PROGRESS IN COLLOID & POLYMER SCIENCE

Editors: H.-G. Kilian (Ulm) and G. Lagaly (Kiel)

Volume 76 (1988)

Trends in Colloid and Interface Science II

Guest Editor: V. Degiorgio (Pavia)

Springer-Verlag Berlin Heidelberg GmbH

ISBN 978-3-662-15929-3   ISBN 978-3-7985-1693-9 (eBook)
DOI 10.1007/978-3-7985-1693-9
ISSN 0340-255 X

© 1988 by Springer-Verlag Berlin Heidelberg

Originally published by Dr. Dietrich Steinkopff Verlag GmbH & Co. KG, Darmstadt in 1988

Softcover reprint of the hardcover 1st edition 1988

Chemistry editor: Heidrun Sauer; Copy edition: Deborah Marston; Production: Holger Frey.

Preface

During the meeting "New Trends in Colloid Science," held in Como, Italy, in September 1987, the European Colloid and Interface Society (ECIS) was founded. The first decision of ECIS was to organize a yearly meeting of European scientists working in the field of colloids and interfaces. The first such meeting was held in Villa Olmo, Como, 2–4 September 1987. It was attended by about 130 participants from 17 European countries. About 100 papers were presented either orally or as posters. Roughly half of the full papers are collected in this volume; the other half appear as abstracts. The papers have been rather arbitrarily subdivided into three sections: Colloids, Amphiphile solutions, and Interfaces. It is a typical and interesting feature of this field that most contributions are on the borderline between physics and chemistry (and sometimes biology) and present, at the same time, a fundamental and a technological side. It will be one task of ECIS to preserve the balance between all these different aspects in future conferences.

On behalf of ECIS, let me thank all the participants for their lively contributions, paper presentations, and discussions; the staff of Villa Olmo for the excellent organization of both the 1986 and the 1987 meetings; the members of the Scientific Committee, Lindman, Aveyard, Vrij, and Henglein for the preparation of the scientific program; my colleagues Laura Cantú and Mario Corti for their effective collaboration in the organization; and the generous meeting sponsors.

Let me finally remind readers that the Proceedings of the 1986 Conference have also appeared as a volume of this series (Vol. 73) which was edited by Prof. Hoffmann, the first Chairman of ECIS.

Vittorio Degiorgio

Contents

Interfaces

Abstracts

Colloids

Adhesive hard-sphere colloidal dispersions. III. Stickiness in n-dodecane and benzene

P. W. Rouw, A. Vrij, and C. G. de Kruif

Van 't Hoff Laboratory, University of Utrecht, Utrecht, The Netherlands

Abstract: Dispersions of silica particles coated with octadecyl chains show different phase separation behaviour when dispersed in benzene and n-dodecane. This difference is shown by measurements of the second virial coefficient as a function of temperature. The difference also follows from dynamic light scattering experiments and measurement of the cloudpoint.

Attraction between silica particles in benzene is attributed to solvent-chain interaction. There seems to be extra attraction between the particles when they are dispersed in n-dodecane.

Key words: Silica particles with octadecylchains, dispersions in benzene and n-dodecane, second virial coefficient, temperature influence, dynamic light scattering, cloudpoint, attractive effects.

Introduction

Silica particles sterically stabilized by octadecyl chains terminally grafted to the surface can be dispersed in a variety of solvents [1]. In marginal solvents, the particles exhibit an effective attraction. As the attractions are temperature-dependent (they increase with decreasing temperature), a phase separation of the gas-liquid type can be induced. There is a strong parallel between polymer molecules dissolved in a marginal solvent and our silica particles dispersed in a solvent.

Vincent et al. [2] have studied silica particles dispersed in linear alkanes and found that phase separation occurred by both raising and lowering the temperature. The description of the interaction between the particles that Vincent et al. gives is in terms of long-range London-van der Waals attraction.

We studied dispersions of small silica particles in n-dodecane and benzene. These dispersions showed phase separation upon cooling. They remained stable with rising temperature up to 100 °C. At room temperature the difference in the refractive index of particles and linear alkanes is small. Indeed, it was shown by Jansen [3] that the decrease in the second virial coefficient, B_2, due to van der Waals forces between silica

cores with radius 50 nm and dispersed in n-hexane, n-octane and n-decane does not exceed 0.5. B_2 was normalized by the hard sphere volume and this reduction should therefore be compared with a hard sphere excluded volume value of 4.0.

In accordance with the results of Vincent et al. we found that a dispersion of silica particles with radius 150 nm in hexane showed a phase transition at 60 °C when the sample was heated from room temperature. At temperatures near the boiling point of the solvent, van der Waals attraction increases due to the decreasing density of the solvent (and dielectric constant). Larger particle radii and closer inter-core distances lead to larger van der Waals forces. For example for a particle radius of 500 nm and an inter-core distance of 0.15 nm the decrease in B_2 is 2. Since a value of -10.6 is needed for B_2 to reach the critical value, the decrease in density of the solvent should induce a more than five-fold increase in the attraction.

Although it might be possible to explain phase transitions in terms of macroscopic London-van der Waals attractions, one should bear in mind that the application of this theory to microscopic separation distances (i. e. 0.15 nm) is of doubtful value.

The variation in the Hamaker constant is much less when dispersions are cooled to temperatures near room temperature. Yet we did observe phase separation upon cooling dispersions in n-dodecane ($\sim 18\,°C$) and benzene ($\sim 30\,°C$), whereas the energy of attraction between the silica particles in both solvents due to van der Waals forces was found to be considerably smaller than kT. In fact, upon cooling dispersions of silica particles in n-dodecane, the difference in the refractive index of particles and solvent diminishes resulting in a decrease in the van der Waals attraction. Therefore attraction between the particles in dispersions having an upper critical solution temperature (UCST), i. e. showing phase spearation upon cooling, cannot be explained solely in terms of macroscopic London-van der Waals forces. The local interaction between segments of the stabilizing chains on the silica surface and of the solvent molecules is considered to be the most important factor for dispersions having a UCST [3]. This model gives a fairly accurate description of experiments that were done on dispersions in toluene [4, 5] and benzene [6].

Within the framework of this "polymer physics" model, however, there may be substantial differences between solvents such as cyclohexane, n-dodecane and benzene because of differences in energy density parameters. The interactions between silica particles dispersed in cyclohexane can be well described with a hard sphere potential. These dispersions do not show phase separation. Dispersions of silica particles in benzene show phase separation into two liquid phases with different particle concentration. Silica particles dispersed in n-dodecane show gelation upon cooling. Even at very low concentrations ($\sim 3\,\%$ vol/vol) these dispersions give a volume-filling gel. Both phase transitions are reversible: raising the temperature after phase transition restores the initial (homogeneous) situation. There is no sign of permanent clusters in the re-heated solutions.

We report here on measurements of the osmotic second virial coefficient, B_2 and on measurements of the dynamic second virial coefficient, k_d, obtained from measurements of the diffusion coefficient. Both coefficients are determined as a function of temperature. Measurements of cloudpoints are discussed as well. The experiments were performed in order to obtain a more quantitative view of the differences in the attraction between silica particles dispersed in n-dodecane and benzene. Measurements of both B_2 and k_d were made at temperatures just above the phase separation temperatures.

Interaction model

A schematic drawing of the coated silica particles is given in Fig. 1. When the solvent is good with respect to the stabilizing chains, the particles tend to repel one another. This repulsion is caused by the fact that chain segments prefer contact with solvent molecules to contacts with chain segments of other particles. This situation occurs when the silica particles are dispersed in cyclohexane. Experiments done on these dispersions show that particle interactions can be well described as hard sphere interactions [7, 8].

When particles are dispersed in solvents, poor with respect to the stabilizing chains, chain-chain contacts become favourable although there is an entropy loss. The interpenetration distance or interaction length is small, since the octadecyl chains are densely packed on the silica surface. There is a small overlap volume in which stabilizing chains interact, giving rise to attraction between the particles. Since solvent quality is temperature dependent, mainly through the entropy term, the attraction between the particles is also temperature dependent.

The interaction between the particles is described by a hard-sphere potential plus an attractive square-well:

$$V(r) = \infty \qquad\qquad r < \sigma$$
$$= -\varepsilon \qquad\qquad \sigma \leq r \leq \sigma + \Delta$$
$$= 0 \qquad\qquad r > \sigma + \Delta$$

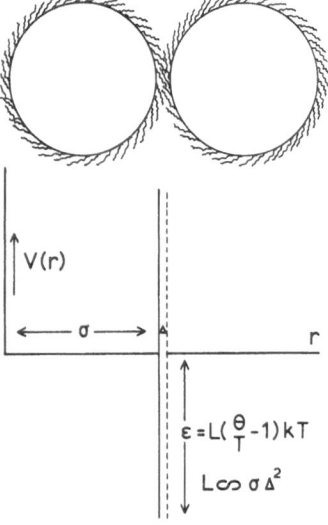

Fig. 1. Schematic drawing of two colloidal silica particles covered with a layer of octadecyl chains

$$\varepsilon = L(\theta/T - 1) \, kT \qquad T \le \theta$$

$$\varepsilon = 0 \qquad\qquad\qquad T > \theta$$

L is a proportionality constant, θ is the theta temperature of the chain-solvent pair, T is the absolute temperature and k is the Boltzmann constant.

ε can be regarded as the free energy of attraction between the silica particles. The equation for ε is similar to the well-known Flory expression used for polymer systems. The expression $L(\theta/T - 1) \, kT$ originates from the theory of polymer solutions (Flory-Krigbaum theory of the second virial coefficient [9]). It consists of an *energy* of mixing term, $kL\theta$, and an *entropy* of mixing term, kLT:

$$\Delta \,(\text{free energy}) = kL\theta - kLT = kTL(\theta/T - 1) \, .$$

The Δ (energy) term resembles the macroscopic London-van der Waals interaction, although it concerns here the process of exchanging solvent with chain segments. The entropic term, however, is missing in the macroscopic theory since the local liquid structure is not present in its formulation.

As in Ref. [10], we will take L to be proportional to the overlap volume $= \pi\Delta^2\sigma/4$, where Δ is the interaction length and σ is the particle diameter.

The particle attraction is now described by three parameters:

1. Δ = interaction-length
2. L = constant proportional to the overlap volume
3. θ = theta temperature of chain-solvent pair.

Following Batchelor [11] we define an attraction parameter related to the second virial coefficient by

$$\alpha = 24/\sigma^3 \int_\sigma^\infty (\exp(- V(r)/kT) - 1) \, r^2 \, dr \, .$$

Substituting the square-well potential and performing the integration we find:

$$\alpha = 24 \, \frac{\Delta}{\sigma} \, (\exp(L(\theta/T - 1)) - 1) \, . \tag{1}$$

Here we used $\Delta \ll \sigma$, which is justified for the following reason. The length of the chains is small compared to the particle diameter. Since the chains can only interpenetrate a small distance, the interaction length is only a small fraction of the chain-length and, therefore, is small compared to the particle diameter. This small value of Δ/σ also implies that the precise shape of the

attractive well is in fact not important in the formulation of α. The width of the well only changes the integral by a constant factor, but the exponential temperature dependence of α is conserved.

For low particle concentrations, the radial distribution function, $g(r)$, can be expressed in terms of the particle pair potential, $V(r)$: $g(r) = \exp(- V(r)/kT)$. We can now rewrite the definition of α as follows:

$$\alpha = 24/\sigma^3 \int_\sigma^\infty (g(r) - g^{HS}(r)) \, r^2 \, dr \, .$$

This shows that α is a volume integral over a distribution function.

We can assign a physical meaning to the interaction parameter α as follows. If ϕ is the particle volume fraction then, the product $\alpha \cdot \phi$ is the mean excess number of particles within a certain volume around a central particle, with respect to the hard-sphere situation. The dimension of this volume must be very large with respect to the range of $g(r)$. Attraction between particles results in a positive value of α (i. e. the number of particles in the neighbourhood of a central particle is enlarged) whereas repulsion between the particles results in a negative value of α (i. e. the number of particles in the neighbourhood of a central particle is reduced). α can also be looked upon as adsorption of particles by a given particle [12].

The temperature dependence of α is governed by Δ, L and θ. Measurement of α as a function of temperature, therefore, provides a way of calculating the parameters Δ, L and θ. Thus we are able to characterize the interaction potential between the silica particles. It is possible to determine the temperature dependence of α both by measurement of the second virial coefficient of the dispersion and by measurement of the diffusion coefficient of the particles in the dispersion. Details of these techniques are discussed in the following sections.

Determination of B_2

Theory

For low particle concentrations deviations of Van 't Hoff's ideal law of the osmotic pressure are described by the second virial coefficient, B_2, which is related to the pair potential, $V(r)$, between the particles by

$$B_2 = \frac{N_A}{M} \, 2\pi \int_0^\infty (1 - \exp(- V(r)/kT)) \, r^2 \, dr \, .$$

Here N_A is Avogadro's number and M is the molar mass of the particles. B_2 is thus related to α:

$$B_2/V_{HS} = 4 - \alpha/2 \qquad (2)$$

where V_{HS} is a specific volume given by

$$V_{HS} = \frac{N_A}{M}\,\pi\sigma^3/6\,.$$

From the equations it is seen that at temperature $T = \theta$, the second virial coefficient attains the hard-sphere-value, $B_2 = 4V_{HS}$. When temperature is lowered, solvent quality becomes poorer and B_2 decreases. Rewriting Eq. (2) and using Eq. (1) leads to

$$\ln\left(1 + \frac{\sigma}{12\Delta}\,(4 - B_2/V_{HS})\right) = -L + \frac{L\theta}{T}\,. \qquad (3)$$

For a given value of Δ, Eq. (3) can be used to evaluate L and θ from the measured B_2 as a function of temperature.

The second virial coefficient can be measured with static light scattering. With this technique one measures a (time-averaged) light intensity as a function of scattering angle. The Rayleigh ratio, $R(K)$, is defined by $R(K) = \dfrac{I(K)\cdot d^2}{I_0 \cdot V_s}$ ($I(K)$ is the intensity of scattered light at the detector, I_0 is the intensity of incident light, d is the distance from sample to detector and V_s the volume of the illuminated part "seen" by the detector). The wave vector K is equal to $(4\pi n/\lambda_0)\sin(\theta/2)$, where n is the refractive index of the sample, λ_0 the wavelength in vacuum and θ the angle between scattered and transmitted light. From the Rayleigh-Gans-Debye theory of light scattering for unpolarized light it follows that

$$R(K) = K^*cMP(K)S(K)(1 + \cos^2\theta) \quad \text{with } K^*$$

$$K^* = \frac{2\pi^2 n^2 (dn/dc)^2}{\lambda_0^4 N_A}\,. \qquad (4)$$

Here c is the weight concentration of the particles. The form factor $P(K)$ accounts for the interference of light scattered from different parts within one particle. The structure factor $S(K)$ accounts for the interference of light scattered from different particles. For not too large values of Ka, the Guinier approximation holds $P(K) = 1 - K^2a^2/5$, where a is the optical radius of the spherical particles. In terms of the radius of gyration of the particles, R_g, $P(K) = 1 - K^2R_g^2/3$.

The structure factor $S(K)$ is the Fourier transform of the total correlation function and is given by

$$S(K) = 1 + 4\pi\varrho \int_0^\infty (g(r) - 1)\,\frac{\sin(Kr)}{Kr}\,r^2 dr$$

where ϱ is the number density of particles and $g(r)$ the radial distribution function, which for low concentrations is $\exp(-V(r)/kT)$. Following Jansen [4] we may use the square-well potential to perform the integration and to find the K-dependence of the structure factor up to terms that are linear in the particle concentration and quadratic in K

$$S(K) = 1 - \phi(8 - \alpha) + \phi\,\frac{K^2\sigma^2}{10}\left(8 - \frac{5}{3}\,\alpha\right).$$

The hard sphere volume fraction $\phi = \pi\varrho\sigma^3/6 = \dfrac{N_A c}{M}$ $\pi\sigma^3/6 = V_{HS}c$. The second term reduces to $2B_2c$ and so for $K \to 0$

$$S(K = 0) = 1 - 2B_2c\,.$$

From Eq. (4) it follows that

$$\frac{c}{R(K = 0)} = \frac{1}{K^*M}\,(1 + 2B_2c)\,.$$

Plotting $c/R(K = 0)$ versus c should yield a straight line (if concentration is not too high). From slope and intercept it is possible to calculate B_2. From the intercept it is possible to determine M. This requires, however, a tedious absolute calibration of the proportionality constant K^*.

Measurement of B_2 by static light scattering is restricted by multiple scattering. Multiple scattering makes interpretation of the data very difficult [13] and is therefore better avoided. Jansen showed that multiple scattering does not affect turbidity measurements [4]. He also described a method for determining the second virial coefficient by means of turbidity measurements. The measurement is closely related to the determination of B_2 from static light scattering.

Turbidity, τ, is defined by the Lambert-Beer relation: $\tau = l^{-1}\ln(I_t/I_0)$ where I_0 is the incident intensity, I_t is the transmitted intensity and l the length of the light path in the sample. When reduction in transmitted intensity is not caused by absorption of radiation but solely by scattering of radiation, a simple conservation law then relates scattering and turbidity

$$\tau = 2\pi \int_0^\pi R(K) \sin\theta \, d\theta.$$

We can substitute the expression for $R(K)$ as given in Eq. (4), and use the relations for $P(K)$ and $S(K)$ given above. After integration, it follows that

$$\tau = K^* c M \frac{16}{3} \pi \left(1 - \frac{8}{3} (\pi n/\lambda_0)^2 R_g^2\right)$$
$$\cdot \left(1 - \phi(8 - \alpha) + \phi \frac{4}{5}\left(8 - \frac{5}{3}\alpha\right)(\pi n\sigma/\lambda_0)^2\right).$$

This expression is correct to order $(\pi n/\lambda_0)^2$. By plotting c/τ versus c and determining the slope, an apparent second virial coefficient is obtained, which is defined by:

$$B_{2,\text{app}}(\lambda_0) = V_{HS}\left(4 - \frac{\alpha}{2} - \frac{4}{5}\left(4 - \frac{5}{6}\alpha\right)(\pi n\sigma/\lambda_0)^2\right).$$
$$(5)$$

This $B_{2,\text{app}}$ is wavelength dependent. By extrapolation at a given temperature, T, of $B_{2,\text{app}}$ to $\lambda_0^{-2} = 0$, B_2 is obtained:

$$B_2(T) = V_{HS}\left(4 - \frac{\alpha}{2}\right).$$

We can also determine α form the slope of the plot of $B_{2,\text{app}}(\lambda_0)$ verus λ_0^{-2} and this should give the same value for B_2 as that found by extrapolation to $\lambda_0^{-2} = 0$. In Eq. (5) three cases can be distinguished depending on the value of α:

1. $4 - 5/6\alpha = 0$; $\alpha = 24/5$.

There is no wavelength dependence for $B_{2,\text{app}}$. The plot of $B_{2,\text{app}}$ versus λ_0^{-2}, therefore, shows a horizontal line.

2. $5/6\alpha < 4$; $\alpha < 24/5$.

The slope of the plot of $B_{2,\text{app}}$ versus λ_0^{-2} is negative and attractions are weak. For $\alpha = 0$ the particles behave effectively as hard spheres.

3. $5/6\alpha > 4$; $\alpha > 24/5$.

The slope of the plot of $B_{2,\text{app}}$ versus λ_0^{-2} is positive and attractions are strong. For these values of α, phase separation will occur.

Measurement of the turbidity of silica dispersions as a function of concentration, wavelength and temperature thus provides a method of determining B_2 as a function of temperature. By plotting the B_2 values according to Eq. (3) we are able to evaluate the square-well parameters.

Determination of k_d

Theory

Another way to determine the development of attractions in particle dispersions is to analyse the temperature dependence of the diffusion coefficient of the particles as measured by dynamic light scattering. Details of this technique are given elsewhere [6] and we will confine ourselves here to the resulting equations.

The apparent radius of the particles is related to the diffusion coefficient by the Stokes-Einstein relation: $a_{\text{app}} = kT/6\pi\eta D$. Here k is the Boltzman constant, T the absolute temperature, η the viscosity of the solvent and D the mutual diffusion coefficient. In the Batchelor treatment of the diffusion coefficient [14] it is found that

$$D/D_o = 1 + k_d\phi$$
$$k_d = 1.45 - 0.56\alpha.$$

If we model the interaction potential between the particles as the square-well potential we find the following expression:

$$k_d = 1.45 - 13.44 \frac{\Delta}{\sigma}\left(e^{L\left(\frac{\theta}{T} - 1\right)} - 1\right).$$

If the Stokes-Einstein equation is applied, the apparent radius of the particles at a given temperature is given by [6]

$$a_{\text{app}} = \frac{a_o}{1 + 1.45\phi - 13.44\phi \frac{\Delta}{\sigma}\left(\exp\left(L\left(\frac{\theta}{T} - 1\right)\right) - 1\right)}$$
$$(6)$$

where a_0 is the hydrodynamic particle radius at infinite dilution and σ is the particle interaction diameter at infinite temperature.

It is evident from this equation that when temperature is decreased (below θ) the apparent radius will increase ($L > 0$). This increase is governed for the most part by L and θ since they are in the exponential argument, whereas Δ does not influence this increase very much if $|\exp| > 1$ since it is in the pre-exponential factor. Rewriting this equation shows that it is possible to determine L and θ for a given Δ by measuring the diffusion coefficient as a function of temperature at a certain volume fraction. The following equation is used:

$$\ln\left(1 + \frac{\sigma(1 + 1.45\phi - D/D_0)}{13.44\Delta\phi}\right) = -L + \frac{L\theta}{T}. \quad (7)$$

L and θ can be calculated from the intercept and slope of the least-squares fit of the expression on the left hand side of Eq. (7) as a function of $1/T$.

We can compare the temperature dependence of D for two samples by plotting k_d as a function of temperature

$$k_d = \frac{1 - D/D_0}{\phi}. \quad (8)$$

For particles that behave as hard spheres, k_d has been shown to be 1.45 [15, 16]. Attraction between the particles reduces k_d [6, 14].

Determination of B_2

Experimental

Measurements were made on two silica systems coded SP 23 and SP 45. Table 1 gives some characterization results for the silica particles. The values for the optical radius in Table 1 are found for dispersions in cyclohexane, but somewhat different values can be found for dispersions in other solvents since the optical radii are sensitive to contrast variations.

Silica particles in cyclohexane were dried at 70 °C under nitrogen flow. A weighed amount of silica was put in a 6 ml tube and dissolved in the solvent. Silica particles coded SP 23 were dissolved in benzene (Baker analyzed) and silica particles coded SP 45 were dissolved in n-dodecane (Baker grade). Concentrations ranged from 2 to 20 % by weight. Dispersions of silica in benzene were kept at 50 °C, i. e. above the phase separation temperature. Dispersions of silica in n-dodecane could be kept at room temperature (25 °C) since phase separation (gelation) occurs at about 18 °C.

As dust particles hardly influence turbidity measurements (they scatter mainly in the forward direction), the samples were not filtered before use. Prior to each measurements, the sample was equilibrated to the measuring temperature by placing the tube in a thermostatically controlled bath. Water was circulated from this bath through the cell holder of the spectrophotometer. The sample was put in a 1 cm glass cuvette (Hellma) and temperature was again allowed to equilibrate. It was assumed that equilibrium was reached

Table 1. Characterization results for silica particles dispersed in cyclohexane

	SP 23	SP 45
Electron microscopy		
r_{EM}/nm	23.5	32.8
standard deviation/nm	2.8	4.2
Dynamic light scattering		
D_0/m^2 s^{-1}	$6.4 \cdot 10^{-12}$	$5.8 \cdot 10^{-12}$
r_{hy}/nm	32	41
Static light scattering		
r_{opt} ($\lambda = 436$ nm)/nm	33.8	45.3
r_{opt} ($\lambda = 546$ nm)/nm	35.6	47.5
Mass density		
ϱ_d/g $\cdot$ ml^{-1}	1.61 ± 0.02	1.63 ± 0.03

when the transmission signal did not change any more. Temperature was measured with a platinum resistance thermometer placed in the cuvette.

Transmission measurements were made with a Shimadzu (Spectronic 200 UV) double beam spectrophotometer. The turbidity is obtained from $\tau = l^{-1} \ln (I(\text{sample})/I(\text{solvent}))$ by measuring the transmission of a cuvette filled with pure solvent and of the same cuvette filled with sample. The wavelength ranged form 400 to 600 nm. Above 600 nm the transmission was close to 100 % and therefore inaccurate. For wavelengths less than 400 nm the transmission was close to 0 % and this also leads to a large error in the turbidity.

Determination of k_d

Experimental

Measurement of the diffusion coefficient as a function of temperature was done on two samples:

SP 23 dispersed in benzene, volume fraction = 0.0128
SP 45 dispersed in dodecane, volume fraction = 0.012.

An Argon ion laser (Spectra Physics, series 2000) was used as a light source ($\lambda = 514.5$ nm). Autocorrelation functions of the scattered light were obtained with a correlator (Malvern Industries K 7025) with 128 channels and analyzed using the method of cumulants. Details of the measurements performed with this technique are given elsewhere [6].

Both samples were filtered through millipore filters (diameter: 1 μm) before measuring. Measurements were done at scattering angle of 90 °. For the sample SP 23/benzene the lowest temperature at which data were taken was just below the cloudpoint. The lowest temperature at which data were taken for the sample SP 45/n-dodecane was well above the cloudpoint. For this sample gelation occurs below the cloudpoint and this causes a dramatic increase in apparent radius.

Apparent radii were calculated form the measured diffusion coefficient using the Stokes-Einstein equation. Viscosity of the solvent was corrected for its temperature dependence using the values in Ref. [17].

The effects of changing the small volume fractions has been studied in detail in Ref. [6].

Determination of B_2
Results and discussion

In Fig. 2. some representative curves of c/τ versus c for the sample SP23/benzene are shown for various temperatures at wavelength = 500 nm. The plots are linear throughout the concentration region, although some scatter is present in the data at high temperature. This scatter is most apparent at high temperatures and

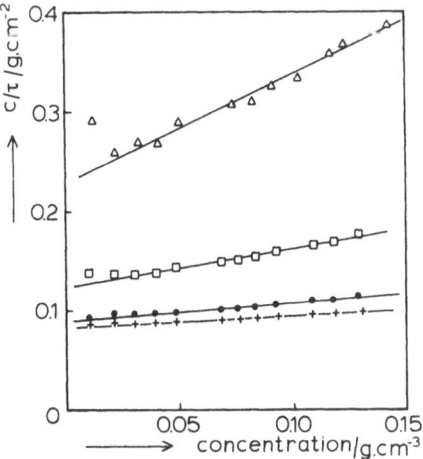

Fig. 2. Reciprocal specific turbidity (c/τ) plotted as a function of concentration (c) for SP23/benzene. Wavenlength = 500 nm; (+) 310.4 K; (●) 313.5 K; (□) 323.2 K; (△) 342.5 K

high wavelength since the contrast is low and transmission values are high ($\sim 90\%$). Intercept and slope of the curves are used to calculate B_2.

Plots of $B_{2,\mathrm{app}}$ versus λ_0^{-2} at each temperature are shown in Fig. 3. Both the ordinate and the slope of $B_{2,\mathrm{app}}$ against λ_0^{-2} depend on α: compare with Eq. (5). From the experimental data in Fig. 3 it may be seen that there is only a slight variation in the slope of the plots. The change in slope is qualitatively consistent with the variation in intercept. As temperature increases, slopes decrease and intercepts increase. Both trends indicate that attraction between the particles is diminished as temperature increases. Determination of the slope, however, is not very accurate, especially at higher temperatures. We may extract from the experimental data two values for B_2: on from the intercept, the other from the average $B_{2,\mathrm{app}}(\lambda_0)$.

The intercept in Fig. 3 gives the value $B_2(i, T)$. From the slope we can calculate α (see Eq. (5)). We used $V_{HS} = 0.63$ g · cm^{-3}, $n = 1.50$ and $\sigma = 64$ nm. Because the slopes of B_2 versus λ_0^{-2} are small and therefore difficult to evaluate, we might just as well extrapolate by horizontal lines. In this way we find an "average" $B_2(a, T)$.

Values for B_2 and α at each temperature are given in Table 2. It is obvious from these data that B_2 decreases and α increases as temperature decreases, indicating that attraction between the particles is developing. For $B_2(i, T)$ and $B_2(a, T)$ we find that at high temperature, B_2 approaches the hard sphere value. The data of B_2 and α against T are shown in Fig. 4 where both the reduced second virial coefficient, B_2/V_{HS}, and $-\alpha/2$

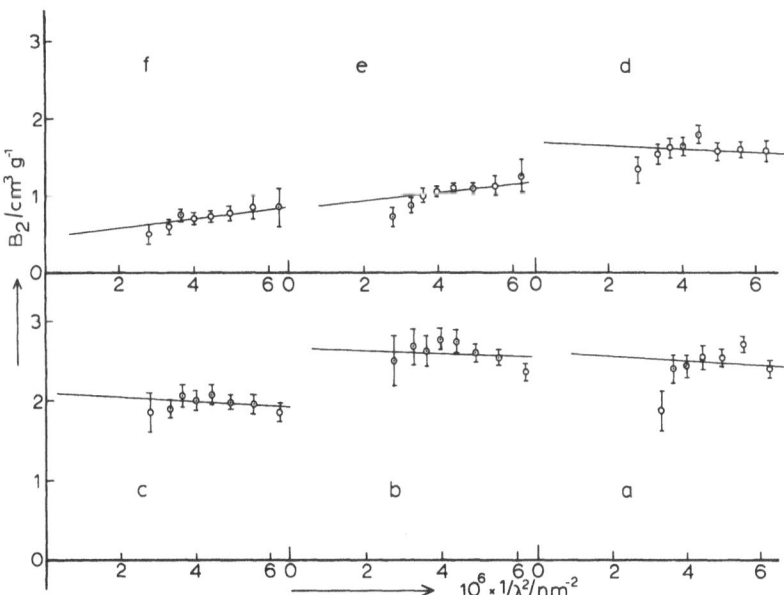

Fig. 3. $B_2(\lambda_0)$ as a function of λ_0^{-2} for SP23/benzene. (a) $T = 342.5$ K, (b) $T = 332.4$ K, (c) $T = 323.2$ K, (d) $T = 318.6$ K, (e) $T = 313.5$ K, (f) $T = 310.4$ K

Table 2. Virial coefficients and interaction parameters from turbidity measurements on SP 23 dispersed in benzene

Temperature (K)	$B_2(i, T)$ ($cm^3 \cdot g^{-1}$)	α	$B_2(a, T)$ ($cm^3 \cdot g^{-1}$)
310.4	0.47	6.22	0.78
313.5	0.83	6.00	1.10
318.6	1.65	4.25	1.58
323.2	2.10	4.16	2.00
332.4	2.63	4.44	2.58
342.5	2.55	1.71	2.45
L	40.5 ± 2.0		26.8 ± 1.1
θ (K)	342.9 ± 6.9		356.7 ± 6.6
T_B (K)	310.1		307.7

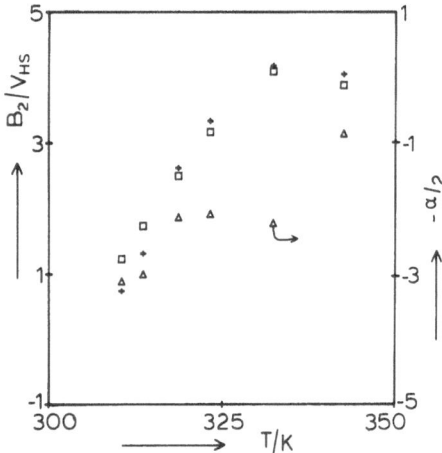

Fig. 4. B_2/V_{HS} and $-\alpha/2$ as a function of temperature for SP 23/benzene. $B_2(i, T)$ (+), $B_2(a, T)$ ($\square$), $-\alpha/2$ ($\triangle$). Symbols are explained in the text

data to $B_2 = 0$ according to theoretical curves. These curves were calculated from Eq. (2) using $V_{HS} = 0.63$ $g \cdot cm^{-3}$. L and θ were calculated from the data according to Eq. (3) using $\Delta = 0.3$ nm. Values for L and θ are given in Table 2. In order to compare the temperature dependence of the B_2 values we normalized temperature by the Boyle temperature and B_2 by the hard sphere specific volume. The data are shown in Fig. 5 where we have plotted the dimensionless second virial coefficient, B_2/V_{HS}, against reduced temperature T/T_B. Values for the Boyle temperature are given in Table 2. The solid curve is calculated from L and θ values as determined by SANS [19], $L = 32$ and $\theta = 346$ K. From Fig. 5 scaling of B_2 data, as determined by turbidity measurements and SANS on a master curve, is observed, which means that the temperature dependence of B_2 is the same in all cases.

In Fig. 6 representative plots are shown for the relations between c/τ and c for the sample SP 45/n-dodecane at wavelength 500 nm. The curves are linear over the whole concentration range. At high wavelength there is some scatter around the line; here transmission values are high ($> 90\%$) and therefore less precise. This certainly holds for low concentrations and low temperatures.

From the slope and intercept of these curves we evaluate $B_{2,app}$ and plot these values as a function of λ_0^{-2}. These plots are shown in Fig. 7 at each temperature. $B_2(i, T)$ is the intercept of the plot of B_2 versus λ_0^{-2}. α is calculated from the slope using $V_{HS} = 0.61$ $cm^3 \cdot g^{-1}$,

are plotted. Both parameters should exhibit the same temperature dependence but shifted by 4 units (see Eq. (2)). From Fig. 4 it can be concluded that $B_2(i, T)/V_{HS}$ and $-\alpha/2$ do not quantitatively agree, since there is not a constant difference between them. We should bear in mind that the uncertainty in slopes and therefore in α is quite large. There is qualitative agreement, however, as can be seen from Fig. 3 since slopes are positive for intercept < 1 (i. e. $\alpha > 24/5$, $B_2(i, T) < 1$) and negative for intercept > 1 (i. e. $\alpha < 24/5$, $B_2(i, T) > 1$) in accordance with theory.

The Boyle temperature is defined for $B_2 = 0$. Just as for simple gases, it is interesting to normalize experimental results by the Boyle temperature. For simple gases universal scaling behaviour is observed [18]. The Boyle temperature was found by extrapolation of the

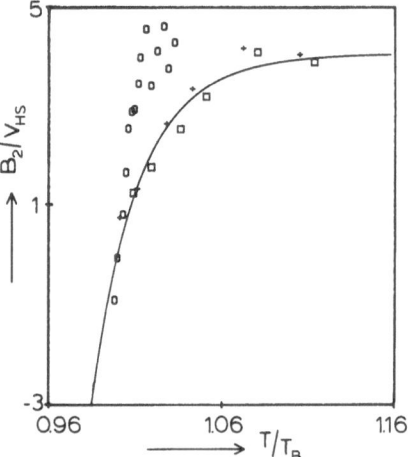

Fig. 5. B_2/V_{HS} as a function of T/T_B for SP 23/benzene. (+) $B_2(i, T)/V_{HS}$; ($\square$) $B_2(a, T)/V_{HS}$; (O) B_2/V_{HS} as determined from dynamic light scattering. The solid curve is based on SANS experiments. Symbols are explained in the text

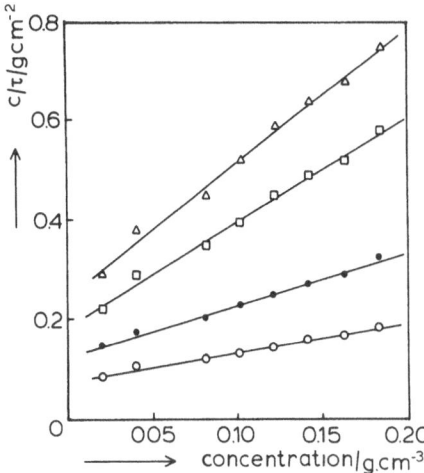

Fig. 6. Reciprocal specific turbidity (c/τ) plotted as a function of concentration (c) for SP 45/n-dodecane. Wavelength = 500 nm; (O) 324.8 K; (●) 314.0 K; (□) 299.0 K; (△) 293.3 K

Table 3. Virial coefficients and interaction parameters from turbidity measurements on SP 45 dispersed in dodecane

Temperature (K)	$B_2(i, T)$ (cm$^3 \cdot$g^{-1})	α	B_2(500 nm) (cm$^3 \cdot$g^{-1})
293.3	1.5 (0.94)	21.1	5.5
299.0	3.3 (2.03)	16.2	5.7
303.0	3.9 (2.44)	10.0	5.4
314.0	4.0 (2.50)	5.7	4.3
324.8	3.8 (2.38)	5.0	3.9
334.2	3.5 (2.19)	3.7	3.2
L	57.2 ± 1.3		
θ/K	313.9 ± 2.6		
T_B/K	290.9		

$n = 1.42$ and $\sigma = 82$ nm. As an "average" B_2, $B_2(a, T)$, we take the value of B_2 at $\lambda = 500$ nm. All values are given in Table 3.

It is remarkable that nearly all $B_2(i, T)$ values are larger than the hard-sphere value $= 4V_{HS} = 2.44$ cm$^3 \cdot$g^{-1} whereas there will certainly be attractions just above 20 °C. There is a plateau value of B_2 ($= 3.90$ cm$^3 \cdot$g^{-1})

at high temperature, therefore the data for $B_2(i, T)$ are scaled so that the plateau value yields the hard sphere value. These scaled values are also given in Table 3 (i. e. values between brackets).

The data of B_2(scaled)/V_{HS} and $-\alpha/2$ against T are shown in Fig. 8. We can see from this figure that the agreement between B_2 and α is only qualitative, i. e. the sharp decrease in the data occurs roughly at the same temperature. The Boyle temperature that can be found from B_2(scaled) is given in Table 3. In Fig. 9 we have plotted B_2/V_{HS} versus reduced temperature

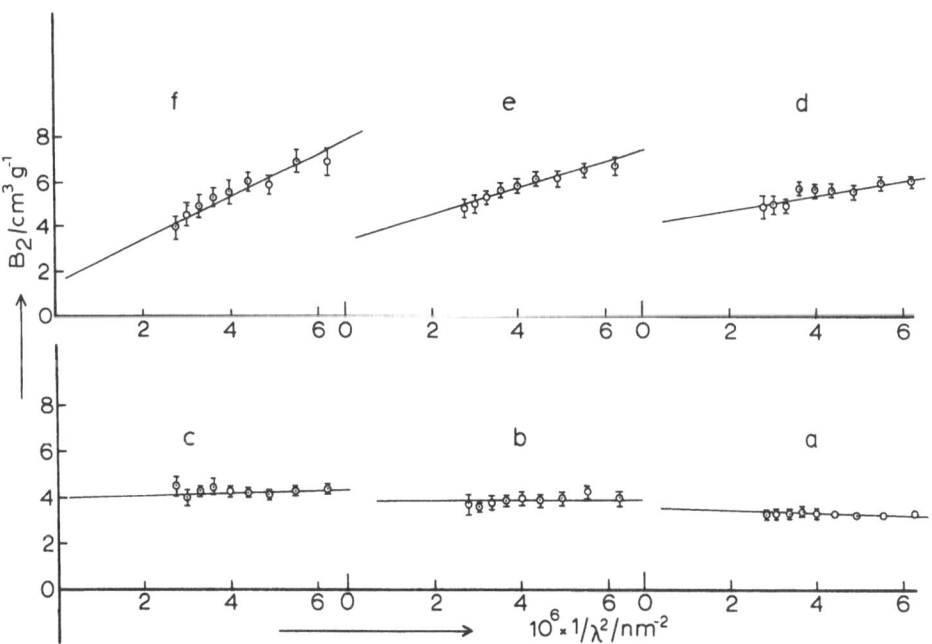

Fig. 7. $B_2(\lambda_0)$ as a function of λ_0^{-2} for SP 45/n-dodecane. (a) $T = 334.2$ K, (b) $T = 324.8$ K, (c) $T = 314.0$ K, (d) $T = 303.0$ K, (e) $T = 299.0$ K, (f) $T = 293.3$ K

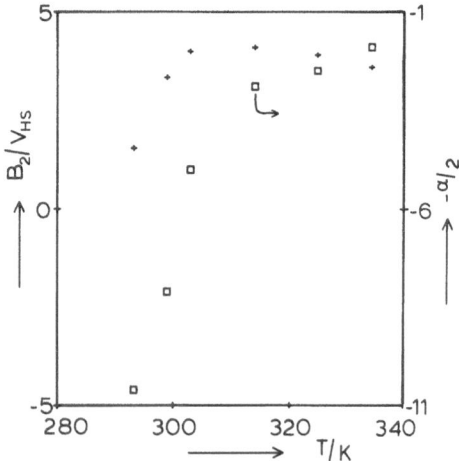

Fig. 8. B_2/V_{HS} and $-\alpha/2$ as a function of temperature for SP 45/n-dodecane. B_2(scaled) (+), $-\alpha/2$ (□). Symbols are explained in the text

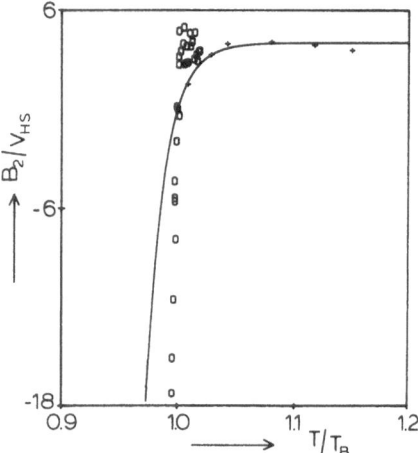

Fig. 9. B_2/V_{HS} as a function of T/T_B for SP 45/n-dodecane. (+) B_2(scaled)/V_{HS}; (O) B_2 as determined from dynamic light scattering. The solid curve is a least-squares fit of the applied square-well model. Symbols are explained in the text

T/T_B for B_2(scaled). L and θ are calculated using $\Delta = 0.3$ nm. We found $L = 57.2 \pm 1.3$ and $\theta = 313.9 \pm 2.6$ K. The solid curve in Fig. 9 is calculated using these values. The average value of $B_2(\lambda)$ (taken at $\lambda = 500$ nm) shows an increase as temperature decreases. This disagrees with the results of turbidity measurements in toluene [4] and benzene (this paper).

The slopes of $B_2(\lambda_0)$ versus λ_0^{-2} are clearly positive at low temperature. This means that α is greater than 24/5 at these temperatures, corresponding to strong attraction. Similar measurements of B_2 in toluene did

not show positive slopes at all. Slopes for the measurement of B_2 for SP 23/benzene, when positive, have only small values. This is in part caused by differences in particle size and refractive index between SP 23/benzene and SP 45/n-dodecane. According to Eq. (5) these differences introduce a factor ≈ 1.5. However, after correction for these differences, the slopes for the measurement of B_2 for SP 45/n-dodecane are still much larger than the slopes for the measurement of B_2 for SP 23/benzene. This also follows from the values of α in Tables 2 and 3, suggesting that attraction between the particles in dodecane is much stronger than either in toluene or benzene.

The measurements of B_2 for SP 23/benzene and SP 45/n-dodecane presented here do not give the same values for B_2 when one compares intercept, i. e. α, and slope of B_2 against λ_0^{-2}. The temperature dependence of these B_2 values, however, seems to be similar. It is emphasized that the most direct value for B_2 is obtained by extrapolation: $B_2(i, T)$, $B_2(a, T)$. The slope of B_2 against λ_0^{-2} and thus α depends on the precise form of the interaction potential and should therefore be treated with caution. There is a clear trend in the temperature dependence of B_2 when one compares silica dispersions in benzene, n-dodecane and toluene. For comparison we plotted the intercept data for SP 23/benzene and SP 45/n-dodecane in Fig. 10. This figure also contains a plot of the intercept data for SJ 9/toluene (silica particles with radius 35 nm Ref. [4]). Using $\Delta = 0.3$ nm, L and θ were calculated from the data given by Jansen [4]. We found [6] $L = 21.1$ and

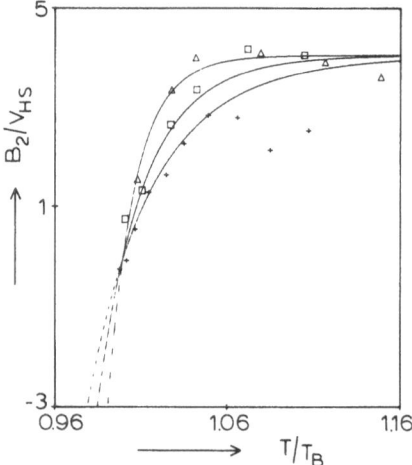

Fig. 10. B_2/V_{HS} as a function of T/T_B for SP 23/benzene (□), SP 45/n-dodecane (△) and SJ 9/toluene (+). Solid lines are least-squares fits of the applied square-well model

$\theta = 342.4$ K. The decrease in B_2 as a function of temperature for SP 45/n-dodecane is very steep, whereas this decrease for SJ 9/toluene is much more gradual. Data for SP 23/benzene are in between. Attraction between the silica particles in n-dodecane evolves rather abruptly, leading to gelation. Attraction between the silica particles in benzene and toluene increases more gradually and this causes phase separation in two liquid phases with different particle concentration.

Determination of k_d

Results and discussion

In Fig. 11 data of a_{app} as a function of temperature are shown for SP 23/benzene. The data were taken at scattering angle of 90° and particle volume fraction of 0.0128. There is a plateau at high temperature where a_{app} attains the hard sphere value of 31.4 nm. As temperature decreases, attraction between the particles evolves and the apparent radius increases, since diffusion is diminished. At low temperature, phase separation occurs and the dispersions become very turbid. This results in multiple scattering that leads to a decrease in a_{app}. By plotting the data according to Eq. (7) and taking $\Delta = 0.3$ nm, we find $L = 141.6 \pm 7.4$ and $\theta = 317.7 \pm 3.8$ K. In the determination of L and θ the data at low temperature where a_{app} decreases were not taken into account.

In Fig. 12 results of the measurements of the apparent radius for SP 45/n-dodecane are shown. These data were taken at scattering angle = 90° and particle volume fraction = 0.012. There is a broad temperature

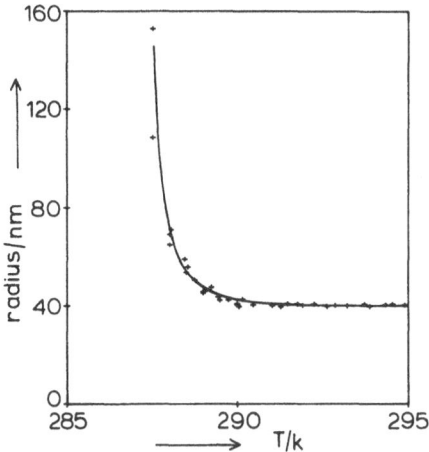

Fig. 12. Apparent dynamic light scattering radius as a function of temperature for SP 45/n-dodecane, $\phi = 0.012$, scattering angle = 90°

region in which a_{app} hardly changes and attains the hard sphere value = 40.3 nm (i. e. the value as measured in cyclohexane). There is a very abrupt increase in a_{app} when the temperature is lowered. This increase is much steeper than the increase observed in the SP 23/benzene sample. Using $\Delta = 0.3$ nm in Eq. (7) we found $L = 293.6 \pm 17.6$ and $\theta = 294.5 \pm 3.3$ K.

From the data shown in Figs. 11 and 12 we can calculate k_d as a function of temperature using Eq. (8). In order to be able to compare the results as found for SP 23/benzene and SP 45/dodecane we plotted k_d as a function of the dimensionless temperature T/T^*. This plot is shown in Fig. 13. T^* is the temperature at which a_{app} equals a_o, i. e. $k_d = o$. T^* (SP 23/benzene) = 310.6 K and T^* (SP 45/n-dodecane) = 290.1 K.

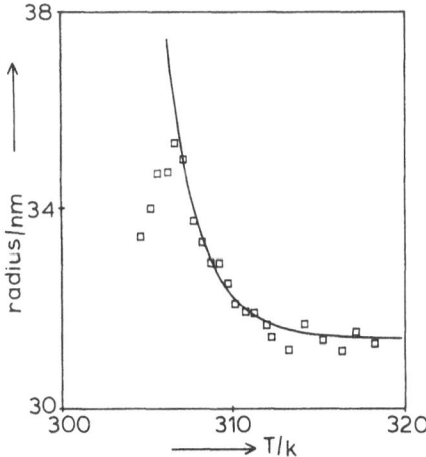

Fig. 11. Apparent dynamic light scattering radius as a function of temperature for SP 23/benzene, $\phi = 0.0128$, scattering angle = 90°

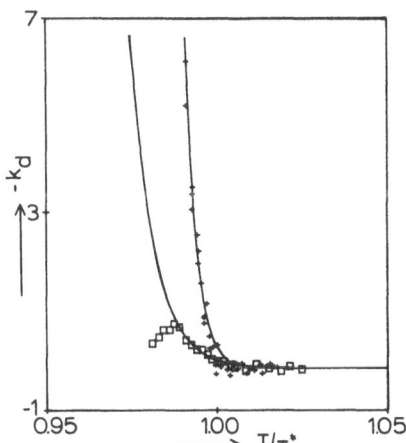

Fig. 13. k_d as a function of T/T^* for SP 23/benzene (□) and SP 45/n-dodecane (+). Solid curves are least-squares fits of the square-well model

From Fig. 13 it is clear that k_d for the dodecane sample decreases much faster with temperature than k_d for the benzene sample. This indicates that attraction between the particles in dodecane evolves rather abruptly, whereas the development of this attraction in benzene is much more gradual.

The values for the square-well parameters: L and θ as determined by turbidity measurements and dynamic light scattering differ. The difference is illustrated in Figs. 5 and 9 where we plot B_2 data as determined from turbidity and dynamic light scattering. α was obtained from the dynamic light scattering data and B_2 was calculated using this value of α. These calculated B_2 values show a much steeper decrease with temperature than B_2 values that were determined from turbidity measurements. This is probably caused by the fact that in the turbidity measurements a concentration range from 0 to 20% by weight was used whereas diffusion measurements were done at low concentration ($\approx 2\%$ by weight). Analysis of data in terms of the second virial coefficient might be incorrect at concentrations near 20% despite the fact that curves of $c/\tau \sim c$ were straight lines up to the highest concentration. Higher virial coefficients might be important here and they will influence the value of B_2 if they are not properly accounted for. It is unlikely that higher order virial coefficients will be important at concentrations as low as 2%; therefore the analysis of the diffusion data in terms of B_2 is correct.

Although the values for L and θ as measured by turbidity and dynamic light scattering are not the same there is certainly a qualitative agreement as far as the temperature dependence of attraction is concerned. In both turbidity and dynamic light scattering measurements on SP 45/n-dodecane there is a broad temperature region for which the particles seem to behave as hard spheres. When the temperature is decreased one observes a rapid increase in attraction between the particles. It follows from both turbidity and dynamic light scattering experiments on SP 23/benzene that the increase in attraction between the particles as a function of temperature is much more gradual than in SP 45/n-dodecane.

Determination of cloudpoint

Results and discussion

Differences in the attraction between silica particles dispersed in benzene and n-dodecane are obvious not only from the resulting phase separation, i.e. fluid

phases (benzene) or gel(n-dodecane), but also from measurement of the cloudpoint. The cloudpoint of a dispersion is the temperature at which the homogeneous phase starts to form a two-phase system. This phase separation is accompanied by a sharp increase in the turbidity of the sample. To determine the cloudpoint, the dispersion must be cooled very slowly so that the system can equilibrate. It was found that cooling dispersions of silica particles in benzene at cooling rates of 1 °C/h or less did not result in different cloudpoints. This means that cooling at 1 °C/h is sufficiently slow for the samples to reach equilibrium. For dispersions of silica in n-dodecane, however, cooling at a rate of 0.2 °C/h still resulted in a somewhat lower cloudpoint than did a cooling rate of 1 °C/h. This shows that dispersions in n-dodecane reach equilibrium much more slowly than dispersions in benzene. The slowest cooling rate we could achieve experimentally was 0.2 °C/h.

In Figs. 14 and 15, the dependence of the cloudpoint on particle radius is shown. Dispersions of silica particles (concentration = 0.05 g · ml^{-1} $\approx$ 0.03 vol. fraction) were cooled at a rate of 1 °C/h. It can be seen from Fig. 14 that there is no clear dependence of cloudpoint on particle size for dispersions in n-dodecane. Particles with radii between 20 and 200 nm all showed a cloudpoint at about 18 °C. There is a strong dependence of cloudpoint on particle radius for dispersions in benzene, as can be seen from Fig. 15. The temperature at which phase separation occurs increases with increasing particle radius. This trend is in accordance with our model since the overlap volume increases with particles radius. This means that L is larger for large par-

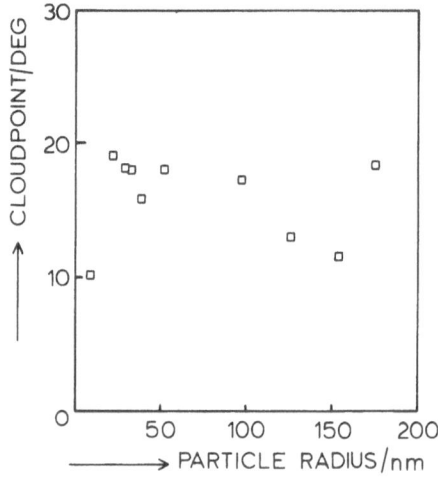

Fig. 14. Cloudpoint as a function of particle radius for silica dispersions in n-dodecane

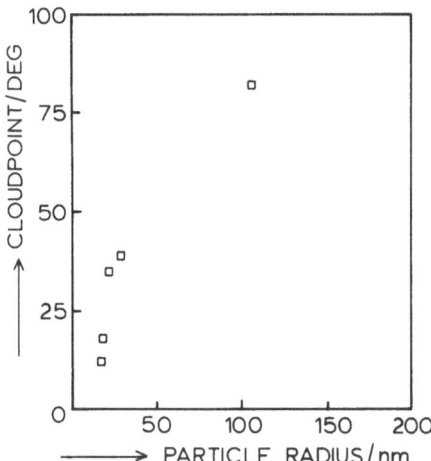

Fig. 15. Cloudpoint as a function of particle radius for silica dispersions in benzene

ticles and therefore phase separation can occur at higher temperatures or smaller values of $\theta/T - 1$.

The absence of dependence of particle radius on cloudpoint for dispersions in n-dodecane conflicts with our model and this indicates that another mechanism might be responsible for phase transition here. We should take into account that the cloudpoints for the n-dodecane samples may not be correct since the cloudpoint still depends on cooling rate below 1 °C/h. We expect, however, that this introduces only a small systematic error in the cloudpoint and does not influence the particle radius dependence.

General discussion

It is clear from Figs. 10 and 13 that the decrease in B_2 and k_d as a function of temperature is much more gradual for SP 23/benzene than for SP 45/n-dodecane. This indicates that attraction between the particles in benzene evolves much more slowly as a function of temperature than attraction between the particles in n-dodecane. One might argue that this difference could be due to the difference in particle radius, since SP 45 is larger than SP 23. Experiments on SP 23 dispersed in n-dodecane, however, showed the same trends.

Since attraction between the particles can become quite strong we wanted a confirmation that the particles did not form aggregates at the temperatures at which turbidity was measured. The wavelength dependence of the specific turbidity, $(\tau/c)_{c=0}$ provides a way to do this since it is strongly affected by aggregation of the particles. The reason is that as aggregation

proceeds structuring effects change the wave vector dependence of the scattered light intensity. The change in wavelength dependence of $(\tau/c)_{c=0}$ as a function of particle size has been studied theoretically by Heller et al. [20]. Long et al. [21] use this change as a demarcation of the flocculation point in polystyrene latices.

We calculated $(\tau/c)_{c=0}$ and plotted its logarithm as a function of $\ln \lambda_0$. The plots are shown in Figs. 16 and 17. In Tables 4 and 5 the slopes of the above-mentioned relations calculated with the least-squares method are given. It can be seen that the wavelength dependence of $(\tau/c)_{c=0}$ is hardly affected by a change in temperature (= attraction). The small differences in slope are probably due to a change in contrast. A decrease in contrast results in an increase in slope. Significant

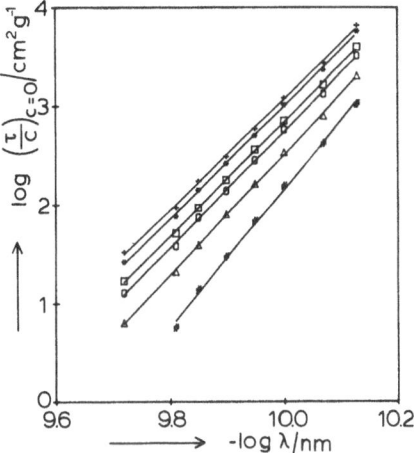

Fig. 16. $\log[(\tau/c)_{c=0}]$ versus $\log \lambda_0$ for SP 23/benzene; (✦) 342.5 K; (△) 332.4 K; (○) 323.2 K; (□) 318.6 K; (✳) 313.5 K; (+) 310.4 K

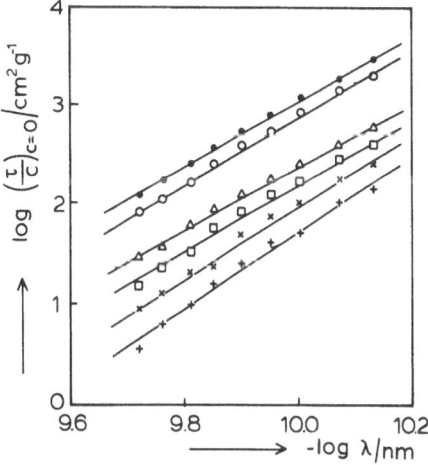

Fig. 17. $\log[(\tau/c)_{c=0}]$ versus $\log \lambda_0$ for SP 45/n-dodecane; (+) 293.3 K; (✕) 299.0 K; (□) 303.0 K; (△) 314.0 K; (○) 324.8 K; (●) 334.2 K

Table 4. Power law behaviour of specific turbidity at different temperatures for SP 45/n-dodecane $(\tau/c)_{c=0} \sim \lambda_0^{-n}$

Temperature/K	n
293.3	3.8 ± 0.2
299.0	3.6 ± 0.2
303.0	3.3 ± 0.1
314.0	3.2 ± 0.1
324.8	3.5 ± 0.1
334.2	3.4 ± 0.1

Table 5. Power law behaviour of specific turbidity at different temperatures for SP 23/benzene $(\tau/c)_{c=0} \sim \lambda_0^{-n}$

Temperature/K	n
310.4	5.6 ± 0.1
313.5	5.7 ± 0.1
318.6	5.8 ± 0.1
323.2	5.9 ± 0.1
332.4	6.1 ± 0.1
342.5	6.9 ± 0.2

permanent aggregation would lead to a much larger difference in slopes between the temperatures. The small differences in slope confirm that permanent aggregation does not play a role a the temperatures at which turbidity data were taken.

There seems to be a fundamental difference in the phase transition behaviour of dispersions of silica particles in n-dodecane and benzene. Attraction between the particles in benzene is due to local interactions between stabilizing chain segments and segments of solvent molecules and can be described quite well by the interaction model. Attraction between the particles in n-dodecane starts very abruptly and we think that this cannot be attributed to the same chain-solvent interaction alone. Ordering of "freezing" of the solvent molecules near the particle surface might occur and thus cause the total attraction to increase. The stabilizing octadecyl chains on the silica surface are very similar to the linear dodecane molecules and might, therefore, induce this freezing effect, leading to an extra "bridging-type" of interaction between two particle surfaces. Experiments with highly branched alkane solvents showed much lower cloudpoints than did linear isomers. This sustains our conclusion that there must be an extra attraction term in linear alkanes due to ordering of the solvent molecules. Such "ordering effects" point to a larger entropic contribution to the

free energy of interaction, which indeed matches with the larger value of L found in the dodecane samples. The alignment of n-dodecane molecules near the particle surface leads to an increase in the effective particle radius, since light scattering behaviour of ordered n-dodecane will differ from light scattering behaviour of randomly oriented n-dodecane molecules in the bulk. It is possible that the large values of B_2 as measured by means of turbidity are due to this increase in particle radius. This ordering effect is absent for the oblate benzene molecules. Phase separation in silica dispersions in which the chain length of the grafted alcohol was varied might also be explained in terms of this freezing effect [22].

In polymer solution theory, the Prigogine-Flory-Patterson (PFP) scheme is used to introduce so-called "equation of state" properties. This makes it possible to explain phase separation in a polymer solution in every (non-polar) solvent when the temperature is sufficiently increased (LCST). This LCST cannot be explained by the simple lattice theory of mixtures which only explains a UCST. It will be clear that the above scheme can also be used to explain the phase separation of coated silica spheres at high temperatures. The physical explanation for solvents that become marginal at higher temperatures is related to the formation of "holes" in a liquid when it approaches the critical density. Such an explanation would be more natural than one which is based on temperature dependence of London-van der Waals forces.

The PFP scheme has also been used to classify interactions in low molecular solvent mixtures. In mixtures of linear alkanes and cyclic alkanes some irregularities are observed which cannot be explained in terms of the PFP theory. The observed effects are attributed to an extra attraction originating from ordering in linear alkanes [23]. We expect that phase separation that occurs upon cooling in dispersions of silica particles in long alkane solvents can be explained along the same lines.

Conclusion

Phase transitions for sterically stabilized silica particles in benzene or n-dodecane are quite different. Attraction between silica particles dispersed in benzene evolves much more gradually as a function of temperature than the attraction between silica particles dispersed in n-dodecane. This leads to gas-liquid phase separation for dispersions in benzene and to gelation for dispersions in n-dodecane.

References

1. Jansen JW, de Kruif CG, Vrij A (1986) J Colloid Interface Sci 114:481
2. Edwards J, Everett DH, O'Sullivan T, Pangalon I, Vincent B (1984) J Chem Soc, Faraday Trans I 80:2599
3. Jansen JW, de Kruif CG, Vrij A (1986) J Colloid Interface Sci 114:471
4. Jansen JW, de Kruif CG, Vrij A (1986) J Colloid Interface Sci 114:492
5. Jansen JW, de Kruif CG, Vrij A (1986) J Colloid Interface Sci 114:501
6. Rouw PW, de Kruif CG (1988) J Chem Phys, in press
7. Vrij A, Jansen JW, Dhont JKG, Pathmamanoharan C, Kops-Werkhoven MM, Fijnaut HM (1983) Faraday Disc Chem Soc 76:19
8. de Kruif CG, Jansen JW, Vrij A (1987) In: Safran SA, Clark NA (eds) Physics of complex and supramolecular fluids. John Wiley & Sons, New York, pp 315–343
9. Flory PJ, Krigbaum WR (1950) J Chem Phys 18:1086
10. Lemaire B, Botherel P, Roux D (1983) J Phys Chem 87:1023
11. Batchelor GK, Wen CS (1982) J Fluid Mech 124:495
12. Vrij A (1985) Proc Kon Ned Akad Wet B 88:221
13. Dhont JKG, de Kruif CG, Vrij A (1985) J Colloid Interface Sci 105:539
14. Batchelor GK (1983) J Fluid Mech 131:155
15. Batchelor GK (1976) J Fluid Mech 75:1
16. Kops-Werkoven MM, Fijnaut HM (1981) J Chem Phys 74:1618
17. Rossini FD, Pitzer KS, Arnett RL, Braun RM, Pimentel GC (1953) Selected values of physical and thermodynamic properties of hydrocarbons and related compounds, Carnegie Press, Pittsburg, Pennsylvania
18. McQuarrie DA (1976) Statistical Mechanics, Harper & Row, New York
19. de Kruif CG, Rouw PW, Briels WJ, Duits MHG, Vrij A, May RP (1988) Langmuir, submitted
20. Heller W (1965) J Chem Phys 42:1609
21. Long JA, Osmond DWJ, Vincent B (1972) J Colloid Interface Sci 42:545
22. Croot R (1987) Bristol, personal communication
23. Heintz A, Lichtenthaler RN (1980) Ber Bunsenges Phys Chem 84:890

Received January 20, 1988;
accepted January 25, 1988

Authors' address:

P. W. Rouw
Van 't Hoff Laboratory
University of Utrecht
Padualaan 8
NL-3584 CH Utrecht, The Netherlands

Progress in Colloid & Polymer Science   Progr Colloid Polym Sci 76:16–19 (1988)

The structure of colloidal suspensions under shear flow: Light scattering experiments

C. Mathis[1]), G. Bossis[1]) and J. F. Brady[2])

[1]) Laboratoire de Physique de la Matière Condensée, Université de Nice, Nice, France
[2]) Department of Chemical Engineering, California Institute of Technology, Pasadena, U.S.A.

Abstract: The structure of a colloidal suspension under shear flow is investigated by light diffusion. The suspension, made of latex spheres ($2a = 750$ Å) dispersed in a water and glycerol mixture, is in a liquid like state. We record the structure factor S(k) in the plane of shear for Peclet numbers ($Pe = \gamma a^2/D$) lower than 10^{-1}. The structure factor presents an asymmetrical deformation when increasing the shear rate: we observe both angular and amplitude shifts of the maximum of $S(k)$; moreover, it depends on the side (compression side or dilatation side) of $S(k)$. These results are compared with a linear approximation and numerical results. This comparison shows that the relevant parameter is $Pe^* = Pe.r^{*2}$, where r^* is an effective radius derived from the interparticle potential.

Key words: Colloidal suspension, shear flow, structure, light scattering, interacting particles.

1. Introduction

The modelling of transport properties in concentrated suspensions rests on a proper understanding of the microscopic change of structure between the solid particles due to an imposed external force. For instance, in the case of the viscosity, we want to know how the particles rearrange when they are submitted to a shear flow. Suspensions composed of monosized polystyrene or silica spheres have been used extensively for studies concerning colloidal crystals [7], coagulation [3], or change of structure due to the effect of temperature [9], ionic strength [4], concentration [8] and shear flow [1]. This work also is devoted to the shear induced structure and we shall focus more precisely on the linearity of this deformation as a function of the shear rate: γ. A more detailed study, both by numerical simulation and by light scattering, is presented elsewhere [2, 6]. We mainly intend to emphasize here the experimental results which demonstrate the non linearity of the deformation even at low Peclet numbers ($Pe = \gamma a^2/D_0$; a: radius of the spheres) $D_0 = kT/6\pi\eta a$: diffusion coefficient of a sphere in a suspending fluid of viscosity η.

In the following section we will describe the experimental method; in part III we will present our results and compare them with a linear theory, and with the angular structure obtained by numerical simulation. We will conclude with a discussion of the results and our prospects for new experiments.

2. Experimental apparatus

2.1 The particles

The sphericity of the particles, observed by electron microscopy, was excellent, as was their monodispersity ($\sigma = \pm 3\%$); the charge surrounding each particle as a double layer stabilizes the suspension by means of a repulsive Debye-Hückel potential; the number of charges at the particle's surface is taken as 1000 e [4] per particle.

The mean diameter was measured in three different ways (electron microscopy, photon correlation, form factor) and we found $d = 750$ Å, whereas Dow Chemical Company indicates $d = 910$ Å: we have kept $d = 750$ Å in our calculations.

2.2 The solvent

The solvent was a mixture of glycerol (53%) and deionised water: (viscosity: $\eta = 9.14$ mPa, refractive index $n = 1.407$, density $\varphi = $

1.14 g cm^{-3}, dielectric constant $\varepsilon_r = 68$). This kind of solvent allows a rather good index matching with the cell (made of quartz), decreases the multiple scattering and avoids Taylor instability in the flow.

2.3 The suspension

The suspension was deionised with an ion exchange resin and maintained under argon during the experiment in order to avoid contamination by CO_2. The volume fraction is $\Phi = 1.32\%$.

2.4 The Couette cell

The measuring cell was made of transparent fused silica. The inner cylinder rotated and the index matching was made with a mixture of glycerol and water (Fig. 1). At the bottom of the cell, a neutral dense liquid avoided secondary flow. The velocity profile was linear ($\pm 5\%$).

2.5 The normalization method

A rotating photodiode measured the total intensity spectra $I(\theta_d)$. We divided each spectrum by a background spectrum: $I^{BG}(\theta_d)$ obtained after adding some salt (NaCl) to the suspension which was therefore in a gaseous-like state. The normalized spectrum: $S(\theta_d) = I(\theta_d)/I^{BG}(\theta_d)$ is free from errors from geometrical defects and no longer includes the form factor.

3. Results

3.1 Measurements

At zero-shear rate we obtained the mean interparticular distance $D_{exp} = 2\pi/k^o_{max} \approx 2460$ Å (k^y_{max}, θ^y_{max} respectively, denote the wave-vector and the scattering angle, respectively, at which the intensity S is maximum at a shear-rate of intensity y). By comparison, the Debye-Hückel length for a completely deionised suspension is approximately $\varkappa^{-1} \approx 282$ Å. Two other parameters are relevant in this system: The Taylor number Ta (always lower than the critical value) and the Peclet number Pe, (by a few per cent for the shear rates investigated) which measures the relative importance of brownian and hydrodynamic motions: see Table 1.

At rest the suspension is isotropic and the pair distribution function has a spherical symmetry: $g(r) = g_o(r)$. In the presence of a flow, the structure is distorted and, relative to a reference particle, we can distinguish between a compressional region where the particles are pushed towards each other and an extensional side where the particles are pulled apart. When the inner cylinder rotates, as indicated in Fig. 1, the positive, or

Table 1. Shear rates and dimensionless numbers of the flow. The zero-shear rate is denoted (o)

y(Hz)	Pe	Ta
6.39 (1)	1.43 10^{-2}	65
12.79 (2)	2.86 10^{-2}	261
25.58 (3)	5.72 10^{-2}	1044

respectively negative, values of θ_d correspond to a scattering wave-vector $k = k_i - k_d$ in the compressional respectively extensional region. Of course, these regions are permutated when the rotation of the cylinder is inverted. We can see in Fig. 2 the evolution of the structure as a function of the shear rate. The most striking feature is the large difference in behaviour between

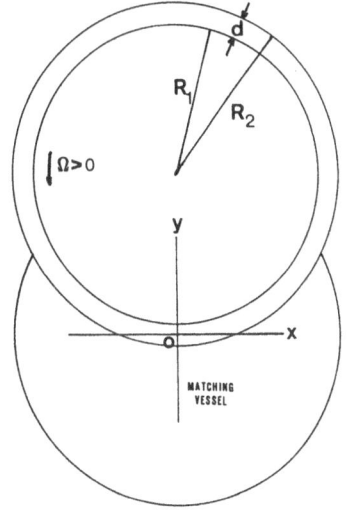

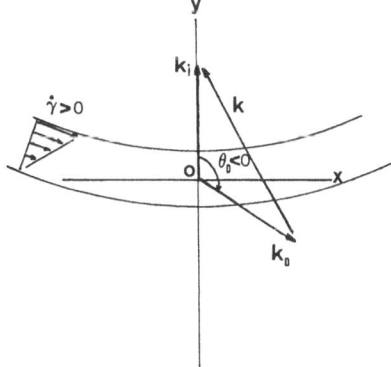

Fig. 1. The rotating cell and the sign convention

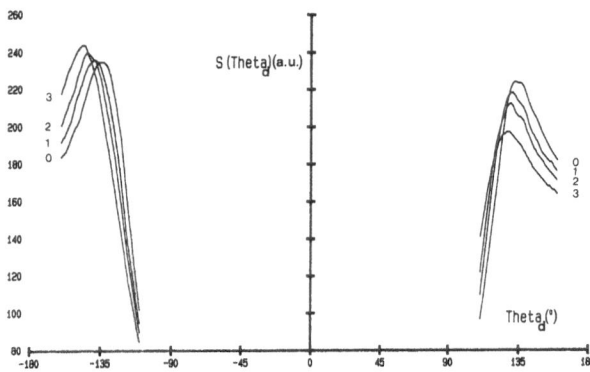

Fig. 2. Structure factors with positive shear rates

the two sides: the intensity of the maximum falls much more rapidly on the extensional side than it increases on the compressional side.

3.2 Comparison with a linear theory

At low shear rates, the deformation of the structure must be linear in the Peclet number (which is the non-dimensionalized shear rate), and so we can write [2, 6]

$$g^y(r) = g^o(r) \left[1 - Pe/2 \, f(r) \, xy/r^2\right]. \tag{1}$$

The function $f(r)$ is the solution of a stationary conservation equation for the pair distribution function.

Following Ackerson and Clark [1], we can apply the simple model in which we assume that the structure is just elliptically deformed, so that:

$$g^y(r) = g^o \left(r(1 - xy/r^2 \, \tau \, y)\right) \tag{2}$$

where τ can be assimilated to a Maxwell time obtained by equating the viscous and elastic stresses: $\tau_M = 1/y$ $\Delta r/r = \eta/G$. Using a first order development we obtain an expression similar to Eq. (1):

$$g^y(r) = g^o(r) \left[1 - xy/r^2 \, \tau_M \, yr \, 1/g^o \, dg^o/dr\right] \tag{3}$$

or, with a Fourier transform:

$$S^y(k) = S^o(k) + k_x k_y/k^2 \, k \, \tau_M \, y \, dS^o(k)/dk. \tag{4}$$

By writing Eq. (4) as $S^y(k) \approx S^o \, (k(1 + k_x k_y/k^2 \, \tau_M \, y))$ we can readily obtain the value of τ from the shift in the position of the maximum intensity of $S^y(k)$ for different shear rates. Actually, this simple model only works on the compressional side where we get $\tau = 8.4$ ms. Equation (2) does not fit the experimental data on the extensional side, where the calculus of τ is impossible. At this stage it appears than Eq. (3) is not experimentally verified, but it could be due to the crude approximation of the deformation function $f(r)$. A direct test of the linearity of the structure deformation can be realized whatever the function $f(r)$, by using the symmetry relative to the shear rate: the expression $[S^{+y}(\theta_d) + S^{-y}(\theta_d)]/2 - S^o(\theta_d)$ should be zero if the deformation was linear with the shear rate. Figure 3 shows that it is not the case: the difference is already noticeable for the lower shear rate and increases almost proportionally with the shear rate.

4. Discussion

It is noteworthy that a linear expansion does not properly describe the deformation of the structure,

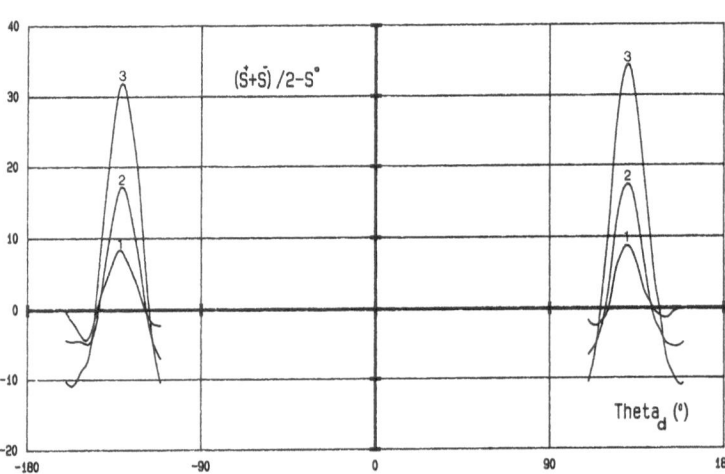

Fig. 3. The difference $[S^{+y}(\theta_d) + S^{-y}(\theta_d)]/2 - S^o(\theta_d)$

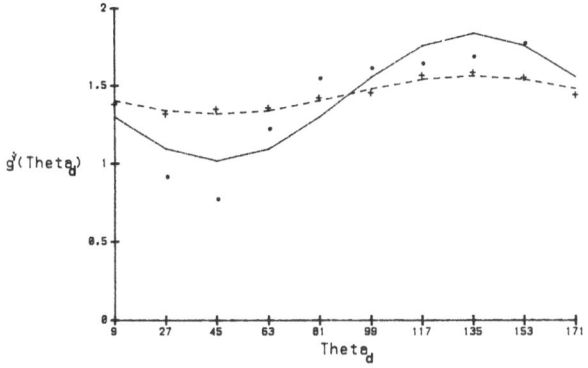

Fig. 4. The angular dependence of $g(r)$: $6d < r < 7d$. Pe $= 0.01$; Pe* ≈ 0.5: (+++) experimental, (---) theoretical. Pe $= 0.1$; Pe* ≈ 5: (···) experimental, (·−··−·) theoretical

even for Peclet numbers as low as a few per cent. In particular, deformation on the extensional side is much larger than on the compressional side, a feature which is also observed by numerical simulation [2], as can be seen in Fig. 4. In this figure we have plotted the angular variation of the pair distribution function obtained by numerical simulation of a monolayer of soft spheres for Pe $= 0.01$ and Pe $= 0.1$. The angular symmetry xy/r^2 is well recovered for Pe $= 0.01$ whereas for Pe $= 0.1$ we observe the same kind of asymmetry than on the experimental spectra at high shear rates. In fact, we have shown that in the case of long-ranged repulsive potential, the relevant expansion parameter was Pe* $=$ Pe$.r^{*2}$ where r^* is the effective radius obtained from the first peak of the pair distribution function. The two curves in Fig. 5 then correspond to Pe* ≈ 0.5 and 5, respectively, and the experimental spectra correspond to $0.15 <$ Pe* < 0.60. We are now faced with a discrepancy between numerical simulation and experimental results since, in the first case, we

have a linear behaviour for Pe* $= 0.5$, whereas in the light scattering experiments the linear theory is not well verified even for the lower value: Pe* $= 0.15$. A possible explanation could be due to the deformation of the ionic cloud during the rotation of two spheres around each other. This deformation must give rise to a dipolar repulsive potential which can contribute to repel the particles on the dilatational side. Other experiments using different concentrations, and so different values of r^*, are needed to clear up this problem. We also plan to use birefringence and dichroism to more easily access the structure deformation in the whole plane of shear.

References

1. Ackerson BJ, Clark NA (1983) Physica 118A:221
2. Bossis G, Brady JF, Mathis C (1988) Colloid Interface Sci, to be published
3. De Rooy N, De Bruyn PL, Overbeek JT (1980) Colloid Interface Sci 75–2:542
4. Grüner F, Lehmann WP (1982) J Phys A, Math Nucl Gen 15:2847
5. Härtl W, Versmold H (1984) Z Phys Chem 139:247
6. Mathis C, Bossis G, Brady JF (1988) Colloid Interface Sci, to be published
7. Pieranski P (1983) Contemp Phys 24:25
8. Pusey PN, Van Megen W (1986) In: Safran SA (ed) Proceeding Symposium on Complex and Supermolecular Fluids. Wiley Interscience, New York
9. Schaefer DW (1977) J Chem Phys 66–9:3980

Received December 14, 1987;
accepted December 30, 1987

Authors' address:

C. Mathis
Laboratoire de Physique de la Matière Condensée
Université de Nice
Parc Valrose
06034 Nice Cedex, France

Progress in Colloid & Polymer Science   Progr Colloid Polym Sci 76:20–23 (1988)

Kinetic and spectroscopic properties of small colloids of metal iodide semiconductors

O. I. Micic, M. T. Nenadovic, T. Rajh, J. M. Nedeljkovic, M. I. Vucemilovic

Boris Kidric Institute of Nuclear Sciences, Vinca, Belgrade, Yugoslawia

Abstract: Small colloidal HgI_2, PbI_2 and AgI particles (with particle diameters in the range 20–50 Å), were prepared in water and acetonitrile, and optical effects due to size quantization were observed. A simple method for the preparation of colloids was used to avoid the possibility of the formation of complexes or other conflicting absorbing species in solution. Colloids of HgI_2 were prepared by simple agglomeration of HgI_2 monomeric molecules. The PbI_2 colloid formation in the solution of crystalline powder dissolved in acetonitrile was studied as well. Small colloids of AgI were prepared by reaction of silver salt with iodide in stochiometric ratio. AgI colloids were also incorporated in transparent silicate glass. Electron transfer reactions from electron donors to AgI particles in aqueous solutions were studied by pulse radiolysis technique.

Key words: Quantization effect, metal iodide, absorption spectra, electron adduct.

Introduction

During the last few years a large number of publications on size quantization effects in colloidal semiconductor particles have appeared [6, 9, 13, 19, 23]. These effects arise from the confinement of charge carriers and lead to an increase of effective band gap as the diameter of the particles decreases below about 5 nm. Recent observations on layered semiconductor clusters on metal-iodide, PbI_2 [17, 24–25, 28] BiI_3 [17, 25] and HgI_2 [16, 17], have shown that very small colloidal particles can be synthetized in acetonitrile and 2-propanol. Though small particles do exist in these systems, other species such as molecular and oligomeric complexes, or I_3^--ions, can be present in the solution as well [16, 17, 28].

In this work we report on the preparation of extremely small colloidal HgI_2, PbI_2 and AgI in water and acetonitrile. Different methods for preparation were used. Our study was also focused on investigation of the nature of excess charges on colloidal particles.

Agglomeration of HgI_2 monomers

Small colloids of HgI_2 can be prepared by reaction of mercury salt with iodide ions in stochiometric ratio in acetonitrile [16, 20] and 2-propanol [17]. However, species such as molecular and oligomeric complexes can be present in the colloidal solution as well [20]. It is known that molecular HgI_2 associates with different anions and organic molecules [10]. An important aspect of the HgI_2 system is the fact that although its solubility product is very small (10^{-28} in water) the solubility is very high (8×10^{-5} mol dm^{-3} in water and $\sim 10^{-2}$ mol dm^{-3} in alcohols) and in that concentration the molecular monomer HgI_2 exists as nearly linear I-Hg-I molecules.

We prepared HgI_2 colloids by agglomeration of HgI_2 monomers in water in the presence of neutral stabilizers. This simple method avoids the presence of starting ions which can form complexes with HgI_2 monomers.

We first studied the effect of stabilizers on absorption spectra of HgI_2 monomers. Figure 1a shows the

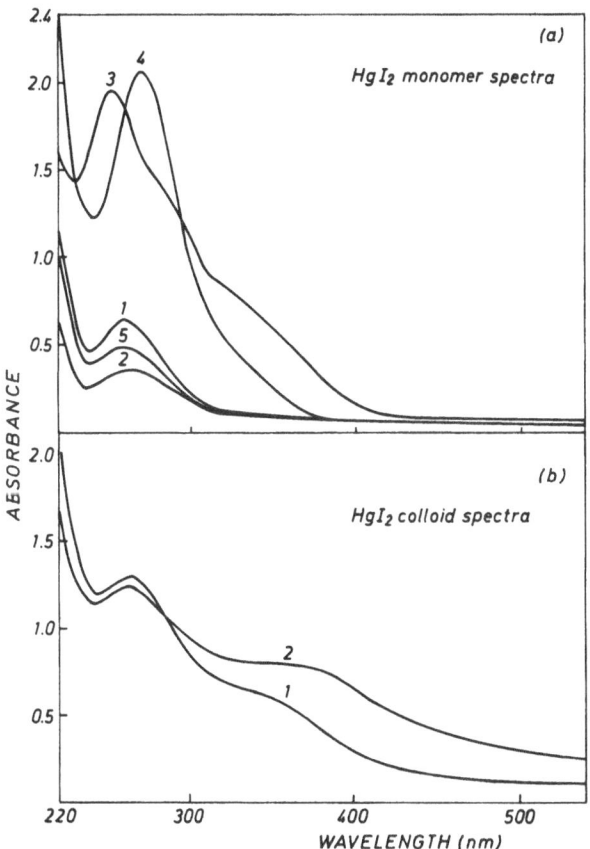

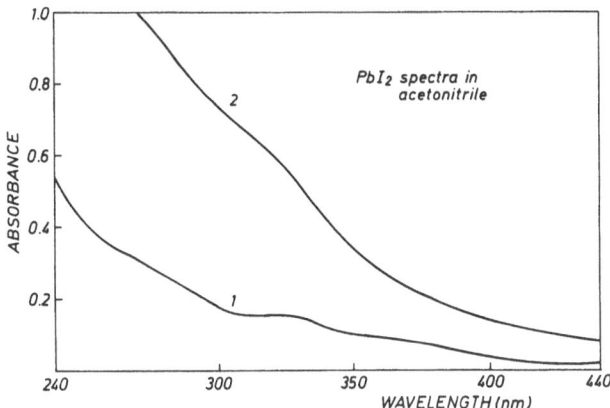

Fig. 2. Optical absorption spectra of PbI_2 in acetonitrile saturated with argon: (1) saturated solution of PbI_2 obtained by dissolution of yellow PbI_2 powder in the presence of 5×10^{-5} mol dm^{-1} $Na_2S_2O_3$ (2) mixing of equal volumes of 4×10^{-4} mol dm^{-3} $Pb(CH_3CO_2)_2$ and 2×10^{-4} mol dm^{-3} $HClO_4$ with 8×10^{-4} mol dm^{-3} in 1% H_2O in acetonitrile

Fig. 1. Optical absorption spectra of HgI_2 in water: (a) Saturated solution of the red HgI_2 in the presence of stabilizers: (1) 0.1% polyethylene glycol; (2) 6×10^{-3} mol dm^{-3} SiO_2; (3) 0.1% poly (vinyl alcohol); (4) 0.1% Polybrene in sulfate form; (5) no stabilizers. (b) Colloidal solution obtained by mixing 3 ml of 2×10^{-2} mol dm^{-3} red HgI_2 in methanol and 100 mol water containing: (1) 0.1% polyethylene glycol; (2) 6×10^{-3} mol dm^{-3} SiO_2

absorption spectra of HgI_2 monomers in the presence of different stabilizers usually used for preparation of colloids. It can be seen that polyethylene glycol (27 000 mol wt) and SiO_2 (diameter 40 Å) do not form complexes with HgI_2 monomers: we used them for the preparation of colloids.

A colloidal solution of HgI_2 (6×10^{-4} mol dm^{-3}) was obtained by mixing concentrated methanol solution of HgI_2 monomers with aqueous solution containing stabilizers. It can be seen that colloidal solutions (Fig. 1b) also have an absorption band with maximum at 260 nm, which corresponds to HgI_2 monomers. Optical absorption spectra of colloidal HgI_2 immediately after mixing show a continuous increase toward the UV region, with a shoulder at about 370–400 nm. On the other hand, the onset of their absorp-

tion is blue-shifted from the band gap of the bulk material (2.1 ev for red HgI_2, onset 595 nm). Such effects can be attributed to the quantization effect that arises from the confinement of charge carriers in the finite volume of distinctly sized particles.

Dissolution of PbI_2 powder

Lead iodide has different properties from HgI_2 materials. PbI_2 is more soluble in water than in organic solvents. Recently it was found that PbI_2 dissolved in organic solvent oxidizes to I_3-ions [17, 28]. In our experiment, PbI_2 powders were immersed in acetonitrile containing 5×10^{-5} M $Na_2S_2O_3$ to suppress the oxidation to iodine; I_3^- does not form under these conditions. Absorption spectra for PbI_2 dissolved in acetonitrile are shown in Fig. 2, curve 1.

Lead iodide is a semiconductor (bulk band gap 2.57 eV) with layered hexagonal structure, where the bonding between adjacent PbI_2 layers is weak [2, 8, 11]. Experiments with MoS_2, also with a structure similar as PbI_2, suggest that this material forms colloids in acetonitrile by cleavage of van der Waals layers by penetrating the solvent [18]. The process is apparently related to that which occurs when clays, also having layered structures, are contacted with liquids to produce colloidal clay dispersions [27].

Formation of PbI_2 colloids in this system was supported by electron microscopy, electron diffraction,

X-ray diffraction and chemical analysis. X-ray powder diffraction patterns of evaporated solution showed lattice parameters that precisely matched hexagonal PbI_2 (c = 4.557 Å, a = 6.979 Å). The same result was obtained for PbI_2 solid material contacted with acetonitrile. Thermogravimetric analysis and X-ray results indicate that PbI_2 does not form an intercalation compound with acetonitrile. PbI_2 was found to spontaneously form colloids upon dissolution in acetonitrile. However, the existence of conflicting absorbing molecular species in solution is not excluded.

Small AgI clusters in liquid and silicate glass

Colloids of AgI are among the most studied systems in colloidal chemistry. We prepared extremely small AgI colloids in water and acetonitrile in the presence of stabilizers. Solutions of Ag^+ and I^- are mixed in stochiometric ratio checked by conductivity measurement. These colloids are stable for days.

Optical absorption spectra are shown in Fig. 3a. In aqueous solution, the absorption spectrum of 100 Å-AgI has an excitonic peak at 3.0 eV and it is similar to that previously obtained for AgI colloids [12, 26]. AgI is a direct gap ionic semiconductor with a band gap of 3.0 eV [4, 5]. In acetonitrile, (Fig. 3, curve 3) particle size was smaller (20–30 Å) than in water prepared with the same stabilizers, and the excitonic peak is shifted towards UV region of about 1 eV. Here again, a quantization effect appears for small particles.

We recently developed a method for incorporating small particles of colloidal semiconductors in transparent silicate glasses [22]. Transparent colloidal glasses were prepared by using AgI aqueous colloidal dispersion to mix tetramethoxysilane and accelerate the polymerization by adding NH_4OH. Gels were then slowly dried in the dark. The absorption spectrum of AgI particles (Fig. 3b) in the glass is identical to those of the starting colloidal aqueous solutions. The luminescence intensity of AgI colloids in silicate glass increases by a factor of about two compared to aqueous sols. This is attributed to passivation of surface recombination sites by the silicate polymer.

Excess electrons on colloidal particles

We used pulse radiolysis technique to characterize excess electrons on colloidal particles. AgI particles have been chosen, since they are stable in aqueous solutions. Radiolytic reduction of colloidal silver halide to the metal has been studied earlier [14]. In our experiments, $(CH_3)_2\dot{C}OH$ was created in the colloidal solution during electron pulse, which then injects electrons into the colloids. The $(CH_3)_2\dot{C}OH$ radicals have a large negative redox potential for electron donation, $\dot{E}\,(CH_3)_2CO/[(CH_3)_2\dot{C}O^-H^+] = -1.23$ V (vs. NHE) and they can react with colloids according to

$$(CH_3)_2\dot{C}OH + AgI_{coll} \rightarrow (e^-)\,AgI_{coll}$$
$$+ CH_3)_2\,CO + H^+ \quad (1)$$

Immediately after the pulse, a transient bleaching was observed (Fig. 4) at wavelengths below 450 nm, with a bleaching maximum around 410 nm. The transient bleaching was also observed in the different small semiconductor colloids in the presence of excess charge carriers [1, 3, 13, 15]. The transient bleaching in Fig. 4 can originate from depletion of AgI particles and from excess electrons accumulated in particles. In the

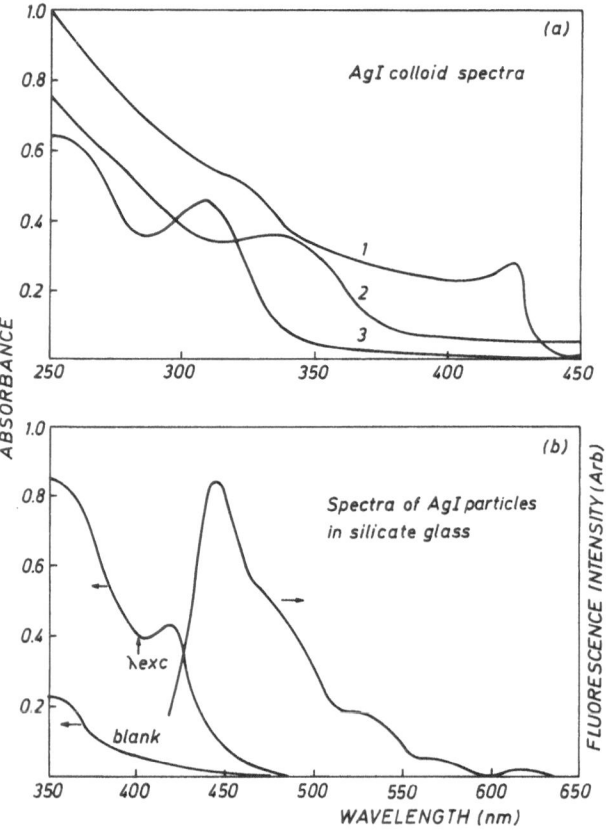

Fig. 3. Optical absorption spectra of AgI colloids: (a): (1) in water (2 × 10^{-4} mol dm^{-3} AgI) without stabilizer; (2) prepared in acetonitrile and dialyzed against water (1 × 10^{-4} mol dm^{-3} AgI) Polybrene (0.02 %); (3) in acetonitrile (10 % H_2O, 2 × 10^{-4} mol dm^{-3} AgI); (b) in silicate glasses

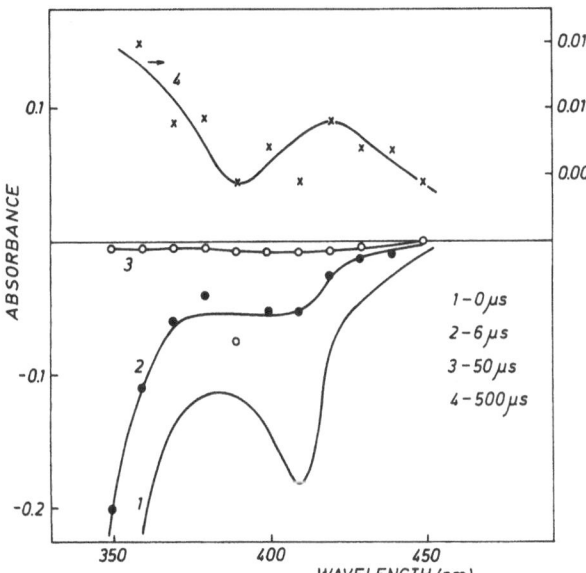

Fig. 4. Transient absorption spectra of 2×10^{-4} mol dm^{-3} AgI in water (containing 3×10^{-4} mol dm^{-3} (NaPO$_3$)$_6$, 0.1 mol dm^{-3} 2-propanol saturated with N$_2$O) recorded immediately after electron pulse (dose 27 Gy) and at different times

latter case, the absorption edge of AgI semiconductor can be shifted to higher energies because of the Burstein effect [7].

The recovery of this bleaching is a fast process followed by a first-order rate law and corresponds to the localization of e^- on silver atom which leads to the reduction of AgI

$$(e^-) \, \text{AgI}_{\text{coll}} \rightarrow \text{Ag}^\circ \, (\text{AgI})_{\text{coll}} + \text{I}^- . \qquad (2)$$

After 500 μs, stable colloidal metallic silver is formed. In Fig. 4 we can see clear evidence of the Ag$^\circ$ formation in a three-step process in the presence of (CH$_3$)$_2$ĊOH radicals. The first step involves electron injection and storage in the AgI particle, followed by localization of electrons at the silver atom and their agglomeration in colloidal silver.

Acknowledgement

We thank Dr. Arthur Nozik for valuable discussions.

References

1. Albery WJ, Brown GT, Darwent JR, Saievar-Iranizad E (1985) J Chem Soc Faraday Trans 1 81:1999
2. Balchin AA (1980) In: Levy I (ed) Crystallography and Crystal Chemistry of Materials with Layered Structures, Vol 2. D Reidel Pub Co, Dorderecht, p 31
3. Baral S, Fojtik A, Weller H, Henglein A (1986) J Am Chem Soc 108:375
4. Bedikyan LD, Miloslavskii VK, Ageev A (1979) Opt Spectros 47:225
5. Berry CR (1967) Phys Rev 161:611
6. Brus LE (1986) J Phys Chem 90:2555 and references therein
7. Burstein E (1954) Phys Rev 93:632
8. DiSalvo FJ (1974) In: Timmerhaus KD, Sullivan WJ, Hammel EF (eds) Low Temperature Physics LT-13, Vol 3. Plenum Press, New York, p 417
9. Fojtik A, Weller H, Koch V, Henglein A (1984) Ber Bunsenges Phys Chem 88:649
10. Griffiths TR, Anderson RA (1979) J Chem Soc, Faraday Trans 2, 75:957 and references therein
11. Harbeke G, Tossani F. (1975) RCA Rev 36:40
12. Henglein A (1983) In: Baxendale Memorial Symposium, Consiglio Nazionale delle Richerche, Instituto di Fotochimica e Radiazioni d'Alta Energia, Centro Stampa, Lo Scarabeo, Bologna, pp 43–58
13. Henglein A, Kumar A, Janata E, Weller H (1986) Chem Phys Lett 132:133
14. Johnston FJ (1978) Rad Res 75:286
15. Kamat PV, Dimitrijevic NM, Fessender RW (1987) J Phys Chem 91:396
16. Micic OI, Nenadovic MT, Peterson MW, Nozik AJ (1987) J Phys Chem 91:1295
17. Micic OI, Zongguan L, Mills G, Sullivan JC, Meisel D (1987) J Phys Chem, in press
18. Nenadovic MT, Rajh T, Herak R, Micic OI, Paterson MW, Nozik AJ (1988) J Phys Chem 92:1400
19. Nozik AJ, Williams F, Nenadovic MT, Rajh T, Micic OI (1985) J Phys Chem 89:397
20. Peterson MW, Micic OI, Nozik AJ (1988) J Phys Chem, to be published
21. Rajh T, Peterson MW, Turner JA, Nozik AJ (1984) J Electroanal Chem 228:55
22. Rajh T, Vucemilovic MI, Dimitrijevic NM, Micic OI (1988) Chem Phys Lett 143(3):305
23. Rossetti R, Nakahara S, Brus LE (1983) J Chem Phys 79:1096
24. Sandroff CJ, Hwang DM, Chung WM (1986) Phys Rev B 33:5953
25. Sandroff CJ, Kelty SP, Hwang DM (1986) J Chem Phys 85:5337
26. Tanaka T, Saigo H, Matsubara TJ (1982) Photogr Sci 26:92
27. Van Olphen H (1977) An Introduction to Clay Colloid Chemistry. J Wiley, New York
28. Wang Y, Herron N (1987) J Phys Chem 91:5005

Received December 10, 1987
accepted December 21, 1987

Authors' address:

O. I. Micic
Boris Kidric Institute of Nuclear Science
Vinca
P.O. Box 522
1101 Belgrade, Yugoslawia

Progress in Colloid & Polymer Science   Progr Colloid Polym Sci 76:24–26 (1988)

Charge carrier dynamics in colloidal semiconductors

H. Weller, M. Haase, L. Spanhel and A. Henglein

Hahn Meitner Institut, Bereich Strahlenchemie, Berlin, F.R.G.

Abstracts: Charge carriers on colloidal semiconductors were characterized by their optical absorption spectra. They were produced in a laser flash or by means of pulse radiolysis whereby the reactions of these charge carriers could be followed. In a biphotonic process, hydrated electrons were formed from CdS. After covering the surface with $Cd[OH]_2$, the fluorescence of the particles appeared at the band edge with a quantum yield of more than 0.5 and the photochemical stability was enhanced by several orders of magnitude to become comparable to that of fluorescence dyes, like rhodamine. Colloids of CdS or Cd_3P_2 could be attached to TiO_2 or ZnO, respectively. In these sandwich colloids an electron transfer occurred from the excited CdS or Cd_3P_2 to the TiO_2 or ZnO part, whereas the positive holes remained in the excited part of the sandwich.

Key words: Photochemistry, colloidal semiconductors, charge carriers, surface modification, sandwich-colloid.

1. Introduction

In the last few years, colloidal semiconductors have been studied by many groups. The method of preparation as well as the size of the particles plays a crucial role in determining their photochemical and photophysical properties [1–7].

If the particles consist of only a few 10 to 100 molecules then the photogenerated charge carriers "feel" the particle walls, which results in an increase in the energy of the excited levels [5, 8]. The absorption and emission spectra are thus blue shifted. This "Size Quantization" effect is observed in many semiconductor colloids. We call such materials "Q-particles". Besides this size effect, the characteristics of the surface essentially determine the photochemical behavior of the particle. In the following we report how, by controlled surface modifications, the properties of the charge carriers can be influenced and how these changes can be demonstrated with the help of spectroscopic and kinetic methods.

2. Results and discussion

When an electron is excited into the conduction band (CB) of a semiconductor by absorption of light ($h\nu_1$), a positive hole (h^+) remains in the valence band (VB).

Following their formation, both charge carriers move to traps of lower energy on the surface of the particle from where either fluorescence ($h\nu_2$) (broad band, stongly Stokes-shifted) or radiationless deactivation (shaded arrow) occurs (Fig. 1a). The latter process is normally very fast in colloidal CdS resulting in very short lifetimes of the charge carriers (smaller than 1 ns) and very low fluorescence quantum yields. Unsaturated sulfide valencies on the surface of the particle, on which the holes are trapped as S^- radicals, are probably responsible for the radiationless deactivation. Trapped holes can be produced and studied spectroscopically by the reaction of pulse radiolytically generated OH radicals with the particles [9]. Typical absorption spectra of the holes, i. e. the difference spectrum to

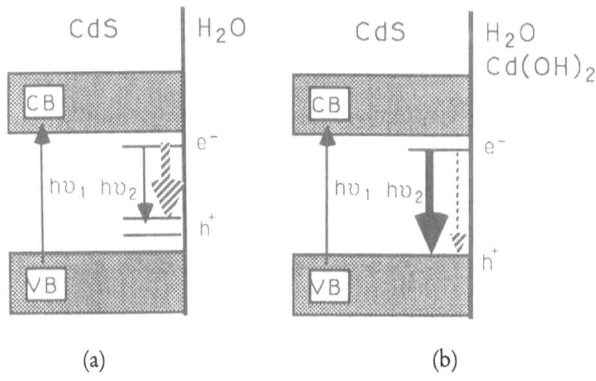

Fig. 1

pure CdS, as well as the absorption spectra of the corresponding CdS sols, are shown in Fig. 2 for large (a) and small (b) particles. The absorption spectrum and the chemical properties of the positive holes are entirely different from that of the free S$^-$ radical in homogeneous solution. The spectrum of the holes in small particles is distinctly blue-shifted in comparison to that of the large particles (size quantization).

Positive holes are the precursors for the usually undesired photocorrosion of CdS:

$$CdS + 2 O_2 \xrightarrow{h\nu} Cd^{2+} + [SO_4]^{2-}.$$

If the surface of CdS is covered with cadmium hydroxide, the sulfide valencies are saturated and the

holes can no longer be trapped (Fig. 1b). The electrons then combine directly with the holes in the valence band. This has dramatic consequences for the photochemical properties of the colloid [10]: (1) the fluorescence band becomes very narrow and appears at the absorption edge of the colloid; (2) the radiationless deactivation is strongly suppressed leading to a rise in the lifetime of the charge carriers to ca. 50 ns and an increase in the fluorescence quantum yield of up to 0.7; and (3) the photocorrosion is practically completely suppressed and the photostability of the colloid is comparable to or better than that of commercial fluorescence dyes.

Variation of the particle size enables the position of the emission maximum to be shifted over a wide spectral region.

The trapped electrons can also be studied spectroscopically in the same way as the positive holes can. The easiest method for doing this is via pulse radiolysis measurements in which the time resolved changes in the optical absorption, after the hydrated electrons have reacted with the particles, are followed. As a consequence of the storage of an electron on the surface of a particle, the absorption of the latter at the band gap is bleached with a $\Delta\varepsilon$ of up to 100,000 M^{-1} cm^{-1} [11]. The trapped electron polarizes the excitonic state (electron-hole pair) resulting from light absorption, whereby the optical transition is blue-shifted in comparison to the "normal" particle. This effect is also observed in Laser-flash photolysis experiments [12, 13]. During the lifetime of the charge carriers, generated by the flash, bleaching of the colloidal solution occurs at the band edge. If, as is normally the case, many photons are absorbed per particule during the flash, emission of electrons into the solvent water is observed [13, 14]. These hydrated electrons are formed in a biphotonic process for which the following mechanism is proposed:

$$CdS \underset{k_1}{\overset{h\nu_1}{\rightleftharpoons}} CdS(h^+/e^-) \underset{k_2}{\overset{h\nu_2}{\rightleftharpoons}} CdS(h^+/e^-)_2$$

$$\xrightarrow{k_3} CdS(h^+) + e^-_{aq}.$$

The absorption of the first photon ($h\nu_1$) creates an electron-hole pair which can recombine with a rate constant k_1. If, during the lifetime of this excited state, the particle absorbs a second photon ($h\nu_2$), then the energy of the second electron-hole pair can result in the electron being at an energy level high enough to leave the particle (k_3). In this process, a positive hole,

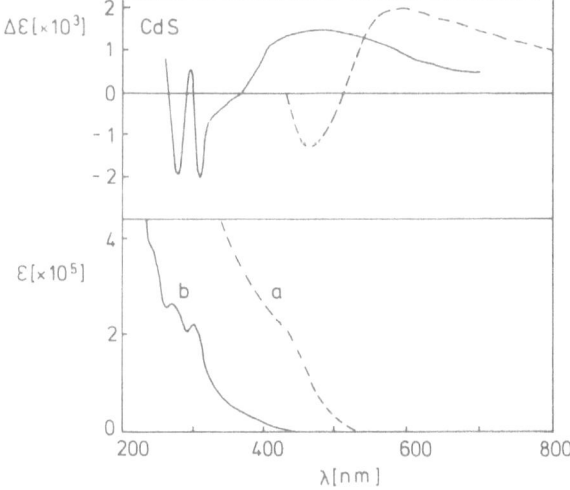

Fig. 2

particle I   particle II

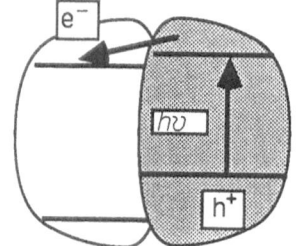

Fig. 3

which can be spectroscopically evidenced, remains on the particle. In competition to electron emission is the very fast recombination (k_2, $\tau < 50$ ps) of the "doubly excited" particle. The hydrated electrons can be identified by their characteristic absorption and typical chemical reactions. Hence they can, in the same manner as the pulse radiolytically generated hydrated electrons, diffuse to an intact CdS particle where they become trapped and bleaching at the band edge is observed.

Using appropriate techniques it is possible to prepare "sandwich colloids" from two different semiconductor materials [15, 16]. Their energy scheme is shown in Fig. 3.

Such a sandwich consists of particle I, having a large band gap and an energetically low-lying conduction band (TiO_2 or ZnO), and of particle II having a small band gap and an energetically high-lying conduction band (CdS or Cd_3P_2). If particle II is excited by long wavelength light, then its fluorescence is completely suppressed and the electron is transferred to particle I. For energy reasons the simultaneously generated positive hole remains on particle II. As a result of this charge separation, the lifetime of the charge carriers becomes very long which leads to a dramatic increase in the photocatalytic activity.

References

1. Henglein A (1982) Ber Bunsenges Phys Chem 86:301
2. Henglein A, Fojtik A, Weller H (1987) Ber Bunsenges Phys Chem 91:441
3. Ramsden JJ, Grätzel M (1984) J Chem Soc Faraday Trans I 80:919
4. Nozik AJ, Williams F, Nenadovic MT, Rajh T, Micic OI (1985) J Phys Chem 89:397
5. Brus L (1986) J Phys Chem 90:2555
6. Nedeljkovic JM, Nenadovic MT, Micic OI, Nozik AJ (1986) J Phys Chem 90:12
7. Kuczynski J, Thomas JK (1983) J Phys Chem 87:5498
8. Weller H, Schmidt HM, Koch U, Fojtik A, Baral S, Henglein A, Kunath W, Weiss K, Dieman E (1986) Chem Phys Lett 124:557
9. Baral S, Fojtik A, Weller H, Henglein A (1986) J Am Chem Soc 108:375
10. Spanhel L, Haase M, Weller H, Henglein A (1987) J Am Chem Soc 109:5649
11. Henglein A, Kumar A, Janata E, Weller H (1987) Chem Phys Lett 132:133
12. Albery WJ, Brown GT, Darwent JR, Saievar-Iranizad E (1985) J Chem Soc Faraday Trans I 81:1999
13. Haase M, Weller H, Henglein A (1988) J Phys Chem, in press
14. Alfassi Z, Bahnemann D, Henglein A (1982) J Phys Chem 86:4656
15. Spanhel L, Weller H, Henglein A (1987) J Am Chem Soc 109:6632
16. Spanhel L, Weller H, Henglein A (1987) Ber Bunsenges Phys Chem 91:1359

Received December 22, 1987;
accepted January 4, 1988

Authors' address:

H. Weller
Hahn-Meitner-Institut
Bereich Strahlenchemie
Glienicker Straße 100
D-1000 Berlin 39, F.R.G.

Progress in Colloid & Polymer Science   Progr Colloid Polym Sci 76:27–31 (1988)

Clay chemistry – colloidal chemistry

R. Söderblom

Swedish Geotechnical Institute, Linköping, Sweden

Abstract: The changes in properties and behaviour of clays when exposed to the influence of various substances are known to a very limited extent, in particular with regard to clays in undisturbed condition. Most of the colloid chemical laws established for diluted colloid solutions are unsuitable for explaining the behaviour of undisturbed clays. Thus the Schultze-Hardy rule for predicting colloid stability when adding electrolytes, and which has hitherto been regarded as a background for different stabilizing processes, is not valid for undisturbed clays. One important factor in this connection is the clay structure. Experiments have been carried out to establish a clay structure by means of dielectric dispersion, comparing clays with artificial lamellar model systems. The investigations of the undisturbed clays were carried out both in the laboratory on extracted samples and in situ by pumping different chemicals into the ground with subsequent sampling. It was found that rheopexy occurred in some clays and caused difficulties in measuring the strength properties. Model studies of rheopexy were carried out. Various experiments regarding the importance of the double layer for stability properties were also initiated.

Key words: Clay colloid stability, clay structure, model systems, infiltration experiments, electric double layer.

Abbreviations

τ_f is the shear strength of a clay in unremoulded state (f is for failure)

nF   is an abbreviation for nanofarad (one thousand million part of a farad). C means capacitance.

[mS] G is the conductivity in millisicmens

f is the frequency

Introduction

Knowledge of the changes in properties and behaviour of clays when exposed to the influence of various substances is very limited, in particular with regard to clays in an undisturbed condition. The laws of colloidal chemistry, which have been established in the case of diluted colloidal solutions and which have hitherto been considered applicable to undisturbed clays, have proved unsuitable in such cases. The investigations show, for example, that the Schultze-Hardy rule [15] is inapplicable to undisturbed clay systems, thereby invalidating the theoretical background for lime stabilizations of undisturbed clay masses. In order to explain the reactions of an undisturbed clay mass, it is necessary to form a suitable picture of the structure of the undisturbed clay. Most of the structures discussed assume a homogeneous structure which cannot, however, explain the brittle structure demonstrated by highly sensitive clay specimens. The fragmentation into small crystalloids that is easily produced demands a different structure, e.g. domains linked together to form clusters and aggregates. This paper discusses, among other things, whether dielectric dispersion and a comparison with artificially created lamellar systems can reveal a resemblance between the clays and lamellar domain systems of this type. Agglomerations of this type give rise to completely different reactions with electrolytes when compared with individual flake-shaped particles, where Onsager's theory of small platelets is applicable. The results of this work show that changes in the double layer of the clay in treatment with various substances influence the strength properties of the clay, so that the system becomes either more fluid or more solid at the same moisture content. The experiments performed so far indicate that there is, in principle, no reactional difference between increases

Fig. 1. Photograph of a landslide near Lake Como, Italy

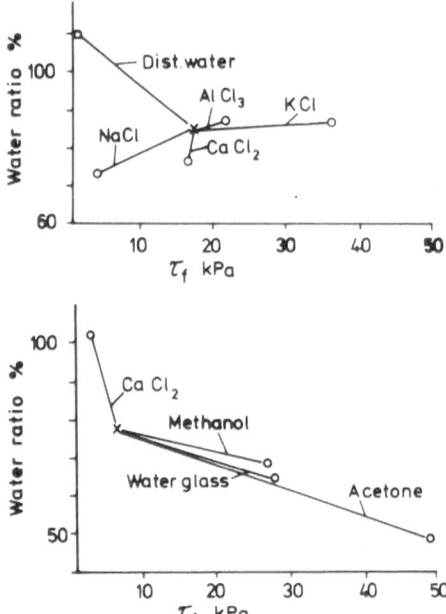

Fig. 2. Influence of different chemicals on the strength of undisturbed clay samples

and decreases in the strength of a clay. Here, the double layer composition constitutes a very important parameter, whose significance for the mechanical properties of clays has hitherto been studied only to a limited extent.

Some time before the actual conference, large areas of Northern Italy not far from Lake Como were exposed to severe landslides which caused great damage. Several towns and villages had been evacuated because of the risk of new landslides. It was heard that two villages had been buried by clay masses sliding down the slopes. An illustration of the extent of these debris masses is given in Fig. 1 in the centre part showing the destruction of houses.

One of the causes of the landslides was intensive rain. Other factors may also be involved. One important question is the extent to which colloid chemical changes in the clay system contributed by reducing its strength and increasing the wettability of the soil.

Knowledge of these types of processes in soil, especially with regard to undisturbed clay systems, is very limited and is practically non-existent in the case of such clays in situ. Almost all reported works are based on studies of remoulded clay samples, clay suspensions etc. where the normal colloid chemical laws are valid. In order to understand the processes at work in nature it is necessary to make a basic study of the behaviour of clays (a) in situ, (b) in the form of "undisturbed" samples and (c) in the remoulded form. It is also necessary to verify the extent to which the normal colloid chemical laws can be used to describe the behaviour of clays. The laws of colloid chemistry were originally worked out to suit the case of diluted colloid solutions [1, 2]. It has been often assumed without any proof that these laws are applicable to undisturbed clays [3–5].

Experiments with undisturbed clay samples indicate that the problem is not as simple as at first thought. A great number of experiments have been carried out with undisturbed clay samples by the present author, treating them with different chemicals. As seen from Fig. 2, the Schultze-Hardy rule cannot be valid for such samples. Ca^{2+} ions give a looser material than sodium and potassium ions. This is in accordance with results obtained by Swedish Railways [6] who have tried to stabilize areas by introducing $CaCl_2$ into the ground. As seen from Fig. 2, Ca^{2+} ions have no stabilizing effect on undisturbed clays. The conclusion from the experiments is that the Schultze-Hardy rule cannot be used as a background for the stabilization of clays.

Fig. 3. Quick clay sample broken into pieces, indicating a nodular structure

Clay structure and mechanical behaviour

In order to explain the reactions of an undisturbed clay mass, it is necessary to form a suitable picture of the structure of undisturbed clay. Hitherto, almost all theories about the properties of clays have been based on an interaction between individual flaky mineral grains, as stated in conventional colloid chemistry dealing with diluted colloid solutions [1, 2]. However, there is a growing opinion that both artificial and natural clay soils are formed of plates, creating aggregates such as packs, clusters or domains.

Many clays have a pattern of microinhomogeneities and can therefore be broken down into small fragments indicating a quasi-crystalline nature. The individual crystallites are separated from each other by thin films of water. Thus there are very weak planes in a clay and these create a risk of instability (Fig. 3).

The domain structure in question has been studied by Aylmore and Quirk [7], Olsen [8], Stepkowska [9] and other authors. The structure has been investigated in several ways, for example, by water sorption, water retention and thermal weight loss. Experiments have been carried out by the present author to verify the clay structure proposed by making comparisons between substances with a proven lamellar domain structure and clays [10]. In this particular case, the lamellar phase of sodium caprylate, decanol and water was used. The clays exhibit approximately the same changes in capacitance and conductance of different frequencies as the model samples. One example of temperature dependence is shown in Fig. 4. The curves are similar to those for the model systems. Measurements were made when the temperature was both increased and decreased.

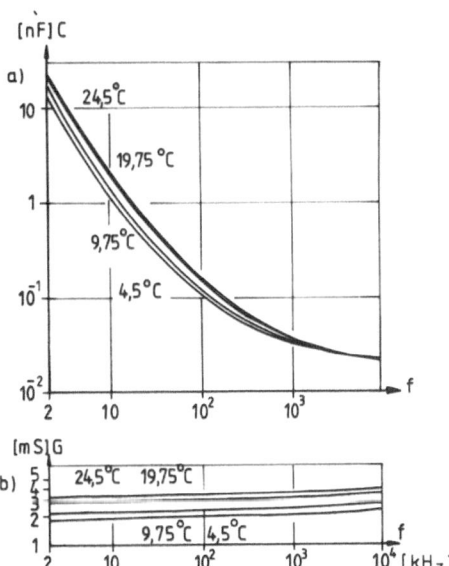

Fig. 4. Capacitance and conductivity as a function of temperature, Lake Roxen 8 m, Sweden

The results obtained show that the behaviour of the clay in chemical treatment must be quite different when working with remoulded systems compared to undisturbed systems. Owing to difficulties with present sampling techniques, the clays must be investigated in situ.

Experimental

The behaviour of remoulded clays is so well-known and unimportant to this work that no results from such systems are reported here. The investigated clay samples were extracted with a 22 mm piston sampler [11]. Samples from many places, including Stockholm, Lidingö, Karlskoga and Norrköping, were investigated. The clay samples were treated with different chemicals and their strength was determined. The reason for mentioning sewage here is that this is supposed to have played an important part in a number of landslides.

On the strength of results obtained from remoulded samples, the Schultze-Hardy rule has been accepted as a principle. Thus, monovalent ions are assumed to coagulate an untreated clay mass, divalent ions and three-valent ions having a successively increased effect.

The results indicate that this is not the case in undisturbed Swedish clays. Sodium ions make the system looser, as do divalent Ca^{2+} ions. Al^{3+} ions also have little effect. On the other hand, the monovalent K^+ ions have a strong effect, similar to NH_4Ac. NH_4OH makes the system much looser. (In Norway, KCl has been used as a stabilizing agent for quick clays [12]). There are also substances, such as water glass, acetone, ethanol, methanol etc., that greatly increase the strength of the clay and decrease its water ratio.

Field experiments

The experiments mentioned above have been carried out on extracted clay samples. There is, however, no proof that such treatments reflect the behaviour in the ground. In order to study behaviour, a device has been constructed which makes it possible to pump different substances into the ground. The first experiments were made with water and sewage. An inspection trench was dug 1 m from the infiltration pipe. It was observed that the water moved relatively fast in the clay and it took only about 15 min for the water to reach the trench. Water was seen to appear in local veins. The rapid flow can be explained by water following the structure and the cracks in the clay.

An inspection showed that tap water passed through the clay without markedly affecting the system. It was also possible to see how the tap water formed small spheres on the clay surfaces, indicating non-wettable conditions, i. e. too large a contact angle between liquid and solid (cf. [13,14]). It may be directly concluded that in a chemical context, the danger from heavy rain depends on the wettability of the soil. According to Chen and Schnitzer [13], humic substances, especially fulvic acids, increase the wettability of soils. The same applies to certain leaf extracts [13] e. g. from poplars and maples. Other extracts, e. g. from citrus trees [14], make a soil non-wettable. If the vegetation on a slope is removed or changed, the wetting conditions may be altered considerably, leading to erosion, dispersion and sliding of the soil.

Sewage, having a lower surface tension, attacks the aggregates and washes out a flowing mass by infiltration in the soil. The liquid is not a dispersion but contains aggregates several millimetres in diameter, with an irregular form. The same occurs with infiltrating sodium diphosphate, for example. It should be noted that similar aggregates can be found in flow masses in large landslides.

Sampling and rheopexy

It was planned that the changes in mechanical properties of the clay around the infiltration pipe would be investigated by means of sampling and a subsequent investigation in the laboratory. As mentioned above, the introduced substance travels through veins and consequently it is difficult to find a suitable site for a borehole. Samples were therefore extracted near the trench wall in a position where liquid was transported. It was found that it is very difficult to determine the

Table 1. Model studies on sampling

Iron (II) oxalate	τ_f (kPa)	water ratio (%)
Untreated	0.4	33.0
Sampled	> 1.1	8.4

actual conditions in the ground. If the clay is composed of aggregates and a cylinder with a sharp edge is used to extract a sample, the edge must also split aggregates. In some clays, this will result in rheopexy, giving a much harder sample than is representative of the in situ conditions. This may be one of the explanations why it has been almost impossible in several cases to find slide-susceptible areas by conventional sampling and why slides may occur in areas judged to be stable.

In order to investigate the disturbances, a number of experiments were carried out with a typical rheopexic model substance, a gel iron (II) oxalate. The gel was precipitated from iron (II) sulphate and was dried to about 30 % water ratio, at which value it showed typical rheopexic behaviour. An example of the results is shown in Table 1. The gel was investigated intact and after "sampling" with a piston tube 22 mm in diameter. It can be seen that changes occur both in the water ratio and in strength. This very important behaviour of some clays must be studied further and the only practical way appears to be to work with model substances. Otherwise, working with clays involves using clay samples that are already disturbed.

Double layer composition and mechanical behaviour of clays

Applying double layer composition in the structure proposed above is a very complicated task, involving double layers of simple particles and double layers of aggregates. An investigation is planned to study the influence of the double layer on the structure of, firstly, lamellar model substances and secondly, clays. The method to be used will be dielectric dispersion technique. In addition, comparative rheological experiments will be carried out with model substances and clays. However, experiments reflecting the behaviour in situ soils can be very difficult to perform, owing to the disturbances mentioned previously. In spite of this, colloid chemistry is an essential aspect in understanding the processes at work in the ground which influence the properties of clay with regard to landslide development.

References

1. Verwey EJW, Overbeek JThG (1948) Theory of the Stability of Lyophobic Colloids. Elsevier, New York
2. Langmuir I (1938) J Chem Phys 6:873
3. Lambe TW (1958) Proc Am Soc Civ Eng 84:1654
4. Moum J, Löken T, Torrance JK (1971) Géotechnique 21:329
5. Woo SM, Moh ZC (1977) Proc 9th Intern Conference Soil, Mech Found Engng, Special session 11, MAA Publishing Company, Taipei, 451:464
6. Jerbo A (1972) Stat Järnv Centralf Geot Kontoret Medd 28
7. Aylmore L, Quirk JP (1960) Nature, London 187:1046
8. Olsen HW (1962) Proc 9th Nat Conf Clays Clay Miner 19:131-161
9. Stepkowska ET (1982) Arch Hydrotechn 29:469
10. Söderblom R, Sandelin CJ, Thornér K, Sjöblom J, Lundström I (1984) Proc Nord Geot m 1984, Swedish Geotechnical Institute, Linköping, 2:949-955
11. Adestam L (1981) Proc 10th Int Conf Soil Mech Found Eng, Balkema A A, Rotterdam 2:413-418
12. Madshus PA, Janbu N (1984) Proc Nord Geot m 1984, Swedish Geotechnical Institute, Linköping 2:921-930
13. Chen Y, Schnitzer M (1978) Soil Sci 125:7
14. Letey J, Osborn J, Pelishek RE (1962) Soil Sci 93:149
15. Jirgensons B, Straumanis ME (1962) A short textbook of colloid chemistry. Pergamon Press, New York

Received December 2, 1987;
accepted December 11, 1987

Author's address:

R. A. Söderblom
Swedish Geotechnical Institute
S-58101 Linköping, Sweden

Progress in Colloid & Polymer Science   Progr Colloid Polym Sci 76:32–36 (1988)

Image analysis of fractal aggregates of precipitated silica

A. Tuel, P. Dautry, H. Hommel, A. P. Legrand and J. C. Morawski[1])

Laboratoire de Physique Quantique, UA CNRS 421 ESPCI, Paris, France
[1]) Rhône-Poulenc Films St. Maurice de Beynost, Miribel, France

Abstracts: Amorphous silica usually appears as a set of more or less spherical particles of a few hundred angstroms in diameter, building up extended aggregates. To understand how the shape of the aggregates observed by electron microscopy is related to the synthesis process, we have simulated by DLA and CCA models the way in which the elementary particles connect to each other. We study in particular how two-dimensional images inform us of the actual three-dimensional structure of the fractal aggregates. Results demonstrate the possibility of describing the morphology of precipitated silica using these models. A better understanding and a better control of the aggregation step seems possible by using these methods.

Key words: Precipitated silica, image analysis, aggregates, fractal dimension, numerical models.

Introduction

The study of aggregation is of both practical and fundamental interest in material science. Silica formed in aqueous solution is a material with a wide range of technological applications and it is the fact that particles are aggregated into ramified structures that often makes it useful.

Typically, it is made by the action of an acid on aqueous alkaline silicate, such as sodium silicate, at pH higher than 7. The precipitated silica produced is made up of aggregates of elementary particles with a specific surface area lower than 350 m^2/g [3].

The size and structure of these aggregates seem to be very important parameters in industrial applications, such as rubber reinforcement. An example is given in Fig. 1, in which we have plotted the tear strength of filled rubber versus the specific surface of the silica measured by CTAB (cetyl trimethyl ammonium bromide) adsorption [7].

Since mechanical properties largely depend upon structure of the aggregates, different methods have been developed to characterize them. In particular, electron microscopy combined with image analysis provides information on their two-dimensional projections. In order to quantify these projections, differ-

ents size parameters such as the projected surface area S, the perimeter P, the longest length L and the smaller with l have been used (Fig. 2) [3].

From these parameters we can also define two factors of form, the circularity $C = P^2/4\pi S$ and the anisotropy $E = L/l$. Although these differents factors only characterize the projections of the aggregates, it has been observed empirically that they could be related to the behaviour of the material after incorporation of the silica [3]. For instance, aggregates with large surface

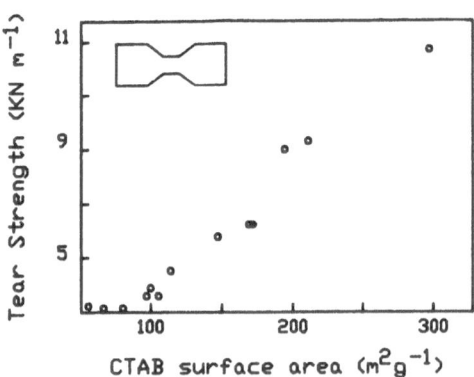

Fig. 1. Tear strength of filled rubber versus the specific surface area of the precipitated silica

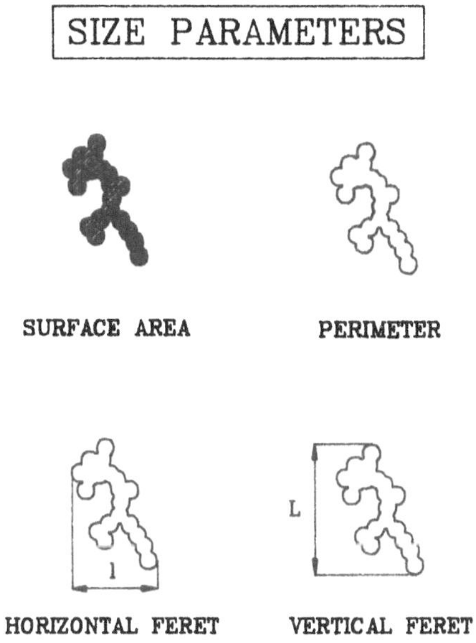

SIZE PARAMETERS

SURFACE AREA   PERIMETER

HORIZONTAL FERET   VERTICAL FERET

Fig. 2. Size parameters used to describe the two-dimensional projections of aggregates

projection areas allow a better dispersion of the silica in the elastomer, which improves its mechanical properties [1].

Surface projections can easily be enlarged by increasing the duration of the precipitation reaction [3]. Concurrently, other measurements have demonstrated that an increase in the time of the precipitation process can also modify the morphology of the aggregates.

Many other correlations could be made between the operation conditions and the morphology of the aggregates themselves. By using these correlations it is possible to predict the morphology of the aggregates in order to obtain specific applications.

Numerical models

In recent years numerous computer simulations and several experiments have revealed that clusters formed by aggregation processes may be characterized as fractal structures on length scales, between the size of the single particle units and the typicall overall size of the cluster.

Witten and Sander [9] have shown that a simple diffusion-aggregation process in which particles are added, one at a time, to a growing cluster of particles via random-walk trajectories leads to scale-invariant structures with a fractal dimensionality, D, which is considerably smaller than the Euclidian dimensionality of the space in which the growth process is occuring. Three-dimensional simulations lead to structures with a fractal dimension of about 2.5.

This model, called DLA model, has in fact been invented to study the morphology of aerosols. However, it did not do so

sucessfully. The strongest evidence for this is probably the fractal dimension itself. For three-dimensional simulations, the clusters have a fractal dimension which is higher than those generally observed in growth experiments.

A more adapted model has been developed by Meakin [6] and Julien [4] in which clusters can diffuse and stick together to form larger clusters. The result of this model, called CCA (Cluster-Cluster Aggregation), is a more open structure with smaller fractal dimension than the DLA clusters. Three-dimensional simulations lead to structures with fractal dimensions of about 1.8.

To compare fractal dimensions measured on electron microscopy images with those from computer models, we examined the two-dimensional projections of three-dimensional clusters built by a CCA aggregation process. Many authors state that the projection of a three-dimensional aggregate to two dimensions will not affect the value of the fractal dimension D for values of D less than 2 [5]. In fact, this reasoning is correct for infinite aggregates, and for D less than, but close to, 2, finite size effects will affect the projection and hence the value of D [8]. But, in the present case, aggregates built by computer simulations and images of silica have been studied using the same method and a direct comparison is possible.

The value of the fractal dimension D, obtained from about 100 projections of three-dimensional simulations, has been evaluated to:

$$D = 1.8 \pm 0.1.$$

The size distribution function of three-dimensional clusters obtained by a CCA process can be fitted by a Botet distribution [2]:

$$N = \alpha^2 n e^{-\alpha n}$$

where n is the number of n-particles clusters. The parameter α depends upon the aggregation time and decreases with it. The same mathematical function has been used to characterize the two-dimensional projections of the aggregates. The correlation between the size distribution function obtained on the projections and the mathematical distribution is still good for small aggregates containing less than 50–60 particles. For a larger aggregate, the projection has a higher probability to superpose on to another one and it is impossible to distinguish them from the projection of a single aggregate. Nevertheless, the size of the aggregates of silica did not exceed 30 particles in projection, so we did not take this problem into account. From these projections it was possible to measure the evolution of the parameter α with the aggregation time.

We also characterized the evolution of the perimeter of the projections with their radius of gyration. We have shown that this evolution can be described by a law of the type:

$$P \sim R^\nu.$$

The value of the parameter ν, measured on about 100 clusters, is

$$\nu = 1.2 \pm 0.1.$$

Knowledge of the parameter ν and of the evolution of α with the aggregation time allows calculation of the circularity of the aggregates at differents steps of the simulation. Since we have defined C

Progress in Colloid and Polymer Science, Vol. 76 (1988)

as the ratio $P^2/4\pi S$, it is easy to see that C varies with the number of steps p of the simulation, following the law:

$$C \sim p^y$$

where $y = 2\nu/D - 1$.

The mean circularity of the aggregates for a given value of p can be given by:

$$\langle C(p) \rangle = \sum C(n) \cdot N_p(n)$$

where $N_p(n)$ is the number of clusters of n particles after p steps of aggregation.

In Fig. 3 we have plotted $\langle C(p) \rangle$ versus p for a number of steps from 1000 to 30 000. The evolution is practically linear.

Image treatment and calculation of the fractal dimension

Electron microscopy images show that silica powder consists of aggregates of spherical particles. The particles have a size of about 250 Å in diameter. We also note that particles overlap frequently. Images were recorded at a magnification of 200 000. In order to calculate the fractal dimension, 32 images were digitalized. The digitalization apparatus consisted of a video camera (Hitachi CCTV) connected to the interface card DGI1 (from MID). Digitalized images were stocked in 256 × 256 point arrays. These images, once recorded were treated in the following manner. The computer program written for this study found the outer limit of the aggregate (one per frame). If there are several aggregates in the frame this routine can distin-

guish them as long as they do not interpenetrate. The average intensity of the image outside the aggregate was then measured and this value taken to zero. Conversely, all the other pixels have an intensity of 1, so that we do not take into account fluctuations of intensity. The images of aggregates were analysed to yield the fractal dimension by finding the scaling relation between mass and size for all the aggregates. Thus we simply counted all the digitalized pixels inside the aggregate and related this to a measure of the size of the aggregate. We chose to use the value $(L + l)/2$ for the size, where L and l are respectively the longest length and the smaller width of the aggregate defined above. From a log-log plot of mass versus size for all the aggregates measured we obtained a set of points which can be fitted to a straight line, the slope of which gives the value of the fractal dimension D.

To calculate the exponent ν which characterizes the perimeter/size scaling, we used the following method: from a digitalized image we kept only those pixels which had two or three nearest neighbors (with an intensity of 1) on the 256 × 256 square lattice. In this way we can get rid of all the points of the surface (which have four nearest neighbors) and the points which stick of the surface by a single side.

Circularity was evaluated using two differents methods. In the first, the mean circularity was measured by taking the mean value of the circularities measured directly on each aggregate. This method gives a relatively low value because it does not take into account the size distribution of the aggregates. The second method consists in measuring the average radius Rm of all the aggregates and correlating it with a size distribution function. We chose to use a Botet distribution so that we could directly access the parameter α. The mean circularity (C) is then calculated with the relation.

$$\langle C \rangle = \sum P_a(n) C(n)$$

where $P_a(n)$ is the probability of having a cluster of n particles.

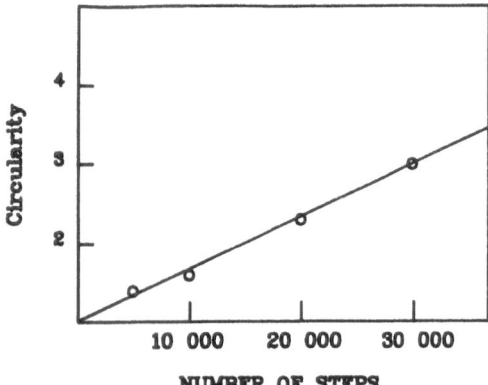

Fig. 3. Mean circularity of the aggregates built by the CCA model as a function of the number of steps in the simulation

Results of the measurements on silica

32 aggregates were digitalized and studied. The result of the total/mass size scaling method for all aggregates is reported in Fig. 4. The point scattering is quite notable but a straight line fit is reason-

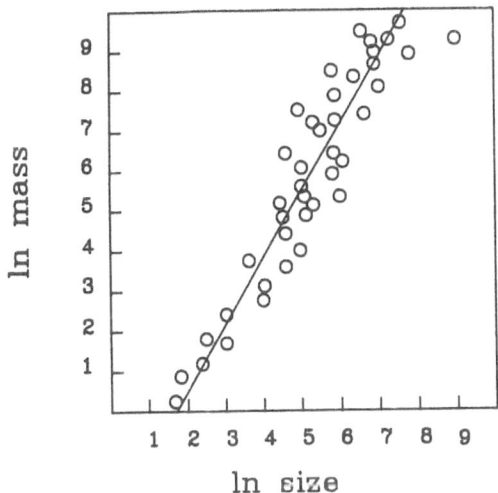

Fig. 4. Scale relation between mass and size for aggregates built by CCA model

able and leads to a value of the fractal dimension D:

$$D = 1.8 \pm 0.1.$$

A similar plot is obtained for the perimeter/size scaling and gives a value of the exponent ν

$$\nu = 1.3 \pm 0.2.$$

The mean value of the radii measured on all the aggregates was found to be equal to 8-particle size units. Considering a Botet-like size distribution function, the corresponding value of the parameter α is approximately 1/4.

For circularity, the first method gives a value of about

$$\langle C \rangle = 4 \pm 0.05$$

and the second gives a higher, and more reasonable, value of

$$\langle C \rangle = 5.25 \pm 0.05.$$

Discussion and comparison with numerical models

We compared these experimental values with those obtained on clusters built with the Cluster-Cluster Aggregation (CCA) model. We have chosen one of the simplest CCA models, the "brownian trajectories" model [5,6]. To compare visually the simulated aggregates with the experimental images, Fig. 5 shows projections of clusters constructed in three dimensions and a typical example of precipitated silica (Rhône-Poulenc) aggregates micrography. This simulation was carried out using brownian trajectories on a square lattice. The similarity of the pictures is clear.

For this model, the fractal dimension of two-dimensional projections, evaluated by averaging over 100 clusters, is estimated to be:

$$D = 1.8 \pm 0.1.$$

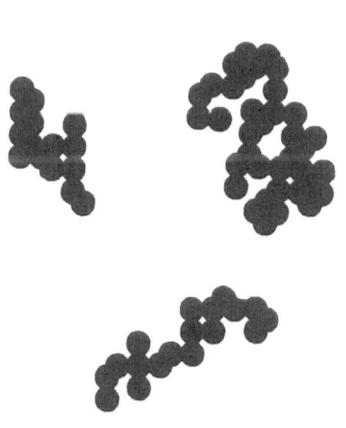

a)

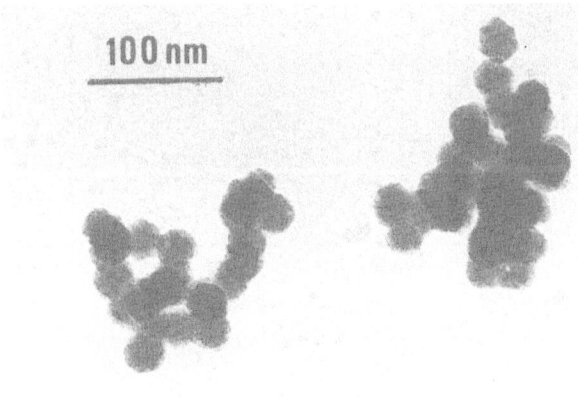

b)

Fig. 5. Projections of three-dimensional aggregates built by CCA model (a) compared to microscopy images of a precipitated silica (b)

The same value has been found for the experimental aggregates of precipitated silica.

Moreover, if we calculate the circularity of simulated aggregates corresponding to a size distribution function with parameter $\alpha = 0.25$, we find that it is equal to:

$$C = 4.75 \pm 0.05.$$

This value is in good agreement with that of 5.25 ± 0.05 found experimentally by using the same size distribution function.

Finally, the exponent ν obtained on simulations was estimated to be:

$$\nu = 1.2 \pm 0.1.$$

This value is slightly lower than the one obtained on experimental silica images, but the precision of the method is such that this difference is not very significant.

Comparing all the experimental values with the results obtained by simulations on a CCA model suggests that aggregation of precipitated silica is, by random walk process, appropriate for aggregation in a liquid solution.

Conclusion

We digitalized and analysed 32 different electron micrographs of precipitated silica aggregates. Fractal dimension was calculated using the average mass/size method. This method gives a value of

$$D = 1.8 \pm 0.1.$$

We also calculated the perimeter/size scaling which can be described by an exponent ν. This exponent was found to be

$$\nu = 1.3 \pm 0.2.$$

The comparison of these values with the results obtained by simulations of theoretical models suggests that the process of aggregation is a random walk process appropriate for aggregation in a liquid solution.

References

1. Bomo F, Morawski JC, Lamy P, Le Thiesse JC (1985) Quantitative determination of the morphology of precipitated silicas: relationship between morphology, specific surface area and mechanical properties of filled elastomers. ACS Rubber Division, Cleveland USA
2. Botet R, Julien R (1984) J Phys A, Math Gen 17:2517
3. Chevalier Y, Morawski JC (1986) Precipitated Silica: preparation, characterization European Chemistry: Silica. Le Bischenberg-Obernai, September 22–24, France
4. Kolb M, Botet R, Julien R (1983) Phys Rev Lett 51:1123
5. Mandelbrot B (1975) Les objects fractals, Flammarion, Paris
6. Meakin P (1983) Phys Rev Lett 51:1114
7. Morawski JC (1982) An exciting future for silica I: Surface area and tear strength. Int Rubber Conf, Paris, France
8. Tence M, Chevalier JP, Julien R (1986) J Phys 47:1989
9. Witten TA, Sander LM (1981) Phys Rev Lett 47:1400

Received December 4, 1987;
accepted December 14, 1987

Authors' address:

A. Tuel
Laboratoire de Physique Quantique
UA CNRS 421 ESPCI
10 rue Vauquelin
F-75231 Paris Cedex 05, France

Progress in Colloid & Polymer Science   Progr Colloid Polym Sci 76:37–41 (1988)

Non-intrusive determination of particle size distribution in a concentrated dispersion

C. Carter, D. J. Hibberd, A. M. Howe, A. R. Mackie, and M. M. Robins

AFRC Institute of Food Research, Norwich Laboratory, Norwich, U.K.

Abstract: Concentration profiles of a 19% oil-in-water emulsion were determined non-intrusively (from a measure of the velocity of ultrasound through the emulsion at a series of heights) over a period of 63 days. The oil concentration gradually increased with height until a cream layer of ~ 70% oil was reached. Such profiles are typical of a dispersion of polydisperse particles with very weak interactions. The size distribution (by weight) was calculated from the rise velocities of contours of constant volume fraction. The sizes ranged from 200 nm to 12 μm. The size distribution of the small particles was compared with fractionated sections of the creaming emulsions sized by dynamic light scattering. The large particles were compared with laser light diffraction studies. In both cases the particle size distribution by creaming agreed well with the optical studies. A delay in the creaming for the small droplets (diameter < 0.5 μm) is observed and discussed. The implications of such a delay for size determination from a single concentration profile are also discussed.

Key words: Size distribution, concentrated dispersion, emulsion, creaming, concentration profiles.

Introduction

Many properties of concentrated colloidal dispersions are strongly dependent on the particle size distribution (e.g. maximum packing fraction, stability to coalescence or gravity) [5]. In our experience it is common for particle diameters to range from 100 nm to 10 μm in suspensions and emulsions. Determination of the complete particle size distribution in this range is very difficult but can be made by electron microscopy or non-intrusively by techniques such as light scattering when a sophisticated analysis using full Mie theory is required. Light scattering techniques require dilute dispersions to avoid multiple scattering. Such conditions may not be appropriate for (originally concentrated) dispersions of colloidal particles of low, but significant, solubility.

Gravitational sedimentation may be used to determine the size distribution in concentrated dispersions from a measure of the concentration profile and application of Stokes' law [3]. The attenuation of X-rays is used to measure the concentration profile in concentrated (opaque) dispersions where there is a large difference in (electron) density between the continuous and disperse phases. However, such a technique is not applicable for systems where the phases are aqueous and organic. Turbidimetric detection of concentration profiles is again limited to dilute dispersions and has the added complication that specific turbidity varies with particle size.

In this study the concentration profiles of a creaming alkane-in-water emulsion are determined from a measure of the velocity of ultrasound at selected heights over the sample. From a study of the complete creaming behaviour of the emulsion a detailed size distribution (by weight) is obtained. The size distribution is used to predict a mean value of the diameter at each height after significant creaming had occurred and the results are compared with dynamic light scattering measurements on sections of the emulsion extracted from those heights. The size distribution of the large particles is compared with light diffraction measurements.

Experimental

Materials and emulsion perparation

Emulsions of n-alkane-in-water (Fisons, AR) were made in a constant-power Waring Commercial Blender using a set programme of shear cycles. The oil was a mixture of nine parts by volume heptane (BDH), 99.5 %) to one part hexadecane (Aldrich, 99 %). By using a mixture of high and low alkanes Ostwald ripening was avoided [4] while obtaining a substantial density difference (307 kgm^{-3}) between the disperse and continuous phases. The emulsion droplets were stabilised to coalescence by 0.28% w/v nonionic surfactant Brij 35 (nominally C12E23, Sigma). The composition of the emulsion was 19 % v/v oil, with an aqueous phase containing 0.20 % w/v Brij 35 (from a phenol/sulphuric assay [6] on the centrifuged sub-cream layer) and 0.20 % w/v sodium metabisulphite as preservative.

Part of the emulsion was placed in a parallel-sided cell, thermostatted, and the concentration profiles determined at 20.0 °C. The remainder of the emulsion was placed in cylindrical glass tube which had a glass tap at the base. The height of both samples of emulsion was 152 mm. 18 samples of 10 cm^3 of the creaming emulsion were removed from the base of the glass cylinder after 9 days. The samples were well shaken, diluted to ~ 10^{-4}%, placed in cylindrical glass sample cells and thermostatted to 20.0 °C for dynamic light scattering measurements.

Methods

The technique to measure volume fraction of a dispersion has been described in detail elsewhere [8]. The velocity of ultrasound is directly related to the relative concentrations of the disperse and continuous phases. Provided the particles are smaller than the wavelength of the ultrasound (~ 200 μm) a simple formula may be used [9]:

$$V^2 = \frac{V_c^2}{\left(1 - \phi(1 - \varrho_d/\varrho_c)\right)\left(1 - \phi(1 - (V_c/V_d)^2 \varrho_c/\varrho_d)\right)}$$

(1)

where V is the velocity through the dispersion of volume fraction ϕ, and ϱ_c (1001 kgm^{-3}), ϱ_d (694 kgm^{-3}), V_c (1484.9 ms^{-1}), V_d (1174.0 ms^{-1}) are the densities of and ultrasound velocities through the continuous and disperse phases. From measurements of the velocity at a series of heights a concentration profile may be determined. Collection of profiles at appropriate time intervals allows creaming/sedimentation and compaction processes to be followed in detail [7].

Densities were measured in a Paar DMA 602 measuring cell.

The dynamic light scattering of argon-ion laser light ($\lambda = 488$ nm) by the separated, diluted emulsion samples was monitored at 90°. The autocorrelation function was optimised for a single sample time over 64 channels. The data were processed to give a z-average particle size [10].

Light diffraction studies were carried out on a diluted sample of the whole emulsion (to an obscuration of ~ 0.2) using a Malvern 2600 HSD Laser Particle Sizer. The data were analysed using a model-independent fit.

Results and discussion

Concentration profiles

The concentration profiles for the creaming emulsion to 63 days are shown (with contours of constant ϕ) in Fig. 1. Two menisci were present. The lower meniscus was very diffuse. The base of the sample did not begin to clear optically for approximately 12 days, although using the ultrasonic technique creaming was detected within 2 h. The sharp upper meniscus descended from the top of the sample and corresponded to a build-up of disperse phase to form a cream layer. The droplets creamed to the top of the sample to an oil concentration of ~ 70 % which changed little with time. Such behaviour is consistent with creaming of an emulsion of discrete polydisperse droplets.

Calculation of droplet size distribution

The rate of rise under gravity U_s of a spherical droplet in a liquid medium is given by Stokes' law:

$$U_s = \frac{d^2 g \Delta \varrho}{18 \eta}$$

(2)

where g is the acceleration due to gravity, $\Delta \varrho$ is the density difference between the disperse and continuous phases, η is the medium viscosity and d is the droplet diameter. The medium (i. e. dispersion) viscosity is greater than the continuous phase viscosity because of the hydrodynamic interactions resulting from the

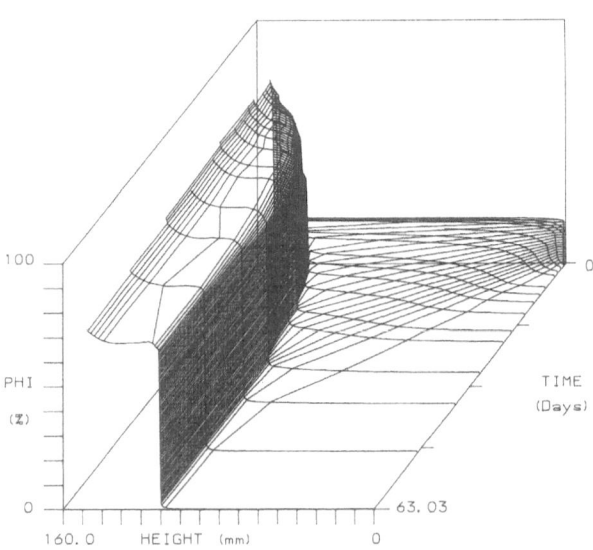

Fig. 1. Concentration profile of the creaming emulsion

presence of the droplets. The viscosity η of a concentrated dispersion of non-interacting, hard-sphere particles may be estimated using a mean-field approach [1]

$$\eta = \eta_o (1 - \phi/\phi_M)^{-2.5\,\phi m} \qquad (3)$$

where ϕ_M is the close-packed volume fraction, taken here to be 0.7 (the volume fraction in the non-compacted cream) and η_o the continuous phase viscosity (at low shear rate). For emulsions of 19 % oil η/η_o is 1.7.

The hazy (diffuse meniscus is characteristic of the creaming of individual polydisperse particles which rise at different velocities. In Fig. 2 the heights of contours of constant ϕ are shown during the creaming process. The changes in contour height with time are linear – the rise velocity of each contour is constant throughout the creaming process, and the velocity increases with contour ϕ. From the rise velocities a series of droplet diameters d_m may be calculated from

Eqs. (2) and (3) (Table 1). At any given height and time the emulsion contains droplets of all sizes up to a maximum diameter d_m corresponding to the local ϕ. Thus a cumulative size distribution (by weight) may be obtained from the velocities of contours at a series of concentrations from $\phi = 0.28$ % (the lowest measurable) to ϕ_o (Table 1). The droplet diameters range from < 0.2 μm to 12 μm, with a weight mean of 1.4 μm (a weight mode of 0.6–0.8 μm and a median of 0.85 μm). The size distribution thus calculated is plotted in Fig. 3.

Table 1. Cumulative size distribution from velocity of contours of constant ϕ. N is the number of data points in each correlation

ϕ/%	% under	U_s/nms^{-1}	corr. coefft	(N)	η/mPas	d_m/μm
0.28	1.47	8.44	0.984	(7)	1.027	0.23
0.30	1.58	9.67	0.985	(7)	1.028	0.24
0.40	2.11	15.2	0.996	(10)	1.030	0.31
0.50	2.63	18.3	0.998	(11)	1.033	0.34
0.60	3.16	21.7	0.997	(11)	1.035	0.37
0.70	3.68	25.0	1.000	(10)	1.038	0.39
0.80	4.21	27.3	0.9993	(9)	1.041	0.41
0.90	4.74	29.3	0.9992	(10)	1.043	0.43
1.00	5.26	31.1	0.9995	(11)	1.046	0.44
1.20	6.32	34.0	0.9999	(8)	1.051	0.46
1.40	7.37	35.0	0.9995	(10)	1.057	0.47
1.70	8.95	37.4	0.993	(8)	1.065	0.49
2.00	10.5	40.7	0.9999	(8)	1.073	0.51
2.50	13.2	43.7	0.9998	(8)	1.087	0.53
3.00	15.8	44.9	0.9996	(9)	1.101	0.54
4.00	21.1	50.5	0.9996	(9)	1.131	0.58
5.00	26.3	55.4	0.9999	(8)	1.161	0.62
6.00	31.6	59.4	0.9996	(8)	1.193	0.65
7.00	36.8	65.6	0.9998	(8)	1.227	0.69
8.00	42.1	72.1	0.9990	(7)	1.261	0.74
9.00	47.4	83.3	0.9997	(7)	1.298	0.80
10.0	52.6	94.9	0.9999	(5)	1.336	0.87
11.0	57.9	115	0.9999	(5)	1.376	0.97
12.0	63.2	141	1.0000	(4)	1.417	1.09
13.0	68.4	162	0.981	(5)	1.461	1.19
14.0	73.7	225	0.9999	(4)	1.507	1.42
15.0	78.9	325	0.9999	(4)	1.556	1.74
16.0	84.2	493	0.9999	(4)	1.606	2.18
16.5	86.8	636	1.0000	(3)	1.633	2.49
17.0	89.5	936	0.9999	(3)	1.660	3.05
17.5	92.1	1390	0.9997	(3)	1.687	3.74
18.0	94.7	2310	0.9987	(4)	1.716	4.87
18.2	95.8	3040	0.9996	(4)	1.728	5.60
18.4	96.8	4040	0.9995	(4)	1.739	6.48
18.6	97.9	6450	0.996	(3)	1.751	8.21
18.7	98.4	7510	0.9999	(3)	1.757	8.88
18.8	98.9	11000	–	(2)	1.763	10.7
18.9	99.5	14000	–	(2)	1.769	12.2

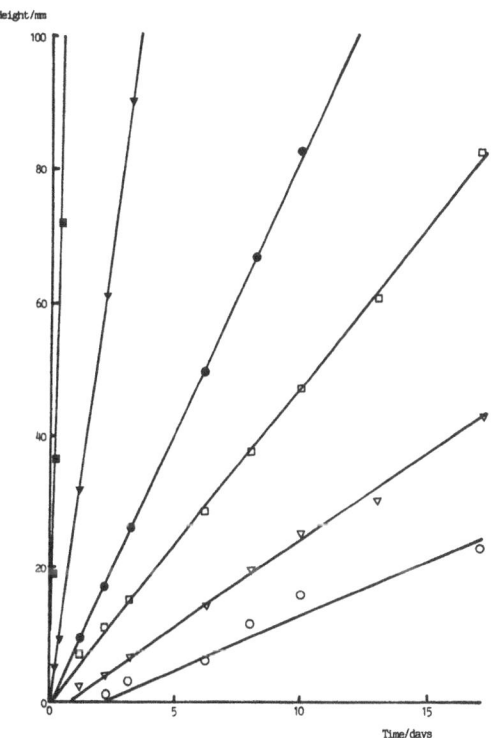

Fig. 2. Height of contours of constant volume fraction ϕ at various times during the creaming process. The gradients are the rise velocities of the contours, which correspond to particles of maximum size d_m. Key: (■) ϕ 18.0 %, $d_m = 4.7$ μm; (▼) $\phi = 15$ %, $d_m = 1.7$ μm; (●) $\phi = 10.0$ %, $d_m = 0.87$ μm; (□) ϕ 5.0 %, $d_m = 0.62$ μm; (▽) $\phi = 1.0$ %, $d_m = 0.44$ μm; (○) $\phi = 0.5$ %, $d_m = 0.34$ μm

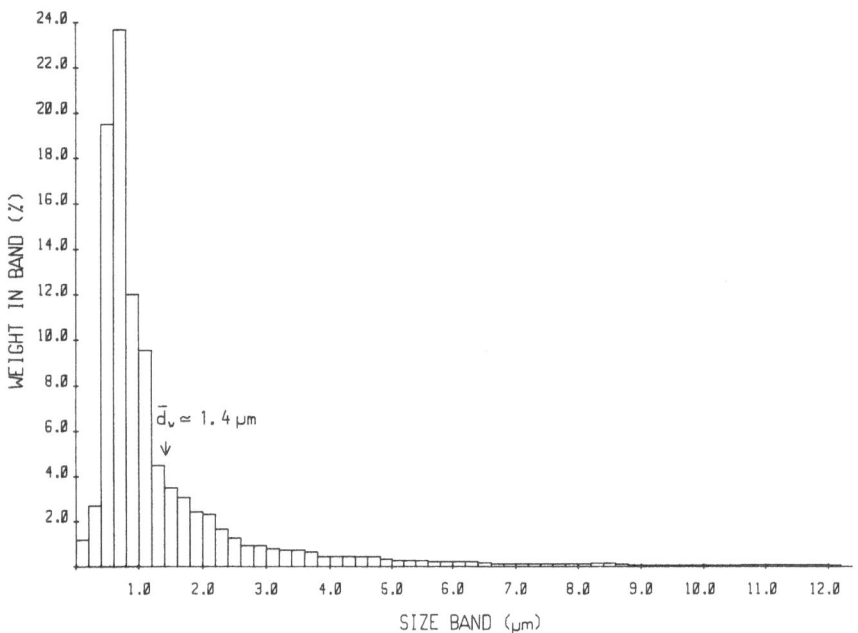

Fig. 3. Size distribution by weight (in bands of width 0.2 μm) from concentration profiles

Comparison with light scattering

It is difficult to compare the wide size distribution determined from the creaming behaviour, with measurements using any other single particle sizing technique. The size distribution of the small particles (those which are often the most difficult to detect in a wide size distribution) were determined independently by dynamic light scattering. Sections were removed (to 109 mm) from a parallel emulsion sample after 9 days as described. The weight-mean droplet diameter at any height and time in the emulsion may be calculated from the full size distribution. In Fig. 4 the mean sizes (determined from light scattering and creaming) are plotted against height. At 109 mm $\phi = 12.4\%$ after 9 days, thus Fig. 4 shows a detailed comparison of the lower part (65% by weight) of the size distribution. The results are in excellent agreement.

The size distribution of the large particle ($d > 2.4$ μm) was determined by laser light diffraction. (In this case the technique is unsuitable for detection of smaller particles because the refractive indices of the continuous phase (1.330) and disperse phase (1.393) are too close.) However, if it is assumed (from Table 1) that 13.9% of the oil is in particles with $d > 2.4$ μm then the relative size distribution of the large particles may be calculated and the results for the two methods are given in Table 2. The results are in agreement, even though the number concentration of the large particles is very small.

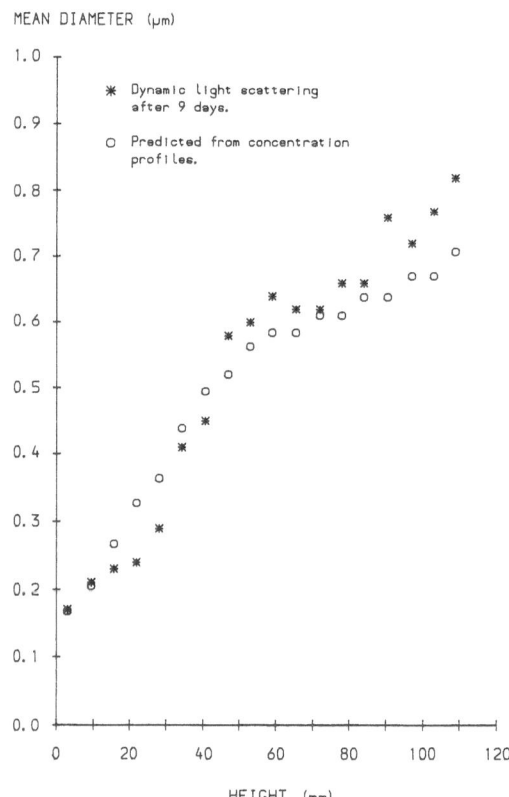

Fig. 4. Mean particle diameter at different heights in a creaming emulsion after 9 days

Table 2. Comparison of size distribution of large particles from creaming and light diffraction studies

$d/\mu m$	2.4	3.0	3.9	5.0	6.4	8.2	10.5
wt% < d (creaming)	86.1	89.2	92.5	95.0	96.7	97.9	98.9
wt% < d (diffraction)	86.1	88.5	93.2	95.9	98.1	98.6	98.9

Comments on size distributions from gravitational settling

Inspection of Fig. 2. shows that contours of low ϕ, corresponding to droplets of diameter < 0.5 μm, are not immediately apparent and delays in creaming of several days are observed as ϕ (and d_m) decreases to < 1% (< 0.4 μm). The delays before creaming of the low-ϕ contours may arise through diffusive, convective or filtration processes. Incorporation of diffusive backflow into a model of sedimentation of sub-micron monodisperse particles has been shown to broaden the meniscus, but not affect the sedimenting velocity [2]. In polydisperse systems there is a backflow of small particles, previously swept up by larger ones, by diffusion and filtration. The higher velocities of the larger particles may also cause local convection currents. Such processes lead to a delay in creaming but do not affect the rise velocities.

The delay in creaming of the small particles implies that, unless corrections are applied, the determination of sizes of sub-micron particles form gravitational settling will be in error if a single concentration profile is used. It is also clear from Fig. 2 that to determine accurately the full size distribution (in which the diameters range over a factor of 55), the concentration profile should be measured on more than one occasion (even in the absence of the delay for the small particles).

Conclusions

The size distribution of a creaming 19% oil-in-water emulsion has been determined non-intrusively from a study of the concentration profiles determined at intervals over 63 days. The droplet diameters ranged from 200 nm to 12 μm with a weight-mean value of 1.4 μm. The lower part of the size distribution (65% by weight) was found to predict mean values of the droplet diameter as a function of height after 9 days, in good agreement with those measured by dynamic light scattering. The size distribution of droplets larger than 2.4 μm (14% by weight) were in agreement with laser light diffraction measurements. A significant delay in the creaming of particles of diameter < 0.4 μm was noted and the implications of such a finding for predictions of size distribution from a single concentration profile discussed.

Acknowledgements

We thank Dr. Gary Barker for helpful discussions concerning the delay in the creaming of the small droplets; and Sarah Gouldby, Paul Gunning and Leonardo Silbert for technical assistance. This work was funded by the Ministry of Agriculture, Fisheries and Food and the Department of Education and Science.

References

1. Ball RC, Richmond P (1980) Chem Phys Liq 9:99–116
2. Barker GC, Grimson MJ (1987) J Phys A 20:305–312
3. Bunville LG (1984) In: Barth HG (ed) Modern Methods of Particle Size Analysis. Wiley, New York, pp 1–42
4. Buscall R, Davis SS, Potts DC (1979) Colloid Polym Sci 257:636–644
5. Dickinson E, Stainsby G (1982) Colloids in Food. Applied Science, Essex, UK
6. Dubois M, Gilles KA, Hamilton JK, Rebers AA, Smith F (1956) Anal Chem 28:350–356
7. Gunning PA, Hibberd DJ, Howe AM, Robins MM (1988) ???????? 2:119–129
8. Howe AM, Mackie AR, Robins MM (1986) J Dispers Sci Technol 7:231–243
9. Urick RJ (1947) J Applied Phys 18:983–987
10. Weiner BB (1984) In: Barth HG (ed) Modern Methods of Particle Size Analysis. Wiley, New York, pp 93–116

Received December 10, 1987;
accepted December 14, 1987

Authors' address:

A. M. Howe and M. M. Robins
AFRC Institute of Food Research
Norwich Laboratory
Colney Lane
Norwich NR4 7UA, Norfolk, United Kingdom

Progress in Colloid & Polymer Science   Progr Colloid Polym Sci 76:42–46 (1988)

Determination of the coagulation kinetics in silica hydrosols by photon correlation spectroscopy

P. Ludwig[1]) and G. Peschel[2])

[1]) Mütek GmbH, Essen, F.R.G.
[2]) Institut für Physikalische und Theoretische Chemie, Universität Essen, Essen, F.R.G.

Abstract: The coagulation kinetics of four silica hydrosols, each monodisperse with respect to discrete silica particle sizes (44.8 nm, 101 nm, 180 nm, and 515 nm) were investigated in the presence of LiCl, KCl and CsCl, respectively, by photon correlation spectroscopy. The concentration range of the electrolytes varied from 10^{-2} up to 1.5 mol dm^{-3}, thus covering the region of slow coagulation followed by that of rapid coagulation. The stability ratio as dependent on the electrolyte concentration was determined. The dispersion with the smallest particles was found to be the most stable.

Key words: Photon correlation spectroscopy, coagulation kinetics, silica particles, stability ratio, lyotropic series.

Introduction

The Derjaguin-Landau-Verwey-Overbeek (DLVO) theory is commonly applied in describing the stability of lyophobic colloids [1, 2]. Its subject is the interaction of colloid particles by repulsive double layer forces and attractive van der Waals forces, respectively.

The stability of colloidal systems is generally affected by potential energy barriers between colliding particles extending beyond the average kinetic energy of the particles. This electrostatic barrier practically disappears at high electrolyte concentrations [2], so that the attractive energy prevails and coagulation of the particles is observed. This process is strongly hindered if the electrolyte concentration is rather low which, in view of the large extension of the electrical double layers, involves the existence of significantly high potential barriers.

A quantitative measure of colloid stability is the kinetic constant of the coagulation process which was mathematically treated by Smoluchowski [3, 4], assuming particle diffusion to be an important prerequisite. Thus Smoluchowski arrived at the definition of "rapid coagulation". In the low concentration range of an electrolyte present, "slow coagulation" prevails which differs from rapid coagulation by a factor W, treated in detail by Fuchs [5] who arrived at a theory of particle diffusion in a force field.

Studies of coagulation kinetics had been carried out in the past by employing a number of methods, such as direct counting in the ultramicroscope [6, 7], light scattering [8, 9], and turbimetry [7, 10]. Automatic particle counting techniques were used by Hatton et al. [11]. Light scattering methods are applicable according to the regions of particle size: Rayleigh region (particle size $< \lambda/10$) or Mie region (particle size $> \lambda/10$), respectively.

A modern tool for scrutinizing coagulation kinetics is dynamic light scattering carried out by photon correlation spectroscopy. Important work in this field traces back to Versmold and Härtl [12] and Barringer et al. [13].

The aim of this paper is to show how the coagulation kinetics of a hydrosol can be systematically investigated by changing the type and concentration of the electrolyte present and the particle size.

Kinetics of coagulation

Rapid coagulation

According to Smoluchowski [3], the kinetics of rapid coagulation are determined by diffusional particle movement.

For the decrease of the number n_1 of primary particles Smoluchowski wrote

$$-\frac{dn_1}{dt} = 16\pi a_1 D_1 n_1 = k_1 n_1. \tag{1}$$

a_1 is the radius of primary particles, D_1 their diffusion coefficient. It is quite clear that in the initial period of coagulation doublets will form, followed by triplets, quadruplets and, to a small extent, even larger aggregates. For the sake of simplicity, Smoluchowski assumed that the rate constants of all kinds of aggregation are equal and, hence, finally arrived at the expression

$$n_i = \frac{n_o \left(\frac{t}{\tau}\right)^{i-1}}{\left(1+\frac{t}{\tau}\right)^{i+1}} \tag{2}$$

exhibiting the time dependence of the number of different aggregates. τ is the half-time of coagulation and given by the relation

$$\tau = \frac{1}{8\pi D_1 a_1 n_o} \tag{3}$$

which includes the rate constant

$$k = 8\pi D_1 a_1. \tag{4}$$

Slow coagulation

Slow coagulation is characterized by the fact that, because of prevailing repulsion forces [14], not every particle encounter results in coagulation.

In treating this, Smoluchowski [3] introduced a factor α reflecting the efficiency of particle collisions and wrote, in analogy to Eq. (3)

$$\tau = \frac{1}{8\pi a_1 D_1 n_o \alpha}. \tag{5}$$

A first quantitative explanation of slow coagulation is due to Fuchs [5] who introduced the stability ratio W as the ratio between the rate constant of rapid and of slow coagulation, respectively:

$$W = \frac{1}{\alpha} = \frac{k_r}{k_s} = 2a \int\limits_{2a}^{\infty} \exp\left(V_{tot}(H_o)/k_b T\right) \frac{dH_o}{H_o^2} \tag{6}$$

k_b is the Boltzmann constant and H_o the minimum distance between the particles having a radius a. V_{tot} is the total potential energy of interaction.

McGown and Parfitt [15] took account of long-range attractive forces and found for W when including viscous interactions [16–18]:

$$W = \frac{\int\limits_{2a}^{\infty} \beta(u) \dfrac{\exp\left(V_{tot}(H_o)/k_b T\right)}{H_o^2} dH_o}{\int\limits_{2a}^{\infty} \beta u \dfrac{\exp\left(V_A(H_o)/k_b T\right)}{H_o^2} dH_o}. \tag{7}$$

The correction term is represented by

$$\beta(u) = \frac{6\, u^2 + 13\, u + 2}{6\, u^2 + 4\, u} \tag{8}$$

with

$$u = \frac{H_o - 2a}{a}.$$

V_A is the potential energy of attraction.

Plotting log W against the logarithm of the electrolyte concentration yields the stability diagram of the colloidal dispersion. Two linear sections are found intersecting at the critical coagulation concentration (CCC) which demonstrates the transition of slow to rapid coagulation.

Coagulation mostly in the primary potential minimum is assumed, but there is ample evidence for the occurrence of coagulation in the secondary potential minimum [6, 19, 20].

Particle interaction

The total potential energy V_{tot} of interaction between two particles is typically given by

$$V_{total} = V_R + V_A. \tag{9}$$

The form of the repulsive electrostatic term V_R depends on whether, during the particle collision, constant charge or constant potential is assumed [21, 22]. In a subsequent paper [23] we shall employ an accurate relation, given by Honig and Mul [24].

According to Hamaker [25], the van der Waals attraction energy between two spheres with a radius a if $a \gg H_o$ is

$$V_A = -\frac{Aa}{12 H_o}.$$ (10)

A is the well-known Hamaker constant, which is composed of that of the material of the particles A_1 and that of the fluid medium A_o. We particularly have

$$A = (A_1^{1/2} - A_o^{1/2})^2.$$ (11)

Photon correlation spectroscopy

Coagulation kinetics can be determined by photon correlation spectroscopy, as was successfully demonstrated by Versmold and Härtl [12]. A detailed description of this modern method is given otherwise [26–28]. The essential procedure is dynamic light scattering from the colloidal particles. From the scattered field amplitude, the normalized correlation function [13] $g^{(1)}(t, t + \tau)$ can be derived.

In particular:

$$g^{(1)}(t, t + \tau^*) = \exp(-\Gamma\tau^*)$$ (12)

Γ being the half width of the Rayleigh line and τ^* the correlation time.

From Γ the diffusion coefficient can be derived according to

$$D = \frac{\Gamma}{K'^2}.$$ (13)

Specifically K' is the light scattering vector, which reads

$$K' = \frac{4\pi n_o}{\lambda_o} \sin \theta/2$$ (14)

where n_o is the refraction index of the dispersion medium, θ the scattering angle and λ_o the wave length of the incident light.

D, on the other hand, can be calculated from Stokes' well known relation

$$D = \frac{k_b T}{6\pi\eta a}$$ (15)

where η is the viscosity of the dispersion medium.

In reality, the test sample becomes polydisperse through the coagulation process, so that an average diffusion constant $\bar{D}$ must be taken account of. This quantity, with the progress of coagulation, is, of course, time-dependent and can be obtained by PCS experiment.

The strategy of Versmold and Härtl [12] is now to calculate $\bar{D}(t)$ in terms of theories given by Smoluchowski [3] and Pusey et al. [28] and to compare this function with that found by the PCS experiment, in order to find the specific value for τ and from this the corresponding rate constant k (k_r or k_s, respectively) by employing the general relation

$$k = \frac{1}{\tau n_o}$$ (16)

as deduced from Eqs. (2), (3) and (4).

Versmold and Härtl [12] arrived at the relation for $\bar{D}(t)$

$$\bar{D}(t) = \bar{D}_o \frac{\overline{p^{5/3}(t)}}{\overline{p^{6/3}(t)}}$$ (17)

when

$$\overline{p^{n/3}(t)} = \sum_{p=1}^{r} p^{n/3} N_p(t)/N_o; n = 5, 6.$$ (18)

Specifically, the expression

$$N_p(t) = N_o t^{(p-1)} (1 + t)^{-(p+1)}; p = 1, 2, 3 \ldots r$$ (19)

holds. D_o the diffusion constant of the original singlet particles can be determined by Eq. (13) when a is obtained, e.g., by electron microscopy. It appears to be convenient to vary p from 1 to 25.

In particular, Versmold and Härtl [12], in their evaluation procedure, replaced $\bar{D}$ and t by $D_{\text{Rel}} = \bar{D}/D_o$ and $t_{\text{Rel}} = t/\tau$, respectively. The function $\bar{D}_{\text{Rel}}(t_{\text{Rel}})$ attained by PCS experiment can be brought to congruence with the calculated one, if, in the latter procedure, τ is appropriately adjusted. According to Eq. (16) k can then be evaluated.

Experimental

Apparatus

Determination of the mean diffusion coefficient was performed using a photon correlation spectrometer (NICOMP Laser Particle Sizer Model 200). The beam of a 5 mW HeNe laser (632.8 nm) was focused in a test cell containing 0.6 ml of the dispersion sample. The scattered light beam was measured at a constant angle of 90°.

Silicic acid particles

Spherical silicic acid particles in the size range between 45 nm and 515 nm were chosen as the disperse phase. They were prepared by controlled hydrolysis of tetraalkylsilicates in ethanol by addition of water containing highly concentrated ammonia [30, 31]. The concentration of the ester was kept at 0.17 mol dm^{-3} to ensure the formation of accurately spherical particles.

In order to obtain hydrosols from the alcosols prepared in this way, the samples were diluted with bidistilled water to three times the initial volume. This dispersion was subject to vaporization in water-jet vacuo until one tenth of the volume remained. Then this sample was filled up again with bidistilled water and the procedure repeated (about ten times) until only traces of alcohol could be detected in the dispersion with the aid of gas chromatography. Details will be reported in a subsequent paper [23].

The size distribution of the different batches of the silica particles obtained (diameter: 44.8 ± 0.4 nm, 101 ± 0.8 nm, 180 ± 1.0 nm, and 515 ± 5 nm) was determined using PCS.

The alkali chlorides (LiCl, KCl, and CsCl) added to the dispersion to study the coagulation behavior were purified before use by employing a particular flotation method [31].

Results and Discussion

In Fig. 1 the logarithm of the stability ratio is plotted against the electrolyte concentration for 515 nm particles. The ranges of slow and rapid coagulation are well separated. The CCC typically shifts according to the lyotropic series from Li$^+$ to Cs$^+$ to lower values. For smaller particles, similar linear sections appeared, but shifted to larger stability ratio values. The smallest particles (44.8 nm) quite unexpectedly showed the highest ordinate values (Table 1). This unusual pattern of coagulation behavior [32] will be extensively discussed in a following paper [23]. Further tests will be applied to the pH dependence of the stability ratio.

While the curves in Fig. 2 referring to the addition of LiCl or CsCl, respectively, ascend with increasing concentration, both in a similar way in the pH range from 2 to 7, the curve obtained without addition of

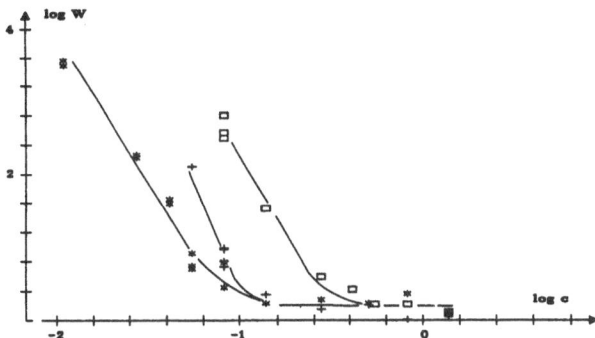

Fig. 1. The stability ratio as dependent on the electrolyte concentration. Particle diameter 515 nm. $T = 293$ K; LiCl (-□-); CsCl (-✶-); KCl (-+-)

these electrolytes steeply rises just in the range of the PZC, which indicates that a considerable share of the repulsive forces should be ascribed to hydration forces in this region [32] which are introduced as "structural forces" in colloid chemistry.

A series of further tests are concerned with the values of the rate constants for rapid coagulation as dependent on particle size. These values, which according to Smoluchowski's theory [3] should not differ when changing the particle size, are presented in Table 2. This fact clearly shows that the coagulation of particles is a highly complex process which evades a simple treatment. Thus Healy et al. [33] state that the stabilizing effect of Li$^+$ and K$^+$ counter ions is attributed to a hydration barrier at the interface. This concept is corroborated by Peschel and Ludwig [34] who found, in the presence of these ions, prevention of ion exchange on a silica surface in a distinct concentration range. Zimehl and Lagaly [35] reported an increase of coagulation power when passing from Li$^+$ to Cs$^+$.

Table 1. CCC/mol dm^{-3} for different sizes of silica particles and alkali metal cations

Diameter (nm)	Li$^+$	CCC K$^+$	Cs$^+$
44.8	1.5	–	0.7
101	0.73	0.39	0.25
180	0.63	–	0.16
515	0.43	0.21	0.12

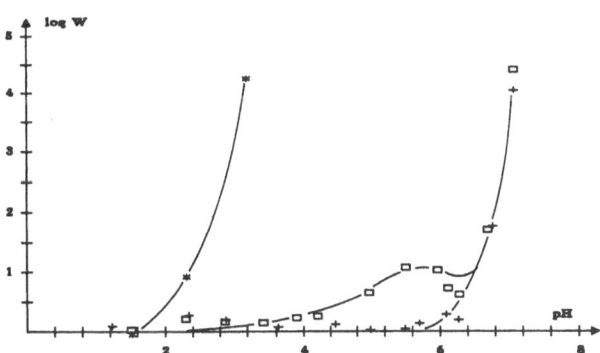

Fig. 2. The stability ratio as dependent on the pH-value. Particle diameter 515 nm. $T = 293$ K. In the absence of electrolyte (-✶-); in 0.2 M LiCl-solution (-□-); in 0.8 M CsCl-solution (-+-)

Table 2. Rate constants of rapid coagulation for different sizes of silica particles

Diameter (nm)	Rate constants/m^3 s^{-1}
44.8	$(1.73 \pm 1.13)\ 10^{-24}$
101	$(2.83 \pm 0.26)\ 10^{-20}$
180	$(1.90 \pm 0.54)\ 10^{-18}$
515	$(3.23 \pm 0.79)\ 10^{-18}$
theoretical value (3)	$5.38 \cdot 10^{-18}$

References

1. Derjaguin BV, Landau LD (1941) Acta Phys Chim 14:633
2. Verwey EJW, Overbeek JThG (1948) Theory of the Stability of Lyophobic Colloids. Elsevier, New York
3. Smoluchowski MV (1916) Phys Z 17:585
4. Smoluchowski MV (1917) Z Physik Chem, Leipzig 92:129
5. Fuchs N (1934) Z Phys 89:736
6. Overbeek JThG (1952) In: Kruyt HR (ed) Colloid Science. Vol 1, Elsevier Amsterdam, pp 282–283, 292–295
7. Watillon A, Romerowski M, Runderbeek F (1959) Bull Soc Chim Belges 68:450
8. Lips A, Smart C, Willis E (1971) Trans Faraday Soc 67:2979
9. Lips A, Willis E (1973) J Chem Soc Faraday 69:1226
10. Lichtenbelt JWTh, Pathmamanoharan C, Wiersma PH (1974) J Colloid Interface Sci 49:281
11. Hatton W, McFadyen P, Smith AL (1974) J Chem Soc Faraday Trans I 70:655
12. Versmold H, Härtl W (1983) J Chem Phys 79(8):4006
13. Barringer EA, Novich BE, Ring TA (1984) J Colloid Interface Sci 100:584
14. Sonntag H (1977) Lehrbuch der Kolloidwissenschaft. VEB Deutscher Verlag der Wissenschaften, Berlin
15. McGown dNL, Parfitt GD (1967) J Phys Chem 71:449
16. Derjaguin BV, Muller VM (1967) Dokl Akad Nauk SSSR 176:738
17. Spielmann LA (1970) J Colloid Interface Sci 33:562
18. Honig EP, Roebersen GJ, Wiersema PH (1971) J Colloid Interface Sci 36:97
19. Hogg R, Yang KC (1976) J Colloid Interface Sci 56:573
20. Marmur A (1979) J Colloid Interface Sci 72:41
21. Frens G (1968) Thesis, Utrecht, The Netherlands
22. Wiese GR, Healy TW (1970) Trans Faraday Soc 66:490
23. Peschel G, Ludwig P (1988) Colloid Polym Sci
24. Honig EP, Mul PM (1971) J Colloid Interface Sci 36:258
25. Hamaker HC (1937) Physica 4:1058
26. Berne BJ, Pecora R (1976) Dynamic Light Scattering. Academic, New York
27. Chu B (1974) Laser Light Scattering. Academic, New York
28. Pusey PN, Fijnaut H, Vrij A (1982) J Chem Phys 77:4270
29. Stöber W, Fink A, Bohn E (1968) J Colloid Interface Sci 26:62
30. Ludwig P (1987) Dissertation, Essen
31. Lemlich R (1972) Adsorptive Bubble Separation Technique. New York, London
32. Allen LH, Matijević E (1969) J Colloid Interface Sci 31:287
33. Healy TW, Homola A, James RO, Hunter RJ, Disc Chem Soc 65:156
34. Peschel G, Ludwig P (1987) Ber Bunsenges 91:536
35. Zimehl R, Lagaly G (1986) Progr Colloid Polym Sci 72:28

Received December 10, 1987;
accepted December 16, 1987

Authors' address:

G. Peschel
Institut für Physikalische und Theoretische Chemie
Universität Essen
Universitätsstraße 5–7
D-4300 Essen, F.R.G.

Small angle neutron scattering studies of polymer latices under sheared conditions

P. Lindner[1]), Ivana Marković[2]), R. C. Oberthür[1]), R. H. Ottewill[2]) and A. R. Rennie[1])

[1]) Institut Laue Langevin, Grenoble, France
[2]) School of Chemistry, University of Bristol, U.K.

Abstract: Results are reported of small angle neutron scattering experiments on a sheared dispersion of a poly(methylmethacrylate) latex in dodecane. By means of a two-dimensional detector the structural arrangement of the particles was determined in directions parallel and perpendicular to the directions of flow. These data suggested that strings of particles were formed in which the particles moved closer together in the direction normal to the flow direction as the shear rate was increased.

Key words: neutron scattering, polymer latices, shear, microstructure.

Introduction

Over the past decade or so, the study of concentrated colloidal dispersions has been greatly facilitated by the use of scattering techniques such as small angle neutron scattering [1] and photon correlation spectroscopy [2]. Moreover, the availability of monodisperse spherical polymer particles at high volume fractions has greatly enhanced the interpretation of the results obtained from such studies. A major interest in research of this type is to relate the structural arrangement of the particles determined by diffraction (microscopic behaviour), under both static and dynamic conditions, to the bulk behaviour of the system (macroscopic), for example, rheology [3].

In this preliminary communication we report the examination of a concentrated dispersion of polymer particles in dodecane by small angle neutron scattering under shear gradients which varied from 0 s^{-1} to $10,000$ s^{-1}. The particles were composed of a spherical core of poly(methyl methacrylate) stabilised by an outer layer of poly(12-hydroxy stearic acid) grafted chemically to the core particle. This is illustrated schematically in Fig. 1.

Experimental

The polymer latex used, SPSO90, was prepared by the dispersion polymerisation technique previously described [4]. The num-

ber average particle diameter determined by electron microscopy was 0.20 μm and the weight average determined by small angle neutron scattering was 0.22 μm. This latex, dispersion in dodecane, had a weight/weight concentration of 44 %.

The small angle scattering measurements were carried out using the neutron diffractometer D11, at the Institut Laue Langevin, Grenoble [5], with a sample-detector distance of 35.7 m and a neutron wavelength, λ, of 0.80 nm. For these elastic scattering measurements, the magnitude of the scattering vector, Q, was defined by,

$$Q = 4\pi \sin(\theta/2)/\lambda$$

with θ = angle between the incident and the scattered beams.

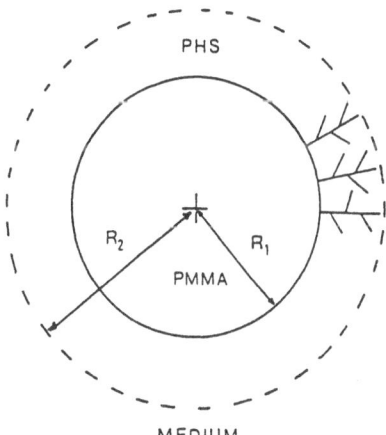

Fig. 1. Schematic illustration of a poly(methylmethacrylate) – poly(12-hydroxy stearic acid) particle

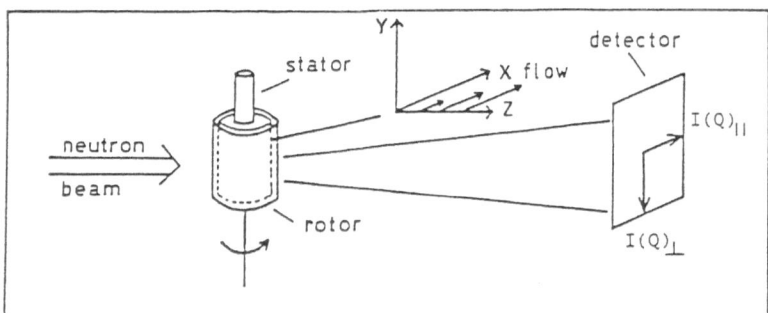

Fig. 2. Shear cell used for scattering measurements

The measurements were carried out using the quartz Couette shear cell designed and constructed by Lindner and Oberthür [6]. This was composed of a static internal cylinder (the stator) and a rotating outer cylinder (the rotor). It is illustrated in Fig. 2. The gap between the stator and the rotor was 0.5 mm giving a total neutron path length of 1 mm through the dispersion. The scattered neutrons were measured using a two-dimensional detector, a square matrix of 64 × 64 elements. This allowed either two-dimensional contour plots to be obtained showing lines of equal intensity or three-dimensional plots showing the intensity across the detector.

Results and discussion

The 44 % dispersion was examined at a number of shear rates and the results obtained at zero, 300 s^{-1}, 3,000 s^{-1} and 6,000 s^{-1} are shown in Fig. 3 in the form of contour plots. Some three-dimensional plots are also shown in Fig. 4 for shear rates up to 10,000 s^{-1}. In all these figures, the velocity of flow was in a direction parallel to the x-axis.

From these figures it can be seen that at zero rate of shear (Figs. 3a and 4a) there is a peak in the intensity which is essentially symmetrical about the centre of the detector. As discussed previously [1] this peak in the scattered intensity, $I(Q)$, as a function of the scattering vector, Q, can be interpreted on the basis of a "liquid-like ordering" of the particles. Under sheared conditions, however, the intensity peak moves to lower scattering vectors in a direction parallel to the flow direction.

Since the scattering information was obtained from a two-dimensional detector pattern it was possible to obtain sector averages over different parts of the detector. For example, taking the x-axis, and the direction of flow as being 0°, a radial average plot of $I(Q)$ against Q was obtained for the sector $0 \pm 15°$. In the direction normal to the flow, the sector taken was $90 \pm 15°$. The results obtained at shear rates of 0 s^{-1} and 10,000 s^{-1} are shown in Fig. 5. At zero shear rate the intensity curves

are identical in the two directions and Q_{max} occurs at 0.0026 Å^{-1}. However, at 10,000 s^{-1}, in a direction parallel to the flow, Q_{max} decreases to ca. 0.0014 Å^{-1} and the intensity curve begins to resemble the behaviour expected from a form factor, $P(Q)$, unmoderated by a structure factor, suggesting in fact a more dilute dispersion. In the direction normal to the flow direction, however, the peak moves to a higher Q value, $Q_{max} = 0.0034$ Å^{-1} and the intensity decreases. This is the behaviour expected for a more concentrated dispersion where an increase in the structure factor, $S(Q)$, moderates the scattering behaviour so that the peak moves to a higher Q value and at the

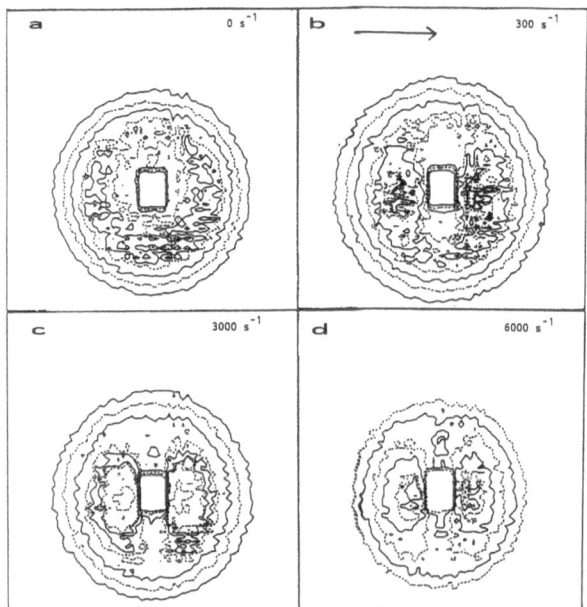

Fig. 3. Contour plot of constant intensity $I(Q)$ lines against Q for a PMMA-PHS latex at a concentration of 44 % w/w at various shear rates: (a) 0 s^{-1}, (b) 300 s^{-1}, (c) 3,000 s^{-1}, (d) 6,000 s^{-1}. (⟶) direction of flow

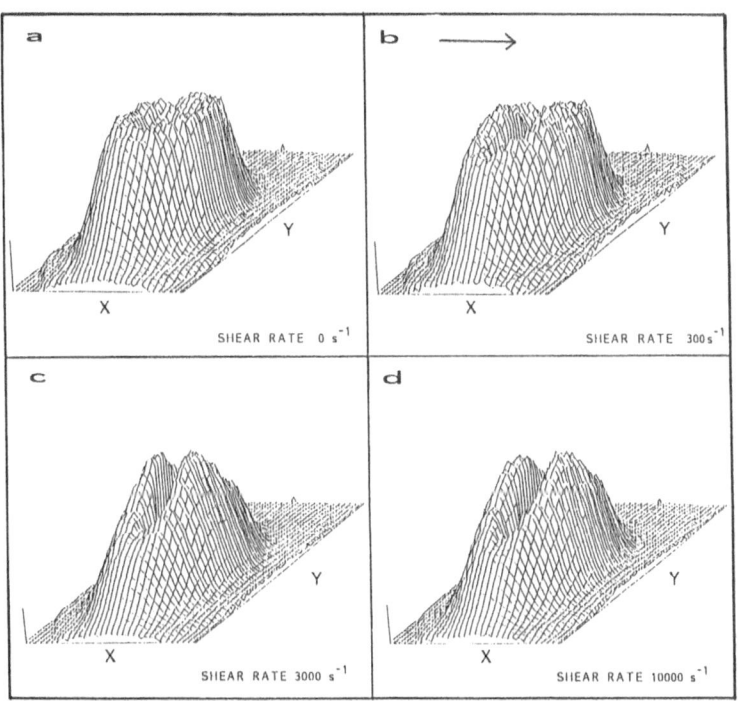

Fig. 4. Three-dimensional plots of scattered intensity $I(Q)$ for a PMMA-PHS latex at a concentration of 44 % w/w at various shear rates: (a) 0 s^{-1}, (b) 300 s^{-1}, (c) 3,000 s^{-1}, (d) 10,000 s^{-1}. (⟶) direction of flow

same time the intensity decreases at the lower Q values [1].

In the direction normal to the flow, it can be deduced from the curves of $I(Q)$ against Q (Fig. 5) that the peak in the structure factor moves to higher Q values and increases in magnitude as the shear rate increases. This suggests that the system is becoming more concentrated in the normal direction. Similarly, parallel to the flow direction the system appears to be becoming more dilute.

The time taken for a particle to diffuse one particle radius is ca $3\pi\eta a^3/kT$ and hence for a 10 nm radius particle this is of order 10^{-6}s. Diffusion for these particles occurs so rapidly that very high shear rates, of order 10^6 s^{-1}, would be required to maintain a shear induced structure. Consequently, the experiments that we carried out on 25 nm radius particles failed to show any structural changes even at shear rates of 10^4 s^{-1}. However, with latex particles of radius 0.1 μm, as used in the present work, the time taken to diffuse one radius is of order 1 ms. In this case, as anticipated, structural changes were observed with the shear rates used.

A schematic representation of the behaviour is shown in Fig. 6. In Fig. 6a, the "liquid-like ordering" of the particles is represented in two-dimensions, i.e. in a plane perpendicular to the beam direction and parallel to the detector; in three-dimensions this system is also liquid-like. Under shear, however, in the schematic two-dimensional illustration (Fig. 6b) we note that the

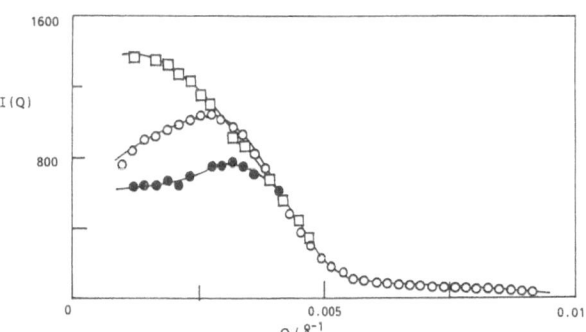

Fig. 5. $I(Q)$ against Q for a 44 % w/w PMMA-PHS latex: (O) Static conditions; (□) in direction of flow at 10,000 s^{-1}; (●) perpendicular to the direction of flow at 10,000 s^{-1}

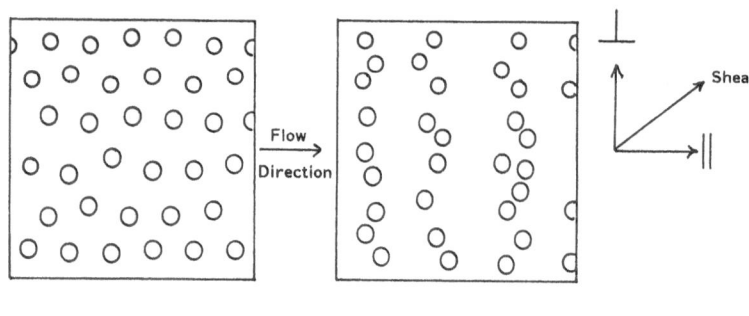

Fig. 6. Schematic diagram illustrating changes in particle packing on going from (a) static, to (b) flowing conditions

particles are moving closer together in the direction normal to the flow. Further diffraction experiments are required with the beam in the direction of the shear gradient which would allow a determination of the interplanar spacing in the z-direction and help confirm the particle arrangement.

A more detailed report of this work covering a range of volume fractions with a more detailed analysis will be published elsewhere.

Acknowledgements

Our thanks are due to ICI PLC for support of this work under the joint Research Scheme and to Dr. Th. F. Tadros for many discussions. We also wish to thank SERC for support and ILL for neutron beam facilities.

References

1. Marković I, Ottewill RH, Underwood SM, Tadros TF (1986) Langmuir 2:625
2. Ottewill RH, Williams NSJ (1987) Nature 325:232
3. Ackerson BJ, Hayter JB, Clark NA, Cotter L (1986) J Chem Phys 84:2344
4. Antl L, Goodwin JW, Hill RD, Ottewill RH, Owens SM, Papworth S, Waters J (1986) Colloids Surfaces 17:67
5. (1983) Neutron Beam Facilites at the High Flux Reactor. Institute Laue Langevin, Grenoble
6. Lindner P, Oberthür RC (1984) Rev Phys Appl 19:759

Received December 17, 1987;
accepted January 4, 1988

Authors' address:

Prof. Dr. R. H. Ottewill
School of Chemistry
University of Bristol
Cantock's Close
Bristol BS8 1TS, U.K.

Weighing monolayers of colloidal silica particles with a quartz crystal microbalance

S.-A. Nordin and B. Kasemo

Department of Physics, Chalmers University of Technology, Göteborg, Sweden

Abstract: A new method of monitoring the deposition of single layers of colloidal particles (10–200 nm in diameter) on a flat surface is described. We demonstrate that with a quartz crystal microbalance (QCM) it is possible to measure directly the addition of one colloidal monlayer of SiO$_2$ particles onto an oxidized Al film, or onto a previously formed multilayer. This technique holds promise for the accurate modelling and control of catalyst preparation, e.g. for the deposition of washcoats, and other applications of monodispersed colloids.

Key words: Monodispersed colloids, quartz crystal microbalance, colloidal silica particles, desposition of colloids, colloids.

1. Introduction

There is currently considerable interest in colloids both from scientific and technological points of view (for a recent short review see [1]). Today it is possible to prepare a variety of monodispersed colloids of different chemical compositions, different geometrical particle shapes and over a large range of particle dimensions from the micrometerscale down to the nanometer scale. It is possible to produce monodispersed systems, the particles of which are very uniform in shape and size [1]. In the present paper we describe, to our knowledge for the first time, the application of the quartz crystal microbalance (QCM) weighing technique to the monitoring of deposition of single and successive monolayers of such particles on surfaces.

2. Experimental

The quartz crystal microbalance

The principle of the quartz crystal microbalance (QCM) is simple. A piezoelectric crystal has an eigenfrequency which is dependent on the oscillating mass. If mass is added to the surface of the crystal, this reasonance frequency decreases.

The QCM is an extremely sensitive weighing apparatus. It easily measures mass in the nanogram region and under ideal conditions a few picograms. It has mainly been used in surface and vacuum

science and technology [2–4]. Recently, several successful attempts have been made to utilize the great sensitivity of the QCM outside the vacuum realm [5, 6]. The technical construction of the QCM is shown in Fig. 1. The 8 MHz AT-cut quartz crystal has a diameter of 9 mm and a thickness of ≈ 0.17 mm.

Metal electrodes are deposited on each side. The circular region in which they overlap is the active, mass sensitive region of the QCM.

The crystal is excited to oscillate by an oscillator circuit and the output resonance frequency is fed into a D/A converter and displayed on a strip chart recorder.

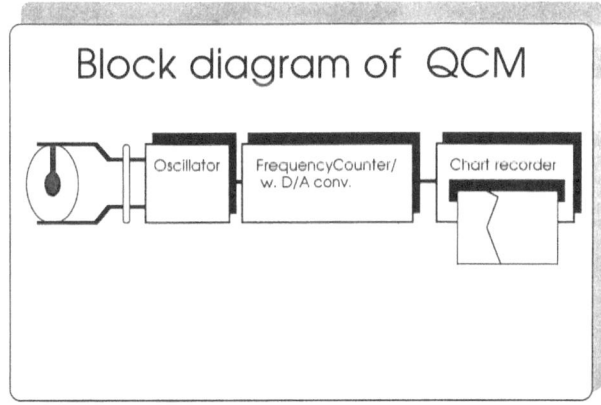

Fig. 1. Block diagram of quartz crystall microbalance

If mass is deposited on either electrode it will induce a frequency decrease. For small changes in mass the following relation holds,

$$\Delta f = - C \cdot \Delta m$$

where Δf is the frequency change induced by the deposition of the mass, Δm, and the proportionality constant, C, depends on the actual crystal in use.

For the so-called AT-cut crystals (supplied by Kvartselektronik AB, Stockholm) of 8 MHz fundamental frequency used in this work, $C \approx 7 \cdot 10^{-9}$ g/Hz $\cdot$ cm^2.

Sample preparation

The sample preparation is a slight variation of a procedure described by Iler [7]. Iler demonstrated monolayer thickness of different colloidal particles on glass surfaces using his method.

Application of one layer of colloidal particles is a wet process and consists of several steps:

1. The crystal is rinsed in pure water.

2. The crystal is dipped in an 0.5 % solution of Chlorhydrol for about 15 s and rinsed in water.

3. The wet crystal is dipped in a 2 % sol of colloidal silica (Ludox TM), and rinsed very carefully with water.

4. The sample is dried and the frequency shift recorded.

For additional layers, steps 2 to 4 are repeated. In Iler's work, the surfaces were dried after step 2. The silica sols had pH $\geq$ 3.3 adjusted with HCl. All sols were filtered before being applied to the crystals.

3. Results

After each sequence of silica particle deposition, the frequency shift (= mass increase) induced by this deposition was measured. Figure 2 shows the measured frequency shifts vs. the number of monolayer depositions of 22 nm SiO$_2$ particles, for four different crystals with Al electrodes. There is some spread in the data but the general tendency is an approximately linear increase with the number of depositions.

The dotted line represents the average of these measurements while the slope of the full line gives the maximum mass increase per deposition if a close-packed layer of 22 nm SiO$_2$ particles are assumed to be deposited at every deposition. The mean of the experimentally observed values (dotted line) corresponds to a packing $\approx$ 54 % of the maximum close-packing.

The maximum weight increase per deposition corresponding to maximum close-packing in each deposited monolayer was calculated as follows: One close packed layer of 22 nm particles corresponds to $\sim 2.4 \cdot 10^{11}$ particles/cm^2, and the layer thickness is 22 nm. The density of SiO$_2$ is 2.2 g/cm^3. This gives that one layer weighs $\sim$ 2.9 µg/cm^2. Since each preparation cycle gives one layer on each side of the crystal, we

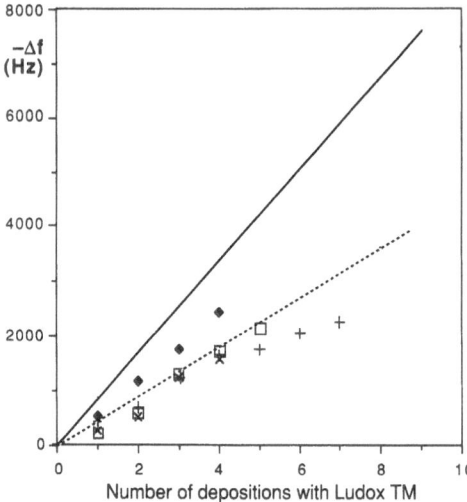

Fig. 2. Measured frequency shifts versus the number of monolayer depositions of 22 nm SiO$_2$ particles

expect a maximum deposition of twice this value. By use of the sensitivity constant, C, given above, we obtain the maximum value $\Delta f_{max} \approx$ 840 Hz/deposition.

4. Discussion

The measured mass uptake per deposition (Fig. 2) clearly illustrates that of the order of one monolayer is deposited for each deposition. The somewhat smaller uptake than the one corresponding to maximum close-packing in the monolayer can have several causes:

Firstly, there is no a priori reason to expect close packing in the deposited layers. The basic packing may very well be more open, and defects are probably also incorporated (vacancies, etc).

Secondly, the relatively large spread in data from crystal to crystal and from deposition to deposition indicates a variation in the quality of the crystals or in the electrode films. We cannot yet exclude the possibility that crystal contamination contributes to this spread.

It should be mentioned that single depositions occasionally resulted in very large frequency shifts for no apparent reason[1]. This was followed by breakdown

---

[1] If the crystals were dried after step 2 the recorded frequency shift was $\sim$ 10 Hz. Thus this preparation step does not contribute to the observed mass increases. Likewise, if the crystals were dipped in the Ludox sol without pretreatment in Chlorhydrol Δf was $\sim$ 20 Hz.

of the oscillation and we suspect that this behaviour is due to agglomeration of particles on the surface, i.e. 3D deposition.

In spite of the spread in the data, they clearly demonstrate that the basic deposition principle is a monolayer by monolayer growth. By optimizing the measurement procedure and improving the control over crystal surface quality, we expect to be able to bring down the experimental spread considerably. This will then make more quantitative measurements possible.

It should also be mentioned that aluminium is perhaps not the ideal substrate for these colloidal layers. Aluminium electrodes tend to flake off in small spots, which usually happens after several depositions. Measurements are under way on different electrode materials and different particle sizes.

Acknowledgements

We are grateful to Professor J.-E. Otterstedt, Dept. of Engineering Chemistry, Chalmers, for introducing us to the technique of monolayer deposition of silica sols. This work is supported by the National Swedish Board for Technical Development, contract no. 835405 and the Micronics program.

References

1. Matijević E (1986) Langmuir 2:12–20
2. Wolsky SP, Zdanuk EJ (eds) (1969) Ultra Micro Weight Determination in Controlled Environments. Wiley, New York (this reference contains a thorough orientation on QCM in vacuum environments)
3. Lu C, Czanderna AW (eds) (1984) Applications of Piezoelectric Quartz Crystal Microbalances. Elsevier, New York
4. Kasemo B, Törnquist E (1980) Phys Rev Lett 44:23, 1555
5. Kaufman JH et al. (1984) Phys Rev Lett 53:2461 (this refrence mentions the work of Nomura et al, in 1981, who operated a QCM immersed in a liquid)
6. Roberts GG et al (19??) Br J Polym Phys, to be published
7. Iler RK (1966) J Colloid Interface Sci 21:569–594

Received January 4, 1988;
accepted January 7, 1988

Authors' address:

S.-A. Nordin
Department of Physics
Chalmers University of Technology
S-412 96 Göteborg, Sweden

Progress in Colloid & Polymer Science

Progr Colloid Polym Sci 76:54–58 (1988)

Amphiphile solutions

Interaction between amphipathic peptides and phospholipid membranes: a deuterium NMR study

E. J. Dufourc[1]), J. H. Freer[2]), T. H. Birkbeck[2]), and J. Dufourcq[1])

[1]) Centre de Recherche Paul Pascal, CNRS, Domaine Universitaire, Talence
[2]) Department of Microbiology, University of Glasgow, Glasgow, Scotland

Abstract: The actions of bee venom melittin and of delta-lysin from *Staphylococcus aureus* on membranes have been monitored by solid state deuterium NMR. Despite inducing quite similar effects on dipalmitoylphosphatidylcholine, significant differences can be shown depending on temperature and on the lipid to peptide molar ratio, R_i.

In the fluid phases, for moderate amounts of toxins and at temperatures $T \gg T_c$ (T_c gel to fluid transition temperature), analysis of the quadrupolar splittings in terms of chain ordering indicates that both melittin and delta-lysin similarly disorder the membrane. The addition of greater amounts of toxins ($R_i = 4$), results in the occurrence of very small structures.

At temperatures above but close to T_c, melittin preferentially orders the center of the bilayer while delta-lysin promotes ordering throughout the entire bilayer thickness. These effects are interpreted as reflecting different locations of the peptides with respect to the membrane surface.

In the gel phase, for lipid to peptide molar ratios > 15, melittin induces isotropic lines reflecting the presence of small discoidal structures whereas delta-lysin does not.

The ability of both toxins to induce bilayer fragmentation is proposed as a possible step in the direct lysis of membranes.

Key words: Deuterium NMR, model membranes, melittin, delta-lysin.

Abbreviations used: NMR: nuclear magnetic resonance, DPPC: dipalmitoylphosphatidylcholine, R_i: lipid to protein molar ratio, T_c: temperature of the gel to fluid transition.

Introduction

Melittin and delta-lysin are two peptide toxins of 26 amino acids from bee venom and secreted by *Staphylococcus aureus*, respectively. Although originating from different species, these amphipathic peptides promote very similar effects on biological membranes, i.e. they lyse a wide variety of cells or subcellular compartments [1, 2]. However, they have different efficiencies, for example on sheep and cod erythrocytes melittin is more active than delta-lysin in the release of hemoglobin [3].

Both toxins bind to phospholipids and are able to induce the release of molecules trapped inside vesicles [2, 4, 5]. They have also been demonstrated to restructure lipid bilayers by fragmentation and fusion for melittin [6, 7], mainly by fusion for delta-lysin [8].

Here the effects of these two toxins on synthetic model membranes are followed and compared by means of deuterium (^{2}H) nuclear magnetic resonance (NMR).

Material and methods

Chain deuterated lecithins ([sn−2−^{2}H$_{31}$]DPPC) were synthesized according to already published procedures [9]. Perdeuterated palmitic acid was a generous gift from Dr. B. Perly (CEN Saclay, France). Deuterium-depleted water was purchased from Aldrich Chem. Co. (USA). Melittin was bought from Bachem A.G. (Switzerland) and delta-lysin was isolated according to already published procedures [10].

Model membranes (multilamellar dispersions) were prepared by hydrating lipids with buffer (20 mM Tris acetate, 2mM EDTA, 100 mM NaCl, pH = 7.5) on a vortex.

Toxins were dissolved in the same buffer and added to the membranes at the desired lipid-to-peptide molar ratio (R_i). Except where otherwise mentioned, samples were incubated for at least 30 min above T_c (the gel to fluid phase transition temperature of the pure lipids) prior to performing the experiments [11]. Before the NMR signal was acquired, the samples were allowed to equilibrate for at least 30 min at the desired temperature which was regulated to ± 1°C.

NMR signals were acquired on Bruker spectrometers, either a WH-270 adapted for solid state studies (B. Clin and E. J. Dufourc, unpublished), or a MSL200, with quadrature detection and by means of the quadrupolar pulse sequence [12]. Data were analyzed on a VAX/VMS 8600 Computer. Spectral deconvolution ("de-Pake-ing") was done as described by Bloom et al. [13] except that we chose to calculate oriented-like spectra with the bilayer normal at 90° with respect to the magnetic field direction; the quadrupolar splitting was then directly measured.

Results

Changes induced on DPPC by the toxins as followed by NMR of deuterium

Figure 1 shows ^{2}H-NMR powder spectra of [sn—2—^{2}H$_{31}$] DPPC dispersions in the presence of melittin (R_i = 20), as a function of temperature. For this experiment, the toxin was added at 25 °C, i.e. in the gel phase, without any incubation above T_c (T_c of DPPC is 41 °C). At 25 °C, the characteristics of the spectrum of pure lipids were not altered after the addition of melittin. This is in contrast to what was obtained in the same temperature range but after incubation above T_c (Fig. 1: $T = 38°$, 30 ° and 0 °C); in all these cases an isotropic line was observed. This line was superimposed on the broad spectrum which represents gel-phase lipids unperturbed by melittin. The presence of this sharp line correlates with the existence of very small melittin-lipid complexes, already characterized by freeze fracture electron microscopy, quasi-elastic light scattering and gel filtration [6, 7, 14]. When the temperature of the system rose above T_c, this sharp line disappeared, only the powder spectra being detected. This again correlates with the presence, under these conditions, of large vesicular structures, as clearly demonstrated by other techniques [6, 7, 14].

When the amount of toxin in the system was increased (R_i = 4), spectra consisted only of the isotropic lines both below and above T_c. This indicates that the bilayers were quantitatively converted into very small structures capable of fast isotropic tumbling, as has already been described by other direct methods [6, 7].

The changes induced by delta-lysin on the same DPPC system are much more subtle (Fig. 2). For the

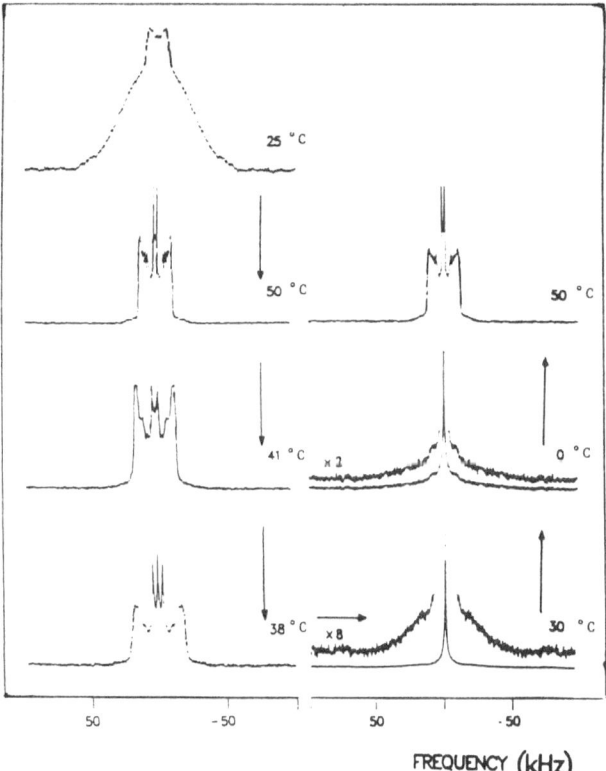

FREQUENCY (kHz)

Fig. 1. Temperature variation of ^{2}H-NMR powder spectra of aqueous dispersions of [sn—2—^{2}H$_{31}$] DPPC in the presence of melittin (R_i = 20). Temperatures (°C) are indicated on the right side of each spectrum. The arrows show the direction of temperature variation during the experiment, after toxin addition at 25°C. Typical experimental parameters: spectral window, 250 kHz; 90° pulse width, 5 µs; delay between 90° pulses, 40 µs; recycling time, 1–1.5 s; 1800 scans

same lipid-to-peptide molar ratio (R_i = 20), no isotropic line was detected in the gel phase. At $T < T_c$, spectra showed little evidence of perturbation, whereas at $T > T_c$ the spectra were modifed markedly. With high amounts of toxin in the system (R_i = 4), isotropic lines were observed, but only in the fluid phase at $T \gg T_c$ [15].

Changes induced in the lipid ordering in the fluid phase

From deuterium NMR spectra of fluid phase [sn—2—^{2}H$_{31}$] DPPC lipids, measurements of quadrupolar splittings allow estimation of the orientational ordering for each labelled position along the sn—2 acyl chain [16]. One may thus represent the ordering of the membrane as a function of the bilayer depth, i.e. on going

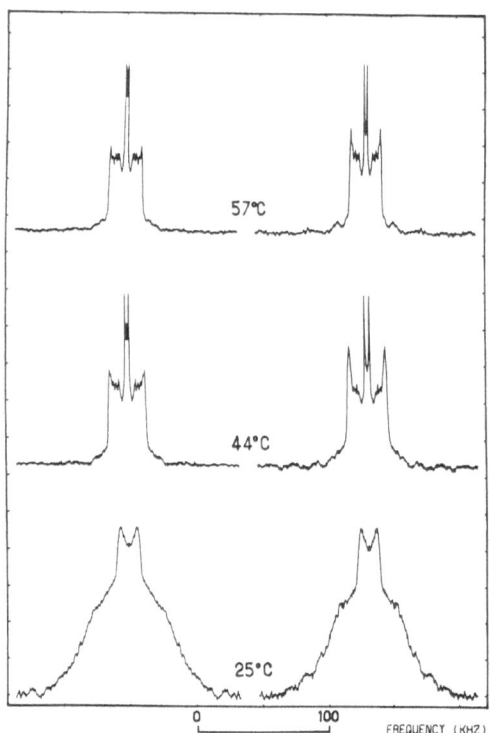

Fig. 2. Temperature variation of ^{2}H–NMR powder spectra of aqueous dispersions of [sn–2–^{2}H$_{31}$] DPPC in the presence of delta-lysin ($R_i = 20$, right side spectra) and in its absence (left hand side spectra). Experimental parameters: spectral window, 166 kHz; 90° pulse, 5 µs; delay between 90° pulses, 40 µs; recycling time, 2s; 3000 scans

from positions C2—C3, close to the glycerol backbone, to the methyl terminal position at the center of the bilayer [17].

Figures 3 and 4 illustrate such variations and show how the ordering is modified by the presence of the toxins. Data are obtained from systems containing moderate amounts of toxins, i.e. situations where powder spectra are still observed in the fluid state and, in the case of experiments involving melittin, the systems proved to be homogeneous and constituted by large unilamellar vesicles [7]. It is notable that both toxins induce a decrease in quadrupolar splitting, i.e. decrease the local orientational order, for all labelled positions, at high temperatures (65 °C).

Near T_c, Fig. 3 shows that melittin preferentially orders the positions near the center of the bilayer ($T = 41°$ whereas delta-lysin promotes an increase in the ordering for all the positions (Fig. 4, $T = 45$ °C).

Discussion

The perturbations detected herein on deuterium NMR spectra and induced by amphipathic peptides might be discussed at two levels: (i) how changes in the shape or symmetry of the spectra are related to stabilization of new lipid phases, (ii) how changes in the quadrupolar splittings in a lamellar phase reveal local

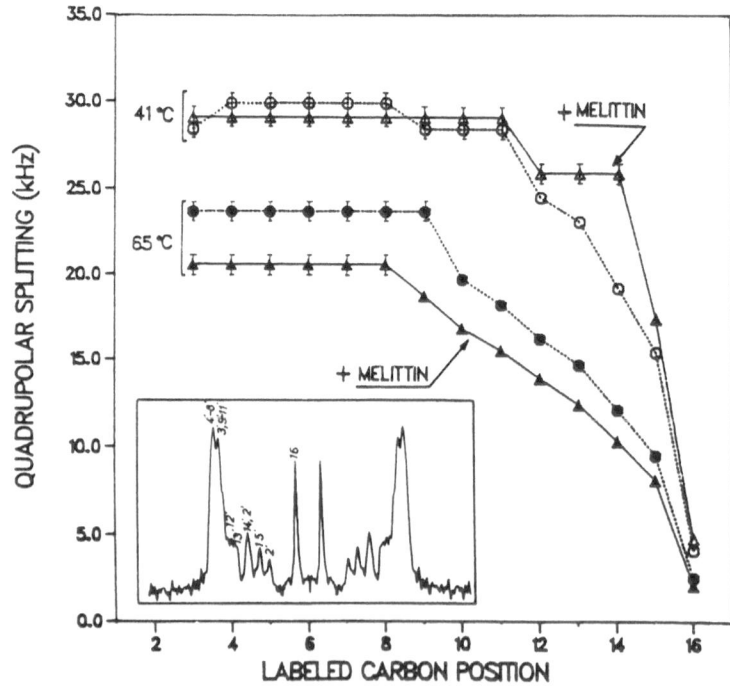

Fig. 3. Quadrupolar splittings from de-Paked spectra (see inset) of [sn–2–^{2}H$_{31}$] DPPC in the presence ($R_i = 20$) and in absence of melittin at 41 °C and 65 °C, as a function of the labelled carbon position along the sn–2 acyl chain. Primed numbers in the de-Packed spectrum in the inset represent the assignment of labeled positions

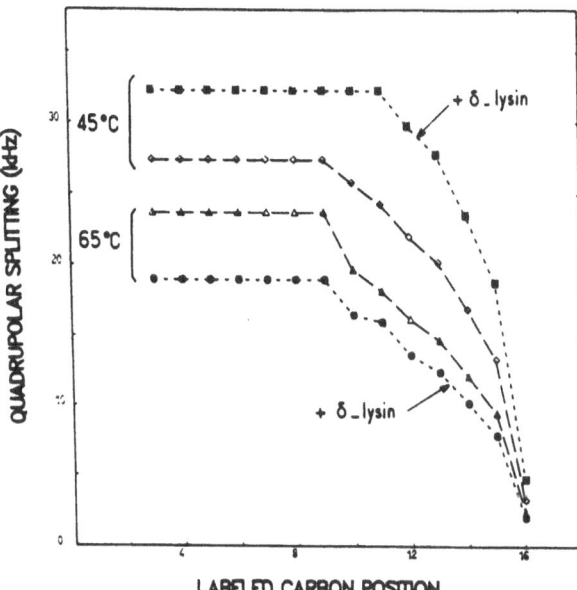

Fig. 4. Quadrupolar splittings from de-Paked spectra of [sn−2−^{2}H$_{31}$] DPPC in the presence of delta-lysin [R_i = 10] and in its absence at 45 °C and 65 °C, as a function of the labeled carbon position along the sn−2 acyl chain

changes at the molecular or submolecular level in the order of the lipid chains.

The drastic effect induced by both toxins on deuterium spectra leading to the total averaging of the quadrupolar splitting and appearance of isotropic lines implies that lipids are involved in new structures.

In the case of melittin, this correlates with the stabilization of new supramolecular lipid-toxin complexes already characterized by electron microscopy, gel filtration and light scattering [7,14]. There is a strict correlation between the conditions necessary to detect sharp isotropic lines and observation of the small micellar species with lipids in the fluid phase.

In the case of delta-lysin there are as yet only indications from negative staining electron microscopy that the toxin is able to induce discoidal complexes with lipid mixtures [2]. However, taking into account the overall behaviour of both toxins, one can speculate that for pure DPPC, the delta-lysin also fragments the fluid bilayers into small complexes. However, when compared to the behaviour of melittin, this fragmentation occurs only at higher doses and higher temperatures.

In the gel state the two toxins differ since isotropic lines are observed only with melittin. This is interpreted as a stabilization of the discoidal complexes charac-

terized by a lipid to toxin molar ratio of about 22 [7], the increase in the amount of the isotropic line in the NMR spectra reflects a parallel increase in the fraction of lipids in discoidal structures. The fast tumbling of the small discs allows the motional averaging of the quadrupolar splitting since the lipid chains within each disc are ordered in a gel-like bilayer array [19], and have no fast local motion.

Regarding perturbations induced in the fluid bilayers, as detected by the changes in quadrupolar splittings, it is apparent that very similar effects are induced by both toxins at high temperatures. This reflects similar position of the toxins with respect to the lipids. It is proposed that they partially penetrate into the bilayer, acting as a wedge, or a spacer, which allows an increase in the conformational space devoted to each methylene group all along the acyl chain.

A different picture appears likely when temperature decreases and approaches that of the transition from the fluid to the gel phase. Under these conditions melittin has a very limited ordering effect on the center of the bilayer, delta-lysin promotes a general ordering of the whole chain. Such different perturbations may correspond to different positions of the two peptides in the bilayer; at temperatures near to T_c, melittin appears to penetrate deeply in the bilayer, whereas delta-lysin appears to be excluded from gel-like domains.

Conclusion

Despite considerable similarities, melittin and delta-lysin can differently affect the structure and the dynamics of model membranes. From the changes in the lipid order parameters, a positioning of the toxins is proposed. This depends on the physical state of the lipids and can differ for the two toxins. This difference is more striking for ordered gel phase lipids where melittin, which is the more hydrophobic, appears to be able to penetrate deeper [17] than the bacterial haemolysin and to induce bilayer fragmentation into small species.

However, in the fluid phase, at $T \gg T_c$, both toxins are probably very similarly bound to the bilayer and act as a wedge. At high doses they promote the appearance of very small complexes, undergoing fast isotropic tumbling in the NMR time scale. They are probably similar to the co-micelles already shown to exist with egg lecithin [7].

The formation of these small objects upon interaction of melittin and delta-lysin with bilayer provides a basis for a possible mechanism of the lysis of biological membranes.

Acknowledgements

We thank Dr. J. F. Faucon for helpful discussions and Pr. J. H. Davis for making his "de-Pake-ing" program available.

References

1. Habermann E (1972) Science 177:314–322
2. Freer JH, Birkbeck TH, Bhakoo M (1984) In: Alouf J, Fehrenbach F, Freer J, Jeljaszewicz J (eds) Bacterial Protein Toxins. Academic Press, London, pp 181–189
3. Bhakoo M, Birkbeck TH, Freer JH (1985) Can J Biochem 63:1–6
4. Dufourcq J, Dasseux JL, Faucon JF (1984) In: Alouf J, Fehrenbach F, Freer J, Jeljaszewicz J (eds) Bacterial Protein Toxins. Academic Press, London, pp 127–138
5. Yianni YP, Fitton JE, Morgan CG (1986) Biochim Biophys Acta 856:91–100
6. Faucon JF, Dasseux JL, Dufourcq J, Lafleur M, Pezolet M, Le Maire M, Gulik-Krzywicki T (1986) In: Mittal KL, Bothorel P (eds) Surfactants in Solution. Vol 5, Plenum Press, New York, pp 873–884
7. Dufourcq J, Faucon JF, Fourche G, Dasseux JL, Le Maire M, Gulik-Krzywicki T (1986) Biochim Biophys Acta 859:33–48
8. Morgan CG, Fitton JE, Yianni YP (1986) Biochim Biophys Acta 863:129–138
9. Perly B, Dufourc EJ, Jarrell HC (1984) J Labelled Compds 21:1–13
10. Kreger AS, Kim KS, Zaboretsky F, Berheinmer AW (1971) Infect Immun 3:449–465
11. Dasseux JL, Faucon JF, Lafleur M, Pezolet M, Dufourcq J (1984) Biochim Biophys Acta 775:37–50
12. Davis JH (1979) Biophys J 27:339–358
13. Bloom M, Davis JH, MacKay AL (1981) Chem Phys Lett 80:198–202
14. Dufourc EJ, Faucon JF, Fourche G, Dufourcq J, Gulik-Krzywicki T, Le Maire M (1986) FEBS Lett 201:205–209
15. Dufourc EJ, Dufourcq J, Birkbeck TH, Freer JH, Biochemistry, submitted
16. Seelig A, Seelig J (1974) Biochemistry 13:4839–4845
17. Dufourc EJ, Smith ICP, Dufourcq J (1986) Biochemistry 25:6448–6455
18. Batemburg AM, Hibbeln JCL, de Kruijff B (1987) Biochim Biophys Acta 903:155–165
19. Lafleur M, Dasseux M, Pigeon M, Dufourcq J, Pezolet M (1987) Biochemistry 26:1173–1179

Received December 21, 1987;
accepted January 4, 1988

Authors' address:

E. J. Dufourc
Centre de Recherche Paul Pascal
CNRS
Domaine Universitaire
33405 Talence, Cedex, France

Progress in Colloid & Polymer Science   Progr Colloid Polym Sci 76:59–67 (1988)

A mechanism of liposome electroformation

M. Angelova and D. S. Dimitrov

Central Laboratory of Biophysics, Bulgarian Academy of Science, Sofia, Bulgaria

Abstract: External electric fields can induce or prevent lipid swelling and liposome formation on solid surfaces. These effects depend on the type of lipid and surface, the medium parameters (temperature, osmolarity, ionic strength), the dried lipid layer thickness, the type and parameters of the electric field (*dc* or *ac*, amplitude, frequency, current), the situation of the lipid (on the very electrode surface or on another surface) and the time of exposure. This paper presents new data and theoretical estimates which allow us to suggest a possible mechanism of liposome electroformation. The new experimental results with negatively charged egg lecithin (EggL⁻) and neutral synthetic phosphatidylcholine (PC) are that (1) cholesterol in mixtures with EggL⁻ inhibits liposome formation, (2) sodium chloride leads to a decrease in the size and number of liposomes; liposomes do not form in solution more concentrated than 10 mM NaCl, (3) dextran also decreases the size and number of liposomes; they do not form in solution of dextran concentration higher than 2.5 mM, (4) *dc* electric fields can overcome the effect of dextran and lead to swelling even in 2.5 mM dextran solutions, (5) *dc* fields are most effective if applied in the first 30 s of lipid swelling; the increase of the period of the field action does not lead to a significant effect on the liposome yield, (6) a similar effect was also observed with (PC), but the critical period of time was a bit shorter — 10 s. The balance of forces acting on a lamellae of hydrating lipid shows that a possible pathway of liposome formation can include at least three basic stages: (1) separation of interacting membranes — the hydration and electrostatic interactions yield the main driving forces for this process, (2) instability of bending which can result from negative membrane tension due to surface and line tension and (3) the bending itself and closing of the membranes — the kinetics of this process and the instability of bending largely determine the liposome size distribution. External electric fields can affect any of these stages by at least two mechanisms: (1) direct electrostatic interaction and (2) redistribution of the counter-ions between the membranes. The interplay between the van der Waals, hydration and electrostatic forces is the major determinant of the mechanisms of liposome electroformation.

Key words: L̲iposomes, e̲lectric fields, m̲embranes, h̲ydration.

Introduction

Any mechanism of liposome formation by lipid swelling on metal electrode surface in external electric fields must also include description of the formation of liposomes without field. Therefore, it must explain two facts: (1) formation of liposomes without field and (2) liposome formation induced or hindered by external electric fields. Our basic concept is that the second type of phenomena may help in understanding the mechanism of liposome formation in general. It means that the information, gained by investigating the interplay between the electric fields and the physicochemical parameters of the system, can elucidate those basic phenomena which underly the formation of liposomes.

Regarding the formation of liposomes without field, there are several experimental observations which should be explained (see, e. g. [1–9]): (a) liposomes form only if the temperature is above that of the main phase transition; (b) they do not form in solutions of high osmolarity; (c) very small ionic concentrations practically prevent liposome formation; (d) the lipids or lipid mixtures should be at least slightly charged; (e) different lipids form liposomes in different ways; additives can improve or prevent liposome formation; (f) during formation, part of the liposomes or all of them can internalize neutral solute or even large solid particles, as well as other liposomes; (g) liposomes are heterogeneous in size with an average radius of the order of microns; (h) they are heterogeneous in thickness, too, ranging from unilamellar to having large number of lamellae and (i) liposomes do not form if the thickness of the dried lipid layer is below a certain critical value.

There are several facts which are specific for the effects of external electric fields [10–14]: (a) external electric fields lead to a decrease of the critical thickness of the dried lipid layer, above which liposomes form; with increasing voltage, this thickness decreases strongly to reach a value of several bilayers; (b) charged lipids form liposomes only if the sign of their charge is the same as that of the electrode; (c) very thin walled liposomes form only when the voltage is relatively high and respectively the thickness of the dried lipid layer is about a value of the order of 10 bilayers; (d) the increase of osmolarity of the solution can be compensated by increasing the voltage and (e) AC fields can induce liposome formation on both electrodes, independently of the sign of the lipid charge.

This communication presents several new facts and a theoretical model which trys to explain some of the experimental observations. It is based on a balance of electric and intermembrane forces and on the conditions for bending instability of membranes, as predicted by the fluctuation wave mechanism (see, e. g. [15–17]).

Materials and experimental methods

We used $L - \alpha$ phosphatidylcholine (EggL$^-$) (Sigma P-5394) which contains 71 % PC, 21 % phosphatidylethanolamine, and 8 % phosphatidylserine, i. e., it is negatively charged, and synthetic $L - \alpha$ phosphatidylcholine (PC) (Sigma P-5763), which is neutral. The cholesterol was from porcine liver, Sigma CH-PL, 99 %. Mixtures of cholesterol/EggL$^-$ of molar ratios 1 : 1 and 9 : 1 were used. The sodium chloride was from Merck. The dextran had molecular weight 80 000.

This lipids were dissolved in chloroform-methanol (9 : 1) mixture. Two drops of this solution (1 μl each) were deposited on two parallel platinum electrodes (diameter 0.48 mm, separation distance 0,5 mm) (see Fig. 1). The solvent was then evaporated under nitrogen. Electric fields were applied and distilled water or water solutions added. The observations were performed under phase contrast. In some cases, Ficoll was added to improve visualising of the thin-walled liposomes. The average number of bilayers was calculated from the amount of lipid, the surface area of the electrode this lipid occupies and data for the area of one lipid molecule. The mean error in determining the number of bilayers was 20 %.

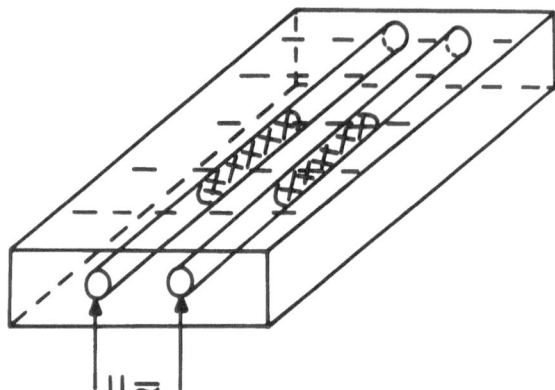

Fig. 1. Sketch of the device used (see text)

Experimental results

Figures 2–5 show swelling of the negatively charged lipid mixture EggL$^-$ (the thickness of the dried lipid film is 120 bilayers) and formation of liposomes without fields (Fig. 2–4) and in DC fields (Fig. 5). EggL$^-$ formed cell-size liposomes in distilled water (Fig. 2a). Adding cholesterol in the lipid mixture inhibited liposome formation (Figs 2b, c). At cholesterol/EggL$^-$ ratio 9 : 1, liposomes did not form. Sodium chloride leads to a decrease of the liposome size (Figs. 3a, b). Liposomes did not form when the NaCl concentration was higher than 10 mM (Fig. 3c). The effect of dextran was even stronger (Figs. 4a, b, c). Liposomes in 0.125 mM dextran (Fig. 4a) were smaller and somewhat different than the liposomes in distilled water. Liposomes almost did not form in 1.25 mM dextran solution (Fig. 4b). The lipid did not swell and liposomes did not form when the dextran concentration was higher than 2.5 mM (Fig. 4c). In this case, the electric field was able to overcome the osmotic forces on the negative electrode (Fig. 5a) and to swell the lipid, but liposomes did not form. On the positive electrode, as expected, the electric field did not cause any observable changes.

Figures 6 (a–e) show the effect of the duration of the electric field on the lipid. In all experiments, the pictures were taken after 30 min from the beginning of the lipid swelling and at this time Ficoll was added to imporve the contrast. Electric fields of different duration were applied at the beginning of swelling.

For this thickness of the dried lipid layer (30 bilayers) the EggL$^-$ did not form liposomes without field (Fig. 6a). For duration 10 s there were no visible changes (Fig. 6b). The time of 30 s was critical for

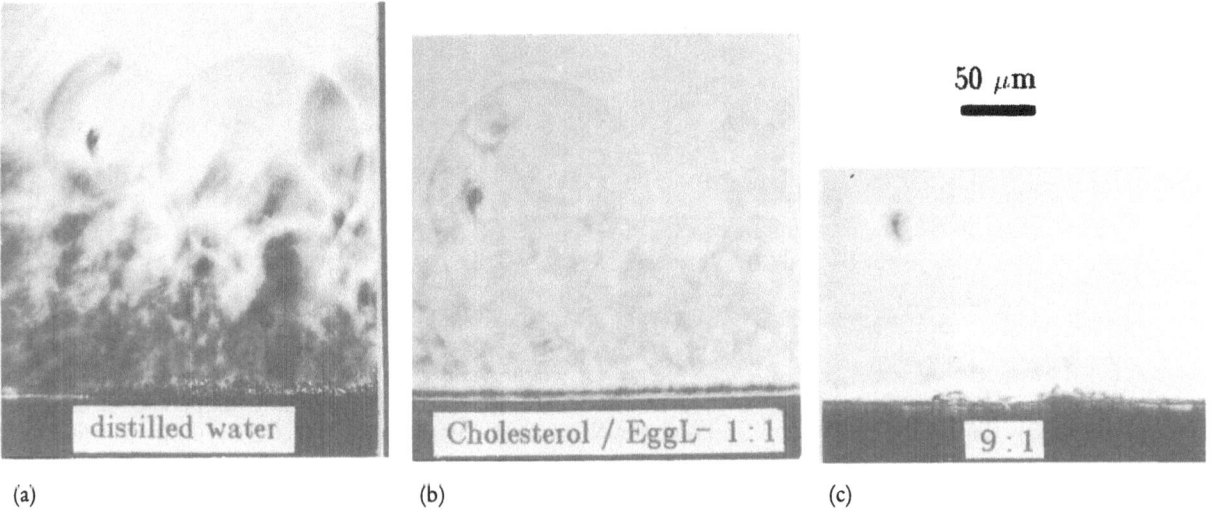

(a) (b) (c)

Fig. 2. Swelling and formation of liposomes from a mixture of EggL⁻ and cholesterol in distilled water (the thickness of the dried lipid layer is 120 bilayers, time of swelling 30 min). (a) Without cholesterol; (b) cholesterol/EggL⁻ ratio − 1:1; (c) cholesterol/EggL⁻ ratio − 9:1

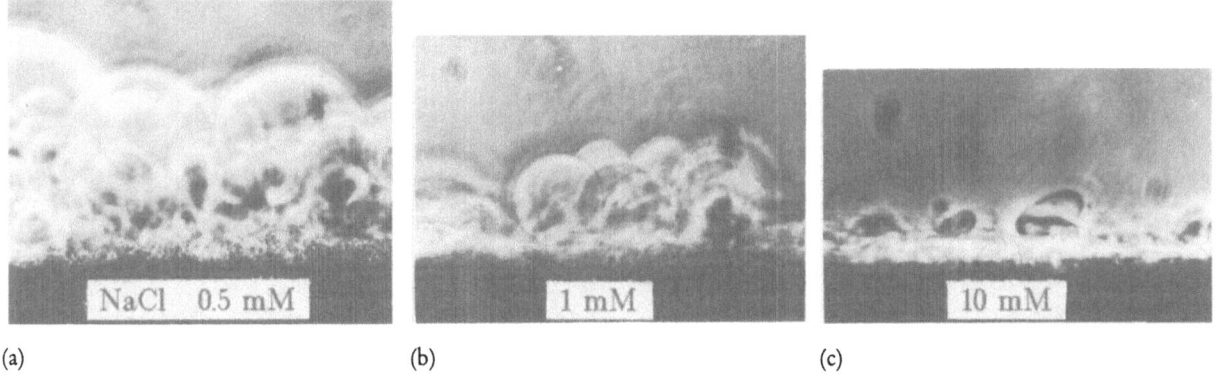

(a) (b) (c)

Fig. 3. Swelling and formation of liposomes from EggL⁻ in sodium chloride solutions (the thickness of the dried lipid layer is 120 bilayers; time of swelling, 30 min). (a) In 0.5 mM NaCl; (b) in 1 mM NaCl; (c) in 10 mM NaCl

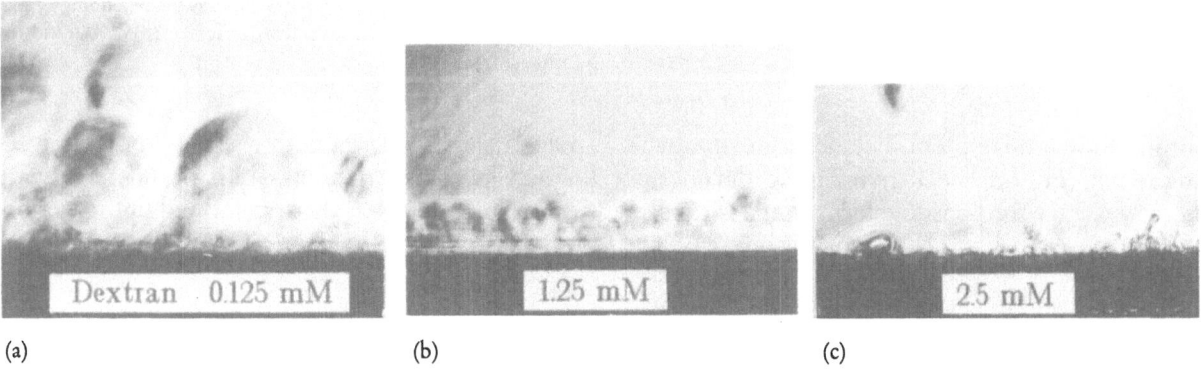

(a) (b) (c)

Fig. 4. Swelling and formation of liposomes from EggL⁻ in dextran solutions (the thickness of the dried lipid layer is 120 bilayers, time of swelling 30 min). (a) In 0.125 mM dextran; (b) in 1.25 mM dextran; (c) in 2.5 mM dextran

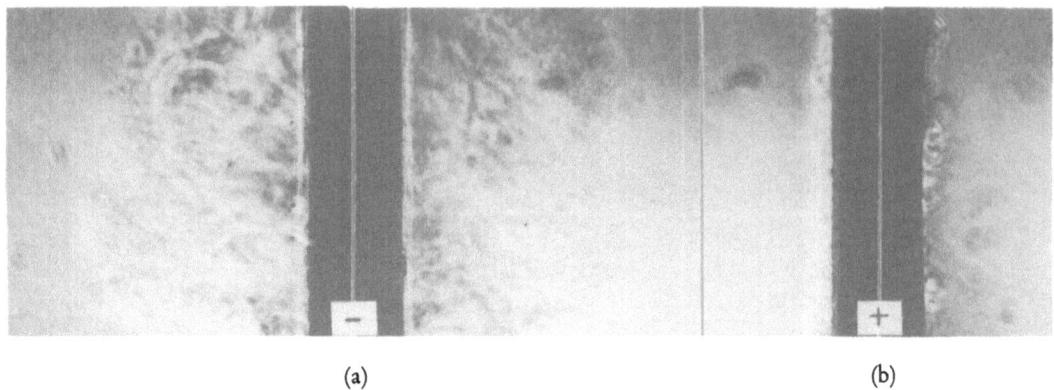

Fig. 5. Swelling and formation of liposomes form EggL⁻ in 2.5 mM dextran solutions under DC electric fields (the thickness of the dried lipid layer is 120 bilayers, time of swelling 30 min, DC field voltage 1 V). (a) On the negative electrode and (b) on the positive electrode

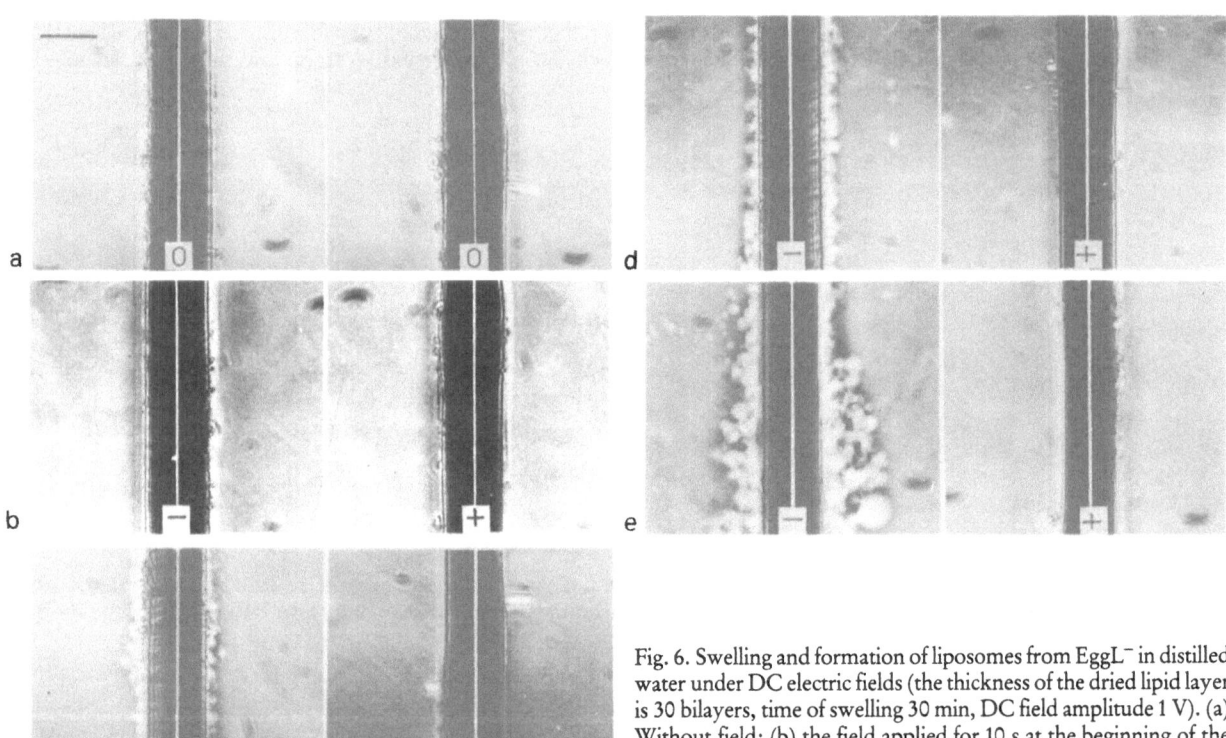

Fig. 6. Swelling and formation of liposomes from EggL⁻ in distilled water under DC electric fields (the thickness of the dried lipid layer is 30 bilayers, time of swelling 30 min, DC field amplitude 1 V). (a) Without field; (b) the field applied for 10 s at the beginning of the swelling; (c) the period of field application is 30 s; (d) the duration is 5 min and (e) 30 min duration

liposome formation (Fig. 6c). The increase of the duration to 5 min (Fig. 6d) and 30 min (Fig. 6e) did not significantly change the liposome formation. As expected, there were no visible changes on the positive electrode.

The critical duration for the neutral lipid PC was 10 s (Figs. 7a, b). Figure 7a shows PC swelling without fields. The field acting for 10 s caused liposome formation (Fig. 7b). The increase of the duration of the applied field did not significantly change the lipid swelling and liposome formation. The basic difference with the case of EggL⁻ was that PC formed liposomes on both electrodes.

The geometric model

We may consider a hydrating lipid layer of thickness l on a semi-infinite metal electrode in an external

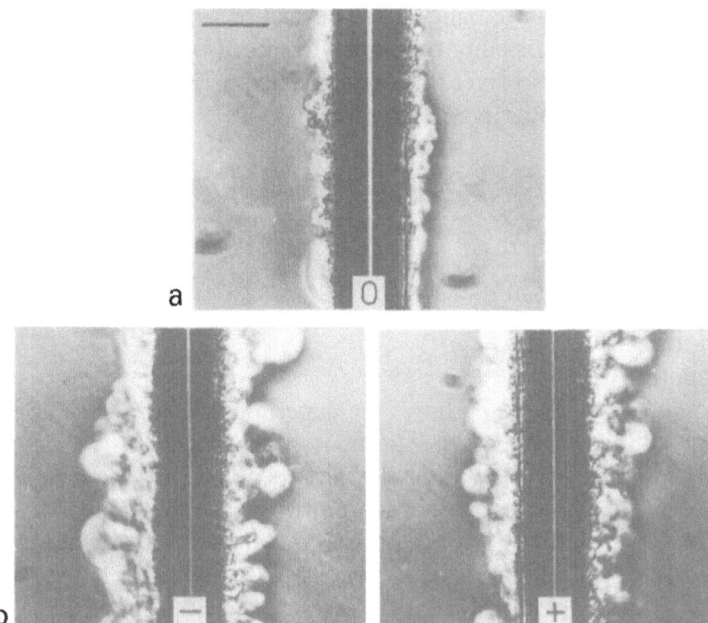

Fig. 7. Swelling of PC and formation of liposomes in DC field in distilled water (the thickness of the dried lipid layer is 50 bilayers, time of swelling 30 min, DC field amplitude 1 V). (a) Without field; (b) the field applied for 10 s at the beginning of the swelling

electric field of intensity $\vec{E}$ (Fig. 8). The lipid lamella, which can be a single bilayer or a sheet of several bilayers, of thickness d, is separated by a thin liquid layer of thickness h, from the bulk of the lipid layer. This system is placed in distilled water or water solutions, where the concentration of the osmotically active solute is c_o.

The balance of force

The lamella is subjected to repulsive and attractive force:

(a) The van der Waals forces can be separated into two parts — those due to interactions between the lamella and the "bulk" lipid layer, yielding attractive pressure $P_{wb}(h, l)$, and forces due to interactions with the metal electrode, giving rise to attractive pressure

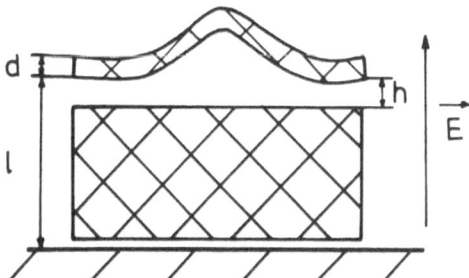

Fig. 8. The geometric model for theoretical estimates

$P_{we}(l)$. In most cases considered here, the thickness l is much larger than the separation h; typical ratio is of the order of 10 to 100. In spite of that, since the metal has a density one order of magnitude larger than the lipid and therefore the Hamaker constant for the metal/lipid interactions is larger than for the lipid/lipid interactions, P_{we} can be of the order or even larger than P_{wb}. P_{wb} decreases rapidly with increasing separation between the membranes, while P_{we} depends only on the distance to the electrode and therefore does not change significantly with increasing separation between the membranes.

(b) The electrostatic interactions are of at least two types — intermembrane interactions characterized by repulsive pressure $P_{ei}(h)$ and electrostatic interactions between the field and the charged membranes which can yield attractive or repulsive pressure $P_{ee}(l)$. Intermembrane repulsion decreases rapidly with increasing the separation h, while P_{ee} does not depend on separation.

(c) The hydration forces, characterized by the pressure P_h are always repulsive. They are dominant for very close separations commonly less than 1 nm. They can act between the membranes and the membranes and the electrode.

(d) The osmotic pressure, P_{osm}, can lead to separation of the membranes. It depends on the solution osmolarity. When the solution osmolarity increases, the osmotic pressure will decrease and can prevent the membrane separation. It decreases rather slowly with increasing separation between the membranes.

(e) The undulation forces, $P_u(h)$, will always repulse the membranes. They depend on the intermembrane separation and not on the separation to the electrode. As was shown by Harbich and Helfrich [18], they can contribute significantly to a pathway of liposome formation.

(f) The dynamic pressure of resistence to motion, $P_d(h)$ (see, e. g. [15]). This can be due to viscous forces, effects of membrane permeability and, in some cases of fast swelling, inertial forces can also contribute. At equilibrium this pressure is zero.

On the basis of the above qualitative review of the basic forces involved in liposome formation, we can write the following approximate balance of forces

$$P_{we}(l) + P_{wb}(h,l) + P_{osm}(h) + P_{ei}(h) + P_h(h)$$
$$+ P_u(h) + P_{ee}(l) = P_d(h); \tag{1}$$

where the dependence on separation from the electrode l and the separation between the membranes h is indicated.

The characteristic thickness of the lipid layer, l_{co}, above which liposomes form without field, can be obtianed from Eq. (1) by assuming $P_{ee} = P_d = 0$ and $l = l_{co}$. Then

$$P_{we}(l_{co}) + P_{wb}(h_o, l_{co}) + P_{osm}(h_o) + P_{ei}(h_o)$$
$$+ P_h(h_o) + P_u(h_o) = 0, \tag{2}$$

where h_o is the separation between the membranes at the characteristic lipid layer thickness without field.

If we assume that the electric field does not significantly affect the other forces, Eqs. (1) and (2) can be transformed into

$$Eq = P_{we}(l_{co}) - P_{we}(l_c) + P_{wb}(h_o, l_{co})$$
$$- P_{wb}(h, l_c) \tag{3}$$

where E is the field intensity and q the surface charge density.

When the lipid layer thickness, l, decreases, the van der Waals pressure P_{we} increases and the electric field intensity E should increase in order to separate the membranes. Therefore, this can explain the observed dependence of the characteristic layer thickness on the field intensity. Unfortuantely, the theoretical analysis of forces in a multilamellar system is rather com-

plicated and does not lead to simple formulae. One possibility of simplifying Eq. (3) is for small critical lipid layer thicknesses. Then the term P_{we} dominates and can be represented as

$$P_{we} = - A/l_c^4; \quad A = (H_w d/2\pi), \tag{4}$$

where H_w is the Hamaker constant for the metal/lipid interactions. The combination of Eqs. (3) and (4) gives

$$Eq = A/l_c^4. \tag{5}$$

Figure 9 shows the experimental data [12] for small thicknesses. The fit is good. Equation (5) well describes the functional dependence on the lipid layer thickness. In addition, if we assume values for $A = 10^{-29}$ J · m and $q = 10^{-1}$ C/m², we get the experimentally measured value 2.4×10^{-28} J · m³/C for the slope of the straight line which describes the dependence $E(l_c^{-4})$. It should be pointed out that the change of the intermembrane electrostatic pressure due to external electric field can contribute significantly to such a dependence. This very complicated process, which may include kinetic effects, is difficult to model theoretically at the moment.

Instability of bending

The membrane, which is exposed on both sides to two different environments, has two different membrane surface tensions σ. The surface tension of the membrane surface which is exposed to the bulk liquid, σ_o, is lower than that of the inner surface, σ_i. This is

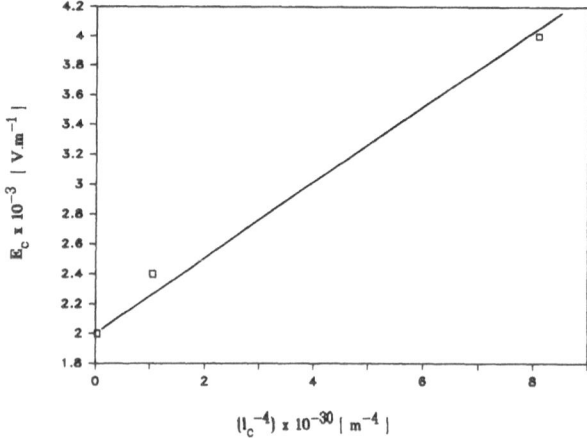

Fig. 9. Comparison of the experimental data from Ref. [12] (redrawn in another scale) with the theoretical estimate given by Eq. (5)

due to the higher surface charge density q_o, because of the higher degree of dissociation of the lipids. The concentration of counter ions in the space between the membranes is high, especially in the initial stages of lipid swelling, which leads to lower surface charge density q_i, respectively higher surface tension σ_i. The decrease in the surface tension due to surface charges can lead to membrane bending. The membrane expands because of the electrostatic repulsion of the surface charges. If the edges of the membrane domain are fixed, the membrane will bend. The difference in surface tensions of both membrane surfaces can lead to bending of the membrane toward blowing a vesicle, like a bubble.

When a external electric field is applied across the membrane, its surface tension changes. This change depends on the transmembrane potential and the wavenumbers k of the fluctuation waves, which always exist due to thermal motion or external constraints. One of the simplest possible cases is when the intermembrane interactions are neglected and the membrane is considered as a single one. For large waves (small wave numbers) an analysis carried out by Leikin [20] led to an expression which gives the following condition for marginal instability

$$T - CU_t^2 + Bk_{cr}^2 = 0, \qquad (6)$$

where k_{cr} is the critical wavenumber, T is an effective membrane tension, which for our model is equal to the sum of both membrane surface tensions, U is the induced transmembrane voltage, C membrane capacitance and B, beding elasticity. It is assumed that in the initial stages of liposome formation the lamella is not exposed to external stretching forces and its edges are fixed.

Each of the surface tensions can be represented as a sum of the surface tension of an uncharged membrane plus that due to surface charges and to the external field, which induces transmembrane potential. We assume that the tension of an uncharged membrane is zero because there are no external stretching forces. Then Eq. (6) can be written in the form

$$Bk_{cr}^2 = C(U_t^2 + \psi^2) \qquad (7)$$

where ψ is the membrane surface potential due to membrane surface charges.

Then the critical wavelength, λ_{cr}, is

$$\lambda_{cr}^2 = 2\pi B/C(U_t^2 + \psi^2). \qquad (8)$$

All wavelengths larger than the critical one will lead to instability. It is seen from Eq. (8) that if the transmembrane potential and the surface potential are zero, the membrane is stable for all wavelengths ($\lambda_{cr} \to \infty$). The increase of the transmembrane potential will decrease the size of the bending membrane area. The increase of the bending elasticity will increase the liposome radius. The increase of the thickness of the lamella also will increase the characteristic length. For $\psi = 0.1$ V, $C = 1$ mF/m^2, $U_t = 0$ and $B = 10^{-17}$ the critical wavelength is of the order of microns. It should be noted that large transmembrane potential can be induced by differences in surface charges due to different environments, as well, as discussed above. In this case it can be included in U_t. However, this will not change the estimate. The difference in the surface tension can determine the direction of the membrane beding. The external electric field can induce very small transmembrane potential of the order of 0.01 mV (for $E = 1$ kV/m). The transmembrane potential induced by external electric fields may be important for non-charged membranes.

The kinetics of the membrane bending and closing after the instability has occurred can be described by taking into account the membrane viscosity, the edge energy and the bending elasticity (see, e.g. [15]). At present, we are performing more refined analysis of this stage.

Discussion and conclusions

This analysis is oversimplified and can serve for a qualitative and in some cases semi-quantitative description of the effects of external electric field. In general, it can explain, with some additional considerations, the basic facts about liposome formation, as presented at the beginning.

The basic driving forces for membrane separation are the electrostatic and osmotic forces. It should be noted that intermembrane electrostatic forces can also be considered as osmotic due to the high concentration of the double layer counterions. Therefore, any external osmotic pressure due to an agent which does not penetrate through the membrane, such as dextran, can prevent membrane separation, and consequently liposome formation, if taken in sufficient concentrations. Our experimental results show that for dextran this critical concentration is about 2 mM. This is in agreement with the observations of Mueller et al. [8] who found that the internal osmolarity of the cell-size liposomes they obtained by swelling, was in this range. In addition, estimates of the concentration of counter ions produced by the lipid dissociation in distilled water in one liposome lead to the same order of magnitude. In addition, the characteristic Debye length of the double layer in distilled water is of the order of 0.1 to 1 µm, which is the lower limit of the size of liposomes obtained by swelling.

The mechanism of preventing liposome formation by ions, such as in our experiments, of sodium chloride, is a bit different. They can enter the intermembrane space (the dextran cannot) and by decreasing the double layer thickness to decrease the intermembrane electrostatic repulsion. In this case, the van der Waals forces dominate and the membranes cannot be separated. The electric field can overcome the effect of the increased osmolarity, as is shown for dextran. However, just separating membranes is evidently not sufficient to induce liposome formation. Other factors, mainly from kinetic origin, should be considered. This may be due to the fact that the membranes cannot bend and close. It seems that the fact that liposomes are formed when the temperature is above the phase transition temperature can also be explained with the impossibility of bending, due to the increased, membrane viscosity, which prevents membrane bending. According to the fluctuation wave mechanism, the membrane will be unstable to bending for all wavelengths above a critical one. Therefore, it may be expected that the liposomes will be heterogeneous in size. In many cases, dominant wave may exist, which has the fastest rate of growth. This wave will give the most expected size of liposomes. Commonly, the length of this wave is of the order of magnitude of the critical wavelength [15]. Hence, we can use the expression for the critical wavelength (Eq. (8)) as an estimate for possible liposome size. It must be pointed that this expression does not take into account the intermembrane interactions, and it therefore cannot explain the dependence of liposome size on the concentration of dextran, NaCl and, to some extent, on the external field. Preliminary estimates, which are not presented here, have shown that the decrease of intermembrane repulsion leads to a decrease of the length of the dominant wave. The kinetic effects in membrane bending and separation seem to be the major causes of the heterogeneity in thickness, too. For very small lipid layer thicknesses, the liposomes are almost unilamellar in thickness because there are no other bilayers to incorporate in the liposome. However, in this case, in order to overcome the strong van der Waals interaction with the electrode, we need to apply external electric fields.

The very good agreement between the theoretical estimates and the experimental data (see Fig. 9) indicates that the main force in this case is the electrostatic interaction between the field and the membrane. In addition, we could not find another way of explaining the fact that charged lipids form liposomes only on the electrodes of the same sign. The experimental data with zwitterionic lipid showed, however, that the redistribution of the counter ions in the intermembrane spaces, due to external electric field, can have a significant effect. It seems that the counter ions are "sucked" from the intermembrane space by the field. There is not enough time for the counterions to diffuse from the double layer around the electrode, where their concentration is high.

Our experimental findings, that the electric field is indeed effective in the first tens of seconds and that longer exposition to the field does not have a significant effect, indicate that the field "helps" to separate the membranes to a certain "critical" distance, after which they can eventually spontaneously bend and form liposomes. The spontaneous bending and closing also takes time. This time can be estimated by a formula derived in Ref. [14] as a function of the edge energy, membrane viscosity and bending elasticity (for the effects of edge energy in liposome formation, see the work of Fromherz [19]).

The type of the lipid can affect lipid swelling and liposome formation by: (1) determining the phase state of the lipid – the principal possibility fo forming bilayer structures, (2) the forces between the membranes and (3) the kinetics of liposome separation and bending. Cholesterol, for example, decreases the membrane permeability to water and increases the membrane viscosity. This inhibits the separation and bending of membranes. In the case of very high cholesterol content, there may also be structural reasons for the lack of liposomes.

It must be again pointed out that due to the many, and very rough, assumptions, this analysis can lead to wrong results in many cases. Therefore, it may be useful to stress again the two basic concepts which should remain valid: (1) the liposome formation requires membrane separation and bending. Therefore, normal forces, which repulse membranes, and tangential forces, which bend the membranes, are involved and (2) external electric field changes both types of forces. They change strongly the normal repulsive forces and therefore they can induce liposome formation or prevent it. The external elelctric field does not change the tangential forces significantly.

Therefore, they cannot directly change the liposome size significantly.

One final point to make is that the above analysis was applied to a concrete system, for our experimental device and data. All the numerical estimates and conclusions are for this particular system. What we are currently doing is to refine the physicomathematical model and to apply the results to other systems.

Acknowledgements

We are indebted to Prof. E. Evans for fruitful discussions and suggestions. We thank Mrs. R. Gadeva for the technical help. This work was supported by the Committee for Science at the Ministerial Council of Bulgaria through Contract No. 189.

References

1. Bangham AD, Standish MM, Watkins JC (1965) J Mol Biol 13:238
2. Szoka F, Papahadjopoulos D (1980) Ann Rev Biophys Bioeng 9:467
3. Szoka F, Papahadjopoulos D (1981) In: Knight (ed) Liposomes: From Physical Structure to Therapeutic Applications. Elsevier/North-Holland Biomedical Press, Amsterdam, pp 51–82
4. Bangham AD (ed) (1983) Liposome Letters. Academic Press, London
5. Gregoriadis G (ed) (1984) Liposome Technology. CRC Press, Inc, Florida
6. Schmidt KH (ed) (1986) Liposomes as Drug Carriers. Georg Thieme Verlag, Stuttgart New York
7. Haydon D (ed) (1986) Faraday Disc Chem Soc 81
8. Mueller P, Chien TF, Rudy B (1983) Biophys J 44:375
9. Dimitrov DS, Li J, Angelova MI, Jain RK (1984) FEBS Lett 176:398
10. Dimitrov DS, Angelova MI (1985) Proc Biotech '85 Geneva 1:655
11. Dimitrov DS, Angelova MI (1986) Studia Biophysica 113:15
12. Angelova MI, Dimitrov DS (1986) Faraday Disc Chem Soc 81:303
13. Dimitrov DS, Angelova MI (1987) Studia Biophyisca 119:61
14. Dimitrov DS, Angelova MI (1987) Progr Colloid Polym Sci 73:48
15. Dimitrov DS (1983) Progr Surface Sci 14:295
16. Dimitrov DS, Jain RK (1984) Biochim Biophys Acta 779:437
17. Dukhin SS, Rulev NN, Dimitrov DS (1986) Dynamics of Thin Films and Coagulation, in Russian. Naukova Dumka, Kiev
18. Harbich W, Helfrich W (1984) Chem Phys Lipids 36:39
19. Fromherz P (1986) Faraday Disc Chem Soc 81:39; 81:347
20. Leikin S (1985) Biol Membrany, in Russian 2:280

Received December 4, 1987;
accepted December 14, 1987

Authors' address:

M. Angelova
Central Laboratory of Biophysics
Bulgarian Academy of Sciences
BG-1113 Sofia, Bulgaria

Progress in Colloid & Polymer Science

Progr Colloid Polym Sci 76:68–74 (1988)

The micellar sphere-to-road transition in CTAC-NaClO₃
A fluorescence quenching study

M. Almgren, J. Alsins, J. van Stam, and E. Mukhtar

Department of Physical Chemistry, University of Uppsala, Sweden

Abstract: The fluorescence decay of pyrene quenched by benzophenone in cetyltrimethyl ammonium chloride solutions has been studied. On addition of chlorate ions to the solution, a transition to long rod-like micelles are known to occur. The fluorescence decay pattern changes in the transition region, from a typical micellar quenching behaviour, following the equation suggested by Infelta, Thomas and Grätzel [22] via the type of decay suggested for polydisperse micelles (Almgren and Löfroth [18]) to a decay pattern which is independent of the chlorate concentration at high concentrations and follows an equation derived for quenching in infinite rods. The weight average aggregation numbers and the widths of the micelle size distributions were determined in the transition region.

Key words: Aggregation number of micelles, fluorescence decay, quenching of pyrene by benzophenone, cetyltrimethylammoniumchloride solution, transtion to rod-like micelles.

Introduction

It was recognized at an early stage that some sort of extended structures form under certain conditions in solutions of ionic surfactants with long hydrophobic chains, even at very low concentrations. The viscosity and viscoelasticity of the solution give a very obvious indication of the presence of interacting aggregates. Hatschek [1, 2] measured the rheological properties of ammonium oleate solutions and was particularly baffled by the irregular behaviour of solutions in the Couette viscosimeter, in which the inner measuring cylinder never came to rest.

A long hydrocarbon chain and a small effective headgroup area are favourable for the formation of long rod-like structures [3]. High salt concentrations and strongly bound counterions reduce the electrostatic repulsion and thus the effective head group area. Cetyltrimethylammonium bromide (CTAB) forms rods at high concentrations [4], whereas the corresponding chloride surfactant, CTAC, does not. Certain hydrophobic counter ions, like salicylate [5] are also known to promote the transition very effectively. Also, some polar aromatic compounds favour the for-

mation of rods. Angelescu et al. [6] studied the effects of phenols on alkali soaps in as early as the 1930s, and a very notable example was found by Nash [7] who showed that CTAB and β-naphthol in equimolar amounts form clearly viscoelastic solutions even at concentrations less than 10^{-4} M.

The more recent development was summarized by Porte [8] who discussed the thermodynamic conditions for the transition. As proposed by Mukerjee [9] the standard free energy of the rod micelle will depend linearly on the aggregation number, N, at large N.

$$G°_N = N\mu°_N = N\mu°_\infty + kT\alpha \qquad (1)$$

The term $kT\alpha$ represents the end-effects, and is assumed to be independent of N at large N. From this assumption, it follows that the average aggregation number should grow linearly with the square root of the amphiphile concentration, and the polydispersity should become very large. These predictions have found experimental support to the extent that a strong increase of N with surfactant concentration occurs in many cases [10, 11]. The experimental results concerning the polydispersity are sparse and uncertain.

The correlation times measured in quasielastic light scattering are often so long that a substantial change in the size of a micelle may take place by monomer exchange during that time. Non-exponential correlation functions have been observed, however, and have been interpreted in terms of size polydispersity [10, 12]. The quantitative significance of the values for the width obtained in this way is not clear to us. Some results from fluorescence quenching, however, do indicate that large, probably rod-shaped micelles, have broad and skewed size distributions [13–15].

In an attempt to bring the analysis one step further, Porte et al. [16] and Eriksson and Ljungren [17] pointed out that the simple theory, even when intermicellar interactions were included to some extent [16], predicts a gradual growth of the micelles with increasing surfactant concentration, seemingly at variance with the observed drastic change of the rheological properties of the solution in a narrow concentration range, the "second CMC". They introduced assumptions, therefore, to explain a sudden sphere-to-rod transition.

We have used a time-resolved fluorescence quenching method, with pyrene as the excitable probe and benzophenone as the quencher, to study micelle growth in the transition induced by sodium chlorate addition to a CTAC solution. The method gives the weight average aggregation number and in addition some information about the width and skewness of the size distribution [14, 18]. The study also reveals a limitation of the method: above a certain size, no further change of the decay pattern can be observed. As elaborated elsewhere [19], this means that the rods are so long that the excited probe and the quencher in their diffusional movement only explore a part of the rod during the lifetime of the excited state. The fluorescence decay curves can then be described as arising from diffusion-controlled quenching in an infinite cylinder.

Experimental

A capillary viscosimeter was used for the viscosity measurements. Fluorescence decay data were collected with the time-correlated single photon counting technique, using a mode-locked argon-ion laser (Spectra Physics Model 171/12) to synchronously pump a cavity-dumped dye laser (Spectra Physics Models 375, 344S) for the excitation. The dye used was DCM operating at 670 nm, frequency doubled in a KDP crystal to 335 nm to match the absorption of pyrene. The plane of polarization of the excitation

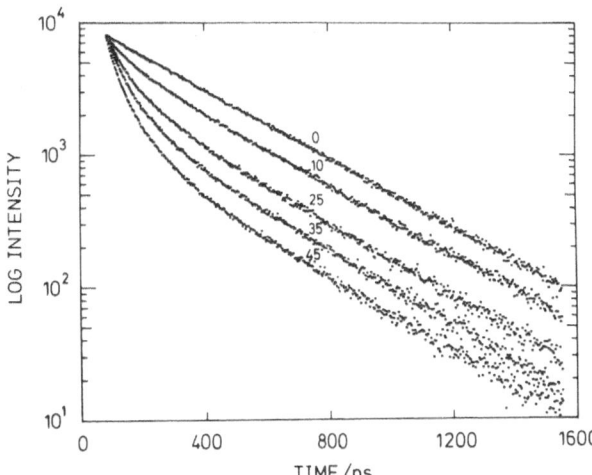

Fig. 1. Fluorescence decay curves for pyrene quenched by benzophenone in micellar solutions of 39.9 mM CTAC and 33.9 mM NaClO₃ at 25 °C. Concentrations of benzophenone in M × 10⁵ are indicated

beam was adjusted to vertical polarization by means of a Soliel-Babinet compensator and the emission was detected with magic angle polarization (54.7°).

The fluorescence emission was measured at 394 nm, selected in a double monochromator (Jobin-Yvon model H10) using 4-nm bandwidth. A channel plate multiplier (Hamamatsu 1564U) was employed, for detection. The multiplier output was conditioned using a Tennelec CFD (TC455) and used as stop pulse in the TAC (Tennelec TC861A). The start pulse was provided by a diode monitoring part of the excitation beam.

The output of the TAC was connected via an ADC to a multichannel analyzer based on a PDP11/23 microcomputer. Data analysis was carried out on a VAX 780 computer, using software based mainly on the programs developed by Löfroth [20].

Materials

Cetyltrimethylammonium chloride (CTAC) was prepared by ion exchange of CTAB (Merck, No 2342, pro analysi) on a Dowex 1–X8 resin. The product was freeze-dried and stored in a desiccator. Pyrene, benzophenone (Aldrich), sodium chloride (Merck) and sodium chlorate (Merck), were samples of high purity and used as received.

The solutions were freed from oxygen by gentle bubbling with argon for 30 min. in a high stem cuvette, usually at room temperature. In the case of the very viscous solutions with a high chlorate content, a temperature of 40 °C was employed.

The fluorescence quenching method

The experimental decay curves shown in Fig. 1 are good examples of what to expect in solutions of small micelles. The decay of the excited probe, pyrene, is a simple exponential when no quenchers are present (and the concentration low enough to prevent excimer formation). With the quencher added, the decay is initially fast (deactivation in micelles with quenchers) but ends with an exponential portion characterized by the same decay constant k_o as the decay when no quencher is present. As will be discussed in the following, benzphenone, which was used as quencher in this study [21] migrates between the micelles sufficiently fast to increase the final decay rate slightly above k_o. Also, in this case, the final decay portion represents the fluorescence decay from micelles which were without quenchers initially; from its relative amplitude, the fraction of micelles without quenchers can be calculated and, assuming a random (poisson) distribution of quenchers over the micelles, the total number of micelles and hence their aggregation number, can be calculated.

If the concentration $[S]$ of surfactant and $[Q]$ of quencher are taken to represent the amounts in micelles, the aggregation number, N, is

$$N \approx [S]/[Q]\ln(F(0)/F_S(0)) \qquad (2)$$

where $F_S(0)$ is the extrapolated value of the fluorescence intensity at the excitation moment according to the final exponential decay, and $F(0)$ the measured intensity.

For polydisperse micelle systems, an unbiased assumption is that the probability of finding a quencher or a probe in a micelle is proportional to its size (aggregation number). Application of Eq. (2) then yields an apparent, or Q-average aggregation number, $\langle N \rangle_Q$, which depends on the quencher concentration as shown in Eq. (3).

$$\langle N \rangle_Q = \langle N \rangle_w - 1/2\,\sigma_w^2 q + 1/6\,\varkappa_3 q^2 - \ldots \qquad (3)$$

Here $\langle N \rangle_w$ is the weight-average aggregation number, and σ_w^2 and $\varkappa_3$ are the second and third cumulants of the weight distribution, $sN_s/\sum sN_s$. The determination of the Q-average aggregation number using Eq. (2), and the subsequent application of Eq. (3), require that the experimental data show a well-developed final exponential portion with the decay constant equal to that of the unquenched decay.

Results and discussion

The viscosity of a 40 mM CTAC solution is almost unaffected by additions of NaCl, whereas the addition

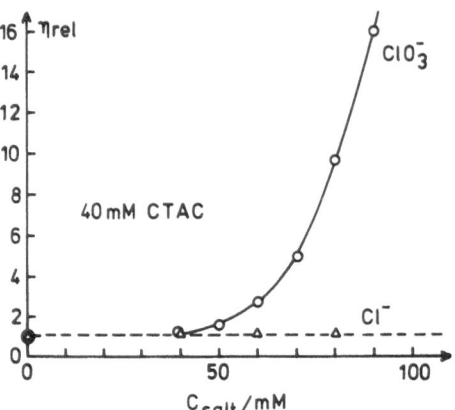

Fig. 2. Relative viscosities at 25 °C for 39.9 mM solutions of CTAC at various concentrations of added salt, NaCl or NaClO₃

of 40 mM NaClO₃ or more induces a rapid increase (Fig. 2). Fluorescence decay curves determined in the transition region for a series of solutions, all with the same pyrene and benzophenone concentrations are shown in Fig. 3. These data are similar to some published previously [19]; the main difference being that the new measurements have been extended to longer times.

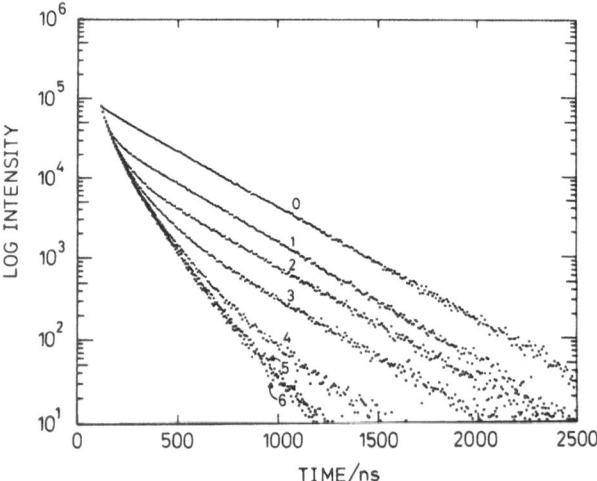

Fig. 3. Fluorescence decay curves for pyrene quenched by benzophenone in micellar solutions of CTAC containing various concentrations of NaClO₃. Pyrene and benzophenone concentrations were kept constant at about 1×10^{-5} M and 4.50×10^{-4} M, respectively, except in solution 0, which contained no quencher. The concentrations of chlorate in the solutions from 0 to 6 were: 0, 0, 32.7, 40.2, 49.2, 70.6, 101 mM

An increase in the micelle size is clearly shown by the decreasing amplitude of the exponential tail, and the increasing importance of the nonexponential initial process. Only the data for solutions without chlorate and with 34 mM chlorate could be analyzed with the conventional Infelta [22] model; these will be discussed first. The data for 42 and 44 mM chlorate, considered next, were more difficult to analyse, presumably because the length and polydispersity of the micelles starts to become important. The decay curves for 70 and 100 mM chlorate, finally, are very similar, although the viscosity data clearly indicate a continued growth in this concentration region. The reason is that when the micelles become long enough, the excited probe and quenchers will not have time to explore the full length of them within the time window available for measurements (which is determined by the lifetime of the excited state and the dynamic range of the measurements). The infinite-rod model [19] is used in the interpretation.

Globular micelles

The decay data shown in Fig. 1 for pyrene quenched by benzophenone in CTAC, 39.9 mM, with NaClO₃, 33.9 mM, and the corresponding data for CTAC without chlorate, were analysed with the Infelta equation [22], Eq. (4), with results as given in Table 1.

$$F(t) = F(0) \exp\{-A_2 t + A_3[\exp(-A_4 t) - 1]\}. \quad (4)$$

A slow migration is indicated by the concentration dependence of the parameter A_2 in Table 1, but the exit rate constant k_- is much smaller than the quenching constant k_q in the micelle. The parameters in Eq. (4) are then

$$A_2 = k_0 + k_- \langle n \rangle = k_0 + k_+[Q]_f$$

$$A_3 = \langle n \rangle \quad (5)$$

$$A_4 = k_q$$

where n is the average number of quenchers per micelle, k_+ the second-order entrance rate constant, and $[Q]_f$ the concentration of quencher in the intermicellar solution.

The results in Table 1 for the solution with the lowest quencher concentration without chlorate, and for the two lowest quencher concentrations of the solutions with chlorate, deviate markedly from the rest by giving larger aggregation numbers and lower values of A_4. Since the accuracy of the estimates at low average occupancy numbers is expected to be low, these results were neglected in the subsequent analysis.

The average of the aggregation numbers estimated for the two systems are 90 and 170, respectively, corresponding to hydrophobic radii of 21.6 and 26.8 Å, showing that the small micelle may be spherical, but the large one has to deviate somewhat from the spherical shape.

The average value of $k_q = A_4$ is obtained as 1.8×10^7 s⁻¹ and 0.78×10^7 s⁻¹ for the small and large micelles, respectively, which is a somewhat larger change than that of the aggregation number. There is also a change in the pyrene lifetime (in the absense of quencher) — it is about 10 % larger in the solution with chlorate.

The A_2-values are plotted vs concentration in Fig. 4. From the slope of the regression lines, values of k_- or $k_+[Q]_f/[Q]$ can be estimated using Eq. (5). The values of the former become 1.7 and 1.3×10^5 s⁻¹, respectively, for the small and the large micelles, and are thus almost inversely proportional to the radii. From the value of the latter, the quencher distribution between micelles and water is obtained if k_+ is estimated as dif-

Table 1. Fluorescence decay parameters, deduced aggregation numbers and rate constants for pyrene quenched by benzophenone in 39.9 mM CTAC and in 39.9 mM CTAC + 33.9 mM NaClO₃ at 25 °C

$[Q] \times 10^5$ (M)	$1/A_2 \times 10^{-6}$ (s⁻¹)	A_3	$1/A_4 \times 10^{-9}$ (s⁻¹)	$x^{2a})$	$\langle N \rangle_Q$
0 mM NaClO₃					
0	303.2	—	—	1.06	—
5.0	305	0.143	90	1.25	113
10.0	303	0.229	60.5	1.01	93
15.0	297	0.334	54.4	0.99	87
25.0	297	0.559	57.2	0.95	89
35.0	294.5	0.645	50.2	0.94	73
45.0	289	1.021	52.5	1.07	90
33.9 mM NaClO₃					
0	333	—	—	1.15	—
4.78	338	0.269	163	1.04	222
9.84	337	0.483	164	1.02	194
14.66	327	0.645	138	0.94	175
25.3	322	1.086	134	1.17	171
35.1	309	1.483	121	1.07	165
45.0	311	1.897	121	1.35	173

a) Reduced x^2-test

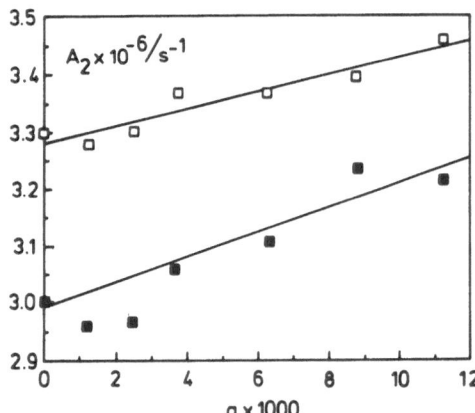

Fig. 4. The decay constant of the final, exponential part of the pyrene fluorescence decay at various quencher concentrations ($q = [Q]/[S]$) in solutions of 39.9 mM CTAC and 0 (■) or 33.9 mM chlorate (□)

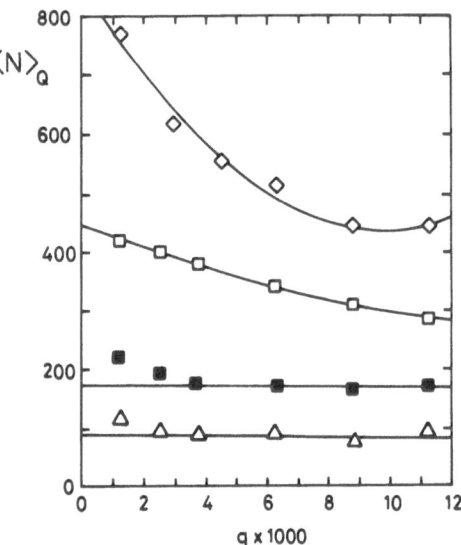

Fig. 5. Q-average aggregation number at 25 °C as a function of quencher concentration for CTAC micelles at the following concentrations of NaClO$_3$: 0, 33.9, 41.4, and 43.7 mM from bottom to up

fusion-controlled [23] with the Smoluchowski equation

$$k_+ = 4\pi RDN_A \qquad (6)$$

where D is the diffusion coefficient of the quencher, assumed to be 5×10^{-9} m^{2}s^{-1}. N_A is Avogadro's constant, and R the encounter radius, which was taken, arbitrarily, as the hydrophobic radius plus 4 Å, leading to an estimated k_+ of 9.5 and 11.0×10^9 M^{-1}s^{-1} in the two cases.

The distribution ratio of the quencher was obtained as 0.037 and 0.042, in satisfactory agreement, which leads to a distribution constant of about 2200, calculated as the ratio of the quencher concentrations in the micellar subphase and in water. This is somewhat larger than the distribution constant of benzophenone between octanol and water [21] which was used for an estimation of the importance of quencher migration in a previous communication [19].

Short polydisperse rods

The small but measurable migration rate of benzophenone between the micelles complicates the analysis of the data in the intermediate size region. The problem is that the exponential tail of the decay is not sufficiently well developed for an uneqivocal determination of the decay constant, in particular not in a case where the polydispersity is also important. We have

chosen to disregard the migration, applied Eq. (2), and tried to estimate the amplitude of a final exponential decay component with the same decay constant as obtained in the absence of quenchers. The method was tested on the data for the solutions with 33.9 mM chlorate, leading to, at worst, 12% higher aggregation numbers than those reported above. The effect is most important at high quencher concentration, and will lead to a slower decrease, therefore, of the Q-average aggregation number with quencher concentration, which in turn leads to a low estimate of polydispersity.

The evaluation of the amplitude of the k_0 mode turned out to be delicate. The results reported in Fig. 5 for the solutions with 42 and 44 mM chlorate were obtained simply by drawing a straight line with slope $-k_0$ in the logarithmic representation of the decay data. Although independent evaluations by different persons using this method gave results that disagreed by at most ± 10%, it is a subjective method, and furthermore, a method that could be suspected to give systematic errors.

An objective method was also used in which an equation similar to Eq. (4), but with an additional term $A_5[\exp(-A_6t) - 1]$ in the exponent (and with $A_2 = k_0$), was fitted to the data. This produced excellent fits and gave systematically higher Q-average aggregation numbers at low quencher concentrations than the first method, in particular at 44 mM chlorate. The method

Table 2. Weight average aggregation numbers and width of the micelle weight distribution at some concentrations of NaClO₃ for 39.9 mM CTAC at 25 °C

[ClO₃⁻] (mM)	$\langle N \rangle_Q$	σ	$\sigma/\langle N \rangle_Q$
0	90.5	29	0.320
34	219	68	0.311
42	442	189	0.427
42[a]	487	229	0.470
44	855	411	0.481
44[a]	1140	572	0.502

[a] Values obtained from fitting with a double exponential Infelta-type of equation, as described in the text.

needs a thorough evaluation by fitting to simulated data; at present the width and the weight-averages from both methods are reported in Table 2 as an indication of the uncertainties involved.

The results in Table 2 show that the micelles start to grow well before the strong increase in the viscosity sets in at around 40 mM chlorate. Above this concentration, however, the increase is much steeper and is also accompanied by an increasing polydispersity, as predicted.

Some calculations performed by Porte et al. [16] showed that a simple theory based on Eq. (1) but with intermicellar interactions taken into account results in prediction of a gradual increase in aggregation number with increasing surfactant concentration. The authors used the results to claim that the simple theory gave predictions which were at variance with experimental observation (from viscosity measurements) of a second CMC, and proceeded to introduce an intramicellar free-energy barrier for the onset of the rod-like growth. Referring to a growth at constant surfactant concentration, the present results are not directly comparable with the theory, but the observed gradual increase of the aggregation numbers nevertheless suggests that the assumption of a substantial free-energy barrier for the rod formation may be superfluous.

The infinite rod

When quenching occurs in infinite rods, the fluorescence decay should follow the equation [19]:

$$\ln F(t)/F(0) = - k_0 t - B_0 c_0 \{\exp(B_1 t) \, \text{erfc}(B_1 t)^{1/2} \\ - 1 + 2(B_1 t/\pi)^{1/2}\}$$

$$B_0 = 3 \, D_m/(k_q R)$$

$$B_1 = (2 \, k_q \, R/3)^2 D_m^{-1} \qquad (7)$$

where k_q is the first-order quenching rate constant in the reaction volume of length $4R/3$ surrounding the excited probe, D_m the sum of the diffusion coefficients of the excited state and the quencher, and R the hydrophobic cylinder radius.

This equation was fitted to the results from quenching experiments with six concentrations of benzophenone in the range from 0.12 to 0.60 mM at a chlorate concentration of 100 mM, in a global analysis with B_0 and B_1 as common parameters. Good fits were obtained with a global x^2-value of 1.63 and individual x^2:s from 0.95 to 1.36, except at the highest concentration where 2.25 was reached. With a value of 21 Å, assumed for the hydrophobic cylinder radius, the following values were calculated from the primarily estimated parameters:

$$k_q = 2.3 \times 10^7 \text{ s}^{-1}$$

$$D_m = 9 \times 10^{-11} \text{ m}^2 \text{ s}^{-1}$$

The quenching constant is close to the value of 1.8×10^7 s^{-1}, obtained for small micelles without chlorate (Table 1) and as a further consistent feature it can be noted that for both the spherical micelle and the rod, the product $R^2 k_q$ has a value close to the estimated diffusion coefficient.

Only a few values of diffusion coefficients for diffusion in rod-like structures have been reported. Johansson et al. [24] measured the diffusion of the surfactants themselves in the rodlike aggregates of two lyotropic nematic phases, formed by sodium decylsulphate and potassium laurate, respectively, by NMR methods, and obtained values of 8.7 and 5.3×10^{-11} m^2 s^{-1}, respectively, which are in the same range as the values obtained in the present study.

The application of Eq. (7) to fluorescence quenching data in long aggregates appears to be a useful route to information about the axial mobilities of small molecules in such structures. The concentration of quenchers that has to be used corresponds to between 0.9 and 5 per 100 Å when pyrene is the excited state. Lower quencher concentrations, and therefore diffusion over longer distances, are accessible with long-lived excited states such as triplets [25].

Acknowledgements

Support from the Swedish Natural Science Foundation and from Knut and Alice Wallenbergs Stiftelse are gratefully acknowledged.

References

1. Hatschek E (1925) Kolloid Z 37:25
2. Hatschek E, Jane RS (1926) Kolloid Z 38:33
3. Israelachvili JN, Mitchell JJ, Ninham BN (1976) J Chem Soc Faraday Trans 2, 72:1525
4. Ulmius J, Wennerström H, Johansson LB-Å, Lindblom G, Gravsholt S (1979) J Pphys Chem 83:2232
5. Reiss-Husson F, Luzatti V (1964) J Phys Chem 68:3504
6. Angelescu E, Popescu DM (1930) Kolloid Z 51:336; Angelescu E, Manolescu T (1941) Kolloid 94:319
7. Nash T (1958) J Colloid Sci 13:134; (1959) J Colloid Sci 14:59
8. Porte G (1983) J Phys Chem 87:3541
9. Mukerjee P (1972) J Phys Chem 79:565
10. Mazer NA, Carey CA, Benedek GB (1976) J Phys Chem 80:1075; Missel PJ, Mazer NA, Benedek GB, Young CY, Carey MC (1980) J Phys Chem 84:1044
11. Porte G, Appell J (1981) J Phys Chem 85:2511
12. Corti M, Degiorgio V (1979) In: Mittal KL (ed) Solution Chemistry of Surfactants. Plenum Press, New York 1:377
13. Löfroth J-E, Almgren M (1984) In: Mittal KL, Lindman B (eds) Surfactants in Solution. Plenum Press, New York 1:736
14. Warr GG, Grieser F (1986) J Chem Soc Faraday Trans 1 82:1813
15. Warr GG, Grieser F, Evans DF (1986) J Chem Soc Faraday Trans 82:1825
16. Porte G, Poggi Y, Appell J, Maret G (1984) J Phys Chem 88:5713
17. Eriksson JC, Ljungren S (1985) J Chem Soc, Faraday Trans 2 81:1209
18. Almgren M, Löfroth J-E (1982) J Chem Phys 76:2734
19. Almgren M, Alsins J, van Stam J, Mukhtar E (1988) J Phys Chem, in the press
20. Löfroth J-E (1985) Eur Biophys J 13:45
21. Russel JC, Wild UP, Whitten DG (1986) J Phys Chem 90:1319
22. Infelta PP, Grätzel M, Thomas JK (1974) J Phys Chem 78:190
23. Almgren M, Grieser F, Thomas JK (1979) J Am Chem Soc 101:279
24. Johansson LB-Å, Söderman O, Fontell K, Lindblom G (1981) J Phys Chem 85:3694
25. Almgren M, Alsins J (1987) Progr Colloid Polym Sci 74:55

Received January 7, 1988;
accepted January 14, 1988

Authors' address:

M. Almgren
Dept. of Physical Chemistry
University of Uppsala
751 21 Uppsala, Sweden

Progress in Colloid & Polymer Science   Progr Colloid Polym Sci 76:75–83 (1988)

Dynamics in microemulsions with nonionic surfactant.
A combined NMR relaxation and self-diffusion study

U. Olsson[1]), M. Jonströmer[1]), K. Nagai[2]), O. Söderman[1]), H. Wennerström[1]), and G. Klose[3])

[1]) Physical Chemistry 1, Chemical Center, Univ. of Lund, Sweden
[2]) Nagasaki University, Faculty of Engineering, Nagasaki, Japan
[3]) Karl-Marx Universität, Sektion Physik, Leipzig, DDR

Abstract: Data on molecular self-diffusion and ^{2}H-NMR relaxation, on deuterons specifically attached to the surfactant molecule, are presented for two ternary microemulsion systems with nonionic surfactant, covering the whole range of water to oil ratios. In one of the systems, the hydrophobic solvent was made up by a 1:1 by weight mixture of cyclohexane and hexadecane. By comparing the diffusion coefficients of the two hydrocarbons, conclusions are drawn concerning the dominating diffusion process of the hydrocarbon molecules. A marked temperature dependence of the self-diffusion properties, as well as in the relaxation parameters, is observed. Of particular interest are the dramatic changes occuring at low oil content within a narrow temperature interval. There, at lower temperatures, swollen micelles are identified that grow in size with increasing temperature. At higher temperatures, data suggest that rapid coalescence and breakup of aggregates provide a mechanism for the macroscopic transport of oil molecules, resulting in a marked increase of the hydrocarbon diffusion coefficient.

Key words: Microemulsions, nonionic surfactant, ^{2}H-NMR-relaxation, self-diffusion study, phase-behavior, transition effects, coalescence of aggregates, break up of aggregates.

Introduction

Over a number of years, microemulsion systems based on nonionic surfactants of the ethylene oxide type have been studied with a large number of experimental techniques such as rheology [1], various scattering methods [1, 2], conductivity [1], freeze fracture electron microscopy [1–3] and NMR methods [4, 5], to mention a few. Special attention has been directed towards the so-called bicontinuous microemulsions [6]. It is generally agreed that in these systems some sort of aggregation of the surfactant occurs to separate the hydrophilic and hydrophobic domains. However, details of the structure and dynamics of these interfacial films are still to a large extent unknown and substantial efforts have been directed towards a description of the microstructure of these systems.

In the present communication we hope to show that a combination of two NMR techniques, vis. NMR self-diffusion and NMR relaxation studies, is particular useful in providing insight into the microstructure in microemulsions in general and in bicontinuous microemulsions in particular.

A special feature of the present study is the use of a mixture of two different hydrocarbons as the oil constituent in the microemulsion. This allows us to take the interpretation of self-diffusion coefficients one step further, and to make some rather definite statements about the microstructure of microemulsions.

In the present communication, data from two microemulsion systems, which we will refer to as system I and II, are presented. System I is composed of 7 % (w/w) pentaethylene glycol dodecyl ether ($C_{12}E_5$), water and a 1:1 (by weight) mixture of cyclohexane and hexadecane. System II contains 14 % (w/w) tetraethylene glycol dodecyl ether ($C_{12}E_4$), water and hexadecane. For system I, only self-diffusion measurements were performed, whereas in system II, both NMR relaxation measurements performed on surfactant deuterons incorporated in the α- position, $CH_3(CH_2)_{10}CD_2(OCH_2CH_2)_4OH$ and self-diffusion measurements were made.

Methods

NMR self-diffusion

The NMR self-diffusion studies were performed as described in several reports. The ratio of two diffusion coefficients were determined directly from the raw data as decribed in Ref. [7].

NMR relaxation

The only NMR relaxation parameter which will be discussed in detail is the transverse relaxation rate, R_2. This parameter was measured from the band widths of the absorbtion peaks at half height after suitable correction for magnetic field inhomogeneities.

Theoretical considerations

In the following, we will give the necessary equations for interpreting the NMR relaxation data. More detailed discussions of the theoretical background for NMR relaxation studies can be found elsewhere [8]. For a CD_2 group, the ^2H transverse relaxation rate is given by [9]:

$$R_2 = (3\pi^2/40) \, \chi^2 \, (3\tilde{J}(0) + 5\tilde{J}(\omega) + 2\tilde{J}(2\omega)) \qquad (1)$$

where χ is the quadrupole coupling constant (typically around 170 kHz) and $\tilde{J}(\omega)$ is the reduced spectral density evaluated at the Larmor frequency of the deuterons. If R_2 is independent of frequency and is substantially larger than the longitudinal relaxation rate, R_1, then only the spectral density at zero frequency contributes to R_2. Thus

$$R_2 \cong (3\pi^2/40) \, \chi^2 \, \tilde{J}(0). \qquad (2)$$

Within the so-called "two-step" model of relaxation [10–12], which has been shown to give a reasonable describtion of NMR relaxation data for aggregated surfactant systems, $\tilde{J}(\omega)$ is given by

$$\tilde{J}(\omega) = (1 - S^2) \, 2\tau_f + S^2 \tilde{J}_S(\omega) \qquad (3)$$

where S is the order parameter analogous to the one measured from quadrupolar splittings, τ_f an effective correlation time for a fast local motion within the aggregate assumed to be in extreme narrowing at the Larmor frequencies studied and $\tilde{J}_S(\omega)$, finally, is the spectral density function describing the dynamics of the entire surfactant aggregate. Combining Eqs. (2) and (3) we obtain

$$R_2 = (3\pi^2/40) \, \chi^2 \, ((1 - S^2) \, 2\tau_f + S^2 \, 3\tilde{J}_s(0)). \qquad (4)$$

If the slow motions assumed to be described by an exponential correlation function and if $(1 - S^2) \, 2\tau_f \ll S^2 \, 6\tau_s$, where τ_s is the correlation time for the slow motion, Eq. (4) reduces to

$$R_2 = (9\pi^2/20) \, (\chi S)^2 \, \tau_S. \qquad (5)$$

It thus follows that for a situation with slow dynamics of the interface, as is the case for all samples of the present study except for the waterfree oil/surfactant systems, R_2 is proportional to a correlation time describing the dynamics of the hydrophilic-hydrophobic interface.

Results and discussion

Phase behaviour

The phase behaviour in three component microemulsion systems stabilized by nonionic surfactants has been examined and described in great detail by Shinoda and coworkers [13–15] and by Kahlweit and coworkers [16–18]. It has been shown that at a constant (and for efficient surfactants low) surfactant concentration there exists an isotropic one-phase solution channel connecting the two binary solutions surfactant/water and surfactant/oil. Such two-dimensional phase diagrams, where the two variables are temperature and the weight fraction (oil/water + oil) $\equiv \alpha$, for the two systems in the present study are shown in Figs. 1 and 2, respectively. The phase diagrams (note that only the extensions of the isotropic solution phase

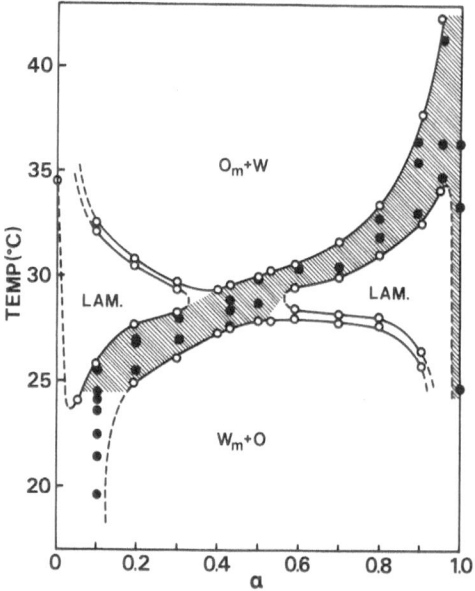

Fig. 1. Schematic phase diagram of the system 7.00 % (w/w) $C_{12}E_5$/water/oil, where the oil is a 1:1 by weight mixture of cyclohexane and hexadecane (system I). (●) Experimental points investigated by self-diffusion measurements. The shaded area within the one phase region shows the extension of the region where $K = 1.69$ was found (see text). Only the phase borders for the extension of the one phase isotropic liquid region are shown in the figure. LAM. denotes a heterogeneous region including a lamellar liquid crystalline phase. (From Ref. [7] with permission)

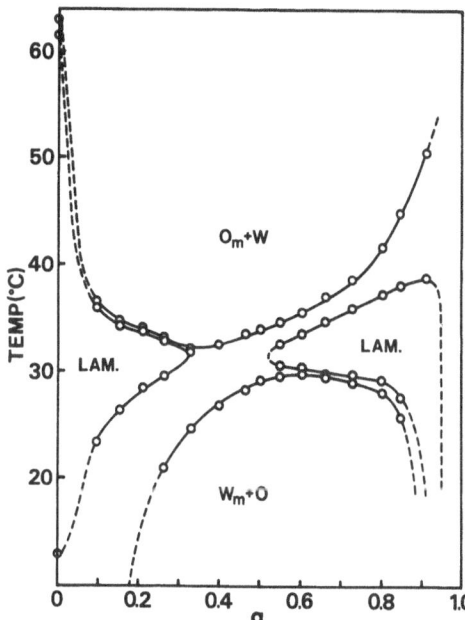

Fig. 2. Schematic phase diagram of the system 14.0 % (w/w) $C_{12}E_4$/water/hexadecane (system II). Only the phase borders of the one phase isotropic liquid region are shown in the figure. LAM. denotes a heterogeneous region including a lamellar liquid crystalline phase

are shown) show the same features as those for similar systems [1, 5]. The isotropic solution is stable only within a narrow temperature range which is shifted to higher temperatures with increasing α. At still higher temperatures, the surfactant is mainly soluble in oil and an oil-rich microemulsion (O_m) and excess water (W) separates. At low temperatures, the surfactant is mainly soluble in water and a water-rich microemulsion (W_m) and excess oil (O) separates. At low and high water content and intermediate temperatures a lamellar liquid crystalline phase (LAM) is found.

General considerations

Normally, microemulsion systems are classified according to their assumed microstructure, into three different classes, namely oil-in-water (O/W), water-in-oil (W/O) and bicontinuous microemulsions. In an O/W microemulsion, oil and surfactant form swollen micelles in a continuous water medium. Analogously, a W/O microemulsion contains swollen reversed micelles in an oil continuous medium. In a bicontinuous microemulsion, however, both oil and water are considered as forming domains, some of which are continuous over macroscopic distances. Several bicontinuous structures have been proposed, which

would take the continuity of both water and oil into account [19–21].

Since the isotropic channels of Figs. 1 and 2 extend over all water-to-oil ratios connecting the binary surfactant/water system ($\alpha = 0$) with the binary surfactant/oil system ($\alpha = 1$), all three types of microemulsions are expected to be present at different α in the channels. In particular, a progressive change O/W → bicontinuity → W/O is expected with increasing α.

Static microemulsion models make clear distinctions between the three classes of microemulsions [22]. The definition of a bicontinuous microemulsion states that there exist sample spanning paths for both oil and water in an instantaneous snapshot, whereas in O/W and W/O microemulsions, sample spanning paths only exist for one of the two components.

From self-diffusion data in the microemulsion systems presently under study, three different regions of the main microemulsion channel can be distinguished, with sharp transitions between them. These three regions differ with respect to the dominating physical processes responsible for the macroscopic transport (diffusion) of water and oil molecules. At low α and low temperatures, the dominating process for the macroscopic diffusion of oil is the diffusion of entire aggregates (swollen micelles) whereas water molecules diffuse as in bulk. At high α and high temperatures, the opposite occurs: Water molecules are trapped in reversed micelles and consequently the dominating diffusion process for water is aggregate diffusion, while oil molecules now diffuse as in bulk. These two regions, readily identified with O/W and W/O microemulsions respectively, are easily identified from self-diffusion data. In between these two regions, the third region, which is distinguishable from the other two with self-diffusion data, can be found. In this region the diffusion process for both oil and water is molecular diffusion in media that are essentially identical to their respective bulk media.

Self-diffusion coefficients as well as NMR relaxation parameters are dynamic observables and consequently give information on the dynamics in the system. Models have to be invoked in order to relate them to a static property such as structure. We will now discuss the three regions, distinguished in terms of their diffusion properties, separately. For simplicity we will refer to these regions by the letters A, B and C, respectively. We will refer to microemulsions where the dominating diffusion process of oil or water is aggregate diffusion as type A and type B, respectively, while the third and final case will be designated as type C.

Even though we agree with the trivial identities (that type A is an O/W and type B is an W/O microemulsion) it appears that the type C microemulsion does not necessarily corresponds to bicontinuity as it is strictly defined by static microemulsion models. Below we will discuss mainly dynamic properties of microemulsions and we feel that we should be consistent in refering our results of dynamic observables to dynamic properties of the system. At the end of this paper we will compare our results with static microemulsion models.

The transition from type A to type C appears to be very sharp and we will address this transition separately below.

The type A microemulsion

The type A microemulsion, which is found at low α and low temperatures, is characterized by $D_{water} \gg D_{oil} = D_{surfactant}$ where D denotes the diffusion constant. In Fig. 3, the diffusion coefficients of both oils and of the surfactant are plotted as a function of temperature for $\alpha = 0.10$ in system I. In Fig. 4 the same plot is made for oil and surfactant at $\alpha = 0.15$ in system II. In Fig. 4 the transverse relaxation rates, R_2, at various temperatures of the surfactant deuterons are also shown. At low

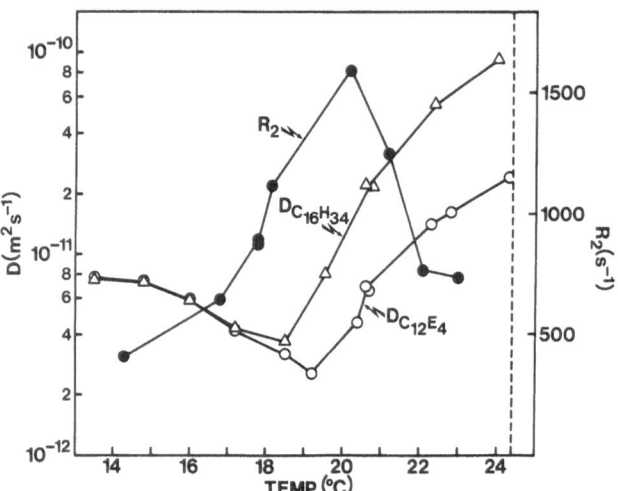

Fig. 4. Temperature dependence of the observed diffusion coefficients (left hand scale) of $C_{12}E_4$ (O) and hexadecane ($\triangle$), and of the transverse relaxation rate, R_2, (right hand scale) of the surfactant deuterons (●) at $\alpha = 0.15$ in system II. Dashed line shows the upper temperature limit of the isotropic one-phase region

temperatures, for both systems, $D_{surfactant}$ practically equals D_{oil} and their diffusion constants are thus a measure of the aggregate diffusion. The presence of closed O/W droplets is evident. Treating the swollen micelles as hard spheres with no hydrodynamic interaction and by applying the Stokes-Einstein equation to the diffusion data (which has been corrected for exuded volume effects through the relation $D = D_o (1 - 2\Phi)^{23}$ where Φ is the volume fraction of surfactant and oil) gives that the diffusion coefficient $8 \cdot 10^{-12}$ m^{2}s^{-1}, found at low temperature ($\approx 15\,°C$) at $\alpha = 0.15$ in system II, corresponds to a diffusion coefficient of a sphere having a radius of about 90 Å. For this sample, the radius estimated from the D-values can be compared with a radius estimated from R_2.

An order parameter of 0.075 was obtained from quadrupolar splittings both at $\alpha = 0.15$ and $\alpha = 0.82$ in their respective lamellar liquid crystalline regions which are indicated in Fig. 2. Using this value and assuming the relevance of applying the "two-step model" of relaxation in the present case, the R_2-value of 400 s^{-1} at 15 °C corresponds to a slow correlation time of about 500 ns. The isotropic reorientation of surfactants in a spherical aggregate is the combined effect of aggregate tumbling and surfactant lateral diffusion. The lateral diffusion coefficient, D_{lat}, of $C_{12}E_4$ is, to our knowledge, unknown. However, in the cubic phase I_1, in the binary system $C_{12}E_5$/water at a surfactant concentration of 60 % (w/w) an observed diffu-

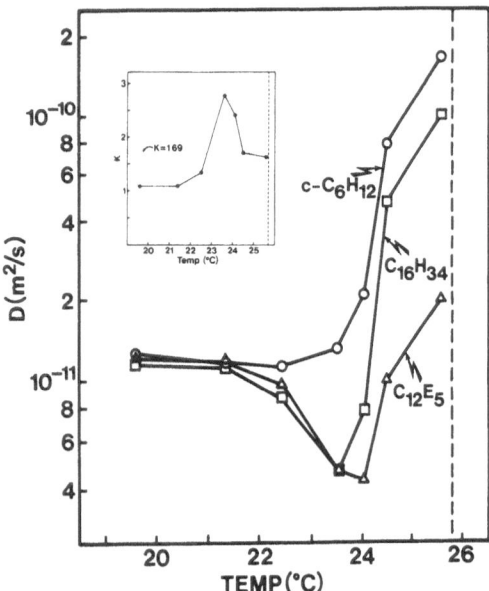

Fig. 3. Temperature dependence of the observed diffusion coefficients of $C_{12}E_5$ ($\triangle$), cyclohexane (O) and hexadecane ($\square$) at $\alpha = 0.10$ in system I. Dashed vertical line shows the upper temperature limit of the isotropic one phase region. Inset: the variation of K with temperature (From Ref. [7] with permission)

sion coefficient of $1.6 \cdot 10^{-11}$ m^{2}s^{-1} was found for the surfactant at 20 °C and I$_1$, data for $C_{12}E_8$ are also available [24]. Assuming in these cases that D_{lat} is three times the observed diffusion coefficients [25] and taking into account the lower molecular weight of $C_{12}E_4$ and the lower temperature, a D_{lat}-value for $C_{12}E_4$ at 15 °C of $2 \cdot 10^{-11}$ m^{2}s^{-1} is estimated. With this value of D_{lat} a correlation time of 500 ns also corresponds to a radius of 90 Å. The agreement is probably coincidental, considering the assumptions made, but it shows that there are no significant disagreements when interpreting the self-diffusion and R_2 data in terms of spherical swollen micelles. It should also be pointed out that in the analysis, a change in D_{lat} by a factor of 2 would affect the extracted radius by a factor of less than $\sqrt{2}$.

The type B microemulsion

In Table 1 some diffusion coefficients of water, surfactant and both oils are given at various temperatures for $\alpha = 0.90$, 0.96 and 1 in system I. In the case of microemulsions with nonionic surfactant of the kind presently used, $D_{water} = D_{surfactant}$ is not found in a type B region. The reason for this is the large solubility of the surfactant in oil. At relevant temperatures, the surfactant and oil are completely miscible and, as has been pointed out earlier [5], the surfactant dissolves as monomers (or very small complexes) in the oil. This is also found in the present system, as is evident not only from the high D-values of the surfactant as shown in Table 1, but also from the longitudianal relaxation rates, R_1, and transverse relaxation rates, R_2, given in Table 2. In this table, R_1 and R_2 of the surfactant deuterons at various concentrations of $C_{12}E_4$ in hexadecane are given. Extreme narrowing conditions are obtained at every concentration and thus any extensive form of aggregation can be ruled out.

A type B situation in the present systems thus has to fulfil $D_{oil} \gg D_{surfactant} > D_{water}$ which is found in several cases, as shown in Table 1. Evidently this condition is not sufficient for producing a type B situation but from comparing the magnitudes of water and surfactant diffusion coefficients at temperatures ≥ 35.5 °C at $\alpha = 0.90$ and for ≥ 36.4 °C at $\alpha = 0.96$, a type B situation can probably be recognized. By applying a two-site model with fast exchange for the surfactant, the fraction of micellized surfactant molecules, $P_{micelle}$, can be calculated from Lindman's first law

$$D_{surfactant} = P_{micelle} D_{micelle} + (1 - P_{micelle}) D_{monomer}.$$
(6)

Table 1. Observed self-diffusion coefficients for water and $C_{12}E_5$ at α = 0.90 and 0.96 in system I

α	Temp. (°C)	Observed diffusion coefficients (10^{-10} m^{2}s^{-1}) water	$C_{12}E_5$
0.90	33.1	1.36 ± 0.07	1.29 ± 0.08
0.90	35.5	0.32 ± 0.01	1.86 ± 0.16
0.90	36.5	0.37 ± 0.02	1.80 ± 0.11
0.96	34.8	0.90 ± 0.04	1.69 ± 0.08
0.96	36.4	0.56 ± 0.03	2.01 ± 0.15
0.96	41.4	0.85 ± 0.05	2.79 ± 0.09

Table 2. ^{2}H-NMR relaxation data for the binary $C_{12}E_4$/hexadecane system

Conc. (% w/w $C_{12}E_4$)	$B_o(T)$	$R_1(s^{-1})$	$R_2(s^{-1})$	Temp. (°C)
0.9	6.0	12.0 ± 1		23.0
4.3	6.0	15.3 ± 0.2	17.2 ± 0.6	23.0
9.3	6.0	18.2 ± 0.2	19.2 ± 0.7	23.0
14.0	6.0	19.5 ± 0.2	21.0 ± 0.5	23.0
14.0	6.0	16.1 ± 0.2	16.6 ± 2.3	27.5
14.0	8.5	15.0 ± 0.2		28.0

Identifying $D_{micelle}$ with D_{water} and $D_{monomer}$ with $D_{surfactant}$ at $\alpha = 1$, $P_{micelle}$ at 36.5 °C for $\alpha = 0.90$ is 0.60. The corresponding value at 36.4 °C for $\alpha = 0.96$ is 0.57. Since the molar ratio of surfactant to water is about twice as large at $\alpha = 0.96$ than at $\alpha = 0.90$, this implies that the reversed micelles are larger at $\alpha = 0.90$, which in turn is consistent with the significantly lower D_{water} at that temperature.

The type C microemulsion

For the purpose of investigating the dominating diffusion process of the oil, a system was selected where the "oil" was a 1:1 by weight mixture of cyclohexane and hexadecane, giving the one-phase solution channel shown in Fig. 1. In the pure binary oil mixture, the ratio

$$D_{cyclohexane}/D_{hexadecane} \equiv K$$
(7)

was determined at five different temperatures at equal intervals between 20° and 38 °C. K showed no significant temperature dependence within this interval and the mean value of the five measurements was 1.69 ± 0.02. Since this neat hydrocarbon mixture serves as a reference system, we will refer to the ratio 1.69 of the two diffusion coefficients in the hydrocarbon mixture as K_o.

According to the results of K in the one-phase channel in Fig. 1, this channel can in principle be divided into three distinct regions, one being where K is practically unity and $D_{\text{oil}} = D_{\text{surfactant}}$. This region is the type A microemulsion that is found at low α and low temperatures and has been discussed above. The second region is where $K = K_o$. This region extends over a major part of the one-phase channel and is indicated by the shaded area in Fig. 1. The third region, which we will discuss in detail below, is the very sharp transition region between the regions of $K = 1$ and $K = K_o$.

The shaded area in Fig. 1 includes 26 individual observations (filled circles in Fig. 1). A completely random variation of K was found with a mean value of 1.69 and a standard deviation of 0.04. It seems unlikely, considering the number of points measured, the variation in conditions and the change in absolute D-values, that the similarity in the K-values is coincidental. We therefore conclude that in the region of $K = K_o$, the diffusion paths that provide the dominating contribution to the long distance diffusion are occuring mainly in an oil medium similar to that of the pure oil mixture.

At this stage it is important to point out that the analogous conditions for the water diffusion have not yet been firmly established. However, it is reasonable that, also in water, the dominating process for the macroscopic transport of molecules can be regarded as a molecular diffusion in the major part of the channels, except at α (≥ 0.9) and high temperatures, where a type B microemulsion is probably present. In the following we will mainly discuss the properties of the systems at lower α (≤ 0.5), where the assumption above has its strongest relevance.

The property $K = K_o$ does not necessarily mean that any oil domain extends over macroscopic distances commparable to the diffusion lengths of the diffusion experiments. In an FTPGSE experiment, one essentially measures the mean squared displacement in one dimension. For a Gaussian diffusion, the one-dimensional mean squared displacement is proportional to the diffusion time, τ, according to the relation

$$\langle\, | \, x(\tau) - x(0)\, |^2 \rangle = 2D\tau\,. \tag{8}$$

At our experimental conditions τ is about 100 ms and D_{oil} varies from $5 \cdot 10^{-11}$ m^{2}s^{-1} at the low α part of the shaded area to about 10^{-9} m^{2}s^{-1} at $\alpha \geq 0.8$. Inserting these numbers into Eq. (8) gives the root mean

squared displacement of oil molecules observed in the experiment which is of the order of 3–14 μm in this α-range, which is significantly longer than the expected structural length scale in the system.

In Fig. 5 the relative diffusion coefficients, D/D_o of water and cyclohexane are plotted vesus α. D denotes the observed diffusion coefficients which, in the case of water, are corrected for the hydration of the surfactant polar headgroup and fast exchange with surfactant hydroxyl protons. D_o is the diffusion coefficient in the pure liquid for water, while for cyclohexane it refers to the 1:1 binary oil mixture at the given temperature. Figure 5 expresses essentially the same diffusion behavior as has been found in a similar system [5]. A strong temperature dependence is found and changes in temperature and α induce opposite changes in D/D_o in D_{oil} and D_{water}. The changes in D/D_o are gradual and in the region of $K = K_o$, D/D_o goes from 0.076 to about 0.9 from low to high α.

For diffusion within parallel planar confinements, a D/D_o of 2/3 is expected, while in a tubular network, diffusion in only one dimension contributes to the long range displacement so that D/D_o is reduced to 1/3 [25]. Further depressions of the ratio D/D_o in a homogeneously continuous system can be explained by invoking folded lamellar structures or similarly coiled tubes. However, such structures seem unlikely and are not

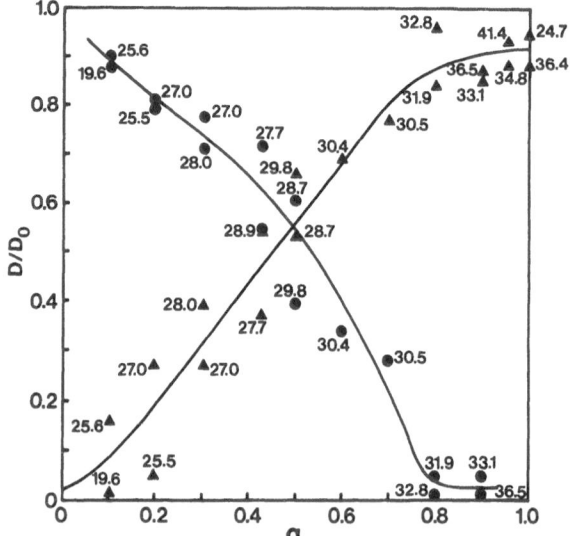

Fig. 5. Relative diffusion coefficients (D/D_o) of water (●) and cyclohexane (▲) as a function of α for experimental points in the main isotropic channel of system I. Solid lines are approximative mean values of D/D_o at each composition. (From Ref. [7] with permission)

supported by the very nature of the transition from type A to type C microemulsions, as will be discussed. A much more appealing explanation of the gradual change in D/D_o of oil from the low α value of 0.076 to the value 0.9 at high α is in terms of a large size polydispersity in the system. This would correspond to the simultaneous existence of very large aggregates, however containing only a finite fraction of the oil, and smaller aggregates. The dominating diffusion paths will be within the large aggregates, whereas the small aggregates will give negligible contributions to the observed diffusion coefficient due to the slow droplet diffusion. For this to be consistent with the observed molecular homogeneity on the experimental timescale (100 ms), a rapid exchange of molecules between the aggregates is necessary.

Since $K = K_o$, the exchange process can neither be selective i.e. favoring one of the two oil molecules present in system I, nor can the actual exchange rates influence D/D_o. Instead, in the polydispersity description the only two properties that influence D/D_o of oil are the fraction of large enough oil domains and the obstruction imposed on the oil molecules by the oil-water interface in these large oil domains. We thus conclude that when D/D_o of oil increases with increasing temperature and increasing α, this is due to the fact that the fraction of larger aggregates increases on behalf of the smaller ones.

The transverse relaxation rate, R_2, of the surfactant deuterons was determined at several temperatures for samples at $\alpha = 0.15$, 0.30, 0.46 and 0.82, respectively. Lorentzian absorption bands were observed in all experiments and the results are shown in Fig. 6 together with the temperature stability of the one-phase isotropic solution at each α. Note that the temperature stabilities given in Fig. 6 differs slightly from those in Fig. 2. The reason for this is probably not an isotopic effect, but rather a result of a slight difference in impurities between the commercial $C_{12}E_4$ and the deuterated compound. There is a very strong temperature dependence in R_2 for the samples at low and intermediate α: a dependence much too strong to be a direct consequence of local molecular changes. Moreover, at $\alpha = 0.15$ and 0.30 a significant maximum in R_2 as a function of temperature is observed, whereas at $\alpha = 0.46$ only a strong decrease in R_2 with increasing temperature is observed. As will be demonstrated, the low α border of the type C region coincides with the maxima in R_2 at $\alpha = 0.15$ and 0.30. The transition region between type C and type B was not located and we will therefore not discuss the high α region further, but

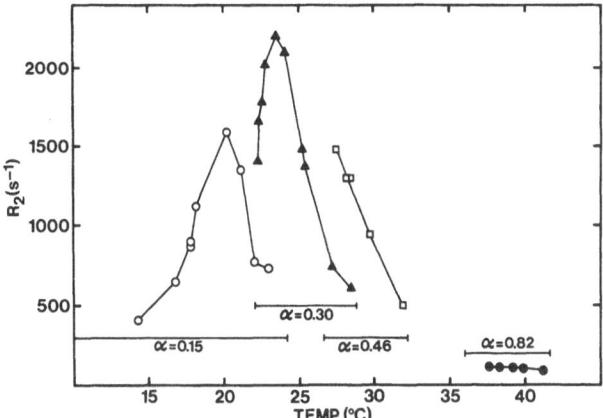

Fig. 6. Transverse relaxation rates, R_2, of the surfactant deuterons as a function of temperature at four different compositions ($\alpha - 0.15$, 0.30, 0.46 and 0.82) in system II. The temperature stability of the one-phase region is shown for each composition

only note that the much smaller R_2 found at $\alpha = 0.82$ is presumably due not only to the presence of smaller aggregates (of the reversed type) but is also influenced by a significant amount of molecular dispersed surfactant in the oil. We will then restrict ourselves to discussing the results from $\alpha = 0.15$, 0.30 and 0.46. In the type C region of these samples, R_2 decreases strongly with increasing temperature. At the same time, the diffusion coefficients of both oil and surfactant increase significantly. If one accepts the explanation given above (the increased diffusion rates are due to an increased fraction of larger aggregates/oil domains) then the value of R_2 cannot be a measure of the average aggregate size. Instead, there must be other dynamic processes, capable of averaging the quadrupolar interaction to zero, that are more rapid than the rate of the isotropic reorientation of the oil-water interface. For reasons that will be given in discussing the transition region between type A and type C, we suggest that for the samples $\alpha = 0.15$, 0.30 and 0.46 a random coalescence and breakup of aggregates give the dominating contributions to R_2 in the type C region, and that the marked decrease of R_2 with increasing temperature is due to an increased rate of the coalescence and breakup dynamics.

The transition from type A to type C

For a given α, this transition occurs within a narrow temperature of only a few degrees. It is signalled by dramatic changes in the diffusion coefficients of oil and surfactant, and in system I it corresponds to a transition from $K = 1$ to $K = K_o$. This transition was followed in detail for sample $\alpha = 0.10$ in system I and sample $\alpha = 0.15$ in system II, and we will again turn our attention to Figs. 3 and 4. In Fig. 3 the diffusion coefficients of both oils and surfactant for $\alpha = 0.10$ in system I are plotted versus temperature. At low temperatures ($\leq 21\,°C$) this sample is a type A microemulsion, whereas above $24.5\,°C$ it becomes a type C microemulsion. The transition region between these two types is, for this sample, then identified as the temperature interval $21° - 24.5\,°C$. As seen by the dramatic decrease of the surfactant diffusion coefficient, there is a substantial growth of the swollen micelles in this region. Moreover, the hexadecane diffusion coefficient equals the surfactant diffusion coefficient up to $23.5\,°C$, whereas for cyclohexane essentially no decrease in its D-value is observed at the onset of formation of these larger aggregates. Clearly an additional diffusional pathway (appart from aggregate diffusion) has been opened for the cyclohexane molecules at about $22\,°C$. This pathway is not available for hexadecane until slightly above $23.5\,°C$ and for the surfactant slightly above $24\,°C$, as seen in Fig. 3. The difference between cyclohexane and hexadecane is further illustrated by plotting K as a function of temperature, as in the inset in Fig. 3. The transition from $K = 1$ to $K = K_o$ occurs in a non-monotonic way. With increasing temperature, K first increases, reaches a maximum of about 3, then falls of and remains equal to K_o. In Fig. 4 the surfactant and hydrocarbon diffusion coefficients at $\alpha = 0.15$ in system II are plotted versus temperature. By comparing this plot with Fig. 3, it is obvious that the knowledge extracted from examining an oil mixture is also applicable in this system. Also, in Fig. 4, R_2 of the surfactant deuterons at several temperatures is plotted for the same sample. Figure 4 thus shows that the minimum in surfactant diffusion coefficient correlates extremely well with a maximum in R_2.

Our interpretation of the data in Figs. 3 and 4 is as follows: At low temperature, well "isolated" swollen micelles are present in these two samples (vide infra). With increasing temperature, there is then a major growth of the micellar aggregates, as seen by the decrease in $D_{surfactant}$ and increase in R_2. At the same time, the ethylene oxide chains of the surfactant headgroup become less hydrophilic [26–28] and the repulsion between adjacent micelles is gradually replaced by an attractive force [29]. As a result, the distance of the closest approach between two adjacent micelles is gradually decreased. Thus at some instant the surfaces of two neighbouring aggregates can come close enough

to each other to allow exchange of individual molecules across a somewhat polar region composed of water and ethylene oxide units. The different solubility in polar media of the two oil molecules in system I thus serves as a selector, favoring the smaller cyclohexane molecule. Finally a point is reached where a complete coalescence of two aggregates will be a significantly occuring dynamic process. In Fig. 4, this point is represented by the correspondning maximum in R_2 and minimum in $D_{surfactant}$. At high temperatures of these extreme points, the random coalescence and, consequently, break ups of aggregates help to create open diffusional pathways for both oil and surfactant and, as seen by the decrease in R_2, serve as a mechanism for the reduction of the quadrupolar interaction to zero: The efficiency of this mechansim increases with increasing temperature. With the naive, but still relevant for order of magnitude calculations, application of the two step model for relaxation, the maximum R_2 in Fig. 6 corresponds to a slow correlation time of about 2 μs.

The dynamic process separating type C from types A and B is not exclusively found in three component systems. Also, in the binary system of surfactant and water, this process is present in the L_1 phase at higher temperatures. For instance, Nilsson et al. reported, in a systematic study of ^1H-NMR bandwidths and self-diffusion, that for 5 % (w/w) $C_{12}E_5$ in water, a significant maximum in the ^1H-NMR bandwith of the surfactant methylene absorption which correlates approximately with a minimum in the surfactant diffusion coefficient [24, 30].

Comparison with static microemulsion models

Since our data carry information on the dynamics in the system, it is obvious that static microemulsion models cannot predict all the details of our data. However, our distinction of type A, B and C microemulsions could in principle be identified with the water continuous, oil continuous and bicontinuous regions of the model [19], or related models [20, 21], if sufficient dynamics is included to account for the molecular homogeneity on the timescale of the diffusion experiment. We note that the Voronoi model is also consistent with the polydispersity suggested above to explain the absolute D/D_o values. The α dependence of D/D_o for oil and water, as shown in Fig. 5, is in qualitative agreement with the Voronoi model, whereas the strong temperature dependence cannot be predicted by the Voronoi model, since its only parameter is the the water-to-oil ratio. The transition from type A to type C could in principle correspond to the percolation predicted by the static microemulsion models. For instance a percolation behaviour as a function of temperature at constant composition could, with the penetrable cylinders model [20], be described in terms of an elongation of the particles. For this model, the percolation threshold is a strong function of the aspect ratio (elongation) of the cylindrical subunits.

The Voronoi model has been extended to include some dynamics by allowing the Poisson points to diffuse [31]. By analyzing the conductivity properties of such a model, it has been shown that even before the percolation threshold coalescence and break up can enhance transport properties [32]. Thus, a microemulsion which is not bicontinuous in the traditional sence, i.e. does not contain sample spanning paths for both oil and water in an instantaneous snapshot, may still have diffusion properties of a type C microemulsion.

In a static model, the strong temperature dependence of D_{oil} at lower α in the type C region could be accounted for by a growth and/or elongation of the structural subunits with increasing temperature. However, the simultaneous strong decrease in R_2 with increasing temperature indicates that this description is at least incomplete. Our data do not explicitly prove, nor do they disprove, the existence of sample spanning paths. In either case, a growth of the oil domains with increasing temperature could be an important mechanism by which the oil diffusion is enhanced.

To summarize, we suggest that significant coalescence and break up of aggregates occur in the type C region and that this is sufficient to provide a dominating molecular diffusion for both oil and water.

Acknowledgement

We thank Arvid Sandell for providing us with the computer program used in the valuation of K and David Anderson for helpful comments concerning the manuscript. The authers or Ref. [2] are thanked for providing a pre-print of their work prior to publication.

References

1. Kahlweit M, Strey R, Haase D, Kunieda H, Schmeling T, Faulhaber B, Borkovec MJ, Eicke H-F, Busse G, Eggers F, Funck Th, Richmann H, Magid LJ, Söderman O, Stilbs P, Winkler J, Dittrich A, Jahn W (1987) J Colloid Interface Sci 118:436
2. Bodet J-F, Bellare JR, Davis HT, Scriven LE, Miller WG (1988) J Phys Chem 92:1898
3. Jahn W, Strey R (1987) In: Meunier J, Langevin D, Boccara N (eds) Physics of Amphifilic Layers, Springer Proceedings in Physics. vol 21, Heidelberg
4. Nilsson P-G, Lindman B (1982) J Phys Chem 86:271
5. Olsson U, Shinoda K, Lindman B (1986) J Phys Chem 90:4083
6. Scriven LE (1976) Nature 263:123
7. Olsson U, Nagai K, Wennerström H (1988) J Phys Chem, in press
8. Lindman B, Söderman O, Wennerström H (1987) In: Zana R (ed) Novel Techniques to Investigate Surfactant Solutions. Dekker, New York, p 295
9. Abragam A (1961) The Principles of Nuclear Magnetism. Clarendon, Oxford
10. Wennerström H, Lindblom G, Lindman B (1974) Chem Scr 6:97
11. Wennerström H, Lindman B, Söderman O, Drakenberg T, Rosenholm JB (1979) J Am Chem Soc 101:6860
12. Halle B, Wennerström H (1981) J Chem Phys 75:1928
13. Kunieda H, Shinoda K (1982) J Dispersion Sci Technol 3:233
14. Shinoda K (1983) Prog Colloid Polym Sci 68:1
15. Shinoda K, Kunieda H, Arai T, Saijo H (1984) J Phys Chem 88:5126
16. Kahlweit M, Strey R (1985) Angew Chem 24:654
17. Kahlweit M, Lessner E, Strey R (1983) J Phys Chem 87:5032
18. Kahlweit M, Strey R, Firman P, Haase D (1985) Langmuir 1:281
19. Talmon Y, Prager S (1978) J Chem Phys 69:2984
20. Anderson DM (1986) Ph D Thesis, Univ of Minnesota, Minneapolis
21. Zemb TN, Hyde ST, Derian P-J, Barnes I, Ninham BW (1987) J Phys Chem 91:3815
22. Bennett KE, Hatfield JC, Davis HT, Macosko CW, Scriven LE (1982) In: Robb ID (ed) Microemulsions. Plenum, New York, p 65
23. Jönsson B, Wennerström H, Nilsson P-G, Linse P (1986) Colloid Polym Sci 264:77
24. Nilsson P-G, Wennerström H, Lindman B (1983) J Phys Chem 87:1377
25. Lindblom G, Wennerström H (1977) Biophys Chem 6:167
26. Karlström G (1985) J Phys Chem 89:4962
27. Lindman B, Karlström G (1987) Z Phys Chem 155:199
28. Ahlnäs T, Karlström G, Lindman B (1987) J Phys Chem 91:4030
29. Claesson P, Kjellander R, Stenius P, Christensson H (1986) J Chem Soc Trans 1 82:2735
30. Nilsson P-G, Wennerström H, Lindman B (1985) Chem Scr 25:67
31. Kaler EW, Prager S (1982) J Colloid Interface Sci 86:359
32. Michels WF (1986) Ph D Thesis, Univ of Minnesota, Minneapolis

Received December 15, 1987;
accepted December 30, 1987

Authors' address:

U. Olsson
Physical Chemistry 1
Chemical Center
University of Lund
P.O. Box 124
S-221 00 Lund, Sweden

Progress in Colloid & Polymer Science   Progr Colloid Polym Sci 76:84–89 (1988)

Relationship between surfactant film bending elasticity and structure and interfacial tensions in microemulsion systems

O. Abillon, B. P. Binks[1]), D. Langevin, and J. Meunier

Laboratoire de Spectroscopie Hertzienne de l'E.N.S., Paris, France
[1]) Chemistry Department, University of Hull, Hull, England

Abstract: We have investigated several microemulsion systems showing Winsor type phase equilibria. We have measured the surfactant film bending elasticity with ellipsometry and the interfacial tensions with surface light scattering experiments. Information about the structure has been deduced from light and X-ray scattering experiments. In order to interpret the ellipsometry measurements, a renormalization of both interfacial tension and bending elasticity has been used. These data are related with the existing theories to structural data of the microemulsion phases. A particular attention has been devoted to bicontinuous microemulsions.

Key words: Winsor microemulsions, structure, X-ray scattering, bending elasticity of films, interfacial tensions, bicontinuous microemulsions.

Introduction and theoretical bases

Microemulsions are dispersions of oil and water stabilised by surfactant molecules [1]. Frequently, the medium, like emulsions, is made of droplets either of oil or water dispersed in a continuous media of respectively water or oil: o/w or w/o microemulsions. Unlike emulsions, however, microemulsions are thermodynamically stable because the droplet size is much smaller (~ 100 Å) and the interfacial tension between oil and water ultralow ($\lesssim 10^{-2}$ mN/m).

The microemulsion structure is determined from an energy balance between several terms: interfacial energy, bending energy of the surfactant monolayers, dispersion entropy, interactions between droplets, etc. [2]. The role of the bending term appears to be dominant in cases where the microemulsion coexists with excess oil or/and water (Winsor equilibria). This bending term, as introduced by Helfrich, may be written [3]:

$$F_c = \frac{1}{2} K (C_1 + C_2 - 2C_0)^2 + \bar{K} C_1 C_2 \text{ per unit area}$$

$$(1)$$

where K and $\bar{K}$ are the splay and saddle-splay bending constants, C_1 and C_2 the local principal curvatures of the surfactant layer and C_0 the preferred curvature of the layer.

In other dispersed systems, the interfacial energy term is much larger than the others. In microemulsions, however, the interfacial tension Γ between oil and water being ultralow, the corresponding energy is small. Γ has to remain ultralow for the system to be stable. This means that the area per surfactant molecule Σ cannot vary very much, because the surface pressure of the surfactant film π cannot vary either: $\Gamma = \gamma_{ow} - \pi(\Sigma)$ (γ_{ow} = bare oil-water interfacial tension) [2]. As a consequence, the droplet radius depends only on the relative proportions of the microemulsion constituents:

$$R = \frac{3\varphi}{C_s \Sigma}$$

$$(2)$$

where φ is the dispersed phase volume fraction, and C_s the number of surfactant molecules per unit volume: $C_s = n_s/V$.

It can be seen from Eq. (1) that the preferred droplet radius is $R_0 = 1/C_0$. The sign of C_0 is by convention positive for oil droplets, negative for water droplets. If one then starts to add oil to a surfactant micellar dispersion in water, the droplets grow (according to Eq. (2)) until they reach the optimum radius R_0. Further addition of oil results in a phase separation: the oil is rejected into an excess phase [4]. The corresponding phase equilibrium is called Winsor I. Similarly, one can explain the equilibrium between a w/o microemulsion and excess water: Winsor II.

When C_0 is small, the surfactant layer has a tendency to become planar and to promote lamellar order. If the bending energy is comparable to thermal energy, the surfactant film will fluctuate strongly. The range of lamellar order is the persistence length of the film introduced by de Gennes [2]:

$$\xi_K = a \exp\left(2\pi K/kT\right) \tag{3}$$

where a = molecular length. If K is of the order of kT, ξ_K is a microscopic distance (~ 100 Å) and the system is macroscopically disordered. The structure is now sponge-like and bicontinuous in oil and water. Simple geometrical space filling models allow us to calculate the mean size (diameter) of the oil and water microdomains:

$$l = \frac{6\varphi_o \varphi_w}{C_s \Sigma} \tag{4}$$

where φ_o and φ_w are, respectively, the oil and water volume fractions [2] (the surfactant volume fraction has been neglected).

The bending constant K depends on the scale at which the surfactant layer deformations are produced: if the scale is larger than ξ_K, the layer is already very rough and easier to bend than a flat layer [5]. Using the renormalized expression for K derived by Helfrich, Safran et al., it is predicted that l will be proportional to ξ_K, the proportionality constant being of the order of unity [6]. If the amount of surfactant is not sufficient to accommodate all the oil and water in the microemulsion, excess oil and water will be rejected in the form of excess phases: Winsor III phase equilibria. Let us note that the Winsor equilibria were also explained by Widom, using a constant value for K, but keeping a non-zero interfacial tension [7].

The interfacial tensions y of the macroscopic interfaces between the microemulsion and the excess phases are also ultralow (Fig. 1). In most cases they are ultralow because the surface pressure of the flat monolayer at the surface pressure $\pi(\sigma)$ almost compensates the bare oil-water interfacial tension:

$$y = y_{ow} - \pi(\sigma)$$

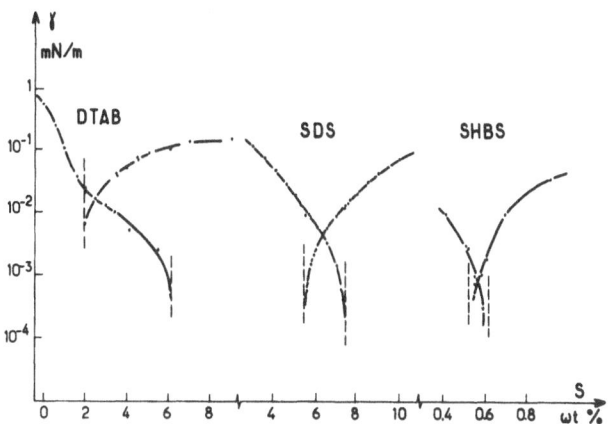

Fig. 1. Interfacial tensions between the microemulsion and the excess oil phase (decreasing values) and the microemulsion and the excess water phase (increasing values) versus salinity for the different systems

σ being the area per surfactant molecule at the macroscopic interface [8, 9]. π, like the free energy, depends on the bending energy, dispersion entropy and microdomain interactions. Assuming again that the bending energy is dominant, $\Gamma = 0$, $\bar{K} = 0$, $R = R_0$, one obtains for droplet microemulsions [2][1]:

$$y = y_c = \frac{2K}{R_0^2} \tag{5}$$

y_c is the energy (per unit area) to unbend the surfactant film. This value of y is independent of the volume fraction φ of the dispersed phase in the microemulsion. This is as observed in several surfactant systems [8].

In some cases, the bending energy is no longer very large compared with the dispersion entropy. Then, as has been shown by Israelachvili [10], $y = y_c + y_e$ with:

$$y_e = -\frac{kT}{4\pi R_0^2} \ln \alpha \varphi \tag{6}$$

α is a numerical constant of order unity depending on the approximations made in calculating the dispersion entropy [11].

This experssion accounts for the y versus φ variation recently reported in some surfactant systems [12].

---

[1]) In Ref. [2], $y_c = K/2R_0^2$ but F_c is taken as $\frac{1}{2} K \left(\frac{1}{R} - \frac{1}{R_0}\right)^2$ so that K is four times larger than the K used here. In order to obtain Eq. (3) for ξ_K, one has to use F_c, as in Eq. (1).

In some other cases, the interactions between droplets are large and attractive. They can promote phase separation between a droplet-rich and a droplet-poor phase beyond a critical point. The coexistence curve being very asymmetrical, the droplet-poor phase contains an appreciable number of droplets only very close to the critical point. Very rapidly its composition approaches that of the continuous phase of the droplets. In the Winsor III phase equilibria, this is the behaviour of the interface between the microemulsion and the excess phase which has the lowest interfacial tension (lowest branches, Fig. 1) [8]. Then, theory predicts that:

$$\gamma = \beta \frac{kT}{\xi_c^2}$$

where β is a numerical constant and ξ_c the correlation length for the critical fluctuations.

In all cases, γ is very small. It follows that thermal fluctuations at the macroscopic interfaces are large, as at the microscopic ones in the bicontinuous microemulsions. Their mean square amplitude is:

$$\langle \zeta^2 \rangle = \int_0^\infty \frac{kT}{\Delta\varrho g + \gamma q^2 + K q^4} \cdot \frac{q\,dq}{2\pi}$$

$$\cong \frac{kT}{4\pi\gamma} \ln \left(\gamma^2 / K \Delta\varrho g \right) \qquad (7)$$

where $\Delta\varrho$ is the density difference between coexisting phases and g the gravitational constant. Typically (far from a critical point) $\Delta\varrho \sim 0.1 \, \text{g/cm}^3$, $\gamma \sim 10^{-2}$ mN/m, $K \sim kT$ and $\langle \zeta^2 \rangle^{1/2} \sim 250$ Å.

This very large roughness can be measured by the reflectivity of light and by ellipsometry [13]. If γ is known, K can be deduced. Again, the problem of the scale variation of K has to be taken into account: K depends on q in the above integral. The renormalization calculation has to be done for both K and γ in this case. As a result, Eq. (3) now becomes:

$$\xi_K = a \exp \left[3.254 \, K(a)/kT \right] \qquad (8)$$

where $K(a)$ is the elastic constant at the scale a [14].

In summary, the theories predict that there is a strong correlation between the behaviour of the macroscopic and the microscopic interfaces in Winsor equilibria. In this paper we will be more concerned with the non-critical interfaces, and present results on the bending elasticity in different model microemul-

sion systems. The results will be correlated with the microemulsion structure and interfacial tensions between excess phases, following the theories.

Experimental results

We have studied three model systems of five component mixtures: oil-water-surfactant-alcohol-salt. Three different ionic surfactants have been used — dodecyl trimethyl ammonium bromide (DTAB), sodium dodecyl sulphate (SDS) and sodium hexadecyl benzene sulphonate (SHBS or Texas #1). The alcohol used was butanol; its addition is necessary because the above surfactants alone cannot form high pressure monolayers at the oil-water interface. The salt screens the electrostatic interactions between surfactant polar heads: it reduces C_0. In this way a continuous evolution from Winsor I-Winsor III-Winsor II can be obtained by addition of salt.

The composition of the systems are (in wt %): dodecane 38.19; brine 56.83; SHBS 1.66; butanol 3.32; or toluene 47; brine 47; SDS/DTAB 2; butanol 4.

The brine was an aqueous solution of sodium chloride for SHBS and SDS, and sodium bromide for DTAB.

The salinities of the Winsor equilibria boundaries are: SHBS: $S_1 = 0.52$, $S_2 = 0.61$; SDS: $S_1 = 5.4$, $S_2 = 7.4$; DTAB: $S_1 = 1.98$, $S_2 = 5.1$ (in wt% in water).

The variations of the interfacial tensions between the microemulsion and the excess phases, as measured with surface light scattering techniques [8, 15, 16], are shown in Fig. 1. They are all ultralow, the lowest corresponding to the SHBS system, the largest to the DTAB one.

The bending elasticities have been measured using ellipsometry [13b, 16]. They are roughly independent of the salinity and are given in Table 1. Since they do

Table 1. Measured and renormalized bending elastic constants K_m and K_r ($a = 10$ Å), persistence length ξ_k, characteristic microdomain size $L = \pi/q_{max}$, correlation length ξ and periodicity d of bicontinuous microemulsions at optimal salinity $S = S^*$, for the three different microemulsion systems

System	K_m/kT	K_r/kT	ξ_k(Å)	L(Å)	ξ(Å)	d(Å)
DTAB	0.40	0.55	63	97	76	177
SDS	0.65	1.00	275	250	170	460
SHBS	0.40	0.86	175	400	247	660

For SHBS microemulsions were the peak at q_{max} is not visible, L has been measured as explained in Ref. [18].

not differ very much, it follows from Eqs. (5) and (6) that to explain the interfacial tension differences, the largest droplet sizes in the Winsor I and II regions must be found in the SHBS system, the smallest in the DTAB one.

We have measured droplet sizes in SHBS and SDS systems with bulk light scattering techniques [8,15], and in DTAB systems with small angle X-ray techniques [17]. Sizes in bicontinuous microemulsions were measured with small angle X-ray and neutron scattering methods in the SDS system by Auvray et al. [18] and de Geyer and Tabony [19], and in SHBS and DTAB systems in collaboration with Auvray [20] and Ober [17]. Typical X-ray spectra corresponding to the three different Winsor equilibria in the DTAB system are shown in Fig. 2. The specturm for $S = 1$ corresponds to oil droplets interacting via a repulsive electrostatic potential and exhibits a peak due to these interactions. The spectrum for $S = 6$ corresponds to water droplets interacting via an attractive van der Waals potential and shows no peak as a result. The spectrum for $S = 2.5$ corresponds to a bicontinuous microemulsion containing equivalent amounts of oil and water. The presence of a broad peak indicates that some correlation between microstructural elements might be present. Recently, Teubner and Strey proposed the following empirical form for the correlation function [21]:

$$g(r) = \frac{d}{2\pi r} e^{-r/\xi} \sin \frac{2\pi r}{d} \tag{9}$$

where d is a domain size and ξ a correlation length: d should scale as l (Eq. (4)) and ξ as ξ_K (Eq. (3)). The Fourier transform of Eq. (9) gives the X-ray spectrum:

$$I(q) \sim \frac{1}{1 + C_1 q^2 + C_2 q^4};$$

$$C_1 = 2C_2 \left(\frac{1}{\xi^2} - \frac{4\pi^2}{d^2}\right); \quad C_2 = \left(\frac{1}{\xi^2} - \frac{4\pi^2}{d^2}\right)^{1/2} \tag{10}$$

which represents the experimental spectra published in the literature very well, including those of Fig. 2.

Empirically, it was found in all the existing studies of this kind that the position q_{max} of the $I(q)$ peak is simply related to l through $l \cong \pi/q_{max}$. Using the Porod's law, Teubner and Strey show that $\xi = 2l/3$. It follows that since $q_{max}^2 = \frac{1}{\xi^2} - \frac{4\pi^2}{d^2}$, then $d \approx 2l$. This is as expected for correlations in a random arrangement of microdomains filled either with oil or water.

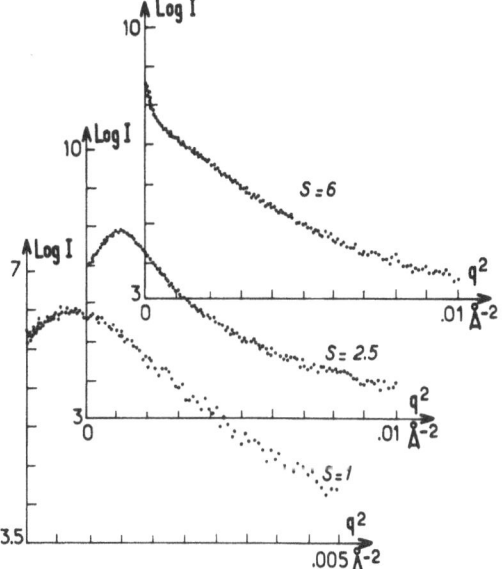

Fig. 2. Scattered X-ray intensity versus scattering wave vector (squared) in different types of phase equilibria for the DTAB system

The values of ξ, d and $L = \pi/q_{max}$ for the DTAB system are reported in Table 2. For the other systems, they are only given in Table 1 at the salinity S^* where $\varphi_o = \varphi_w$ (optimal salinity for oil recovery enhancement with microemulsions). The normalized measured sizes are also shown in Fig. 3. This illustrates the good agreement between the measured sizes and the geometrical formulae 2 and 4.

From this data it can be shown that the quantities $\gamma R^2/kT$ and $\gamma_{sup} l^2/kT$ are the order of unity (γ_{sup} is the largest of the two interfacial tensions in three-phase equilibria) [16,22]. γ_e is negligible compared to γ_c which also scales reasonably well with the data (see Table 2 for the DTAB data).

The persistence length has also been calculated according to Eq. (8) and is reported in Table 1. Again the data are in qualitative agreement with the theoretical predictions: the ratio ξ_K/l is of the order of unity, although it varies from one system to another.

Conclusions

The important role of the bending elasticity of the surfactant monolayers has been illustrated in Winsor equilibria, where microemulsions coexist with excess oil and/or excess water phases. In o/w and w/o microemulsions, the droplet radius is about the spontaneous radius of curvature of the layer, and the interfacial

Table 2. DTAB system. Measured sizes and interfacial tensions and comparison with theory. γ_e has been calculated for $\alpha = 1$ (Eq. (6)). $L = R$ for droplet microemulsions in column 6

S	R (Å)	L (Å)	d (Å)	ξ (Å)	$\gamma_{sup.}$ (mN/m)	$\gamma L^2/kT$	γ_c (mN/m)	γ_e (mN/m)
1	64				0.12	1.22	0.108	0.018
2		120	217	73	0.026	0.93		
2.5		97	177	76	0.020	0.47		
3		103	185	64	0.033	0.87		
4		90	164	57	0.051	1.02		
5		78	140	44	0.091	1.37		
6	51				0.105	0.68	0.171	0.32
7	54				0.138	1.00	0.152	0.03
9	42				0.134	0.58	0.252	0.053

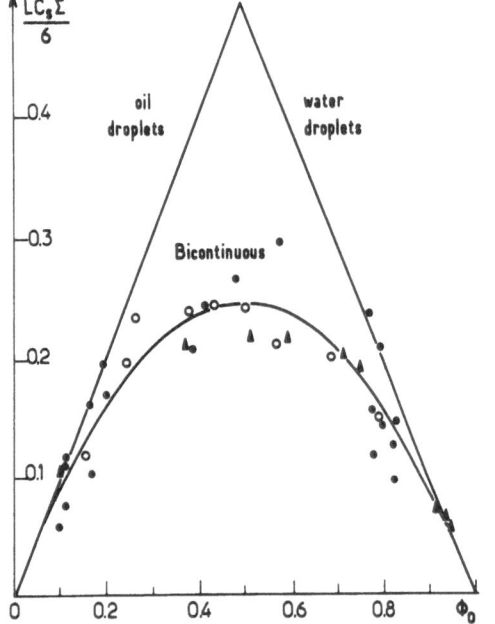

Fig. 3. Reduced characteristic sizes in the three different microemulsion systems versus oil volume fraction: (O) SDS; (●) SHBS; (▲) DTAB

the theories: the role of interactions between the surfactant layers is not taken into account in the calculation of the persistence length, the microscopic surface tension and the saddle-splay bending constant have been neglected. More seriously, the validity of the renormalization expansions for K might be limited. Further improvement of the theories and possibly new experimental developments (measurement of Γ [23], $\bar{K}$, ...) are needed to achieve a complete understanding of these phase equilibria.

Acknowledgements

Dr. Binks wishes to thank the Royal Society for the award of a Postdoctoral Fellowship under the European Science Exchange Programme. We have benefited from partial financial support from C.N.R.S. and P.I.R.S.E.M. (Greco Microemulsions) and from the C.E.E. (contract EN3C-0018F).

tension between the microemulsion and the excess phase roughly corresponds to the energy required to straighten the layer (per unit area) in order to cover the macroscopic interface. In bicontinuous microemulsions, the average size of the microdomains is close to the persistence length of the layer.

Although the agreement with the theories is qualitatively good, the quantitative predictions can differ by factors as large as two or three. The origin of these discrepancies probably arises from oversimplifications in

References

1. (a) Prince LM (ed) (1977) Microemulsions. Acad Press
   (b) Meunier J, Langevin D, Boccara N (eds) (1987) Physics of Amphiphilic layers. Springer Verlag
2. de Gennes PG, Taupin C (1982) J Phys Chem 86:2294
3. Helfrich W (1973) Z Naturforsch C 28:693
4. Safran SA, Turkevich LA (1983) Phys Rev Lett 50:1930
5. Helfrich W (1985) J Phys, Paris 46:1263
6. Safran SA, Roux D, Cates M, Andelman D (1986) Phys Rev Lett 57:491
7. Widom B (1984) J Chem Phys 81:1030; Widom B, Dowson KA, Lipkin MD (1986) Physica 140A:26
8. Pouchelon A, Chatenay D, Meunier J, Langevin D (1981) J Colloid Interface Sci 82:418; Cazabat AM, Langevin D, Meunier J, Pouchelon A (1982) Adv Colloid Interface Sci 16:175
9. Aveyard R, Binks BP, Mead J (1987) J Chem Soc Faraday Trans I 83:2347
10. Israelachvili J (1987) In: Mittal KL, Bothorel P (eds) Surfactant in Solutions. Plenum Press, Vol 4, p 3

11. Overbeek JThG, Verhoeckx GJ, de Bruyn PL, Lekkerkerker HNW (1987) J Colloid Interface Sci 119:422
12. Fletcher PDI (1987) Chem Phys Lett 141:357
13. (a) Meunier J, Langevin D (1982) J Phys Lett 43:L-185
    (b) Meunier J (1985) J Phys Lett 46:L-1005
14. Meunier J (1987) J Phys, Paris 46:1819
15. Guest D, Langevin D (1986) J Colloid Interface Sci 112:208
16. Binks BP, Meunier J, Abillon O, Langevin D, in preparation
17. Abillon O, Binks BP, Otero C, Langevin D, Ober R (1988) J Phys Chem, in press
18. Guest D, Auvray L, Langevin D (1985) J Phys Lett 46:L-1055
19. de Geyer A, Tabony J (1985) Chem Phys Lett 113:83; (1986) 124:357
20. Auvray L, Cotton JP, Ober R, Taupin C (1984) J Phys 45:913; (1984) J Phys Chem 88:4586
21. Teubner M, Strey R (1987) J Chem Phys 87:3195
22. Langevin D, Guest D, Meunier J (1986) Colloid Surf 19:159
23. Huang JS, Milner ST, Farago B, Richter D in Ref [1b], p 346 and Kaler EW in Ref [1b], p 364

Received December 10, 1987;
accepted December 21, 1987

Authors' address:

D. Langevin
Laboratoire de Spectroscopie Hertzienne de l'E.N.S.
24, rue Lhomond
F-75231 Paris Cedex 05, France

Progress in Colloid & Polymer Science   Progr Colloid Polym Sci 76:90–95 (1988)

The disordered open connected model of microemulsions

I. S. Barnes[1][2]), S. T. Hyde[2]) B. W. Ninham[2]) P.-J. Derian[1]), M. Drifford[1]), G. G. Warr[1]), and T. N. Zemb[1])

[1]) Département de Physico-Chemie, Centre d'Études Nucléaires de Saclay, Gif-sur-Yvette, France
[2]) Department of Applied Mathematics, Research School of Physical Sciences, The Australian National University, Canberra, Australia

Abstract: We present further evidence in support of our Disordered Open Connected (DOC) model for microemulsion structure: small-angle X-ray scattering (SAXS) data for the systems didodecyl dimethyl ammonium bromide (DDAB)-octane-water and DDAB-dodecane-water. The models is a simple geometrical approximation to a random interface of constant mean curvature. Using only the composition of the sample and the desired curvature of the oil-water interface, we predict the SAXS spectrum, the position in the phase diagram of the conductivity anti-percolation, and the approximate boundary of the single-phase region.

Key words: Microemulsions, scattering, microstructure, phase diagram, curvature.

Introduction

In a series of recent papers [1–3] we have presented the Disordered Open Connected (DOC) model for microemulsions and evidence in support of it. This model uses an old idea: some microemulsions adopt a bicontinuous structure resembling a random network of interconnecting cylinders [4], but the detailed quantitative analysis of such a structure is new.

Our choice of systems involving the double-chained cationic surfactant didodecyl dimethylammonium bromide (DDAB) is crucial. This surfactant forms isotropic liquid phases (L_2 phases or oil-rich microemulsions) in the ternary systems with water and any of the oils cyclohexane, hexane, hexene, octane, decane, decene, dodecane, tetradecane [5] (and probably many others), without the need for addition of cosurfactant or salt, and is almost completely insoluble in water and in oil. Thus, problems associated with partitioning the surfactant and cosurfactant between oil, water and the interface are eliminated. The bromide counterions give excellent X-ray contrast.

The phase diagrams for these systems, with the exception of tetradecane, have several features in common. The situation for tetradecane, where the chain length of the surfactant is exceeded by that of the oil, is qualitatively different and will be treated separately

[6]. The L_2 phase region is shaped like a drop hanging from the oil corner. Conductivity and self-diffusion NMR measurements show that the phase is bicontinuous at low water content but continuous in oil only at higher water content.

We have attempted to reconcile the data for these systems with a model of water spheres in oil. Once the internal volume fraction Φ and specific surface Σ are known (and these are fixed by composition as described below), the number density n and radius R of spheres are uniquely determined by the relations

$$\Phi = \frac{4}{3}\pi R^3 n \qquad \Sigma = 4\pi R^2 n$$

and one can calculate the full scattering curve from this information without difficulty [1]. At high water content, the small-angle X-ray scattering (SAXS) spectra agree with spheres, with only a hard core interaction. At low water content, where both the shape and the position of the peak are qualitatively different, the structure cannot be a dispersion of water droplets. While the conductivity percolation is quite sharp, the transition from "sphere" to "non-sphere" SAXS spectra is gradual.

The evidence for the hard spheres ("reverse micelles") model is overwhelming at higher water contents, but it is untenable on the other side of the percolation line. Yet this is a single "phase" in the usual macroscopic sense of the word. To resolve this paradox we have developed a gemetrical model which consists of random network with varying connectivity which includes, at one extreme, isolated spheres. The model has no free parameters — all can be measured experimentally — and allows us to predict the position of the percolation transition, the variation of peak position in SAXS and (more approximately) the position of the phase boundaries. It is also possible to calculate the full SAXS curve for any sample composition.

In this paper we set out the underlying assumptions and the predictions of our model, and then present new data for the octane and dodecane systems.

Geometrical constraints on structure

The essential idea of our approach is to distill the chemistry and physics of the system into a small number of geometrical constraints, and then to find a structure which satisfies those constraints. We think of the sample as being divided into two parts: the "water" region, which consists of water, surfactant headgroups and counterions; and the "oil" region, consisting of the oil and the surfactant tails. The interface between these two regions is the interface of contrast seen by X-ray scattering. The actual definition of the position of the interface is open to choice: as long as it is reasonable, the resolution of the experiment will be insufficient to discriminate. We then arrive at three constraints on the geometry:

1. The model must give the correct values for the volume fractions of the two regions. Once the partial molar volumes of the components are known, these fractions can be determined directly from the composition of a sample. As the two fractions must add up to 1, it is sufficient to know the water or "internal" volume fraction Φ_{int}.

2. The model structure must have the correct value for the interfacial area per unit volume Σ. This can be determined in two ways: either from the amount of surfactant in the sample and the headgroup area per molecule (which has been found to remain almost constant at a value of about 68 Å²) or from the Porod limit in SAXS. The two have always been found to agree well [2, 3].

3. The curvature of the interface between the oil and water regions must be such that the surfactant molecules can pack on it. We specify this curvature with the surfactant packing parameter v/al [7], where v is the

effective tail volume (including any oil penetration), a is the head group area and l is the tail length. We assume that this remains constant throughout the L_2 phase region, and obtain a value from the fit to spheres at high water content. As the degree of oil penetration into the tails varies greatly between oils, a new value must be obtained for each oil used.

Geometrical realisation of the structure

Now consider a disordered surface of constant mean curvature (H surface): a disordered version of the infinite periodic minimal- and H-surfaces proposed as models for cubic and hexagonal phase [8]. We approximate this by a network of spheres and cylinders, specified by the density of sphere centres in space n, the average coordination number (number of cylinders meeting at each sphere) Z, the radius of the spheres R, and the radius of cylinders r. To maintain the same value for the mean curvature on the spherical and cylindrical parts of the surface it is necessary to fix $R/r = 2$. This leaves three parameters for the structure, to be determined by the three geometrical constraints above.

For given values of n, Z and R, the structure is constructed as follows. First we generate a random hard-sphere distribution of points in space with minimum separation $2R + 2l$. This condition corresponds to the requirement that the water spheres approach no closer than two tail lengths. The distribution is generated by a simple Monte Carlo procedure. A sphere of radius R is then placed at each of these points. Next we determine the nearest neighbours of each point. This is done by constructing Voronoï polyhedra: two points are neighbours if their Voronoï polyhedra share a face. Finally we order the resulting list of pairs of neighbours by their separation distance, and connect neighbouring spheres by cylinders of radius $r = R/2$, in order of proximity, until the average coordination number for the structure is Z. (Note that $Z = 0$ is just the hard spheres model.)

It is now possible to derive equations relating the parameters n, R and Z to Φ_{int}, Σ and v/al. We have the following formulae for Φ_{int} and Σ, which are exact as long as the cylinders do not intersect:

$$\Phi_{int} = \left[\frac{4}{3} \pi R^3 - \frac{1}{3} Z r^3 (\Omega \varrho^3 - \pi \sqrt{\varrho^2 - 1}) \right] n$$

$$+ \pi r^2 Z \left[\frac{(8.15) n^{2/3}}{(13.4)} - n r \sqrt{\varrho^2 - 1} \right]$$

$$\Sigma = (4\pi - \Omega Z)\, R^2 n$$

$$+ 2\pi r Z \left[\frac{(8.15)\, n^{2/3}}{(13.4)} - nr\sqrt{\varrho^2 - 1} \right]$$

where $\varrho = R/r$ and

$$\Omega = 2\pi[1 - \sqrt{1 - 1/\varrho^2}\,].$$

Finally we have two different expressions for v/al. The first comes directly from the definition and gives

$$v/al = \frac{\Phi(R+l, r+l, Z, n) - \Phi(R, r, Z, n)}{l\Sigma(R, r, Z, n)},$$

but is prone to errors caused by overlaps, especially for large values of Z. The second method comes from differential geometry, and uses the relationship between the variation of surface area as the surface is displaced along its normal and the area-averaged mean (H) and Gaussian (K) curvatures. We have

$$H = \frac{\Sigma_+ - \Sigma_-}{4\delta\Sigma_0} \quad K = \frac{\Sigma_+ + \Sigma_- - 2\Sigma_0}{2\delta^2\Sigma_0}$$

where Σ_0 is just area calculated before, and Σ_+ and Σ_- are the areas of parallel surfaces displaced respectively outside and inside the original surface by a small distance δ. Then

$$v/al = 1 + Hl + \frac{1}{3} Kl^2.$$

This method is strictly only correct for smooth surfaces [9], but it holds in the limit if the cusps are smoothed and the radius of curvature of the smoothing is allowed to go to zero.

We thus have a simple geometrical model with three parameters n, R, and Z which are uniquely fixed by the three experimentally measurable constraints Φ_{int}, Σ and v/al. From the model structure as it varies across different compositions we can now predict various aspects of the behaviour.

Predictions of the model

Firstly, with fixed v/al, corresponding to a fixed oil, consider a water dilution path through the L_2 region, beginning on the surfactant/oil edge of the phase diagram and proceeding in a straight line, at constant surfactant : oil ratio, towards the water corner. Our model

predicts that this will produce steadily decreasing values for the connectivity Z. This can be seen to be intuitively correct by noting that along such a path the internal volume increases and the surface area decreases, and recalling that spheres enclose more volume for a given surface area than do cylinders. We predict an "anti-percolation" at $Z \approx 1.1$; for lower values of Z the water network is not connected and so the sample will not conduct. Our predicted values for the water surfactant ratio at percolation are in good agreement with the measured values [3]. For octane, with $v/al \approx 1.18$, we predict the percolation at a water : surfactant ratio of 1.72, the observed ratio is 2.06; for dodecane with $v/al \approx 1.12$, our predicted ratio is 2.59 and the observed ratio is 2.50.

Next consider the variation of the peak position D^* in SAXS. For a rule of thumb we assume simply that $D^* \approx n^{-1/3}$. This estimate is derived from reasoning based on regular lattices and should be correct to within about 20 % [1]. Predicted and measured values of D^* for octane and dodecane samples are given in Tables 1 and 2.

By using a finite element method which "digitizes" a structure and performs a three-dimensional Fourier transform [10], we are able to predict the full scattering curve from the model structure. Despite some numerical problems, this yields good qualitative agreement with the measured spectra [3, 10], except in those cases where there is clear evidence of an attractive interaction between aggregates. Such an attraction is outside the current scope of the model.

Finally, we are able to give approximate predictions of phase boundaries. These are determined by examining a number of water dilution paths and different surfactant : oil ratios, and determining the maximum and minimum water contents possible along each. The maximum water content is the point at which $Z = 0$. Beyond this the surface and volume constraints require spheres which are too big for the surfactant molecule to pack on them. The system prefers demixion to a change in the surfactant wedge shape which could only be brought about by squeezing some of the oil out of the tail region. The minimum water content prediction is less precise. We postulate that it corresponds to the maximum possible connectivity for the network, which occurs at $Z = 13.4$ [11].

These boundaries are the limits of composition at which a DOC structure can exist with the correct value for v/al. This says nothing about the possible existence of other phases at the same composition with lower free energy or of the possibility of two- or three-

Table 1a. Composition and known values for the DDAB/dodecane/water microemulsions

Sample	Dod. 1	Dod. 2	Dod. 3	Dod. 4	Dod. 5
% Weight H_2O	61.1	52.3	43.2	34.6	25.0
DDAB	27.5	24.3	20.4	16.4	11.8
Oil	11.4	23.4	36.4	49.0	63.2
% Vol. H_2O	58.9	48.4	38.5	29.7	20.7
DDAB	26.5	22.6	18.2	14.1	9.7
Oil	14.6	29.0	43.3	56.2	69.6
C_s (M/l)	0.573	0.488	0.393	0.305	0.210
W	57	55	54	54	54
Φ_{int}	63.2	52.2	54.3	32.0	22.3
$\varrho_{int}(e/Å^3)$	0.343	0.343	0.343	0.343	0.343
Σ ($Å^2/Å^3$) (calculated)	0.023	0.020	0.016	0.012	0.009

C_s is the concentration of surfactant; W the molar surfactant : water ratio; Φ_{int} the volume fraction of the high electronic density "internal" region; ϱ_{int} the electronic density of that region (the electronic density of the "oil" regions is 0.267 e/$Å^3$ throughout); Σ is the interfacial area per unit volume calculated from the compostion.

Table 1b. Measured quantities and values derived from the geometrical model for the DDAB/dodecane/water microemulsions

Sample	Dod. 1	Dod. 2	Dod. 3	Dod. 4	Dod. 5
l_c (Å)			43	55	77
D^* (Å)	120	150	180	217	270
Σ ($Å^2/Å^3$) (measured)	0.03	0.025	0.016	0.013	0.008
Z	3.75	2.5	2	2	1.5
n ($Å^{-3}$) ($\times 10^{-7}$)	4	3	2	1.5	0.7
$(n^{-1/3})$ (Å)	135	150	171	188	242
R_s (Å)	68	71	75	75	85

From the X-ray data we have (without any model) the average chord l_c; the peak position $D^* (= 2\pi/q_{peak})$; and the interfacial area per unit volume Σ calculated from the measured Porod limit. For the DOC model description we have the density of Voronoï centres n; the radius of the water spheres R (the cylinders have radius $R/2$); and the average coordination number for the network Z. The predicted peak position is given approximately by $n^{-1/3}$.

Table 2a. Composition and known values for the DDAB/octane/water microemulsions

Sample	Oct. 1	Oct. 2	Oct. 3	Oct. 4
% Weight H_2O	21.1	30.4	42.7	46.0
DDAB	39.3	34.6	28.5	27.0
Oil	39.6	35.0	28.8	27.0
% Vol. H_2O	18.0	26.4	38.0	41.3
DDAB	33.7	30.2	25.4	24.2
Oil	48.3	43.4	36.6	34.5
C_s (M/l)	0.728	0.652	0.549	0.523
W	14	23	38	44
Φ_{int}	23.6	31.4	42.2	45.3
$\varrho_{int}(e/Å^3)$	0.366	0.355	0.347	0.346
Σ ($Å^2/Å^3$) (calculated)	0.03	0.027	0.022	0.021

Same notation as for Table 1a.

Table 2b. Measured quantities and values derived from the geometrical model for the DDAB/octane/water microemulsions

Sample	Oct. 1	Oct. 2	Oct. 3	Oct. 4
l_c (Å)	25	30	32	40
D^* (Å)	68	90	134	143
Z	≥ 8	5	0.5	0
n ($Å^{-3}$) (10^{-7})	≈ 10	13	4	3.4
$(n^{-1/3})$ (Å)	100	92	136	143
R_s (Å)	≈ 30	32	63	68

Same notation as for Table 2a.

phase equilibria. Nevertheless, the predicted boundaries seem to be in reasonable agreement with those determined experimentally. The phase boundary of the L_2 phase is shown in Fig. 1.

Experimental

We measured SAXS spectra on the D 22 camera at LURE (Orsay, France), using $\lambda = 1.22$ Å. Beam times were 15 min per sample. The absolute scaling was obtained using pure water as a

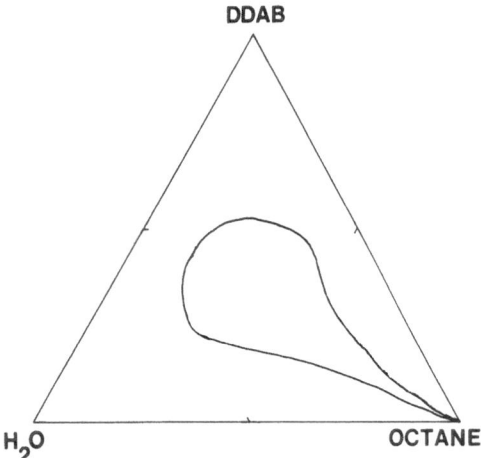

Fig. 1. Phase boundary for the L_2 phase in the DDAB/octane/water system. The predicted boundary is obtained from the variation of the coordination number Z along several water dilution paths: for a DOC structure to be possible, Z must be between 0 and 13.4

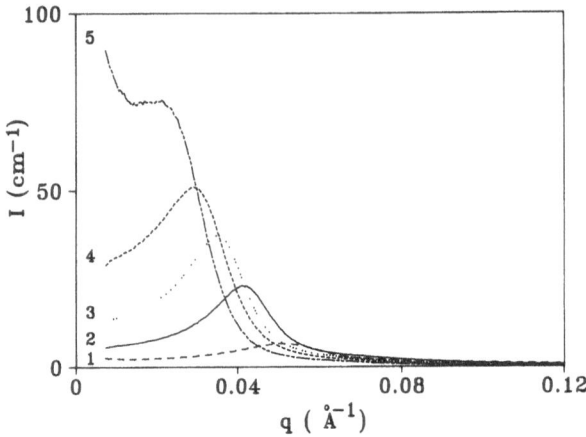

Fig. 2a. Absolute scaled scattered intensity $I(q)$ in cm^{-1} for the DDAB/dodecane/water microemulsion samples. The scattering for sample 5 is affected by the proximity of the critical point in the phase diagram [3]

standard [12]. To obtain a larger range of q-values we used two sample detector distances: 0.75 m and 1.5 m. The two data sets were then merged to form a single scattering curve. As the camera has point collimation, no deconvolution was used. The validity of the absolute scaling was checked using the value of the invariant determined from the composition [3]. For each sample we determined the interfacial area Σ (Å^2/Å^3); the average chord l_c, and the peak position $D^* = 2\pi/q_{peak}$. Plots of the scattering curves for dodecane are shown on in Fig. 2a. Porod plots allowing determination of the specific surface are in Fig. 2b. The scattered intensity at the peak is 300 times higher than in the high-q Porod region. The compositions and other known values for the samples are given in Tables 1a for dodecane and 2a for octane; the values measured form the scattering and predicted by the DOC model are in Tables 1b for dodecane, 2b for octane.

Discussion

Due to the limited q-range available ($0.007 < q < 0.12$), the averaged chord l_c cannot be calculated accurately for dodecane samples 1 and 2. For the remaining samples we find in Tables 1b and 2b that $l_c \approx D^*/3$, which cannot be reconciled with former models for scattering form microemulsions (see discussion in Ref. [1,3]). Comparison of the measured peak position D^* and the rough predicted peak position $n^{-1/3}$ in Tables 1b and 2b show agreement to within about 15%.

The values for the average coordination number Z for the dodecane microemulsion given in Table 1b show that the structure has high connectivity at low oil content, near to the neighbouring bicontinuous cubic phase. We imagine this cubic phase to have a structure

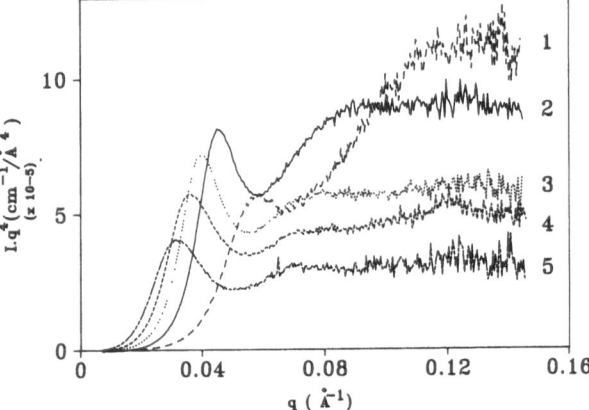

Fig. 2b. Porod plot (Iq^4 vs. q) for the DDAB/dodecane/water microemulsion samples. The constant value at high q is the asymptotic limit from which the interfacial area is determined. The much higher connectivity in sample 1 is seen by the qualitative change in the shape of the scattering curve. For samples 2–5, where the connectivity is around 2, spheres dominate the scattering and the minimum in Iq^4 gives the sphere radius by $R = 4.5/q^*$ where q^* is the position of the minimum. This agrees extremely well with the values predicted by the DOC model

similar to the microemulsion, only with cubic symmetry; to achieve this the coordination number would have to be at least 4. At higher oil content the spectra are very similar to those for spheres, and we find coordination numbers $Z \leq 2$. The sphere radii deduced from the spectra agree extermely well with those predicted by our model. If one makes the calculation for these samples with spheres only ($Z = 0$), however, the radii obtained are different and the interfacial curvatures are unrealistic. Essentially, at relatively low coordination, the cylinders contribute little to the scattering, but absorb excess interface and reduce the interfacial curvature allowing a more realistic structure.

It is interesting to contrast our geometrical approach with the phenomenological model recently proposed by Teubner and Strey [13]. Arguing from statistical mechanics, they postulate that the scattering from microemulsions should be of the form:

$$I(q) = \frac{1}{a_2 + c_1 q^2 + c_2 q^4}$$

where the coefficients a_2, c_1 and c_2 allow the determination of the domain size d ($= D^*$ in our notation) and the coherence length ξ. The preferred curvature of the interface is not included in their model. Further, while the above expression allows the determination of two characteristic distances in order to fit the scattering curve on arbitrary (not absolute) scale, it does not describe or even guarantee the existence of microstructure which satisfies these constraints.

Nevertheless, fits of the Teubner-Strey model to experimental spectra are extremely good. They find consistently that $\xi \approx 0.5\,d$, which is a direct consequence of a Voronoï tesselation such as we have proposed. The two models, arguing by very different methods, arrive at similar values for the "domain size" in a sample.

Recently Kaler and Vonk [14] gave another simple and general expression for the scattering of a distorted lamellar model for microemulsions which also gives excellent fits to measured spectra, even on absolute scale. Their model also has two parameters: the thickness of the lamellae and the persistence length for the orientation. In this case too, the persistence length is about twice the repeat distance, as would be expected for a Voronoï tesselation.

Acknowledgement

The authors thank the staff at LURE, especially Claudine Williams, for their assistance with the scattering experiments.

References

1. Zemb TN, Hyde ST, Derian P-J, Barnes IS, Ninham BW (1987) J Phys Chem 91:3814–3820; (1988) J Phys Chem 92:2286–2293
2. Ninham BW, Barnes IS, Hyde ST, Derian P-J, Zemb TN (1987) Europhys Lett 4:561–568
3. Barnes IS, Hyde ST, Ninham BW, Derian P-J, Drifford M, Zemb TN (1988) J Phys Chem 92:
4. Chen SJ, Evans DF, Ninham BW (1984) J Phys Chem 88:1631–1634
5. Blum FD, Pickup S, Ninham BW, Chen SJ, Evans DF (1985) J Phys Chem 89:711; Fontell K, Ceglie A, Lindman B, Ninham BW (1986) Acta Chem Scand A40.247–256
6. Derian P-J et al, in preparation
7. Israelachvili JN, Mitchell DJ, Ninham BW (1976) J Chem Soc Faraday Trans II 72:1525–1568
8. Scriven LE (1977) In: Mittal KL (ed) Micellization, Solubilization and Microemulsions. Vol 2, Plenum Press, New York, pp 877–893; Hyde ST, Andersson S, Ericsson B, Larsson K (1984) Z Kristallographie 168:213
9. Anderson DM (1986) Ph D thesis, University of Minnesota
10. Barnes IS, Zemb TN (1988) J Appl Crystallogr, in press
11. Hyde ST (1988) submitted
12. Zemb TN, Charpin P (1985) J Phys 46:249–256
13. Teubner M, Strey R (1987) J Chem Phys 87:3195–3200
14. Kaler EW, Vonk CG (1987) presented at the 10th Small Angle Scattering Conference, Prague, July; to be published in J Appl Crystallogr

Received December 22, 1987;
accepted January 4, 1988

Authors' address:

I. S. Barnes
Department of Applied Mathematics
Research School of Physical Sciences
The Australian National University
Canberra, ACT 2601, Australia

Progress in Colloid & Polymer Science   Progr Colloid Polym Sci 76:96–100 (1988)

The influence of hydroxylic group substitution on micellar properties of polyoxyethylene alcohols

I. Miesiac and J. Szymanowski

Technical University of Poznań, Institute of Chemical Technology and Engineering, Poznań, Poland

Abstract: Acetates and chlorides of the polydisperse polyoxyethylene alcohols containing from 6 to 14 carbon atoms in the alkyl group and average number of 2–14 oxyethylene units were synthesized. Their critical micelle concentrations were determined using the surface tension and light scattering methods. The influence of the terminal acetyl group and chlorine atom on the products' hydrophilicities was discussed. It was found that acetates demonstrate similar hydrophilicity as the polyoxyethylene alcohols. Chlorides are much more hydrophobic and depending upon the polyoxyethylene chain length a chlorine atom is equivalent to 2–4 methylene groups.

Key words: Polyoxyethylene alcohol derivatives, hydrophilicity, critical micelle concentration.

Introduction

Polyoxyethylene alcohols, $R(OE)_n OH$ are the most important nonionic surfactants. Commercial products obtained in the reaction of fatty alcohols with ethylene oxide are complex mixtures of homologues with different polyoxyethylene chain length. They also contain some amounts of polyoxyethylene glycols. According to several authors [1, 6] the average length of the polyoxyethylene chain, i.e. the average number of oxyethylene units, is equivalent to the actual length of this chain in well-defined, individual polyoxyethylene glycol monoalkyl ethers.

The physico-chemical and usage properties of polyoxyethylated alcohols can be significantly modified by blocking the terminal hydroxyl group, e.g. by the substitution for this hydroxyl group of an acetyl group or a chlorine atom. Such products exhibit much lower foaming ability [3, 5]. Chlorine derivatives also show high resistance to thermal oxidative destruction in strongly alkaline media [6].

The aim of this work was to determine the influence of substitution of the hydroxyl group in polyoxyethylene alcohols by the acetyl group and the chlorine atom on the critical micelle concentration, as determined from surface tension isotherms and from light scattering data.

Experimental

Polydisperse mixtures of polyoxyethylene glycol monoalkyl ethers were obtained from pure n-hexanol, n-decanol, n-tetradecanol (Reachim) and ethylene oxide in the usual way, by carrying out the reaction under a small pressure 0.3–0.4 MPa and in the presence of sodium hydroxide (0.1 %). These products were characterized by the typical content of the successive homologues and by the typical content of the polyoxyethylene glycols (0.5–4 %) [7, 9]. The average numbers of oxyethylene units were derived from the hydroxyl values [2].

Chlorine derivatives were obtained in the reaction of the polyoxyethylene alcohols with thionyl chloride using 2–3 fold molar excess of this reagent in benzene solution, at the temperature 50°–70°C in the presence of a small amount of active carbon [6]:

$$R(OE)_n{-}OH + SOCl_2 \rightarrow R(OE)_n{-}Cl + SO_2 + HCl$$

where $E = C_2H_4$.

After 1 h, active carbon was separated by filtration, then the diluent and other volatile compounds were evaporated by means of a water bath and nitrogen flow. Light products without $SOCl_2$ flavour were obtained.

Acetates of polyoxyethylene alcohols were obtained by acetylation of the starting materials with acetyl anhydride using 5-fold molar excess of the acetylation reagent at 100 °C, and removing all volatile compounds under reduced pressure [8].

The surface tension was measured using the ring method. The scattered light intensity was measured at an angle of 90° and at a wavelength of 435.8 nm. These measurements were carried out at room temperature (20 ± 1 °C) in the same way as described previously [4].

The molecular area at the interface was determined from the surface excess calculated according to the Gibbs isotherm:

$$A = \frac{1}{N\Gamma} \text{ and } \Gamma = -\frac{dy}{d \ln c} \frac{1}{RT}$$

where N is the Avogadro number, Γ the surface excess, $dy/d \ln c$ the maximum slope of the surface tension isotherm, R the gas constant, T the temperature [K].

Results and Discussion

The exemplary isotherms are shown in Fig. 1. Some effect of the ageing time on surface tension isotherms was observed. This effect depends on the surfactant hydrophobicity and is especially strong for hydrophobic compounds forming milky dispersions. For hydrophilic products for which the CMC is greater than 100 µmol dm^{-3}, this effect is small and constant values of the surface tension are obtained during 0.5–1 h. For all products considered, the ageing time of 24 h was sufficient to determine the proper surface tension isotherm.

The time effect on the scattered light intensity is almost negligible for hydrophobic products, and only a small influence is observed for products having a long polyoxyethylene chain. In this last case, the inflection points shift towards higher concentrations with increasing ageing time. This effect is probably connected with the rearrangements of the micelle structure during dilution. Above the CMC some rearrangements also occur and as a result the intensity of the scattered light significantly decreases (Fig. 1).

In the most cases, similar CMC values were obtained using both the methods considered, and, for calculations, their average values were taken (Table 1). In a few cases, non-typical plots of light scattering without any inflection point were obtained (Fig. 2), especially for hexanol derivatives. Then the CMC values were taken only from the surface tension isotherms.

The acetyl group causes a small decrease of the CMC, and this effect tends to disappear as the length of the polyoxyethylene chain increases. However, for the chlorine derivatives, the effect of substitution of hydroxyl group is much stronger and the values of CMC decrease by about 2 orders of magnitude.

On the other hand, the influence of both types of substitution on the values of molecular area is quite different. For chlorine derivatives, these values are practically of the same order $30-50 \cdot 10^{-20}$ m^2 as for the polyoxyethylene alcohols, and increase in a typical way with increasing length of the polyoxyethylene chain.

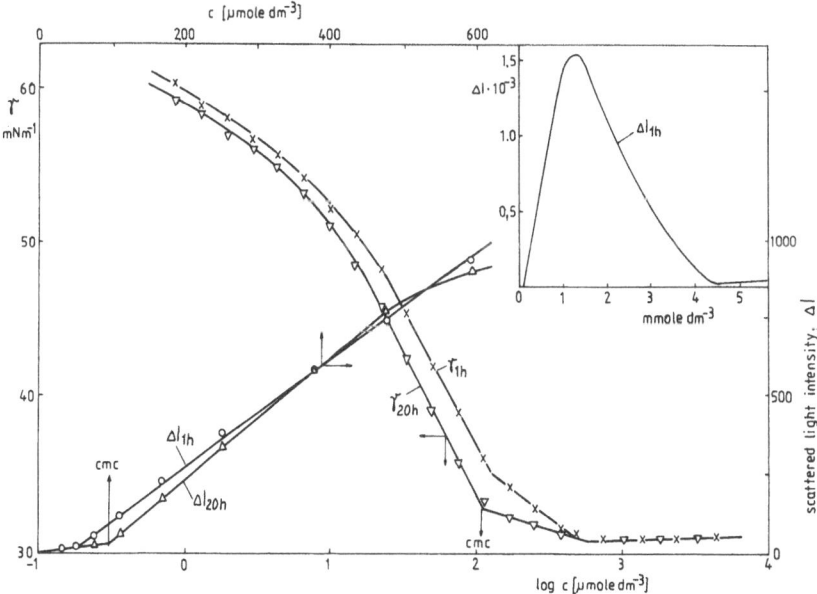

Fig. 1. Determination of critical micelle concentration (CMC) from surface tension and light scattering isotherms (product $C_{10}(OE)_{14}Cl$)

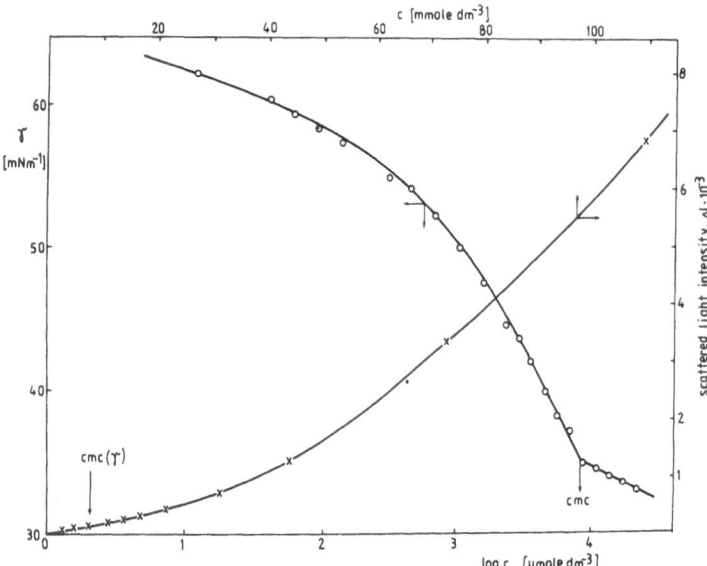

Fig. 2. Non-typical plot of light scattering without explicit inflection point (product $C_6(OE)_{14}Cl$)

The opposite effect is observed for acetyl derivatives, especially for those containing a short polyoxyethylene chain, for which much higher values of the molecular area were obtained. Thus, the introduction of the slightly hydrophobic acetyl group disturbs considerably the well-ordered adsorption layer, but the CMC values do not change. This effect tends to disappear as the length of the polyoxyethylene chain increases.

The surface tension values at the CMC are similar in the range of 30–40 mN m^{-1}, regardless of the kind of terminal group.

Figure 3 shows the relation between the CMC and the average length of the polyoxyethylene chain. For polyoxyethylene alcohols and their acetates, typical straight line relations, with small slope are observed. For chlorine derivatives, two different straights can be derived. Only for $n > 10$ is the slope of the typical order, but for $n <$ the effectiveness of the polyoxyethylene chain is much stronger. The gradients of the obtained straight lines are given in Table 2.

For practical purpose, the effectiveness of one oxyethylene unit can simply be expressed as an equivalent

Table 1. Micellar properties of polyoxyethylene alcohols, their acetates and chlorides [$C_m(OE)_{\bar{n}}X$]

| No. | Product | | X = −OH | | | X = −OCCH₃ (O) | | | X = −Cl | | | |
	m	$\bar{n}$	CMC	A	γ	CMC	A	γ	CMC γ	CMC Δ|	A	γ
1	6	2.1	−	−	−	−	−	−	138	145	73.8	43.6
2	6	6.0	63000	78.9	39.0	13000	53.1	36.4	930	−	93.0	36.4
3	6	10.7	−	−	−	−	−	−	6900	−	66.4	34.5
4	6	14.2	−	−	−	−	−	−	8300	−	65.2	34.6
5	10	0	340	32.2	27.6	−	−	−	−	−	−	−
6	10	2.2	480	30.1	24.2	340	93.8	36.6	6.0	5.0	23.2	38.6
7	10	3.0	−	−	−	−	−	−	8.0	6.0	31.4	36.4
8	10	4.1	620	33.1	24.8	490	71.5	32.1	14.5	12.0	35.8	36.0
9	10	6.1	810	36.2	25.6	610	56.1	29.2	20.9	21.0	37.2	33.8
10	10	8.2	1010	36.5	28.2	750	49.1	29.1	77.0	75.0	40.0	33.6
11	10	10.4	1200	37.8	30.0	870	47.3	29.3	81.0	80.0	50.8	30.3
12	10	14.1	1490	50.0	33.6	1460	53.8	30.3	107.0	95.0	50.0	32.8
13	14	5.9	9.3	35.8	28.8	15.5	76.3	29.4	2.2	2.0	43.4	30.0
14	14	14.1	18.2	47.8	35.8	−	−	−	12.6	12.0	51.6	35.0

CMC = [µmole dm^{-3}]; A = molecular area in [m$^2 \cdot 10^{-20}$]; γ = surface tension at the CMC in [mN m^{-1}]

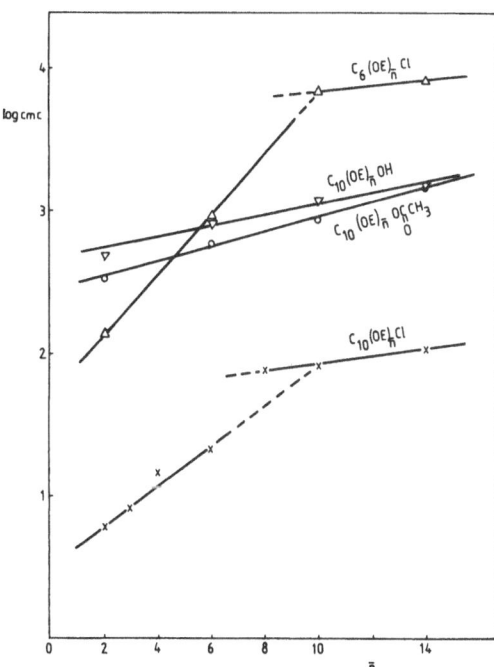

Fig. 3. Effect of the polyoxyethylene chain length on the critical micelle concentration for various product series

Table 2. Effectiveness of one oxyethylene unit for different product series

Products	Gradient $\dfrac{\Delta \log CMC}{\Delta n}$	Equivalent alkyl length Δm_{eq}	$\bar{n}$ range
$C_{10}(OE)_{\bar{n}}OH$	0.038	−0.08	2 ÷ 14
$C_{10}(OE)_{\bar{n}}OCCH_3$ $\overset{\|}{O}$	0.052	−0.11	2 ÷ 14
$C_{10}(OE)_{\bar{n}}Cl$	0.141	−0.30	2 ÷ 6
	0.028	−0.06	8 ÷ 14
$C_6(OE)_{\bar{n}}Cl$	0.212	−0.46	2 ÷ 6
	0.023	−0.05	10 ÷ 14

length of alkyl chain in a standard homologue series, e.g. in hexaoxyethylene glycol monoalkyl ethers for which:

$$\log CMC = A + B\,m$$

where m denotes the number of carbon atoms in the alkyl chain, CMC is given in μmol dm^{-3}, and $A = 7.79$, $B = -0.47$ are the constants calculated from literature data [5].

Similarly, for a homologue series of products investigated containing different number of oxyethylene units, $\bar{n}$:

$$\log CMC = a + b\,\bar{n}$$

where a, b are constants, while b is the gradient given in Table 2.

By combining both the equations in their differential forms, the length of an alkyl chain equivalent to 1 oxyethylene unit can be obtained:

$$\Delta m_{eq} = \frac{b}{B}, \quad [\Delta \bar{n} = 1].$$

The calculated values of Δm_{eq} (Table 2) show that for polyoxyethylene alcohols, their acetates, and for

chlorides containing more than 8–10 oxyethylene units, 1 oxyethylene unit is equivalent to $-(0.05 \div 0.11)$ methylene groups, while for chlorides containing less than 8–10 oxyethylene units, this contribution increases to $-(0.30 \div 0.46)$ methylene groups. The abrupt change of the relative hydrophilicity of the polyoxyethylene chain for $\bar{n} = 8$–10 may be connected with the change of their form from a zig-zag to a meander one.

Considering the same standard series of hexaoxyethylene glycol monoalkyl ethers, and assuming the ordinary gradient of $\Delta \log CMC/\Delta n = 0.047$ as calculated from the literature data [4, 5] the following relation can be obtained:

$$\log CMC = 7.79 - 0.47\,m + 0.047\,(n - 6).$$

Rearranging this equation, the following relation for the equivalent length of an alkyl chain can be obtained:

$$m_{eq} = 15.97 + 0.1n - 2.13 \log CMC.$$

Introducing into this equattion the values of CMC (in μmol dm^{-3}) and average numbers of oxyethylene units, $\bar{n}$, for the products considered, the equivalent length of their hydrophobes was calculated (Table 3). For polyoxyethylene alcohols, these values are near the actual length of the alkyl chain, which supports the validity of the CMC determination. The values for acetates are only slightly higher, for about 0.2–0.3 [CH$_2$]. Thus the acetyl group practically does not change the products' hydrophilicity. However, for chlorine derivatives, much higher values of m_{eq} were obtained, especially for products containing a short

Table 3. Equivalent length of the hydrophobe for polyoxyethylene alcohols, their acetates and chlorides $C_m(OE)_{\bar{n}}X$

No.	Products		$X = -OH$	$X = -OCCH_3$ $\overset{\parallel}{O}$	$X = -Cl$
	m	$\bar{n}$			
1	6	2.1	–	–	11.60
2	6	6.0	6.35	7.80	10.25
3	6	10.7	–	–	8.86
4	6	14.2	–	–	9.04
5	10	0	10.58	–	–
6	10	2.2	10.48	10.80	14.61
7	10	3.0	–	–	14.47
8	10	4.1	10.43	10.65	14.00
9	10	6.1	10.38	10.64	13.76
10	10	8.2	10.39	10.66	12.78
11	10	10.4	10.45	10.75	12.95
12	10	14.1	10.62	10.64	13.11
13	14	5.9	14.35	14.02	15.87
14	14	14.1	14.44	–	15.06

polyoxyethylene chain. Thus, in the last case, the effective length of hydrophobe is greater for about 4–5 methylene groups in comparison to the polyoxyethylene alcohols. This hydrophobic effect of the chlorine atom is lower for products with a longer polyoxyethylene chain ($\bar{n} > 8 - 10$) and is equivalent to approximately 2 methylene groups.

References

1. Becher P (1967) In: Schick MJ (ed) Micelle formation in aqueous and nonaqueous solution, Nonionic Surfactants. Dekker M Inc, New York, p 478
2. DGF Einheitsmethoden C-V 17b
3. Ger Pat 25 44 707 (1975)
4. Miesiac I, Merkwitz H, Szymanowski J, Berger J (1986) J Colloid Interface Sci 114:425–431
5. Mukerjee P, Mysels KJ (1971) Critical Micelle Concentrations of Aqueous Surfactant Systems. National Bureau of Standard 36, Washington
6. Schönfeldt N (1976) Grenzflächenaktive Äthylenoxid Addukte. Hanser, München
7. Szewczyk H, Szymanowski J (1982) Tenside Detergents 19:357–362
8. Szymanowski J, Miesiac I (1985) Tenside Detergents 22:230–239
9. Szymanowski J, Szewczyk H, Atamańczuk B (1984) Tenside Detergents 19:357–362

Authors' address:

Prof. J. Szymanowski
Technical University of Poznań
Institute of Chemical Technology and Engineering
pl. Skłodowskiej-Curie 2,
60-965 Poznań, Poland

Progress in Colloid & Polymer Science   Progr Colloid Polym Sci 76:101–105 (1988)

Quasi-long-range order in swollen lyotropic lamellar phases

J. Marignan, J. Appell, P. Bassereau, and G. Porte

Groupe de Dynamique des Phases Condensées (CNRS UA 233) U.S.T.L., Montpellier, France

Abstract: Information on interactions involved in the stability of lamellar phases are obtained by X-ray, neutron and light scattering investigations. In the phases swollen by brine, stability seems to be driven by thermally induced steric interactions. In the phases swollen by oil, the experimental results indicate that long-range repulsive interactions are responsible for the stability of the stack of lamellae.

Key words: Surfactant, bilayer, liquid crystal, small angle X-ray scattering, light scattering.

1. Introduction

Much attention has been focused on understanding inter-membrane interactions in multilayered systems. Lamellar phases in lyotropic systems involving surfactants can provide a good model for this system.

Indeed, they consist of a periodic stacking of hydrophilic and hydrophobic layers where the mean repeat distance $\bar{d}$ can often be varied to a great extent by addition of the appropriate solvent. There are many examples in the literature of an initial lamellar phase L_α with a moderate pitch incorporating large amounts of water or brine [2, 5–7, 15, 16] or oil [4, 10, 17]. The resulting mixture consists of thin hydrophobic (conversely hydrophilic) layers separated by very thick water (conversely oil) layers.

This experimental situation is especially interesting because the average distance $\bar{d}$ between the membranes becomes much larger than the range of the direct microscopic interactions: namely, the attractive Van der Waals forces which quickly decay as d^{-4}, hydration forces, screened electrostatic repulsion.

In the absence of unscreened electrostatic interactions, it has been demonstrated [9, 12] that the entropically induced undulation forces are dominant in the process of stability of these phases. This long-range interaction, first proposed by Helfrich [8], scales as d^{-2}. On the other hand, in the presence of unscreened electrostatic interactions, on expects [12] this last interaction to dominate for long distances and the swelling of the L_α phase to be no longer governed by thermal fluctuations [16].

2. Water or brine dilution

We have studied three different systems:

a) Cetylpyridinium chloride (CPCl)/1-hexanol/brine (0.2 M NaCl)

b) Sodium octylbenzene sulfonate (OBS)/1-pentanol/brine (0.5 M NaCl)

c) n-Dodecylbetaine/1-pentanol/water.

In the systems (a) and (b), the electrostatic repulsions are strongly screened by the high concentrations of free ions in the solvent (Debye length < 7 Å). In the last one, (c), the surfactant is a neutral zwitterionic molecule. Thus, we suspect that the electrostatic interactions can be neglected at high dilution.

The brine (water) rich corner of the phase diagrams has been determined [1, 13]. As an example, the phase diagram of the system CPCl/hexanol/brine is represented in Fig. 1. We can note that there is a strong analogy between these different systems [15]. Firstly, the lamellar phase characterized by its birefringence at rest appears stable up to very high dilution. The long-range orientational order persists even for large distances between the membranes ($\bar{d} \gtrsim 1000$ Å). Moreover, the radial geometry of the phase diagram indicates that a true dilution of this lamellar phase can be performed with brine (or water for the system (c)). This allows us to study the stability of the stacking lamellae when the interlamellar distance increases, the proper characteristics of the individual membrane remaining unchanged.

These multilayered systems, like smectics, exhibit a singularity in the structure factor which characterizes

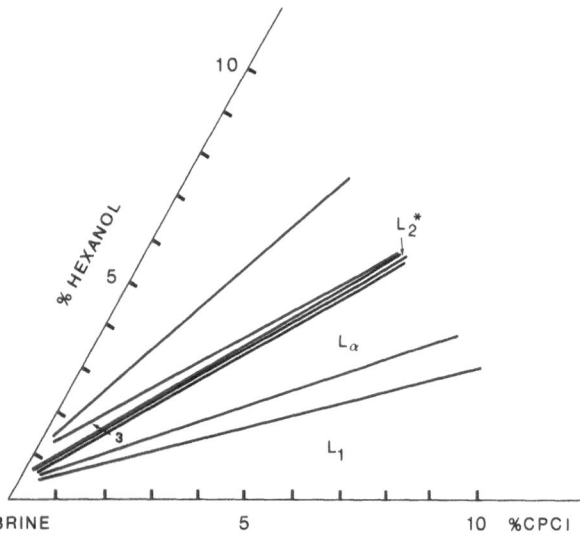

Fig. 1. Brine-rich corner of the phase diagram of the system CPCl/hexanol/brine (0.2 M NaCl) ($T = 30\,°C$). L_1 is a direct isotropic phase, L_α is the lamellar phase and L_2^*, the "anomalous" isotropic phase

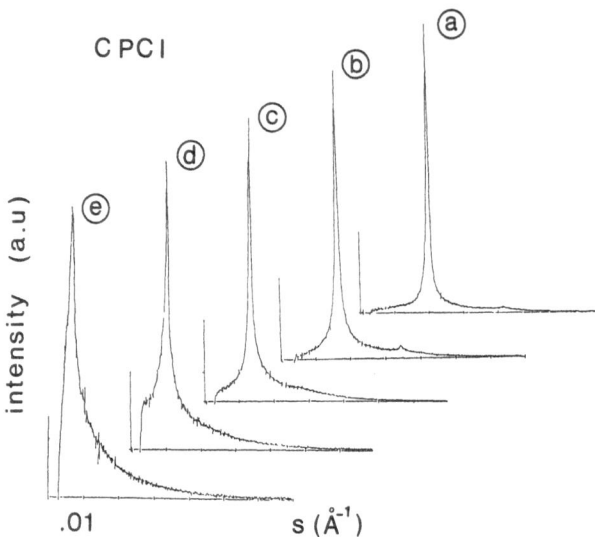

Fig. 2. Evolution of the longitudinal X-ray profiles with the volume fraction of brine ϕ_w in the CPCl system. The molar alcohol/tensioactive ratio is 3.15. (a) $\phi_w = 46\%$, $d = 49$ Å; (b) $\phi_w = 54\%$, $d = 58$ Å; (c) $\phi_w = 63\%$, $d = 64$ Å; (d) $\phi_w = 72\%$, $d = 98$ Å; (e) $\phi_w = 81\%$, $d = 153$ Å

the quasi-long-range translational order. Scattering experiments have been performed to observe the evolution of this positional order with the swelling of the phase. Figures 2–4 give the evolution with the dilution of the X-ray scattering profiles obtained using the standard SAXS set up in our laboratory. For the three systems, we can observe the persistence of the first Bragg singularity and a progessive broadening of the tail of the peak. Such a behavior has also been seen in other systems by Safinya, Roux et al. [17]. It is noticeable that this broadening is less pronounced in the case of the CPCl than for the other systems.

We have performed neutron scattering experiments on the line D 11 at I.L.L. (Grenoble) on the most diluted samples of the CPCl system. The first Bragg peak was observed up to wave vectors corresponding to intermembrane distances of about 1100 Å.

The linear variation of the position of the first Bragg peak ($s_B = 1/d$) with the dilution ϕ_w (Fig. 5) confirms that the individual membrane remains unchanged along the dilution path [1].

At any dilution, the lamellar structure was also indicated by the presence of focal conics in disoriented samples.

More quantitatively, the peak-to-tail intensity ratio can provide information about the interactions between the lamellae. Indeed, as in the smectic-A liquid crystal phase, the structure factor exhibits a power low

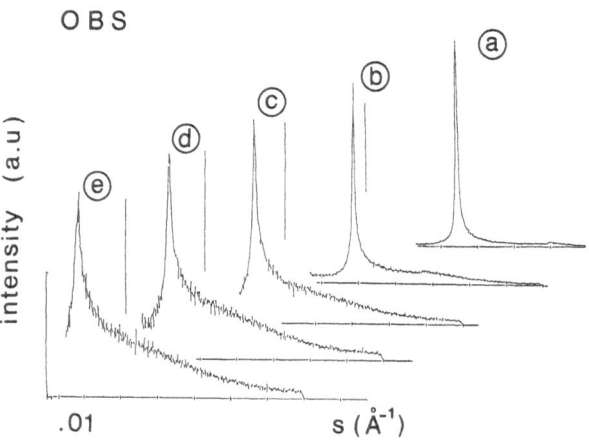

Fig. 3. Evolution of the longitudinal X-ray profiles with the volume fraction of brine ϕ_w in the OBS/pentanol/brine system. The molar alcohol/tensioactive ratio is 3.5. (a) $\phi_w = 51\%$, $d = 40$ Å; (b) $\phi_w = 63\%$, $d = 52$ Å; (c) $\phi_w = 70\%$, $d = 71$ Å; (d) $\phi_w = 75\%$, $d = 84$ Å; (e) $\phi_w = 80\%$, $d = 112$ Å

behavior near the nominal position of the Bragg peak q_B which was first calculated by Caillé [3]:

$$S(0, 0, q_z) \, \alpha (q_z - q_B)^{-(2-x)} \tag{1}$$

where the exponent x is directly related to the effective potential between the membranes.

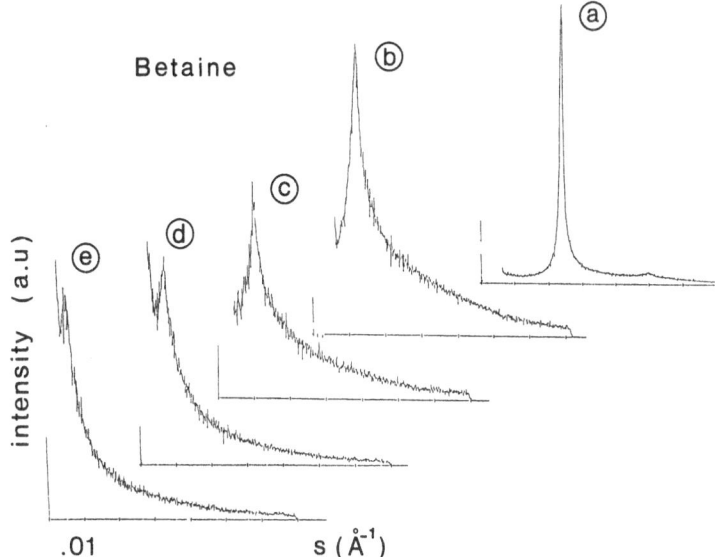

Fig. 4. Evolution of the longitudinal X-ray profiles with the volume fraction of water ϕ_w in the betaine system. The molar alcohol/tensioactive ratio is 3.2. (a) $\phi_w = 50\%$, $d = 41$ Å; (b) $\phi_w = 70\%$, $d = 83$ Å; (c) $\phi_w = 75\%$, $d = 111$ Å; (d) $\phi_w = 81\%$, $d = 131$ Å; (e) $\phi_w = 94\%$, $d = 169$ Å

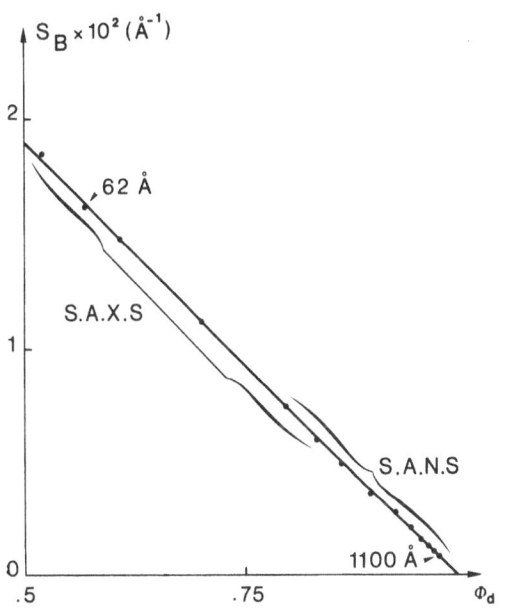

Fig. 5. Position of the Bragg peak ($s_B = 1/d$) as a function of the volume fraction of diluant (brine) in the CPCl system. The same evolution is found for the other two systems

Leibler and Lipowski have demonstrated [12] that in the absence of long-range microscopic interactions (for example, an electrostatic potential), the stability of the phase is governed by the thermally induced fluctuations of the membranes.

In this case, the only relevant potential is similar to the steric potential, first evaluation by Helfrich [8]:

$$V_{\text{und}} \propto \frac{(kT)^2}{k_c} \frac{1}{d^2} \qquad (2)$$

where k_c is the bending rigidity modulus of a single membrane. Then, for a finite thickness δ of the lamella:

$$x = a \left(1 - \frac{\delta}{d}\right)^2. \qquad (3)$$

For large d, the exponent x must behave like a constant independent of the system. Safinya, Roux et al. have obtained an asymptotic value for x of 1.33 for two different systems [16,17], which is in good agreement with the evaluation of Helfrich. Our X-ray results indicate firstly that the intensity scattered around the Bragg position actually has a power law decay; but, on the other hand, x seems to reach a system-dependent asymptotic value for the most swollen samples: for the CPCl system, the constant a is found to be equal to 1.3 but it is quite different for the other two [14]. The discrepancy which is not predicted by the actual theories is not yet understood.

3. Oil dilution

Previous works of our group [10,11] have shown that the lamellar phase of the ternary system — OBS/n-pentanol/water — can be diluted to a great extent with an appropriate mixture of decane and pentanol. In this case, the aliphatic solvent swells the hydrophobic part of the initial liquid crystal and the lamella then consists of a thin water layer embedded between two layers of surfactant and alcohol. This system is noteworthy for the high degree of dilution which can be obtained; indeed, optical birefringence persists even in samples containing only 0.3 % of the initial liquid crystal. It provides a good opportunity to study the evolution of the diffraction pattern of multilayered systems over a large range of interlayer distances $\bar{d}$.

For moderate dilutions (up to a spacing $\bar{d} \simeq 200$ Å), the X-ray scattering profiles are very similar to those of brine swollen systems (Fig. 6). They exhibit one peak only and a broadening of the tail of the peak with the dilution.

On the contrary, for very high volume fractions of the diluant ϕ_d (97.7 % < ϕ_d < 99.7 %), the diffraction pattern is completely different. The light scattering profile exhibits *two* Bragg peaks, up to repeat distances of 1.5 μm (Fig. 7). Moreover, the position of the peaks are in the ratio 1 : 2, which is well in favor of a lamellar structure.

The position of the first Bragg peak varies linearly with the volume fraction of diluant (Fig. 8) for mean interlamellar distances 2000 Å < $\bar{d}$ < 15000 Å, which confirms the unidimensional expansion of this phase, and also proves that the addition of the mixture decane/pentanol that we used does not affect the composition of the membrane over this whole range of dilution. In particular, the thickness of the lamella remains constant; it has been found to be equal to 40 Å, which is not far from the pitch of the initial liquid crystal (35 Å).

The appearance of the second order Bragg peak at large spacing is quite puzzling. If the stability of the

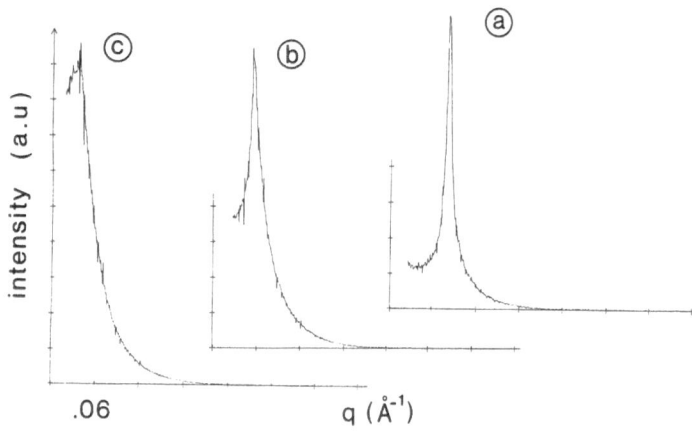

Fig. 6. Evolution of the X-ray profiles with the volume fraction of solvent ϕ_d (decane + pentanol) in the OBS/decane/pentanol/water system ($q = 2\pi S = 2\pi/d$). (a) $\phi_d = 50$ %, $d = 69$ Å; (b) $\phi_d = 65$ %, $d = 102$ Å; (c) $\phi_d = 76.6$ %, $d = 155$ Å

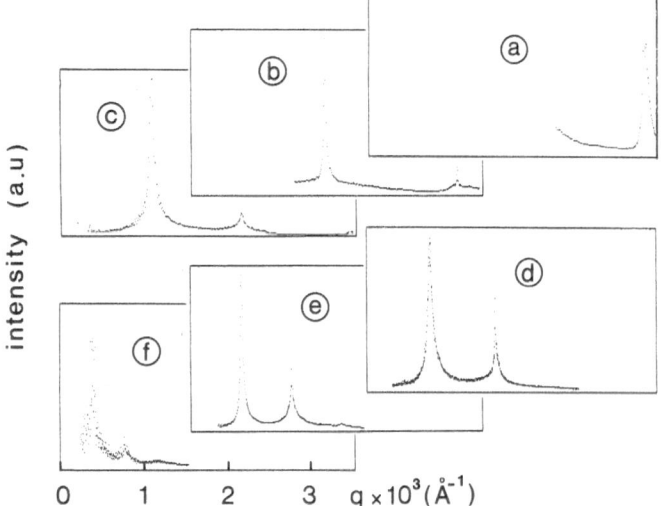

Fig. 7. Evolution of the light scattering profiles with the volume fraction of solvent ϕ_d for the most oily swollen samples of OBS system. ($q = 2\pi S = 2\pi/d$). (a) $\phi_d = 97.72$ %, $d = 1900$ Å; (b) $\phi_d = 98.86$ %, $d = 3950$ Å; (c) $\phi_d = 99.24$ %, $d = 5850$ Å; (d) $\phi_d = 99.43$ %, $d = 8400$ Å; (e) $\phi_d = 99.545$ %, $d = 10400$ Å; (f) $\phi_d = 99.70$ %, $d = 16200$ Å

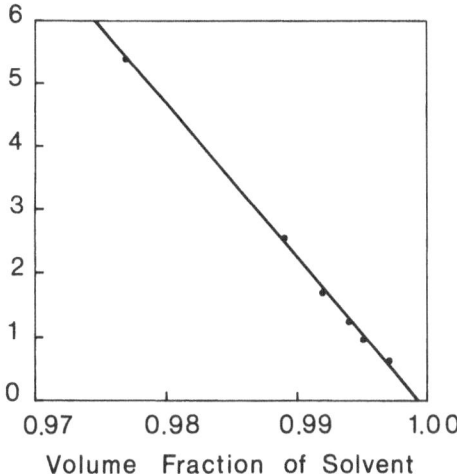

Fig. 8. Position of the first Bragg peak $(s_B = 1/d)$ of the light scattering experiments as a function of the volume fraction of solvent in the OBS/pentanol/water/decane system

phase is governed by the undulation forces, the exponent x_2 of the second order must be equal to $4a$ [17]. (Here, the thickness of the membrane is negligible with respect to the interlamellar distance d.) But, for the predicted a-values, $4a$ is much larger than 2 ($4a$ is equal to 5.3 for Helfrich) which does not permit any divergence of the intensity at $q = 2q_B$.

To account for our results, a type of interaction scaling as d^{-1} must be considered. In this case, the exponent x varies as $d^{-1/2}$ [12] and is thus asymptotically equal to 0; this would allow 2 singularities at large dilution in the diffraction pattern.

Although the membrane should be a priori neutral in systems swollen by oil, electrostatic interactions (scaling actually as d^{-1}) could be dominant here. Indeed, Wennerström [18] proposes that the double layer forces can be induced by lipophilic (OBS) ions present in the apolar medium. In the present case, the solubility of hydrophobic ions in the alkane can be increased both by the presence of substantial quantities of alcohol, which increase the dielectric permittivity ε_a, and by the large Bohr radius of the ion.

Nevertheless, we have as yet no experimental evidence of this process.

4. Conclusion

In summary, we have experimentally demonstrated that the large swelling of classical lyotropic lamellar phases, either by water or by oil, is a unidimensional expansion process if the right solvent is used. The sta-

bility of the stack of non-electrostatic membranes embedded in brine or water is governed by the entropically driven steric interactions. But the specificity between the different chemical systems has not yet been explained.

On the other hand, our results on the decane hyperswollen system cannot be explained by the fluctuations-induced interactions in the actual position of the theories. Their predictions, which are confirmed for moderate spacing ($d < 200$ Å) could be incomplete for enormous dilution ($d \simeq 15000$ Å).

Important experimental work has still to be performed to draw conclusions about the dominant interaction in this system: steric, electrostatic...

Acknowledgements

This research has received partial financial support from a PIRSEM (CNRS) grant number no. AIP 2004.

References

1. Bassereau P, Marignan J, Porte G (1987) J Phys 48:673
2. Benton WJ, Miller CA (1983) J Phys Chem 87:4981
3. Caille A (1972) C R Acad Sci B274:891
4. di Meglio JM, Dvolaitzky M, Taupin C (1985) J Phys Chem 89:871
5. Ekwall P (1975) In: Brown GM (ed) Advances in Liquid Crystals. Academic, New York
6. Ghosh O, Miller CA (1987) J Phys Chem 91:4528
7. Gomati R, Appell J, Bassereau P, Marignan J, Porte G (1987) J Phys Chem 91:6203
8. Helfrich W (1978) Z Naturforsch 33a:305
9. Janke W, Kleinert H (1987) Phys Rev Lett 58:144
10. Larche FC, Appell J, Porte G, Bassereau P, Marignan J (1986) Phys Rev Lett 56:1700
11. Larche FC, El Qebbaj S, Marignan J (1986) J Phys Chem 90:707
12. Leibler S, Lipowsky R (1987) Phys Rev B 35:7004
13. Marignan J, Gauthier-Fournier F, Appell J, Akoum F, Lang J (1988) J Phys Chem 92:440
14. Marignan J, Appell J, Bassereau P, Delord P, Larche FC, Porte G (1987) Mol Cryst and Liq Cryst 152:153
15. Porte G, Gomati R, El Haitamy O, Appell J, Marignan J (1986) J Phys Chem 90:5746
16. Roux D, Safinya CR (1988) J Phys 49:307
17. Safinya CR, Roux D, Smith GS, Sinha SK, Dimon P, Clark NA, Bellocq AM (1986) Phys Rev Lett 57:2518
18. Wennerström H, Progr Colloid Polym Sci, in press

Received December 2, 1987;
accepted December 7, 1987

Authors' address:

J. Marignan
Groupe de Dynamique des Phases Condensées
CNRS UA 233
U.S.T.L., Place E. Bataillon
F-34060 Montpellier Cedex, France

Progress in Colloid & Polymer Science   Progr Colloid Polym Sci 76:106–108 (1988)

From isotropic w/o microemulsions to lyotropic mesophases by differential scanning calorimetry

D. Senatra [1])[2]) and Z. Zhou[1])[3])

[1]) Department of Physics, University of Florence, Florence, Italy
[2]) CISM (of the M.P.I.) and GNSM (of the C.N.R.) groups
[3]) Department of Physics, Hubei University, Wuhan, Hubei, People's Republic of China

Abstract: Results are reported on the thermal analysis of biphasic hexadecane/n-hexanol/ K-oleate and water samples with composition falling outside the domain of existence of homogeneous and isotropic microemulsion systems. The very first appearance of macroscopic phase separation of the samples into an isotropic w/o microemulsion and a birefringent mesophase, was investigated by differential scanning calorimetry (DSC). The partitioning of the hydrocarbon and of the "free" water in the two phases of biphasic samples was analyzed.

Key words: Microemulsions, free water, interphasal water, DSC study, biphasic samples, LC mesophases.

Introduction

A water-in-hexadecane system with potassium oleate and n-hexanol in the 3/5 mass ratio, was investigated by differential scanning calorimetry as a function of water addition. The domain of existence of the transparent isotropic and monophasic solubility area of the w/o type of the above system was investigated along a line starting from the oil-rich side of the pseudoternary phase diagram and moving toward the zero oil vertex, by adding water to samples characterized by a given combined surface active agent to oil mass ratio. The concentration interval studied, by encompassing the w/o microemulsion monophasic domain, entered the region in which macroscopic phase separation occured in the samples.

Particular attention was devoted to the monophasic domain limiting concentration at which pretransitional effects were observed, linked with structural arrangements prefiguring the formation of the lyotropic liquid crystalline phase.

Some preliminary results are reported about the partitioning of both hydrocarbon and water in the different phases of the samples.

Experimental

The experimental path followed in the present study (AB line) is shown in Fig. 1, in which the pseudoternary phase diagram of the actual system is plotted for the mass ratio surfactant-to-cosurfactant of 0.6. Each sample on the AB line is characterized by a constant mass ratio (K-oleate + n-hexanol)/hexadecane = 0.68. The sample water content is expressed by the weight ratio: C = water/ total sample. The concentration interval studied extends up to C = 0.42.

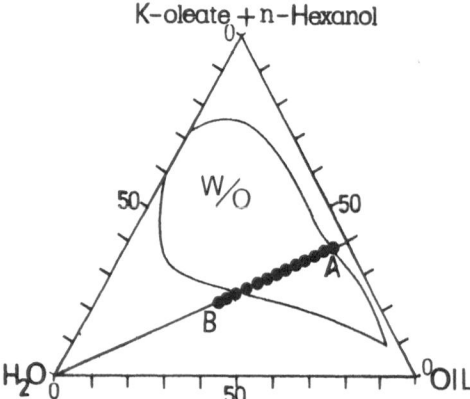

Fig. 1. Pseudoternary phase diagram of the system studied. AB line: experimental path followed in the present study

Two procedures were adopted to formulate the samples, namely: by preparing two twin specimens with the same composition, and by splitting into two subsamples a freshly formulated one immediately after vigorous stirring. The samples were kept into a thermostatted chamber at 301 K for 1 year to allow both the completion of the process of phase separation and the two phases to reach a state of equilibrium.

Since in biphasic samples the proportions of the two phases depend critically upon the temperature, any handling was made inside the thermostatted chamber. The isolation of each of the two phases was achieved by "soaking" the upper isotropic part of the biphasic samples with a syringe equipped with a teflon tube. For homogeneity of procedure with the monophasic w/o systems, the amounts required for the measurements of the isotropic parts were taken from both the divided and the undivided biphasic samples. Observation of the phases was made visually with the help of two crossed polaroids and a white light source.

The study of the thermal properties of the different phases was performed with differential scanning calorimetry (DSC) by means of a Mettler TA 3000 thermal analyzer equipped with a DSC-30 low temperature cell. The heat flow rate (dQ/dt) as a function of increasing temperature was recorded, at constant pressure, during the controlled heating $(dT/dt = 8$ K/min) of previously frozen samples (DSC-ENDO). Greater details on the DSC measuring procedure, as well as on previous findings of the actual system, may be found in Refs. [1–3].

The criterion for checking the accomplishment of a correct division of each of the two phases of a given sample was as follows:

1. The known weight of hexadecane in the "in toto" sample was assumed as the reference datum, $(g_{oil})_{tot.}$, and its partitioning in both the isotropic and the liquid crystalline phase was considered specifically;

2. From the measured latent heat of fusion (enthalpy) associated with the melting of the hydrocarbon, the weight of oil involved in the endothermic process was calculated. For each sample, two such measurements were performed, namely, one for the isotropic $(g_{o, is})$ and, one for the LC mesophase $(g_{o, LC})$.

3. The summation of the weights of oil calculated from the experimentally measured enthalpies, was therefore compared with the "a priori" known weight of hydrocarbon contained in the sample as a whole. If $(g_{oil})_{tot.} = (g_{o, is}) + (g_{o, LC}) \pm 5\%$, the division of the two phases was assumed as correct.

Results and discussion

The behaviour of the enthalpy of both the hexadecane (ΔH_h) and the "free" water (ΔH_w) [1, 2] as a function of increasing concentration is plotted in Fig. 2, in which the values are expressed in KJ/mol of sample for oil and water, respectively. Figure 2 summarizes the results analyzed at the present stage of the research on both the w/o microemulsions [1] and the isotropic upper phases of each sample studied. The lines to the left are the computed regression lines on the experimental points. Bars are standard errors of the mean of six independent measurements.

The very first macroscopic phase separation occured at $C_{tot} = 0.372$ with the formation of an extre-

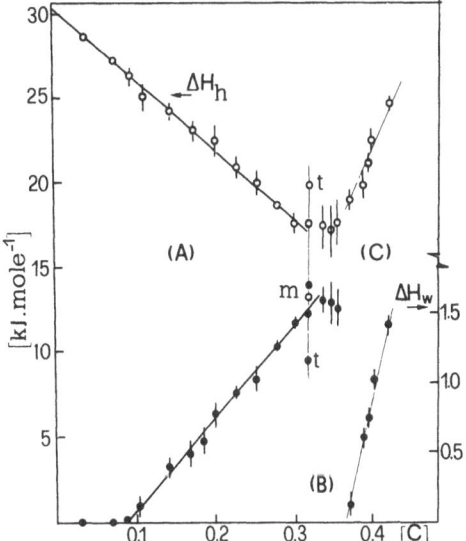

Fig. 2. Enthalpy changes against increasing concentration. ΔH_h-hexadecane (O); ΔH_w — free water (●). The three regions (A) (B) and (C) are characterized as follows: (A) Linear relationship between the measured enthalpy and the concentration. Lines are least squares fit regression lines on the experimental points. (B) Pretransitional concentration-interval within which the linear relationship does not hold. (C) Appearance of biphasic samples. The data refer to the isotropic phases only. Concentrations are C_{tot}

mely thin lens at the bottom of the test tube. The latter, if observed between two crossed polaroids, displays "birefringence".

Pretransitional effects prefiguring the phase separation, were found in the interval $0.316 \leq C_{tot} \leq 0.372$ in which the samples were still transparent and looked homogeneous. However, by taking the amounts required for DSC measurements from regions of the test tubes differing in height, it clearly emerged that within the aforementioned concentration interval, the samples, although macroscopically isotropic, were not homogeneous. In fact, as shown in Fig. 2 by the experimental points marked with "t" (top), "m" (middle) and "b" (bottom), the DSC results depended on the level at which the specimens were taken in the test tube. The top specimens always displayed a less free water and a higher hexadecane content than the bottom ones. The thermal curves of these samples appear of the same type as the ones recorded on homogeneous w/o microemulsions (see 'f.i.' Fig. 3 of Ref. [4]). Any attempt to reach a phase separation in these samples failed, even by temperature processing, centrifugation and sonication.

The DSC analysis of the isotropic upper phase of the very first biphasic sample $(C_{tot} = 0.372)$ showed

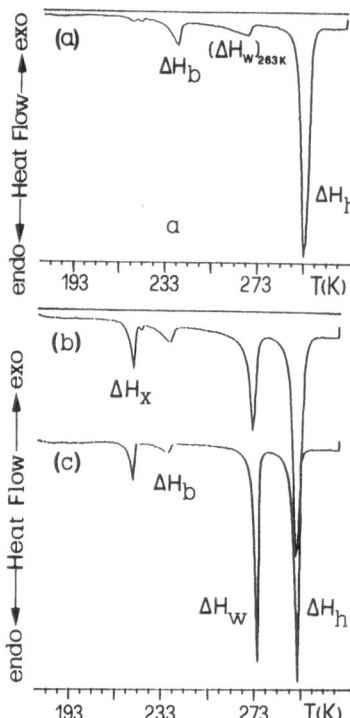

Fig. 3. Thermal curves of isotropic upper phases of biphasic samples. (a) DSC recording of the isotropic part of a sample with $C_{tot} = 0.372$ at which the very first appearance of a macroscopic phase separation was observed. (b), (c) DSC thermograms of upper isotropic phases with higher water content. $C_{tot}^b = 0.388$, $C_{tot}^c = 0.419$

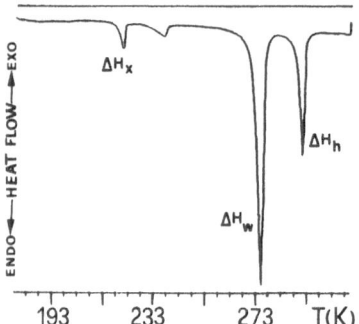

Fig. 4. DSC curve of a liquid crystalline "bottom" mesophase of the sample with $C_{tot}^c = 0.419$

that this system is distinguished by an "interphasal" water endotherm at $T = 263$ K [2] and, practically, by the disappearance of the contribution due to the n-hexanol (ΔH_x) (Fig. 3a). The thermal curves of systems, the isotropic upper phases of which possess a higher water content, exhibit the same characteristics found for ordinary w/o microemulsions (Fig. 3b, c).

A typical example of a DSC spectrum of a liquid crystalline phase of a sample showing two well separated phases, is plotted in Fig. 4. The bottom LC phase was found to contain the 12.22 % of hexadecane and 25.20 % of free water, with respect to the 34.55 % and the 41.93 % contained in the sample as a whole, with $C_{tot} = 0.419$.

In the two divided parts of the biphasic samples, the DSC data of both the isotropic and the liquid crystal-line phases, were found not to depend on the level at which the specimens were taken from the test tubes.

Conclusions

Differential scanning calorimetry was found to be a very effective tool for both the identification of the domain of existence of homogeneous w/o microemulsions and for analysis of the partitioning of different components in either phase of biphasic samples.

References

1. Senatra D, Zhou Z, Pieraccini L (1987) Progr Colloid Polym Sci 73:66
2. Senatra D, Guarini GGT, Gabrielli G, Zoppi M (1984) J Phys Paris 45:1159
3. Senatra D, Gabrielli G, Guarini GGT (1986) Europhys Lett 2:455
4. Senatra D, Garielli G, Caminati G, Zhou Z (1988) IEEE Transactions on Electrical Insulation, in press

Received December 3, 1987;
accepted December 11, 1987

Authors' address:

Prof. S. Donatella
Physics Department
University of Florence
Largo E. Fermi, 2 (Arcetri)
50125 Florence, Italy

Progress in Colloid & Polymer Science

Progr Colloid Polym Sci 76:109–112 (1988)

A fluctuation wave mechanism of membrane electrofusion

D. S. Dimitrov and D. V. Zhelev

Central Laboratory of Biophysics, Bulgarian Academy of Science, Sofia, Bulgaria

Abstract: Fusion requires close approach and destabilization of the membranes and the intervening liquid layer. Due to thermal motion or other reasons the thickness of the membranes and the distance between them fluctuate. External electric fields can change the amplitude of the fluctuations and induce instability. On one hand, the field can induce attractive forces between the membranes. On the other hand, it can destabilize and rupture the membranes themselves. The kinetics of the fluctuation wave growth depends on the viscosities of the membranes and the medium between them, as well as the other physical parameters and the amplitude of the applied electric field. We suggested a simple model to calculate the characteristic times of unstable fluctuation waves which can lead to fusion. The basic conclusion is that the fluctuation-wave mechanism is kinetically favorable for establishing close local contact between the membranes. It can also be responsible for formation of hydrophilic pores in the membranes which can serve as sites for inducing fusion. This mechanism is operating for the initial stages of electrofusion. These results do not disprove other possible mechanisms.

Key words: Membranes, electrofusion, fluctuation waves.

Introduction

Fusion requires close approach and destabilization of the membranes and the intervening liquid layer. Due to thermal motion or other forces, the thickness of the membranes and the distance between them fluctuate. External electric fields can change the amplitude of the fluctuations and induce instability. On one hand, the field can induce attractive forces between the membranes; on the other, it can destabilize and rupture the membranes themselves. The kinetics of the fluctuation wave growth depend on the viscosities of the membranes and the medium between them, the amplitude of the applied electric field and on other physical parameters.

We suggested two types of models to calculate the characteristic times of unstable fluctuation waves which can lead to fusion [1–7]. The more simple model [1] considers the instability of the liquid films between the membranes, which are assumed to behave as two-dimensional bodies, characterized by their membrane tensions and bending elasticities. The more sophisticated development of the fluctuation wave mechanism of membrane fusion [6] describes the membranes and the intervening film as three-dimensional viscoelastic bodies. The fusion occurs as an instability phenomenon of this multilayred system. This model predicts that in electrofusion the most unstable region is that of contacting membranes and therefore fusion of the contacting membranes will take place before the rupture of the non-contacting membranes. This model, however, is too complicated. It requires a lot of mathematical computations and a couple of physical parameters, which are commonly unknown.

Recently, we estimated, both experimentally and theoretically, the order of magnitude of the attractive intermembrane force induced by AC electric fields and its possible functional dependence on the distance between the two membranes for large separations [7,8]. In another work [9] we found experimentally that the dependence of the breakdown voltage on the pulse duration to induce electroporation of single cell membranes is very similar to that for inducing fusion. The theoretical estimates showed that a fluctuation wave mechanism may be responsible for pore formation induced by the electric field [10].

This communication explores the feasibility of a mechanism of membrane electrofusion, where the strong attractive force between the membranes is induced by the high voltage DC pulse. The pulse, on one hand, destabilizes the membranes, and, on the other, induces instability of the liquid layer between the membranes, due to the strong attraction of the polarized cells.

The model

We will consider two fluctuating membranes (Fig. 1). The condition for instability of the liquid film between the membranes has the form [1]

$$\tau^{-1} = \omega = (h^3 k^2 / 24\mu)(2dP/dh - Tk^2 - Bk^4) \quad (1)$$

where ω is the angular velocity, which is inversely proportional to the characteristic time, τ, of fluctuation growth; μ is the liquid viscosity, h the average separation between the membranes, k the wave number, which is inversely proportional to the wave length; T membrane tension, B bending elasticity modulus and P the pressure which approaches the membranes and destabilizes the intervening liquid layer. The value of this pressure is essential for membrane fusion and will be considered in more detail.

The forces

One of the basic problems in understanding mechanisms of membrane fusion is how the forces which

resist membrane approacch can be overcome. The hydration forces are very strong at close approach, but the approaching membranes are deformed and the viscous resistence to motion becomes extremely high at close separation. It is reasonable to suppose that during mutual dielectrophoresis (or any other way of initial membrane approach) membranes can overcome the electrostatic barrier, due to double layer forces, and experience very strong resistance at closer separation, of the order of 1 nm, where hydration forces dominate.

Therefore, we will consider only the interplay between "polarization" forces (induced by the external electric field) and the hydration forces. We found the following expression for the attractive polarization force at close approach [8]

$$F_p = (AE/h)^2 \quad (2)$$

where A is a constant and for a protoplast of radius R_s = 18.4 μm is equal to 10^{-14} $N^{1/2}$m^2/V. The approaching membranes deform and commonly form a region in which the membranes are almost flat. The radius of this region is of the order of R_s especially when the external deforming force is large. In this case, the effec-

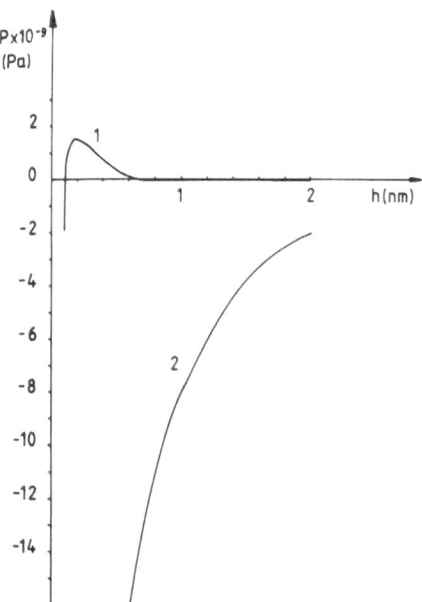

Fig. 2. Dependence of the pressure P on the separation h for two different field intensities: curve (1) $E = 2$ kV/m (which is typical for dielectrophoresis and induces cell aggregation) and curve (2) $E = 20$ kV/m (which is sufficient to induce breakdown and fusion)

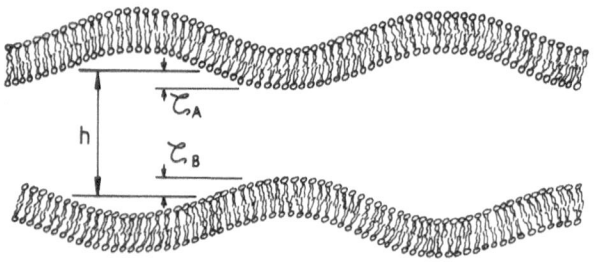

Fig. 1. Sketch of two fluctuating membranes (see text)

tive pressure, due to polarization forces, P_p, which acts on the flat membranes, can be estimated as

$$P_p = F_p/S = F_p/(R_s)^2 \tag{3}$$

where S is the contact area.

Equations (2) and (3) lead to

$$P_p = (AE/hR_s)^2 = C(E/h)^2 (N/m^2) . \tag{4}$$

The total pressure P acting on the film between the membranes is

$$P = P_h - P_p = P_o \exp(-h/l) - C(E/h)^2 \tag{5}$$

where typical values for the constants P_o and l are 10^{10} N/m^2 and 0.193 nm [11] and $C = 2 \cdot 10^{-19}$ Nm^2/V^2. Figure 2 shows the dependence of the pressure P on the separation h for two different field intensities: $E = 2$ kV/m (which is typical for dielectrophoresis to induce cell aggregation), $E = 20$ kV/m (which is sufficient to induce breakdown and fusion). It is seen that for small voltages the hydration barrier is higher than the polarization force. For high voltages, the attractive force dominates for any separation distance. This means that the membranes will approach each other.

The time of continuous membrane approach

The characteristic time, τ_a of this approach can be calculated by using the Reynolds formula, which, after integration, reads

$$\tau_a = \mu R^2 \ln(h_i/h_f)/CE^2 \tag{6}$$

where h_i and h_f are the initial and final separations, respectively. For $E = 10^5$ V/m it yields times of the order of Milliseconds, and times of the order of seconds for lower intensities ($E = 0.2 \cdot 10^5$ V/m).

The time of fluctuation wave growth

The times of continuous membrane approach are rather long. Therefore, there a faster kinetic mechanism can exist. The calculation of the characteristic time of growth of the dominant (fastest) fluctuation wave, τ_f, on the basis of Eq. (1) for small bending elasticities, gives

$$\tau_f = 24\mu T/h^3 (dP/dh)^2 . \tag{7}$$

We assume that the marginal thickness, h_o at which the fluctuation begins to grow, is of the order of 100 nm, $T = 1$ mN/m and $\mu = 1$ mPa.s. Then, for high voltages ($E = 10^5$ V/m), we obtain times of the order of nanoseconds, and times are of the order of microseconds for intermidiate intensities ($E = 10^4$). Therefore, for those values of the electric fields where the breakdown is observed, the characteristic time of fluctuation wave growth is much faster than the continuous uniform thinning of the film between the membranes. For smaller thicknesses, the ratio of the two times is even smaller.

The critical membrane separation

We will use a formula for the marginal and critical thickness of rupture of the film between the membranes, which is based on Eqs. (1) and (6) [1]. For the present model system, this formula reduces to

$$h_o/h_{cr} = 2^{1/3} \tag{8}$$

$$h_{cr}^3 = C(ER)^2/T . \tag{9}$$

For typical values of the field intensity, this formula gives large marginal and critical thicknesses in the range 10 to 100 μm. This means that the fluctuations start to grow immediately after pulsation if the membranes are at close approach.

Discussion

We would like to emphasize two important aspects of this model:

(1) The polarization force was estimated for large separations, but the expression (4) was used also for small separations; there are no reasons why this expression should not be valid; however, there are no rigorous experimental and theoretical verifications.

(2) The suggested mechanism is kinetic — it is based on the competition between the growth of the unstable fluctuation waves and the continuous approach. It assumes that there is no activation barrier between the initial and final states. This means that the resistance is purely of a viscous (diffusion) type and the contact of the two membranes will always occur if the driving force is positive, i.e. tends to cause membrane approach.

One of the basic conclusions is that the fluctuation-wave mechanism is kinetically favorable for establishing close local contact between the membranes. It can also be responsible for formation of hydrophilic pores in the membranes which can serve as sites for inducing fusion. The attraction force, induced by the DC pulses, is sufficient to induce both pore formation and fusion of the membranes. This is one possible explanation for the observed high correlation between the pulse amplitude/pulse length dependences in electric breakdown and electrofusion [9, 12].

The mechanism proposed in this communication describes just one of the possible stages of membrane fusion, that is, the rate of making contact and the type of contact. Therefore it does not necessarily contradict other mechanisms which can describe other stages or the whole event (see, e.g., Refs. [12–15]).

In a previous paper [16], we showed that red blood cells can make very strong contact in conditions characteristic for dielectrophoresis. We even supposed that the membranes may fuse. It seems that in these experiments the polarization force of attraction was sufficient just to overcome the electrostatic barrier. Estimates based on Eq. (4) have shown that the polarization force is of the order of magnitude and larger than the electrostatic repulsion due to the ionic double layers. In this case, fusion would be possible only where there was no hydration barrier. This can occur in sites where large defects of appropriate proteins exist. In addition, the surrounding medium should facilitate this event. This is in agreement with the observation [16] that this tight contact occurred in glycine/glucose, but not in sorbitol solutions. However, even if local fusion had occurred, the mixing of cell contents may have been impossible, because the fusion event is not possible in all the lipid matrix due to hydration forces. Therefore we can conclude that for effective cell fusion, it is necessary to apply fields of higher intensity in order to overcome the hydration barrier.

Acknowledgement

This work was supported by the Committee for Science at the Ministerial Council of Bulgaria through Contract No. 189.

References

1. Dimitrov DS (1982) Colloid Polym Sci 260:1137
2. Dimitrov DS (1983) Progr Surface Sci 14:295
3. Dimitrov DS, Zhelev DV (1984) Colloid Interf Sci 99:327
4. Dimitrov DS, Jain RK (1984) Biochim Biophys Acta 779:437
5. Dimitrov DS, Jain RK (1984b) J Colloid Interf Sci 101:489
6. Dimitrov DS, Zhelev DV, Jain RK (1985) J Theor Bil 113:353
7. Dimitrov DS, Zhelev DV (1985) Studia Biophysica 110:105; (1987) Symposium on Fundamental Mechanisms of Membrane Fusion, Jerusalem
8. Stoicheva N, Tsoneva I, Dimitrov DS (1985) Z Naturforsch 40c:735
9. Zhelev DV, Dimitrov DS, Doinov P (1988) Bioelectrochemistry and Bioenergetics, in press
10. Dimitrov DS (1984) J Membrane Biol 78:53
11. Loosley-Millman M, Rand P, Parsegian V (1982) Biophys J 40:221
12. Zimmermann U (1982) Biochim Biophys Acta 694:227
13. Leikin SL, Kozlov MM, Chernomordik LV, Markin VS, Markin YuA (1986) Chizmadzhey, Biologiheskie Membrani, in Russian 3:1159
14. Sugar IP, Neumann E (1987) Biophys Chem
15. Sowers AE (ed) (1987) Cell Fusion
16. Dimitrov DS, Tsoneva I, Stoicheva N, Zhelev D (1984) J Biol Phys 12:26

Received December 4, 1987;
accepted December 14, 1987

Authors' address:

D.S. Dimitrov
Central Laboratory of Biophysics
Bulgarian Academy of Science
Sofia 1113, Bulgaria

Progress in Colloid & Polymer Science   Progr Colloid Polym Sci 76:113–118 (1988)

Studies on the solubilization of basic polypeptides in reversed micelles by ultracentrifuge measurements

G. Ebert, M. Plachky, M. Senō[1]), and S. Shoji[2])

Fachbereich Physikalische Chemie der Philipps-Universität, Marburg a. d. Lahn, F.R.G.
[1]) Institute of Industrial Science, University of Tokyo, Tokyo, Japan
[2]) Department of Industrial Chemistry, College of Technology, Gunma University, Gunma, Japan

Abstract: The sedimentation behavior of reversed micelles (RM) of the system AOT/ H_2O/iso-octane was studied before and after solubilizing poly(L-lysine).

In contrast to the Schlieren-pattern of the system AOT/H_2O/iso-C_8, those of AOT/ H_2O/(Lys)$_n$/iso-C_8 show two Schlieren peaks. The slowly moving one corresponds to the "empty RM", the other — with a higher sedimentation rate — to the "filled RM". From the size of the area below these peaks, it follows that at a maximum $\approx$ 45 % of the RM are filled with (Lys)$_n$. Also, the diffusion patterns show an overlapping of the diffusion process of these two species. The results obtained by the moving boundary method can be confirmed by equilibrium sedimentation measurements.

As a result of these studies, the number of (Lys)$_n$, AOT, and H_2O molecules in one RM can be determined. In any case, in the filled RM, only one (Lys)$_n$ molecule is solubilized.

Key words: Reversed micelles, sedimentation behavior, polypeptides, solubilization, conformation.

Introduction

Recently Senō et al. [8] have shown that basic polypeptides like poly-L-lysine (Lys)$_n$ poly-L-ornithine (Orn)$_n$, poly-L-arginine (Arg)$_n$, as well as copolymers fo basic α-amino acids with L-leucine, can be solubilized in reversed micelles (RM) of AOT (Bis-(2-ethylhexyl)sulfosuccinate)/water in isooctane (iso-C_8). By circular dichroism measurements it was revealed that in contrast to aqueous solutions containing AOT, in which these polypeptides prefer the α-helical conformation, they attain the β- strucure in the RM mentioned above [6, 8]. Because polypeptides in the β-conformation have the tendency to from n-meric aggregates by forming intermolecular hydrogen bonds, there is the question of whether the polypeptides solubilized in RM also form aggregates or not. In answering this question the determination of the mass of "empty" and "filled" RM by sedimentation measurements with an analytical ultracentrifuge should be very useful.

Materials and methods

Materials

Two samples of poly-L-(lysine) with a $\overline{DP}$ = 600 and 2400 (molar mass $\approx$ 100 000 and $\approx$ 400 000) were obtained by polymerizing the N-carboxyanhydride (NCA) of N-ε-carbobenzoxy-L-lysine (Cbo-Lys) (0.1 mol in 200 ml tetrahydrofuran) with triethylamine as an initiator. The N$_\varepsilon$-cbo-group was removed by treating with HBr/HCl in acetic acid.

Methods

The sedimentation measurements were carried out using an analytical ultracentrifuge (Beckman Instruments, model E) with the Schlieren-pattern device.

Moving boundary

For determining the sedimentation coefficient by the moving boundary method, an aluminium double sector cell placed in a titanium rotor An-H was used. The measurements were carried out at 60 000 rpm and 20 °C.

Diffusion coefficient

For determining the diffusion coefficient by the synthetic boundary method, a capillary-type cell with a centerpiece of carbon-filled epon was used. The synthetic boundary was formed at 4400 rpm and from the diffusion peak 10 exposures were made at 2 min intervals.

Sedimentation equilibrium runs

For determining the mass of RM by the sedimentation equilibrium method, a six-channel centerpiece was used. The height of the solution column was only a few millimeters but in the case of viscous solutions one run may take up to 1 week.

Density measurements

The density of the solutions was determined by the high precision densitometer DMA 10 (Paar, KG, Graz, Austria). The samples were degassed by a supersonic treatment.

Results

As can be seen from Fig. 1, the Schlieren-pattern of a moving boundary run fo RM, containing $(Lys)_n$ shows two peaks. The large peak moving at a slow rate corresponds to that obtained with RM free from $(Lys)_n$

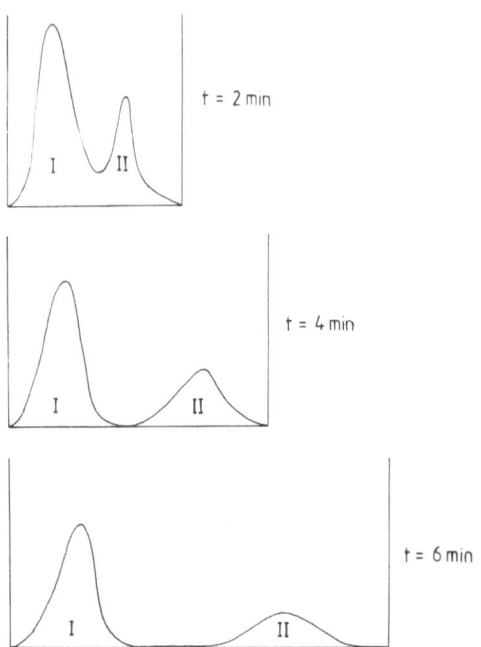

Fig. 1. Schlieren-patterns (schematically drawn) obtained during the sedimentation of the reversed micellar system $(Lys)_n$/AOT/ H_2O/i-C_8 at 2-min intervals. $w_o = 6$; AOT concentration, 0.1 mol/l, 60 000 rpm; (I) "empty" reverse micelles; (II) "filled" reverse micelles

(empty RM); the second one obviously corresponds to the $(Lys)_n$-filled RM. The s_c- c_{AOT}-plots for empty RM shown in Fig. 2 and those of the first slowly moving peak are almost identical to them.

From the ratio of the areas of these peaks, the fraction of filled micelles can be determined. For $w_o = 6$ and an AOT concentration of 0.1 mol/l it reaches $\approx$ 45 %, as is shown in Fig. 3. At higher concentrations of $(Lys)_n$ the polymer precipitates.

Diffusion measurements

By diffusion of the solute, the synthetic boundary broadens as a function of time. The concentration gradient can be expressed by a Gaussian function in the ideal case. This broadening can be used as a measure of the diffusion rate.

The diffusion coefficient

$$D = \frac{k \cdot T}{f} = \frac{dx}{dt} \cdot 1/(dc/dx)$$

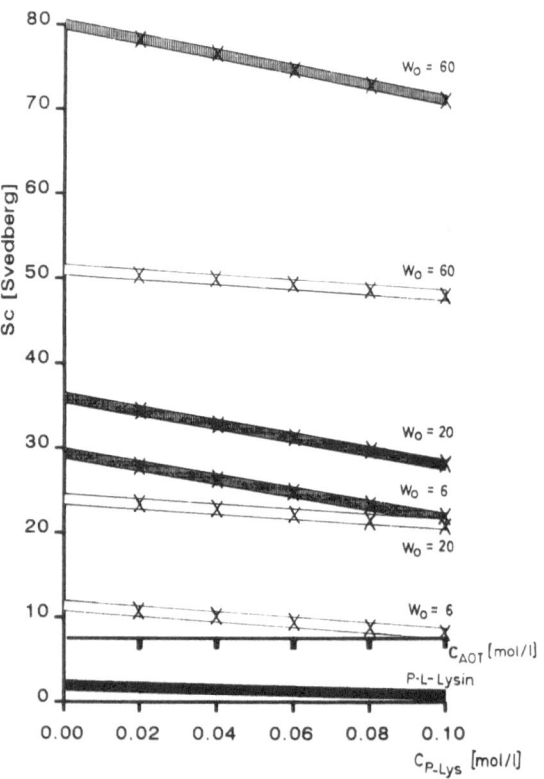

Fig. 2. Sedimentation coefficient (s_c)-concentration plot for empty and filled reversed micelles of the system $(Lys)_n$/AOT/H_2O/i-C_8. The curve below (▬▬) was obtained for $(Lys)_n$; (═══) empty; (▨▨▨) filled RM

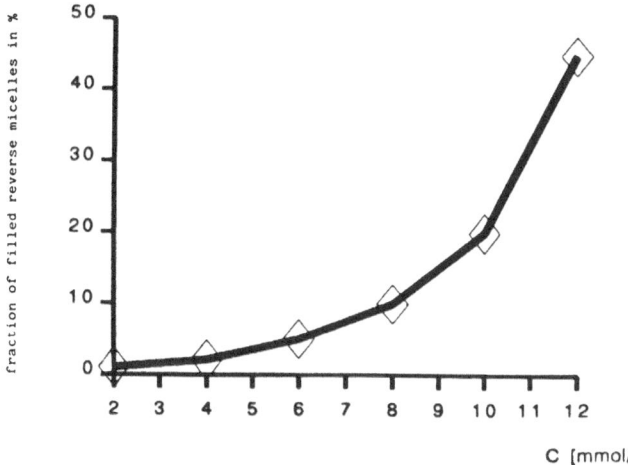

Fig. 3. Fraction of filled reversed micelles of the system $(Lys)_n$/AOT/H_2O/i-C_8 as a function of $(Lys)_n$-concentration. $w_o = 6$; AOT concentration, 0.1 mol/l

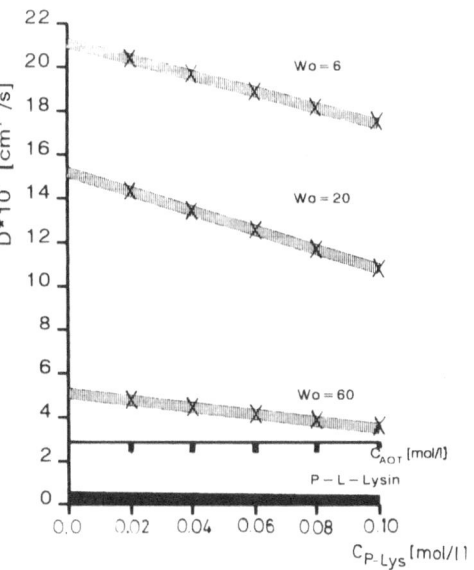

Fig. 4. Diffusion coefficient of empty reversed micelles of the system $(Lys)_n$/AOT/H_2O/i-C_8 as a function of AOT concentration and of $(Lys)_n$ depending on poly(L-lysine) concentration

(where $f \cdot dx/dt$ is the friction force; f the friction coefficient, dc/dx the concentration gradient, dx/dt the migration rate) can be determined by the equation

$$D = \frac{(\Delta c)^2}{4\pi \cdot t \cdot (dc/dr)^2_{max}} \, .$$

The concentration change which is proportional to the area A below the diffusion peak and the maximum of the concentration gradient $(dc/dr)_{max}$ which is proportional to the hight H_{max} of the peak allow one to evaluate the diffusion coefficient. Because of the concentration dependence of D, this was determined at different concentrations c of RM and extrapolated to $c = 0$ (Fig. 4).

In filled RM, however, the usual method for determining D cannot be used because the peak obtained by the Schlieren method is composed of that for empty and for filled RM with different D-values.

For determining the D-values of both the diffusion peak obtained can be separated into two Gaussian curves (Fig. 5a) belonging to each type of RM ("empty" or "filled"). This is shown in Fig. 5b. The separation method has been described in detail by Plachky [7].

The concentration dependence of empty and filled RM for different w_o-values are shown in Fig. 6. The masses of the RM calculated by the Svedberg equation using s_o and D_o are listed in Table 1.

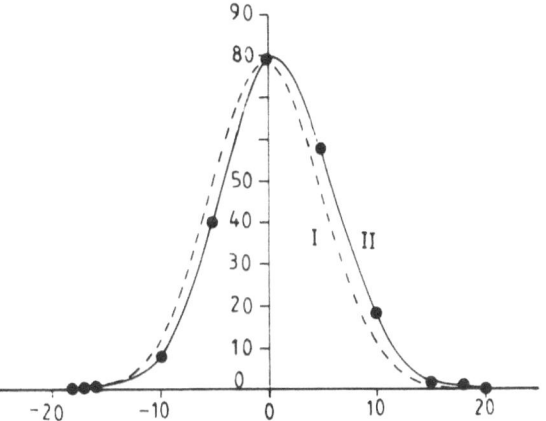

Fig. 5a. Diffusion peak obtained experimentally (–●–) for empty reversed micelles and by calculation (– – –) using a Gaussian function

Table 1. Mass of "empty" and "filled" reversed micelles at various w_o-values obtained by s_o and D_o using the Svedberg equation

System	w_o	s_o (S)	D_o (cm^2/s) $\cdot 10^{-7}$	M
$(Lys)_n$ reversed micelles				
filled rev. micelles	6	27	2.1	880 000
	20	36	1.8	1 380 000
	60	80	1.5	3 670 000
empty rev. micelles	6	21	21.0	37 000
	20	24	15.0	110 000
	60	51	5.0	700 000
$(Lys)_n$ water	–	2	0.35	400 000

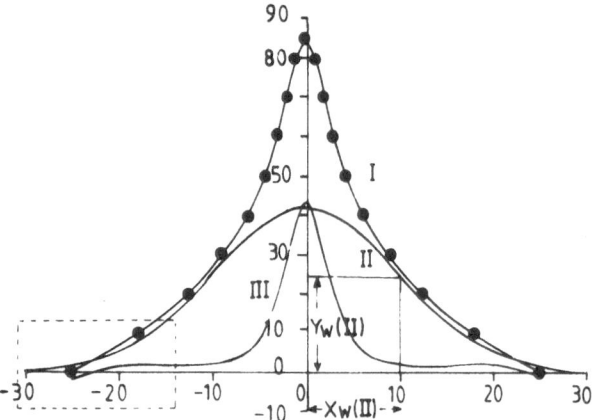

Fig. 5b. Separation of an experimentally obtained diffusion peak (I) of a $(Lys)_n/AOT/H_2O/i-C_8$ reversed micellar system into the curve for filled (II) and empty (III) reversed micelles by computer calculation

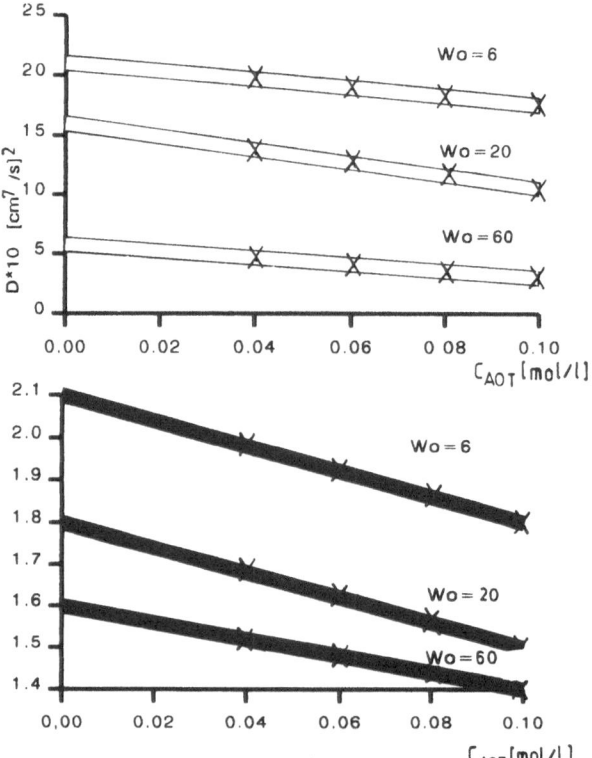

Fig. 6. Diffusion coefficients of empty and filled reverse micelles of the system studied

Sedimentation equilibrium measurements

The sedimentation equilibrium method was used to prove the results obtained on the multicomponent systems above. In this case one can calculate the mass of the particles investigated, without determining the diffusion coefficent, directly from the Schlieren-patterns of the sedimentation diffusion equilibrium obtained by the system studied.

The mass can be obtained from the concentration gradient by the following relation

$$M = \frac{2RT}{\omega^2 \cdot (1 - V^* \cdot \varrho_1)} \cdot \frac{\Delta \ln c}{\Delta r^2}$$

where ω is the angular velocity, c the concentration at r, r the distance from the rotation center, V^* the partial specific volume of the solute, ϱ_1 the density of the solvent.

The concentration c at the distance r from the rotation center can be determined from the distance Y_i between the gradient curve and the base line in the usual way, and $\Delta \ln c/\Delta r^2$ can be obtained directly from the inclination of the straight line $\ln (y_i/r_i)/r_i^2$ calculated with the method of least squares.

In contrast to empty micellar systems $AOT/(H_2O)_i-C_8$, the plot for $(Lys)_n/AOT/H_2O/i-C_8$ consists of two straight lines differing in inclination, as is shown in Fig. 7. The one with the lower inclination is due to the empty micelles, as was confirmed by a comparison with a plot obtained from an empty RM system. The steeper curve is due to the filled RM. In this way the mass of both kinds of RM can be determined by one series of experiments, by plotting the mass measured at various concentrations c of AOT against c (AOT) and extrapolating to $c = 0$.

The masses obtained from these sedimentation equilibrium measurements agree well with those determined by the moving boundary method, especially those for empty RM, as one can see from Table 1.

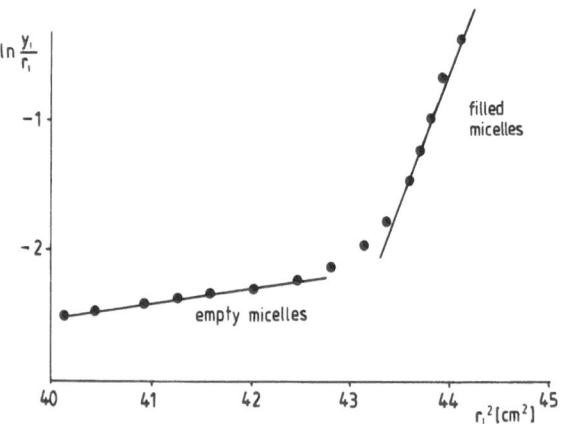

Fig. 7. $\ln (y_i/r_i) - r_i^2$ plot for a sedimentation equilibrium run of $(Lys)_n/AOT/H_2O/i-C_8$ reversed micelles. $w_o = 6$, 20 °C

Solubility of (Lys)$_n$ and (Arg)$_n$ in the water pool of reversed micelles

Surprisingly, the amount of some basic poly(α-aminoacids) solubilized in the water pool of the AOT/H$_2$O/i-C$_8$ reversed micelles decreases with increasing w_o. As can be seen from Fig. 8, it passes a minimum at $w_o = 20$. However, compared with the "solubility" at $w_o = 6$ and 12, it is also rather low at $w_o = 60$.

Discussion

The structure of reversed micelles containing solubilized polypeptide molecules can be described by the so-called water-shell model [1, 2]. In this model the polypeptide molecule is surrounded by a water layer separating the polymer from the ionic headgroups of the surfactant. In polypeptides of high molecular mass and low w_o-values, one can assume that several empty reversed micelles coalesce, the polypeptide molecule forming one filled RM according to the relation [1, 2]

$$a \, | \, s n_o \, (\text{H}_2\text{O})_i \, | \, + P \rightarrow | \, s n P \, (\text{H}_2\text{O})_j \, | \qquad (1)$$

where a is the number of empty RM forming one filled RM, s the surfactant molecule, n_o the number of surfactant molecules forming one empty RM, n the number of surfactant molecules in one filled RM, P the polypeptide molecule, i the number of water molecules in one empty RM, j the number of water molecules in one filled RM.

The mass of one filled reversed micelle m_T is:

$$m_T = m_p + j m_w + n m_s \qquad (2)$$

where m_T is the mass of one filled RM, m_p the molar mass of one polypeptide molecule, m_w the molar mas of H$_2$O, m_s the molar mass of the surfactant.

Substituting j by $n w_o$ one gets

$$n = \frac{m_T - m_p}{w_o \cdot m_w + m_s} \qquad (3)$$

or

$$m_T = m_p + j m_w + (j/w_o) \cdot m_s . \qquad (4)$$

For j, the relation

$$j - \frac{m_T - m_p}{m_w + (m_s/w_o)} \qquad (5)$$

is valid [1, 2].

Because m_w, m_s and w_o are known and m_T and P are experimentally measured, the number of surfactant molecules n and of water molecules in one filled RM can be easily calculated. The corresponding values are listed up in Table 2 together with the masses of empty and filled RM.

From these values one can conclude that at $w_o = 6$ only one polypeptide molecules is solubilized in one RM.

An important problem is how may empty RM form one filled RM because it is obvious that one polypeptide molecule consisting of 600 or 2400 monomer units with a length of 2160 Å or 8640 Å in the all-trans conformation, and of 900 or 3600 Å in the α-helical form, does not fit into a RM with a radius of ≈ 15 Å, if one considers the water pool inside only. Of course, in

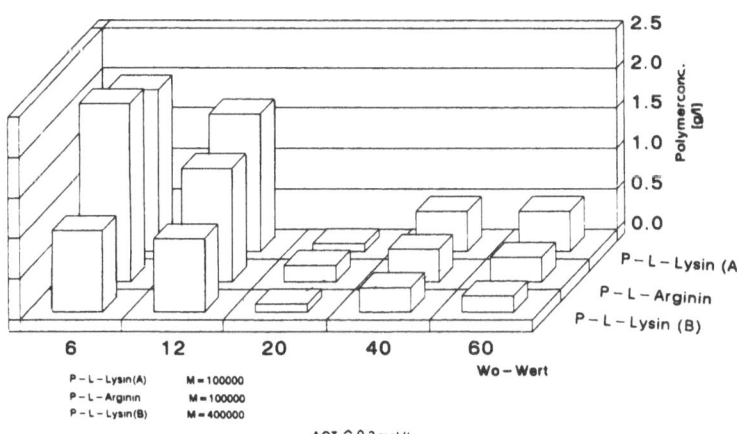

P-L-Lysin(A)   M = 100000
P-L-Arginin   M = 100000
P-L-Lysin(B)   M = 400000

AOT ≙ 0.2 mol/l

Fig. 8. Solubility of poly(L-lysine) and poly(L-arginine) in reversed micelles of the system under consideration

Table 2. Data of "empty" and "filled" reversed micelles containing (Lys)$_n$ at various w_o-vales

$M_p \times 10^3$	w_o	$M_l \times 10^3$	$M_t \times 10^3$	$J_{H_2O} \times 10^3$	n_{AOT}	P_n	a	r_s (empty) [Å]	r_s (filled) [Å]
100	6	37	100	1.1	187	600	2.7	22.9	80
400	6	37	780	4.1	690	2400	9.5	22.9	230
400	12	80	1200	14.5	1212	2400	10.0	–	–
400	20	120	1600	29.8	1490	2400	10.0	32.2	268
400	40	420	2600	75.6	1890	2400	5.2	–	–
400	60	700	3600	125.0	2100	2400	5.3	96.5	322

M_l is the mass of the empty reversed micelles; M_t the mass of the filled reversed micelles; j_{H_2O} the number of water molecules in one RM; n_{AOT} the number of AOT molecules in one RM; P_n the degree of polymerization of the (Lys)$_n$ used; a the number of empty reverse micelles forming one filled RM; r_s the Stokes' radii

the case of native proteins, the molecules are folded forming a tertiary structure which is known in many cases and therefore also their radii are known. But in the homopolypeptide used, the molecular shape is of a linear type. From the results of the ultracentrifuge measurements, it appears (Table 2) that one RM at $w_o = 6$ contains only 67 AOT and 402 water molecules. On the other hand, because the empty RM form one filled RM, together with one (Lys)$_{600}$, there are only 200 AOT present. This means that the ratio of NH_3^+ lysine side groups to SO_3^- groups of AOT is 1:3 and there is a ratio $H_2O/NH_3^+ = 2$. These water molecules are part of the hydration shell of the $-SO_3^-$ as well, and due to these closely neighboured anions a dissociation of the $NH_3^+Br^-$ cannot take place. Therefore, similar to the solid state, the Br^- ions are in close contact with the $-NH_3^+$ ions stabilizing the ordered periodical conformation, as discussed previously [3–5]. The fact that poly(L-lysine) and also poly(L-arginine) also show the CD-spectra of α-helix or β-structure in reversed micelles at $w_o = 60$ demonstrates that a dissociation of the polycation-anion salt leading to a more or less disordered state does not occur.

Acknowledgement

The authors are grateful to Deutsche Forschungsgemeinschaft for financial support.

References

1. Bonner FJ, Wolf R, Luisi PL (1980) J Solid Phase Biochem 5:255–268
2. Luisi PL (1985) Angew Chem 97:449–460
3. Ebert Ch, Ebert G, Werner W (1973) Kolloid Z Z Polym 251:504–505
4. Ebert Ch, Ebert G (1977) Colloid Polym Sci 255:1041–1053
5. Ebert G (1985) In: Boschke F (ed) Solvation and Ordered Structure in Colloid Systems. Topics in Current Chemistry, Vol 128, pp 1–36
6. Noritomi H (1986) Behaviour of Enzymes and Polypeptides in Reversed Micelles, Thesis, University of Tokyo, Japan
7. Plachky M (1987) Untersuchungen der Konformation und Solvatation von Poly-(α-aminosäuren) in inversen Mizellen, Thesis, Philipps-Universität Marburg/Lahn, FRG
8. Senō M, Noritomi H, Kuroyanagi Y, Iwamoto K, Ebert G (1984) Colloid Polym Sci 262:727–733

Received December 10, 1987;
accepted January 12, 1988

Authors' address:

G. Ebert
Fachbereich Physikalische Chemie
Philipps-Universität
Hans-Meerwein-Straße
D-3550 Marburg a. d. Lahn, F.R.G.

Progress in Colloid & Polymer Science

Progr Colloid Polym Sci 76:119–122 (1988)

O/W Microemulsions at low surfactant content

C. M. C. Gambi[1]), L. Léger[2]) and C. Taupin[2])

[1]) University of Florence, Department of Physics, Florence, Italy, CISM (of the MPI) and GNSM (of the CNR) groups
[2]) Laboratoire de Physique de la Matière Condensée, Collège de France, Paris, France, GRECO "Microemulsions" du CNRS

Abstract: Oil-water microemulsions of the five components system water/sodium chloride/toluene/1-butanol/sodium dodecyl sulfate exhibit diffuseness (due to a turbidity gradient) for a very low surfactant content, 0.6 ÷ 0.006 % w/w. The main characteristics of such microemulsions are described and the existence of nonuniformity is discussed.

Key words: Microemulsions with 5 components, water, toluene, 1-butanol, chloride and dodecylsulfate of sodium, Winsor-phases, diffusion effects, influence of temperature.

In this paper we discuss some interesting characteristics of o/w microemulsions, recently investigated [1, 2] for a very low surfactant content (0.6 – 0.006 %), of the five components system: water/sodium chloride/toluene/1-butanol/sodium dodecylsulfate (SDS). Because the aim of the work is a better understanding the role of alcohol, surfactant and temperature in the properties of the amphiphilic compounds, a phase diagram has been determined at constant brine/toluene ratio ($\simeq 2$) and constant salinity (6.5 %); the high salinity value should ensure a practically complete charge screening at the internal interface of the system. The investigation has been carried out in a surfactant concentration range well below that which usually leads to conventional Winsor phases. Under such conditions, a new behavior is exhibited by the samples: transparent upper oily regions coexist with diffuse aqueous regions, the diffuseness being due to a turbidity gradient (see Fig. 1 of Ref. [1]); the interface between the oily and the aqueous regions (which will be simply called 'interface' in the following) is sharp. The phase diagram investigation has been therefore performed to cover the domain for which the new behavior is observed, for different alcohol and surfactant contents and at different temperatures. In Winsor equilibria, a turbid phase of the microemulsion type coexists with an oily phase (Winsor I) or an aqueous phase (Winsor II) or both (Winsor III); the excess phases are always transparent. What distinguishes the samples of this

phase diagram from conventional Winsor equilibria is the turbidity gradient of the aqueous domain. This diffuse domain coexists either with transparent oily regions or with turbid intermediate homogeneous regions of microemulsion type, plus transparent oily regions, depending on the parameters' value.

In the five components system under study, an alcohol/surfactant ratio (w/w) higher than 0.9 develops a Winsor III equilibrium and, for a further increase of that ratio (0.9–9), the value of the active mixture percentage necessary to obtain the Winsor III equilibrium decreases [3]. In our phase diagram investigation [2], the alcohol/surfactant ratio covers the range 4–380, the alcohol and SDS contents are in the range 1.9–2.7 % and 0.006–0.6 %, respectively; for all the samples, the active mixture percentage is 2.8–1.9 % and the investigated thermal range 12°–30 °C. The aqueous regions exhibit diffuseness for SDS and 1-butanol content in the range 0.006–0.3 % and 1.9–2.5 %, respectively, and for $T = 16°$–26 °C. An active mixture percentage 2.3–2.8 % and an alcohol/surfactant ratio > 10 are conditions necessary to observe such a diffuseness. We point out that the alcohol content cannot be decreased below 1.9 %, otherwise the surfactant spreads on the interface in the form of a white structure; the 0.006 % SDS content corresponds to the minimum amount still weighted with a good accuracy. For all the samples studied, outside the thermal range in which diffuseness is shown, the aqueous region exhibits a homogeneous turbidity for $T > 26\,°C$, while the surfactant spreads on the interface for $T \leq 16\,°C$ (the two liquid regions separated by the interface are transparent). This last result suggests that, for diffuse samples, the surfactant is close to the solubilization limit, for the given proportions of the other components. In summary, for all the diffuse samples, the brine/toluene ratio is constant and the active mixture percentage is also quite constant; thus, in the pseudoternary phase diagram brine/toluene/active mixture, the mass composition points of the samples are approximately in the same position, while the alcohol/surfactant ratio varies by three orders of magnitude. By a comparison of this phase diagram with that of Ref. [3], the mass composition point of diffuse samples is found to be on the boundary between Winsor I and Winsor III equilibria for the alcohol/surfactant ratio ~ 9.

We should point out that our phase diagram was performed by visual observation on samples thermally stabilized at $0.1\,°C$ for longer than 1 month. This condition could be insufficient to assure the attainment of the thermodynamic equilibrium for samples which are close to the boundary between Winsor I and III equilibria. Thus, the composition and the structure of a typical sample, composed of brine 65.74%, NaCl salinity 6.5%, toluene 31.90%, 1-butanol 2.30% and SDS 0.04% (w/w), have been studied with a $\pm 0.025\,°C$ thermal stabilization over 1 month; the aqueous domain is still diffuse. The composition of the upper region and the average composition of the aqueous domain, as well as the composition profile inside the aqueous domain, have been evaluated by the index of refraction measurements. A direct estimate of the composition of the upper transparent region has been done by gas chromatography. Concerning the structure investigation, the autocorrelation function of the scattered light intensity, $g(t)$ (QELS analysis), has been measured as a function of height throughout the sample. Because the process is diffusive, the mutual diffusion coefficient has been evaluated for all the heights. Furthermore, the transmitted light intensity has been also measured throughout the sample. Variations of the composition and the structure distributions have been observed only during the first stabilization week, thus more than 1 week's stabilization time has been allowed for the sample in all experiments.

We summarize here the main results:

Upper region: No index of refraction variation is observed, thus the upper region is uniform in composition. From the gas chromatographic spectrum, the region is mainly composed of toluene and 1-butanol, in

good agreement with the index of refraction results. Furthermore, no correlation time of $g(t)$ is detectable in the limit of resolution of the apparatus and no variation of the transmitted light intensity is observed. This implies that no aggregates of supramolecular size are present there. The upper region results in a homogeneous oily solution, similar to the excess oily phase of Winsor I and III equilibria.

Aqueous Domain: An index of refraction profile indicates a progressive variation of the composition from the interface to the bottom of the domain. Different profiles are obtained for different temperatures in the range $18°–24\,°C$. In the aqueous domain, in contrast to the upper region, we detect a correlation time of $g(t)$ for all the investigated heights down to the bottom of the cell. A transmitted light intensity profile is also detected and this profile changes with temperature.

In summary, from the composition and the structure results, we can describe the aqueous domain as composed of three regions: intermediate, diffuse and lower, respectively, from the interface towards the bottom of the sample; the intermediate and the lower regions are uniform in composition and structure, while the diffuse region displays a composition and a structure gradient. The composition and the structure profiles vary with temperature; their trends (they are sharp for $T = 20\,°C$ and $T = 22\,°C$ and smooth for $T = 18\,°C$ and $T = 24\,°C$) agree with the phase diagram observations, despite the better thermal stabilization accuracy of such analyses: in fact, observing with the naked eye, the diffuseness disappears for $T \leq 16\,°C$ and $T > 26\,°C$. Finally, the body of the results allows us to establish that the upper region is an excess oily region, while the aqueous domain is of the o/w microemulsion type, down to the bottom of the cell.

To interpret the nonuniformity of the aqueous domain, a discussion in terms of incomplete phase separation, critical type regime, sedimentation of polydisperse globules due to gravity and globules aggregation has been provided in Ref. [2]. The hypothesis of a critical or precritical regime has been excluded. The diffuse samples relax very slowly after mechanical agitation and are placed on the phase boundaries between two and three phase coexistence domains; thus, a possible explanation for the diffuseness is an incomplete phase separation between the intermediate and the lower regions. Before attaining the studied state (diffuse aqueous region plus transparent oily region), the aqueous domain is a polydisperse emulsion, the concentration gradients being larger than for the studied state; thus, under the hypothesis of a very slow equilibrium attainment, the tendency is clearly towards a Winsor I equilibrium. However, the sample diffuseness could be linked to a segregation due to gravity of dispersed objects having different sizes, the formation of which could be favoured by the peculiar composition of the sample itself. In this last case the sample should be at equilibrium. To test the hypothesis of a gravity segregation, we have estimated the size of the dispersed objects of the sample intermediate and lower regions. In fact, because the 50 % D variation measured between those regions cannot be interpreted in terms of volume fraction variation at constant size, the $D(h)$ profile suggests the existence of two neighbouring regions with scattering objects of different size (the D value is $0.8 \cdot 10^{-7}$ cm^2/s and $1.3 \cdot 10^{-7}$ cm^2/s for the intermediate and the lower regions, respectively). Assuming that globules are dispersed throughout the aqueous domain, from the mutual diffusion coefficient we can give a rough estimate of a hydrodynamic radius R_H, neglecting corrections due to the finite concentration of the sample, assuming $D \sim D_o$ and simply using the Stokes-Einstein formula; assuming as viscosity that of water wet get $R_H \sim 200$ Å. Moreover, we obtain $R_H = 230$ Å and 160 Å in the intermediate and lower regions, respectively. These values are surely overestimated as the second virial coefficient contribution has been neglected, but the ratio between them (~ 1.5) is presumably correct because of the small volume fraction variation of the dispersed phase between the two regions, $\Delta \phi = 0.0014$. If several sizes coexist by aggregation or curvature fluctuations, in both cases, a sedimentation of large enough globules may arise due to gravity. The competition between Brownian motion and gravity, for samples at equilibrium, gives an exponential distribution $\exp(-\Delta\varrho V g h / k_B T)$ with $\Delta\varrho =$ density difference between the continuous and the dispersed phase, where V is the droplet volume and g the gravity accelearation; the characteristic length scale over which sedimentation occurs is $l = k_B T / \Delta\varrho V g$. In our case, supposing the sample at equilibrium, we get $l = 18$ cm for $R = 160$ Å, and $l = 6$ cm for $R = 230$ Å. The size distribution should thus vary with h, larger globules sedimenting towards the oil-microemulsion interface. Of course, a sum of two exponential decays should give a monotonic decreasing function; thus, a gravity segregation of globules of two different sizes cannot explain the whole trend of the profiles. Furthermore, no significant polydispersity has been found in QELS experiments, therefore we cannot deduce a featuring size distribution of the globules. The main ques-

tion about the origin of the diffuseness remains not fully answered. However, it can be demonstrated that a microemulsion is obtained down to a very low surfactant content. Furthermore, the microemulsion interfacial film appears to be very rich in alcohol. An indirect estimation of the interfacial film composition can be obtained as follows. Since under reasonable hypotheses the average volume fraction of the dispersed phase and the average hydrodynamic radius have been evaluated, the composition and the structure results can be compared by means of the formula $R = 3\phi / n_s \Sigma$ where R is the average radius of the globules, n_s the number of surfactant molecules per unit volume and Σ the area per surfactant molecule of the interfacial film ($\Sigma \simeq 60$ Å^2) [4]. Usually the Σ value is related to one surfactant molecule plus one alcohol molecule; however, in the present case, using as R and ϕ values of 200 Å and 0.0275, respectively, the number of alcohol molecules per surfactant molecule at the interface of the microemulsion is ~ 20, one order of magnitude higher than usual. Such an alcohol-rich film is expected to present peculiar mechanical characteristics corresponding to a quite low rigidity coefficient (K) of the interface which could favor strong curvature fluctuations leading to polydispersity or aggregation. We recall that the aqueous domain loses its diffuseness for $T \leq 16\,°C$ and for $T > 26\,°C$. For $T \leq 16\,°C$ the aqueous domain is perfectly transparent; probably the curvature energy is too high, with respect to thermal energy ($K > K_B T$), to give a dispersion of microemulsion type. For $T > 26\,°C$ the aqueous domain is turbid homogeneous; the thermal energy seems to prevail ($K < K_B T$) and such a dispersion is favored. Thus, for intermediate temperatures, an intermediate case is expected: the competition between rigidity and thermal energy could favor large curvature fluctuations, therefore the coexistence of dimetric, trimeric . . . aggregates can be justified, as well as the coexistence of globules of different size. At the present stage of measurements we have no information on the shape of the globules. The 1.5 ratio between the diameters of the globules of the intermediate and lower regions could support the hypothesis of an intermediate region mainly composed of oily dimeric aggregates.

References

1. Gambi CMC, Léger L, Taupin C (1987) Europhys Lett 3(2):213–220
2. Gambi CMC, Léger L, Taupin C (1987) J Phys Chem 91:4536–4544

3. Bellocq AM, Biais J, Clin B, Gelot A, Lalanne P, Lemanceau B (1980) J Colloid Interface Sci 74(2):311–321
4. de Gennes PG, Taupin C (1982) J Phys Chem 86:2294–2304

Received December 10, 1987;
accepted December 18, 1987

Authors' address:

C.M.C. Gambi
University of Florence
Department of Physics
L.E. Fermi 2
I-50125 Florence, Italy

Progress in Colloid & Polymer Science   Progr Colloid Polym Sci 76:123–131 (1988)

Lyotropic nematic phases of double chain surfactants

G. Hertel and H. Hoffmann

Lehrstuhl für Physikalische Chemie I der Universität Bayreuth, Bayreuth, F.R.G.

Abstract: Polarization microscopy, ^{2}H-NMR measurements, surface alignment and alignment in magnetic fields were utilized to look for nematic systems within a selection of N-alkyl-N,N-dimethyl-N-hexadecyl ammonium bromide surfactants. Three binary lyotropic nematic systems are reported which consist of rodlike micelles with positive anisotropy of the diamagnetic susceptibility $\Delta\chi$ and negative birefringence: cetyltrimethyl ammonium bromide (CTAB)/water; N,N-dimethyl-N-ethyl-N-hexadecyl ammonium ($C_{16}C_2$DMABr)/water and N-butyl-N,N-dimethyl-N-hexadecyl ammonium bromide ($C_{16}C_2$DMABr)/water. The ternary systems CTAB/decanol/water and $C_{16}C_2$DMABr/decanol/water were investigated. In each system, two nematic phases were found, one consisting of rodlike micelles with positive $\Delta\chi$ and negative total birefringence and one consisting of disclike micelles with negative $\Delta\chi$ and positive total birefringence. A region of coexistence between the two nematic phases is reported for both systems. A systematic method of preparation of nematic phases in similar systems is proposed.

Key words: Lyotropic, nematic phases, ^{2}H-NMR spectroscopy, magnetic field.

Introduction

Since the first system of a binary lyotropic nematic phase was reported in 1979 [1] only a few binary systems that show lyotropic nematic phases have been discovered. They were found by chance when binary systems were carefully investigated. It still cannot be predicted whether a certain surfactant will form a nematic phase in aqueous solution or not. There is a second reason why only a few lyotropic nematic systems have been found so far. It is often difficult to detect an existing nematic phase because the concentration and temperature ranges are small. Sometimes the phase transitions are slow and it takes a long time until the system is in equilibrium.

A few hydrocarbon detergent binary nematic systems have been reported until now: decyl ammonium bromide/D_2O [2], tetradecyl trimethyl ammonium bromide/H_2O [3], tetradecyl trimethyl ammonium benzene sulphate/H_2O [3], tetradecyl dimethyl aminoxide/H_2O [4]. These systems show a nematic phase consisting of rod-shaped micelles in a concentration range below a hexagonal phase. Previously a binary nonionic nematic system was published which contains disclike aggregates [5].

Binary nematic phases of perfluorosurfactant/water systems have been reported and investigated by Lindman [6], Boden [7], Hoffmann and Kalus [8, 9]. Saupe [10] and Charvolin [11] reported and investigated the biaxial nematic phase in a ternary system.

Looking at known lyotropic nematic phases, one can see that a surfactant which has a strong tendency to build up a hexagonal or a lamellar phase does not build up a nematic one. Nematic phases often occur in systems which lie between these two extremes. Surfactants with short alkyl chains and large headgroups prefer hexagonal phases when the surfactant concentration is high (30 % ww). When the headgroup of the surfactant is small compared with the alkyl chain, lamellar phases are preferred. To modify the tendency to build a certain phase means to modify the headgroup area or the charge of the headgroup or to change the cross section of the alkyl chain.

Here we investigated a selection of N-alkyl-N,N-dimethyl-N-hexadecyl ammonium bromides. The length of the second alkyl chain was varied from methyl, ethyl, butyl and hexyl to octyl. In the following text, the corresponding surfactants are referred to as *CTAB* (cetyl trimethyl ammonium bromide),

$C_{16}C_2DMABr$, $C_{16}C_4DMABr$, $C_{16}C_6DMABr$ and $C_{16}C_8DMABr$. The binary phase diagrams were investigated by polarization microscopy, ^{2}H-NMR spectroscopy, mechanical alignment on glass surfaces and alignment in magnetic fields.

A long-chain alcohol can be looked upon as a surfactant with a very small headgroup compared with the alkyl chain. Adding a long-chain alcohol to a system of ionic surfactants leads to mixed micelles consisting of alcohol and ionic surfactant. Replacing ionic surfactant with alcohol therefore means a reduction of required area for the headgroup. In the investigated ternary systems, we took n-decanol to replace various amounts of the ionic surfactant.

Experimental

CTAB was purchased from Merck and purified by recrystallization from ether/ethanol. The double chain surfactants were a gift from Hoechst Gendorf. They were purified by recrystallisation from ether/ethanol, too. The purity was tested by determining the melting point under the polarization microscope and by determining the CMC with conductivity measurements.

For the ^{2}H-NMR measurements, a JEOL FX 90 Q with external lock was used. All investigated solutions were prepared with a mixture of 5 % ww D_2O and 95 % ww H_2O as solvent. This mixture provides good ^{2}H-NMR spectra within a short accumulation time and the small amount of D_2O does not effect the phase diagram. Larger amounts of D_2O change the phase boundaries.

The experiments of alignment in a magnetic field were made with a Bruker BE 20 magnet. Fields up to 2.1 T were available.

A polarization microscope (Zeiss Standard Pol 16) with a λ-plate and a hotstage (Mettler FP 80) was used to determine the sign of birefringence and the kind of liquid crystalline texture. The liquid crystalline phases were filled in microslides (Camlab) of quartzglass with a path-length of 0.2 mm.

^{2}H-NMR measurements

^{2}H-NMR measurements were used to determine the sign of the anisotropy of the diamagnetic susceptibilities ($\Delta\chi = \chi_{parallel} - \chi_{perpendicular}$) and to determine the phase boundaries of the magnetic phases. The sign of optical and diamagnetic anisotropy is always referred to the symmetry axis (optical axis) of the nematic phase.

Measuring the quadrupole splitting of D_2O in the surfactant system is a good way to distinguish between isotropic, nematic and non-aligned liquid crystalline phase. In the isotropic phase, the movement of the aggregates is very fast compared to the NMR time scale. No quadrupole splitting is visible and the result is

a single sharp signal. In non-aligned liquid crystals, the movement of the aggregates is slow. The O—D axis of the associated water on the surface of the micelles has no preferred orientation. The observable quadrupole splitting is a function of the angle α between the O—D axis and the direction of the magnetic field: $\nu = \nu_{max}\frac{1}{2}$ $(3\cos^2\alpha - 1)$. This function produces values between ν_{max} ($\alpha = 0°$) and $-\frac{1}{2}\nu_{max}$ ($\alpha = 90°$). When there is a statistical orientation, there are many more possibilities for orientation perpendicular to the magnetic field than parallel. The observed spectra is a superposition of all states between these extremes (Fig. 1). The intensity of a certain quadrupole splitting refers to the probability of the corresponding orientation. In a powder pattern the two middle peaks correspond to a perpendicular orientation to the magnetic field. The shoulders can be assigned to the parallel orientation. When a nematic phase is aligned there is only one value of quadrupole splitting. The results is a spectrum consisting of a sharp doublet with a certain quadrupole splitting.

To determine the sign $\Delta\chi$ is to determine whether the director of the nematic phase aligns parallel ($\Delta\chi > 0$) or perpendicular ($\Delta\chi < 0$) to the direction of the magnetic field in the NMR spectrometer. If the alignment of the nematic phase is slow compared to the time of measurement, there are two ways to obtain the sign of $\Delta\chi$:

The first is to record a set of time dependent spectra, starting with a non-aligned nematic phase. This first spectrum is a typical powder pattern. When $\Delta\chi$ is positive, the shoulders of the powder pattern will increase with time. That means the probability of parallel align-

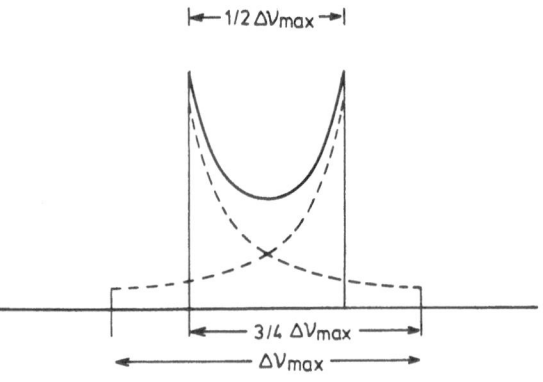

Fig. 1. Powder pattern of a non-aligned liquid crystalline phase

ment is increasing. A nematic phase with negative $\Delta\chi$ shows the same powder pattern at the beginning of the experiment. After some time the shoulders of the powder pattern will vanish and the two middle peaks will become increasingly sharp. For this experiment it is not essential that the magnetic field has to be perpendicular to the axis of the probe tube. Both spectrometers with conventional electromagnets (magnetic field perpendicular to the axis of the probe tube) and spectrometers with cryogenic magnets (magnetic field parallel to the axis of the probe tube) can be used.

The second way to determine the sign of $\Delta\chi$ is to record three types of NMR spectra in a spectrometer where the axis of the probe tube is perpendicular to the direction of the magnetic field. First, a spectrum of the aligned phase is recorded. Then the probe is turned about an angle of 90° round the axis of the probe tube. The third spectrum is recorded with a fast spinning (50 Hz) probe tube. In the case of a positive $\Delta\chi$ the quadrupole splitting is a function of $\frac{1}{2}(3\cos^2 y - 1)$ where y is the angle of rotation of the probe tube against the original position. The ratio of the quadrupole splitting of the three spectras should be $1:0.5:0.25$. In the rotated spectrum the quadrupole splitting is the average of the described function. If the quadrupole splitting is comparable with the frequency of spinning, rotation peaks are visible in the rotated spectrum. The distance of these rotation peaks is equal to the spinning rate. In the case of negative $\Delta\chi$, rotation of the probe tube should not effect the magnitude of quadrupole splitting. All spectras should have the same splitting when there are no wall effects in the nematic phase.

Polarization microscopy

Polarization microscopy was used to distinguish between nematic, hexagonal and lamellar textures, to determine the phase boundaries, the sign of birefringence Δn and to distinguish between disclike and rodlike nematic phases. All investigations under the microscope were carried out in microslides with a pathlength of 0.2 mm (Fig. 2).

Nematic phases usually give the characteristic "Schlieren" textures under the microscope. If the nematic phase is very viscous it can take a long time to get such a characteristic texture. Therefore, great patience of the experimentor is necessary to find such phases with the help of a polarization microscope.

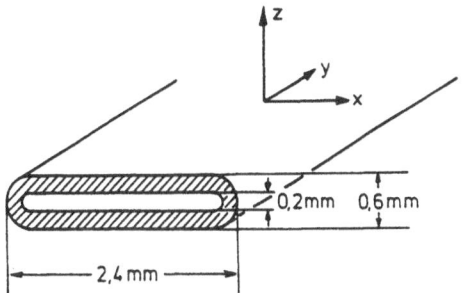

Fig. 2. Dimensions of the microslides

Distinction between rodlike and disclike nematic phases

Both the glass surface of the capillary and the headgroup of the surfactant are polar. Polar interactions force disclike nematic micelles to align with their optical axis perpendicular to the glass surface. Rodlike micelles align with their optical axis parallel to the surface of the capillary. The alignment is slow and may last several hours depending on the viscosity of the solution. Shortly after capillary filling, the nematic phase consisting of disclike micelles is macroscopically disordered. Under the polarization microscope, strong birefringence is visible. 1 h after filling, the sample produces a pseudoisotropic texture. The reason for this behaviour is surface alignment. In the pseudoisotropic texture the director is parallel to the viewing direction. Only at the edges of the capillary, where the glass surface is not perpendicular to the direction of view, there is strong birefringence when the long axis of the capillary is at 45° to the transmission axis of the polarizers.

Rodlike nematic phases undergo flow alignment during capillary filling. The preferred macroscopic orientation, with the director parallel to the long axis of the capillary, remains some time after filling. Compared with disclike nematic phases, it remains birefringent because of surface alignment with the optical axis of the rods perpendicular to the viewing direction.

In view of these facts, it is possible to distinguish rodlike from disclike nematic phases. This result can be checked by comparing the conclusions with the phase diagram. A nematic phase of rodlike micelles has to be followed by a hexagonal phase going to higher surfactant concentrations. After a nematic phase of disclike micelles, there is always a lamellar phase.

Intrinsic and form birefringence

The birefringence is defined as $\Delta n = n_{\text{parallel}} - n_{\text{perpendicular}}$. Parallel and perpendicular refers to the symmetry axis of the nematic phase. There are two parts of birefringence: intrinsic and form birefringence. The intrinsic birefringence is due to the anisotropy of the single surfactant molecule mainly caused by the anisotropy of the C-C bonds in the alkyl chain. But the headgroup and counterion also contribute to the anisotropy. The reason for form birefringence is a regular stacking of particles in a continuous medium, here the aggregation to anisometric micelles. In a hexagonal or a nematic phase of rodlike micelles, the symmetry axis of the surfactant molecule is perpendicular to the symmetry of the phase, if the surfactant molecules are looked upon as cylindrical symmetric. Form and intrinsic birefringence have different signs. In disclike nematic or lamellar phases, the two symmetry axes are parallel. Both parts contributing to the birefringence have equal signs. Therefore, there is no connection between the sign of intrinsic birefringence and total birefringence, the only one which can be determined by polarization microscopy.

The sign of total birefringence

The λ-plate is a quartz plate with a small refractive index n_α and a large one n_γ perpendicular to n_α. The angle between the transmission axis of the polarizers and the optical axis of the quartz plate is 45°. If white light is used, red I (550 nm) is the interference colour of this λ-plate [13]. A macroscopic aligned sample with a small refractive index n'_α and a large one n'_γ can be put under the microscope in two ways: The large refractive indices of λ-plate and sample n_α and n'_α are parallel. The birefringence of the two objects add and the interferrence colour turns blue. If n_α is parallel to n'_γ, there is subtraction and the interference colour turns yellow.

For rodlike nematic phases, flow alignment can be used to achieve macroscopic orientation, because rods align with their optical axis parallel to flow. Discs, however, align with their symmetry axis perpendicular to the direction of flow. There is no defined orientation. In the case of disclike aggregates, surface alignment at the edges of the capillary is used to determine the sign of the total birefringence.

Alignment can also be achieved by a magnetic field (2 T). Discs and rods with positive $\Delta\chi$ align with their optical axis parallel to the direction of a magnetic field. The influence of the magnetic field is much larger than the wall effects which can be observed within a small region near the glass surface. If the sign of $\Delta\chi$ is negative, no definite orientation can be achieved by a magnetic field. All directions perpendicular to the field are energetically equivalent. Only the orientation parallel to the field has a higher level of potential energy. So both wall effects and magnetic field have to be used to investigate such systems. Table 2 gives a summary of the behaviour of nematic phases when a magnetic field

Table 1. Features of the binary nematic phases

	CTAB	$C_{16}C_2DMABr$	$C_{16}C_4DMABr$
Shape of micelles	rodlike	rodlike	rodlike
Sign of total birefringence	negative	negative	negative
Sign of $\Delta\chi$	positive	positive	positive
Maximum range of temperature	25–38 °C	14–30 °C	20–44 °C
Surfactant concentration for maximum range of temperature	25 % ww	32 % ww	41 % ww
	0.92 mol/kg H_2O	1.25 mol/kg H_2O	1.71 mol/kg H_2O
Maximum range of concentration	23–28 % ww	31–35 % ww	40–42 % ww
Temperature for maximum range of concentration	25 °C	14 °C	21 °C
Maximum quadrupole splitting of the aligned phase (magnetic field 2.1 T)	37 Hz	80 Hz	386 Hz

Table 2. Behaviour of nematic phases in microslides and magnetic fields

	Alignment after capillary filling (flow direction parallel to the y-axis)	Alignment after storing the nematic phase in the capillary for several hours (wall alignment)	Alignment of the director when a magnetic field is applied along the		
			x-axis	y-axis	z-axis
Disc $\Delta\chi > 0$	random orientation x-z plane	director parallel to the z-axis (parallel to the viewing direction)	$\parallel x$	$\parallel y$	$\parallel z$
Disc $\Delta\chi < 0$		no birefringence visible except at the edges of the capillary	$\parallel z$	$\parallel z$	random x-y plane
Rods $\Delta\chi > 0$	director parallel y-axis	random orientation of the director within the x-y plane	$\parallel x$	$\parallel y$	$\parallel z$
Rods $\Delta\chi < 0$			random y-z plane	random x-z plane	random x-y plane

of 2 T and microslides with certain dimensions (Fig. 2) are used. The viewing direction under the polarization microscope is always parallel to the z-axis (Fig. 2).

Results

Binary systems

The binary phase diagrams of CTAB (Fig. 3), $C_{16}C_2DMABr$ (Fig. 4) and $C_{16}C_4DMABr$ (Fig. 5) are similar. Beginning at low surfactant concentrations, the sequence of the phases is isotropic/nematic/hexagonal. The region after the hexagonal phase was not investigated further, because it was not possible to prepare homogeneous bulk solutions. The isotropic solution just below the nematic phase is very viscous. The nematic phases of the three systems consist of rodlike micelles and are very viscous, especially the nematic phase of CTAB. This is probably the reason why the

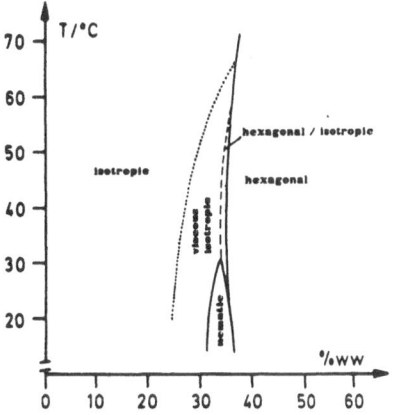

Fig. 4. Phase diagram of the system $C_{16}C_2DMABr$/water

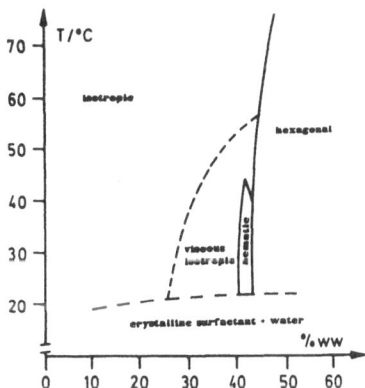

Fig. 5. Phase diagram of the system $C_{16}C_4DMABr$/water

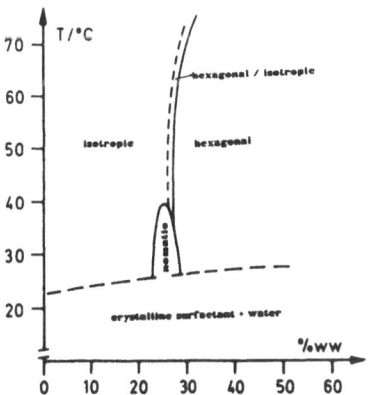

Fig. 3. Phase diagram of the CTAB/water

nematic phase has not been previously discovered in detailed studies of this phase diagram [14]. The phase boundaries between the hexagonal and the nematic phase are difficult to determine and are not precise. Phase separation takes a long time because of the high viscosity. It is not clear whether the system is in equilibrium or not, which is why we marked a two-phase region in the phase diagram (Fig. 3). In the labelled one-phase region, the phases could be clearly assigned using the methods described above. The nematic phases in the binary systems of $C_{16}C_2DMABr$/water (Fig. 4.) and $C_{16}C_4DMABr$/water (Fig. 5) are not so viscous and are easier to detect. A clear boundary between hexagonal and nematic phase can be drawn. The features of the three binary nematic phases are summarized in Table 1.

Figure 6 demonstrates the angular dependence of the quadrupole splitting when the sample tube is rotated at a certain angle in the magnetic field of the NMR spectrometer. A spectrometer with a conventional electromagnet is necessary to perform this experiment.

Figure 7 shows the alignment of the nematic rods in the system $C_{16}C_4DMABr$/water in a set of time dependent ^2H-NMR spectra. The peaks of the final aligned phase result from the shoulders of the initial powder pattern. The director aligns parallel to the magnetic field.

In the binary phase diagram of $C_{16}C_6DMABr$/water, no birefringent liquid crystalline phase was found up to a surfactant concentration of 80% ww. In the region of 80% ww there is perhaps a cubic phase. This has to be investigated by further measurements.

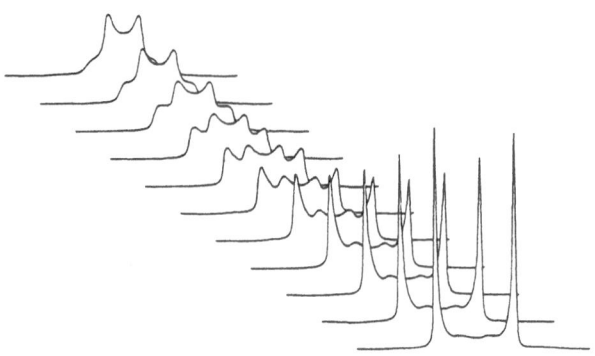

Fig. 7. Alignment of a nematic sample $C_{16}C_4DMABr$ parallel to the magnetic field (time scale 90 min, $T = 44\,°C$, $c = 41\%$ ww)

In the binary system $C_{16}C_8DMABr$/water (Fig. 8) there are two lamellar phases with opposite signs of total birefringence [12]. No nematic phase was found.

Ternary systems

In this investigation, only parts of the ternary phase diagram were studied. Decanol was added stepwise to the isotropic solutions below the nematic phase of the binary systems until a birefringent liquid crystalline phase resulted. The samples (Figs. 9, 10) were then investigated by the methods described.

CTAB/decanol/water

Adding decanol to the isotropic solutions of CTAB/water increases the viscosity. Just below the final birefringent phase, shear induced birefringence can be observed in the isotropic solution — an indication for anisometric micelles.

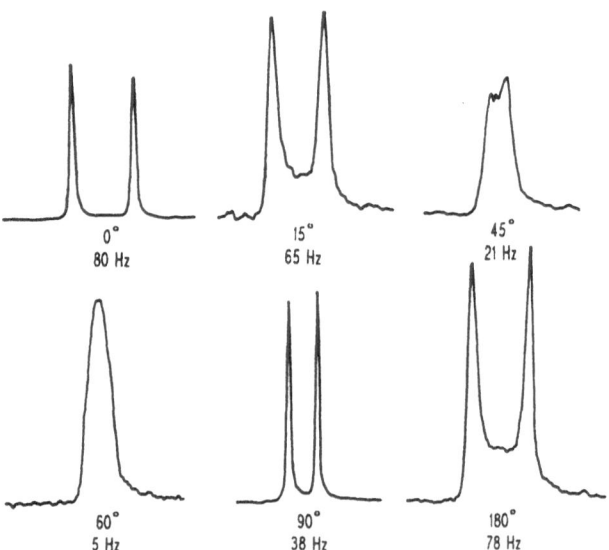

Fig. 6. Quadrupole splitting of an aligned nematic sample $C_{16}C_2DMABr$ as a function of the angle between the director of the sample and the direction of the magnetic field (spectrometer with conventional electromagnet)

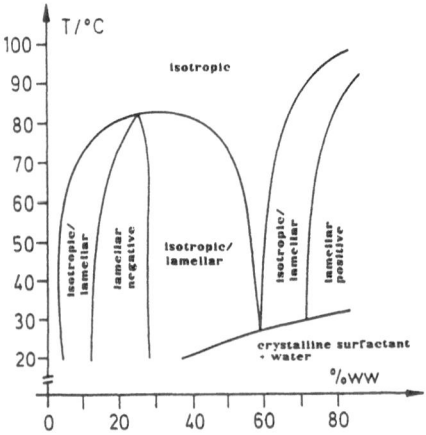

Fig. 8. Phase diagram of the system $C_{16}C_8DMABr$/water

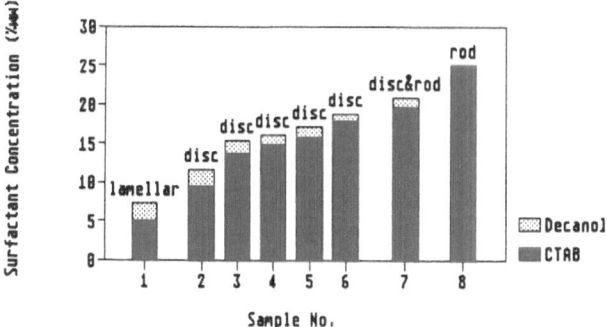

Fig. 9. CTAB/decanol/water − composition of investigated samples

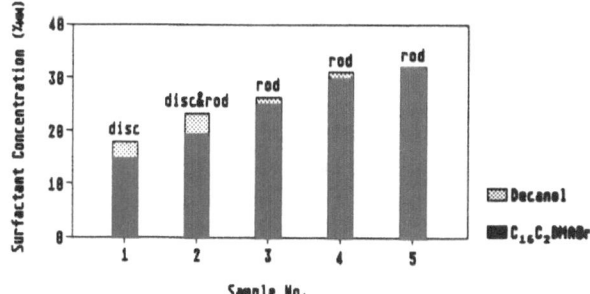

Fig. 10 C$_{16}$C$_2$DMABr/decanol/water − concentration of investigated samples

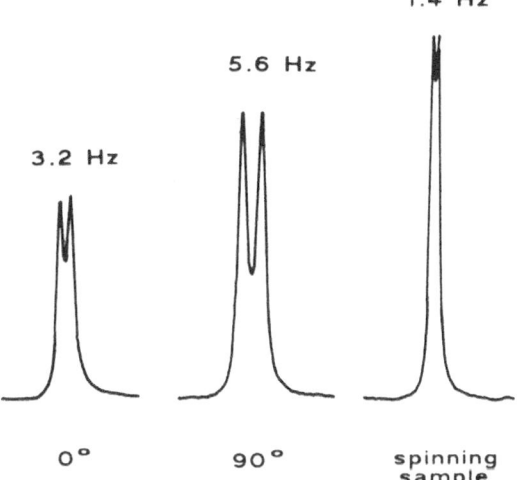

Fig. 11. Nematic phase of the system CTAB/decanol/water consisting of disclike micelles with wall alignment

Adding small amounts of decanol near the binary nematic phase produces a nematic phase of mixed micelles with the same features as the binary one. The quadrupole splitting, however, becomes a bit smaller.

Starting with low surfactant concentrations, much more decanol in necessary to obtain a liquid crystalline phase. This phase has low viscosity compared to the binary nematic phase. Figure 11 shows a set of ^{2}H-NMR spectra of the aligned sample. The quadrupole splitting in the 0° position is smaller than the splitting in the 90° position: the opposite of a nematic phase with positive $\Delta\chi$. How is it possible to find an explanation for this behaviour? The problem was solved by looking at the aligned sample with crossed polarizers and experiments under the polarization microscope. The nematic phase consists of disclike micelles which align with their optical axis perpendicular to a magnetic field ($\Delta\chi < 0$). The total birefringence is positive. The following happens when this sample is placed in an NMR spectrometer with a magnetic field perpendicular to the axis of the sample tube: the director has to align perpendicular to the field. Polar interactions be-

tween glass surface and headgroups produce the second constraint. The director has to align perpendicular to the surface of the probe tube. These two conditions effects a homogeneous orientation. In the 90° position, the director is parallel to the magnetic field and therefore the quadropole splitting is higher than the 0° position. The rotated spectrum splitting is a quarter of the maximum splitting and fits this interpretation.

Starting with a CTAB concentration between these two samples, the resulting sample consists of two coexisting nematic phases. Although phase separation was extremely slow, separated nematic phases were obtained when the samples were equilibrated for 2 months. Figure 12 shows a micrograph of the separated system (top: nematic rodlike; middle: nematic disclike; bottom: isotropic).

C$_{16}$C$_2$DMABr/decanol/water

Decanol was added to the isotropic solutions as described. Figure 10 shows the concentrations of the investigated solutions. The result is the same as in the system CTAB/decanol/water. Near the binary nematic phase is a nematic phase of mixed micelles with the same features. This region of rodlike mixed micelles is much larger than in the system of CTAB/decanol/water. The quadrupole splitting becomes smaller with increasing addition of decanol and with decreasing surfactant concentration (Fig. 13, samples 1, 2, 3).

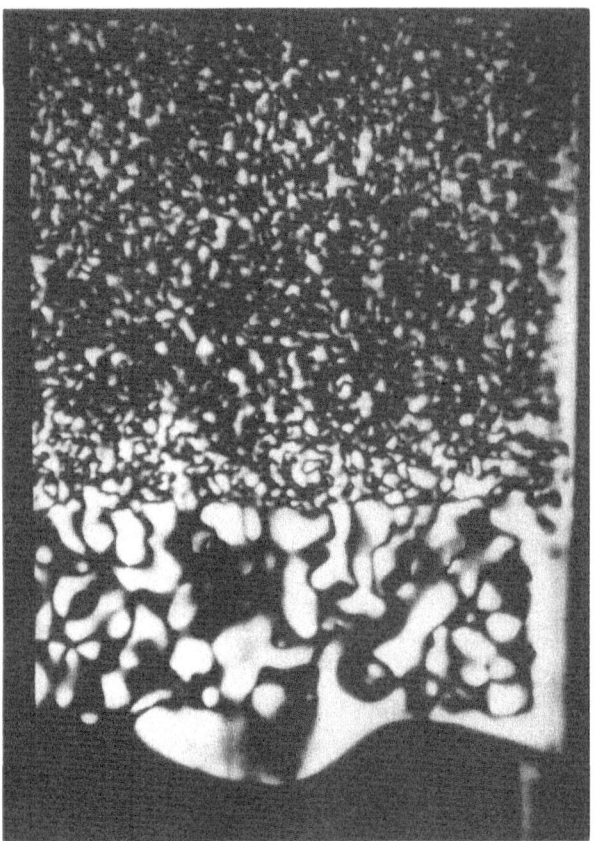

Fig. 12. Macroscopic phase separation of the two nematic phases in the system CTAB/decanol/water (top: rodlike micelles; bottom: disclike micelles)

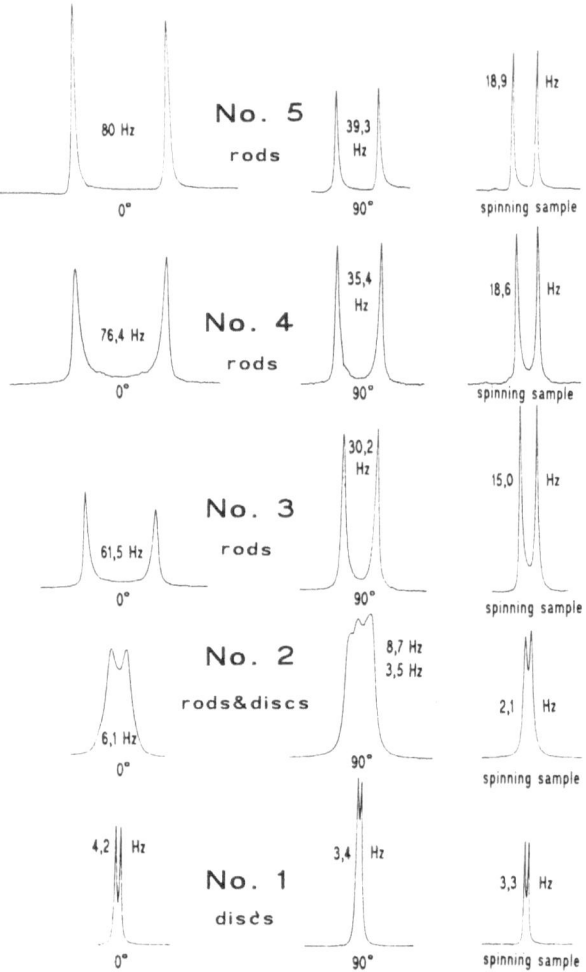

Fig. 13. Transition from nematic rods to nematic discs in the system $C_{16}C_2DMABr$/decanol/water

Sample No. 5 consists of nematic discs with negative $\Delta\chi$ and positive total birefringence. The viscosity is much smaller than that of the rodlike nematics. Looking at the set of ^{2}H-NMR spectra (Fig. 13, sample No. 5) one can see that wall effects, as are found in the system CTAB/decanol/water, are missing. The quadrupole splitting is nearly independent of the rotation angle.

The spectrum of sample No. 4 (Fig. 13) shows the coexistence of nematic rods ($\Delta\chi > 0$) and nematic discs ($\Delta\chi < 0$) without macroscopic phase separation. After two months there was macroscopic phase separation, too, with the rods at the top and the discs at the bottom of the sample tube.

Further addition of decanol always resulted in the formation of a lamellar phase in both ternary systems. When decanol was added to the isotropic solutions of $C_{16}C_4DMABr$ and $C_{16}C_6DMABr$, the product was always a lamellar phase. Perhaps there is a small

nematic region, too, but we did not have enough samples to find it.

Conclusions

All the results taken together suggest that within a certain class of surfactant, nematic phases can be prepared systematically, following a certain rule. The determining factor is the ratio of the cross section of the headgroup to chain. Polar or charge interactions can be looked upon as contributing to the area of the headgroups, which means that repulsive interactions give the same result as big headgroups. Arranging the liquid crystalline phases in order of increasing ratio of area of headgroup to area of chain one finds: reversed phases → lamellar phase → nematic discs → hexagonal phase

→ nematic rods → cubic phase of spherical micelles. This ratio is a function of concentration, too. With increasing surfactant concentration there is a more dense packing of the headgroups, which is the reason why lamellar phases always occur at higher surfactant concentrations if a system builds up both hexagonal and lamellar phases.

Systematic variation of headgroup and alkyl chain is obviously the right method for systematic preparation of nematic phases. The number of surfactants which are actually able to produce nematic phases in the binary system is not very large.

Acknowledgement

We would like to thank the DFG and the Fonds der chemischen Industrie for financial support. This research project was carried out as a part of the SFB 213. We also would like to thank Prof. Wokaun for his assistance in the interpretation of the ^{2}H-NMR data. The surfactants were a gift of Hoechst AG, Gendorf.

References

1. Boden N, Jackson PH, McMullen K, Holmes MC (1979) Chem Phys Lett 48:476
2. Radley K, Tracey AS (1984) Mol Cryst Liq Cryst Lett 1 (3–4):95–103
3. Boden N, Radley K, Holmes MC (1981) Mol Phys 42 (2):493–496
4. Hoffmann H, Oetter G, Schwandner B (1987) Progr Colloid Polym Sci 73:95–106
5. Luehmann B, Finkelmann H (1986) Colloid Polym Sci 264:189–192
6. Fontell K, Lindman B (1983) J Phys Chem 87:3289
7. Boden N, Jackson PH, McMullen K, Holmes MC (1979) Chem Phys Lett 65:476
8. Reizlein K, Hoffmann H (1984) Progr Colloid Polym Sci 69:83–93
9. Herbst L, Hoffmann H, Kalus J, Reizlein K, Schmelzer U (1985) Ber Bunsenges Phys Chem 89:1050
10. Saupe A, Boonbrahm P, Yu LJ (1983) J Chim Phys 80:7–13
11. Charvolin J, Hendrikx Y (1985) Phys Rev B 33 (5):3534
12. Schwandner B (1986) Thesis, Bayreuth
13. Beyer H (1977) Handbuch der Mikroskopie. 2 Aufl, VEB, Berlin
14. Hakemi H, Varanasi PP, Tcheurekdjian ? (1987) J Phys Chem 91:120–125

Received December 10, 1987;
accepted December 21, 1987

Authors' address:

G. Hertel
Lehrstuhl für Physikalische Chemie I
Universität Bayreuth
Postfach 10 12 51
D-8580 Bayreuth, F.R.G.

Progress in Colloid & Polymer Science   Progr Colloid Polym Sci 76:132–139 (1988)

Selective incorporation of dyes with fluorocarbon and hydrocarbon chains into coexisting micellar phases of sodium perfluorooctanoate and dimethyl tetradecyl aminoxide

F. H. Haegel and H. Hoffmann

Lehrstuhl für Physikalische Chemie I der Universität Bayreuth, Bayreuth, F.R.G.

Abstract: Solutions of sodium perfluorooctanoate (SPFO) and dimethyl tetradecyl aminoxide (TDMAO) show a miscibility gap which can be determined by surface tension measurements. Within this demixing region, the initially clear viscous solutions are metastable and split after some weeks into two phases of which the lower one is very turbid and contains a lamellar phase.

In metastable solutions, two kinds of micelles coexist. This is shown by contrast variation in light scattering using water-glycerol mixtures. The necessary premise that there is no basic difference between solutions in water and water-glyerol 50 : 50 w/w was proved by ^{19}F-NMR measurements.

Both kinds of aggregates can be stained selectively by corresponding cyanine dyes. When phase separation occurs, the lower perfluorosurfactant-rich phase contains mainly fluorinated dye, whereas in the upper phase the normal dye is enriched. The phase separation can be accelerated by ultracentrifugation.

Coexisting micellar aggregates can be used for the spatial segregation of two photosystems which are necessary to drive photolytic water splitting by a two-photon process.

Key words: Micelles, sodium perfluorooctanoate, dimethyl tetradecyl aminoxide, miscibility gap, centrifugation.

Introduction

When perfluoro- and hydrocarbon-surfactants are simultaneously present in aqueous solutions it is possible that they form mixed micelles or perfluoro- and normal micelles separately with some mutual solubility.

Examples of both types have been described in the literature [1–14]. There is now strong experimental evidence that demixing on a microscopic level can indeed take place. While perfluoro- and hydrocarbon chains generally tend to demix and binary systems are known which have a miscibility gap, in micelle forming systems the situation is more complicated in as much as the Gibbs free energy for mixing or demixing contains contributions from headgroup and counterion interaction and this makes it difficult to predict whether a particular combination of surfactants will or will not demix. Ottewill et al. [1] have recently reported that in

mixtures of both surfactants, the extension of the chain length by one or two CH_2- or CF_2-groups can make the difference for mixing or demixing.

They also proposed the idea that some aggregation between CF_2- and CH_2-chains might already occur on one and the same micelle without actual demixing of the micelles taking place. The question of demixing is not always easy to answer and some of the experimental results have been discussed controversially [1–3]. With more and more data becoming available on mixed systems, it is now becoming clear that some experiments give a clear cut answer while other results are more difficult to interpret unambiguously.

SANS measurements with contrast variation seem to reveal clearly whether demixing or mixing takes place [1]. In mixed micelles, only one type of micelles is present and the scattering intensity therefore has to disappear when the scattering density of the solvent is matched to the one of the micelles by variation of the H_2O/D_2O ratio. When demixing takes place, it is only possible to match one kind of micelles and the scattering intensity cannot pass through zero. Asakawa et al. [4] have recently presented another clear-cut experiment to prove microscopic demixing. They were successful in separating the two types of micelles by gel filtration. In this investigation we will study mixtures of sodium perfluorooctanoate (SPFO) and dimethyl tetradecyl aminoxide (TDMAO). This combination has some special features which have not been observed in other mixtures. Both individual systems have been studied. SPFO forms rather small globular micelles up to high concentration, whereas TDMAO forms rodlike micelles above 10 mM [15]. We shall present evidence which clearly shows the existence of separate micelles in mixtures of the two surfactants. For sufficiently high surfactant concentrations, the system will also macroscopically demix. Due to the high viscosity of the system, macroscopic demixing is a rather slow process, however.

For comparison, we will also give some results for the mixing system of the cationic fluorosurfactant 1,1,2,2,3,3,4,4-octahydroperfluorooctyl trimethyl ammonium iodide (4CFOTAI) and TDMAO.

Experimental

Materials

Dimetyl tetradecyl aminoxide was received from Hoechst as a 25% solution, freeze dried and recrystallized from acetone [15].

For the preparation of sodium perfluorooctanoate, perfluorooctanoic acid (from Hoechst) was dissolved in water/methanol 3:1 and titrated by sodium hydroxide. Methanol was removed by evaporation and water by freeze drying. The product is washed with ether/acetone.

1,1,2,2,3,3,4,4-Octahydroperfluorooctyl trimethyl ammonium iodide was a gift from Hoechst and used without further purification.

2-Methyl-1-(1,1,2,2-tetrahydro)-perfluorodecyl pyridinium bromide was a gift from Hoechst and used without further purification.

4-Methyl-1-tetradecyl chinolinium bromide was synthesized by the method of Shelton et al. [16]. 4-Methyl quinoline pract. (> 97%) and bromotetradecane purum from Fluka were used.

For the prepatation of 2-(2-(4-dimetyhl aminophenyl)ethenyl)-1-(1,1,2,2-tetrahydro)perfluorodecyl pyridinium bromide (dye 1) 4-dimethyl aminobenzaldehyde z. S. from Merck was condensed with 2-methyl-1-(1,1,2,2-tetrahydro)-perfluorodecyl pyridinium bromide from Hoechst by the method of Clemo and Swan [17].

4-(2-(4-Dimethyl aminophenyl)ethenyl)-1-tetradecyl chinolinium bromide (dye 2) was also synthesized according to Ref. [17].

The water for preparing the solutions was demineralized and distilled. For the water-glycerol mixtures, 87% glycerol p. a. from Merck was used.

Measurements

Surface tension was measured by the method of du Noüy with a Lauda tensiometer. Refractive index increments were determined with a differential refractometer Chromatix KMX-16 at 632.8 nm. Small angle light scattering measurements were carried out with a Chromatix KMX-6 at the same wavelength. NMR spectra were recorded with a JNM-FX90Q Fourier transform NMR spectrometer from Joel. Electric birefringence was measured with an apparatus described elsewhere [18]. Viscosity measurements for solutions with low viscosities were carried out in an OCR-D oscillating capillary rheo- and densitometer from Chempro Paar. For high viscosities (above 10 mPas) a Rheometrics Fluids Rheometer was used. The sedimentation was carried out in a Centrikon T-2050 ultracentrifuge with a fixed-angle rotor. The solutions were centrifuged at 250,000 g (42,000 cpm) for 5 h. 2-ml samples were taken and diluted with 2 ml of acetone to avoid light absorption by scattering of the surfactant aggregates. All measurements were carried out at 25 °C.

Results

Surface tension measurements

For the demixing system SPFO/TDMAO, the surface tension of the mixtures is lower than that of both pure surfactants, whereas for the mixing system 4CFOTAI/TDMAO the surface tension changes continuosly (Fig. 1). Between the molar ratio 0.2 and 0.85 for TDMAO, the surface tension for SPFO/TDMAO mixtures is independent of the composition. As the surface tension is determined by the monomer concentration of both the perfluoro- and the hydrocarbon surfactant, this is an indication that both monomer concentrations are constant in this range. The simplest

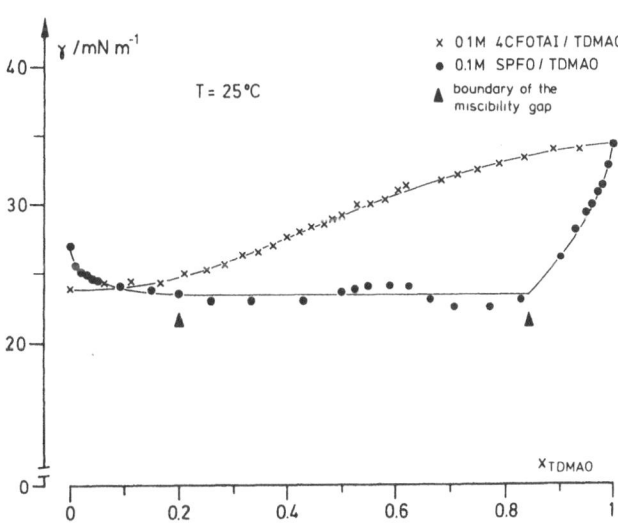

Fig. 1. Surface tension at 0.1 M total surfactant concentrations of SPFO/TDMAO and 4CFOTAI/TDMAO against molar ratio of TDMAO

way to explain this is by assuming demixing of the micelles [2, 9].

The situation can be compared to the total vapor pressure over a mixture of two liquids with a miscibility gap. For constant temperature, the vapor pressure in the miscibility gap is constant and determined by the vapor pressure of the two phases which are in equilibrium with each other. At the concentration of 100 mM (more than three times the CMC of SPFO and 1000 times the CMC of TDMAO) the composition of the micelles is nearly identical with that of the whole solution. This means the mole fractions at which the surface tension becomes independent of the composition also mark the boundaries of the miscibility gap and give the composition of the coexisting micelles.

CMC values

The CMC values measured by surface tension, on the other hand, also show a minimum at a molar ratio of TDMAO 0.7 (Fig. 2). This behavior is typical for surfactants with synergistic effects which mix preferentially and seems to be a contradiction to the result given above. This system, however, is more complicated.

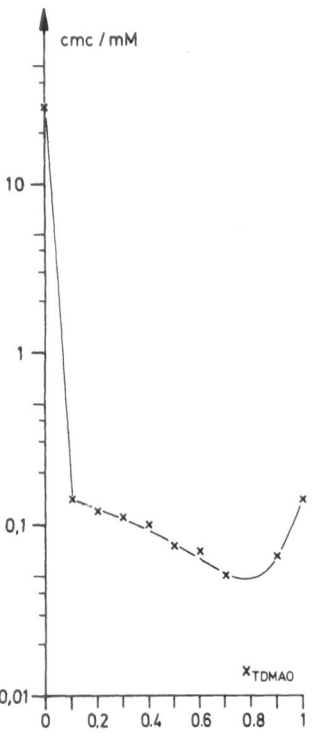

Fig. 2. CMC of SPFO/TDMAO against molar ratio of TDMAO

Most of the changes of surface tension and CMC values occur at rather small fractions of the second component. Obviously, in this range, mixed micelles are formed which are more stable than the micelles of the single components. This behavior is very different from that which has been reported in mixtures of anionic perfluoro- and hydrocarbon surfactants. In those systems it was found that the mixtures have higher CMC than the single components [6–8]. The data look similar to results obtained in mixtures of anionic and zwitterionic surfactants [19].

In combinations of this type, the nonionic surfactant acts to some degree as a cationic surfactant, because the positive charge of the zwitterionic headgroup is closer to the micellar core than the negative charge. This leads to a considerable repulsion between the headgroups and higher CMC values compared to those of nonionics like polyglycol ethers. In the mixed system, the repulsive interaction is reduced and the micelles become more stable.

This stabilization occurs when the zwitterionic surfactant is added to the anionic one and also when the anionic is added to the zwitterionic surfactant. In this respect, the mixtures of the perfluoro- and hydrocarbon surfactant seem to behave in the same way as mixtures of hydrocarbon surfactants. At both extreme sides of the diagram, the charge effects are overpowering any demixing tendency which might be present in the system due to the incompatibility of the perfluoro- and hydrocarbon chains.

pH-effect

pH measurements for different mixing ratios show that the weak base TDMAO becomes stronger when SPFO is added and the pH increases (Fig. 3). This behavior was already known for mixing with hydrocarbon anionic surfactants [20]. A strong interaction of the protonated aminoxide and the perfluorooctanoate anion can be suggested. For 100 mM solutions, however, the amount of protonated TDMAO is low — 25 mM at the pH-maximum. Nevertheless, the solubility of SPFO in TDMAO-rich micelles is higher than for an ideal solution and the CMC is therefore lower than that of TDMAO. Beginning from the pH-maximum, the CMC increases, but always remains lower than for ideal mixing.

Electric birefringence

The high degree of mixing for TDMAO-rich solutions can also be shown by electric birefringence.

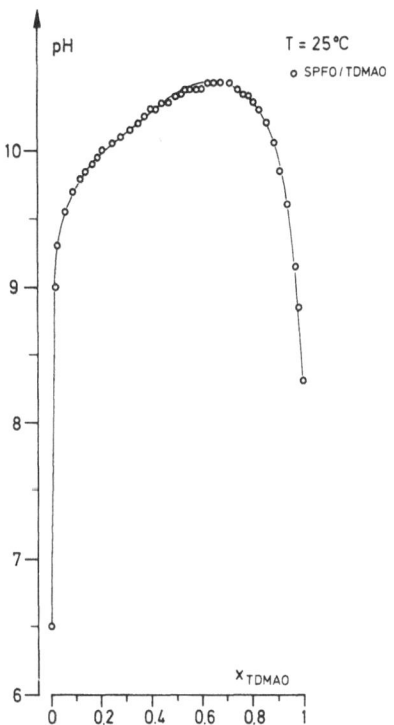

Fig. 3. pH at 0.1 M total surfactant concentration against molar ratio of TDMAO

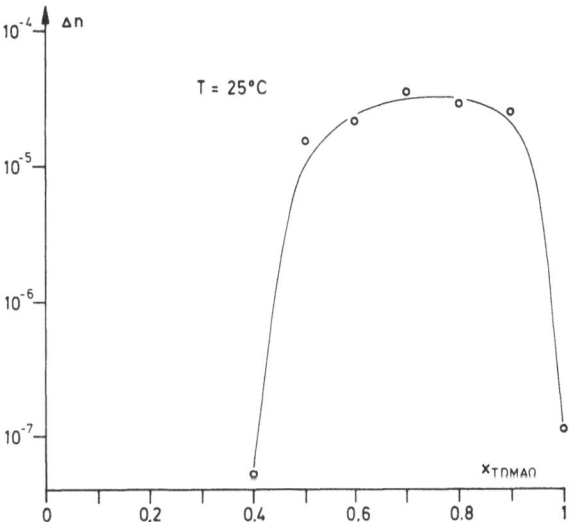

Fig. 4. Birefringence of 0.1 M mixtures of SPFO/TDMAO at 3000 V cm^{-1} and 3 ms pulse length against molar ratio of TDMAO

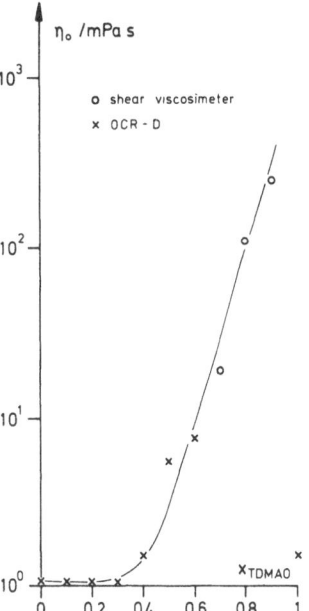

Fig. 5. Zero shear viscosities of 0.1 M mixtures of SPFO/TDMAO against molar ratio of TDMAO

TDMAO forms rodlike micelles and the solutions exhibit birefringence. On mixing with SPFO, the birefringence increases by two orders of magnitude (Fig. 4) and the relaxation times become much longer. This means the interaction or the size of the particles increase dramatically. For molar ratios of aminoxide lower than 0.4, the birefringence disappears and only globular micelles are present, as in pure SPFO.

Viscosity measurements

This behavior can also be detected by viscosity measurements. Whereas TDMAO and SPFO have low zero shear viscosities, mixtures of both components with TDMAO molar ratios higher than 0.3 show higher viscosity (Fig. 5). It is quite certain that the increase of viscosity is mainly due to an increase of interaction, because the effect is strongest at low SPFO concentrations. There, the ionic strength is also low and the shielding of the charged micelles is weak. The considerable increase of the interaction, even for small SPFO contents, can only be understood, if it is incorporated into the TDMAO micelles to a high degree.

Light scattering

Whereas for high TDMAO contents and low concentrations the system is strongly mixing, a miscibility gap is found for TDMAO molar ratios between 0.2 and 0.85 in 100 mM solutions. The coexistence of two kinds of micelles can be shown by light scattering with contrast variation. In water-glycerol mixtures 50 : 50 w/w the refractive index increment for SPFO becomes negative, whereas that of TDMAO decreases to a lower positive value. For a TDMAO molar ratio of 0.4, the refractive index increment vanishes and micelles of this composition have no contrast to the solvent (Fig. 6). So the scattering intensity for those micelles should also be zero. This is, however, not the case. The Raleigh ratio is not zero at this point nor in the vicinity (Fig. 7), which means that micelles of this composition do not exist. The scattering intensity is nearly independent between the aminoxide molar ratio of 0.2 and 0.8. This behavior would be expected for coexisting micelles.

NMR measurements

The change of the solvent may affect the solution behavior of the surfactants and the measurements in water-glycerol mixtures may not be transferable to water solutions. The effects of solvent change, however, seem to be small. This can be shown by ^{19}F-NMR. The chemical shift for the CF_3-group changes considerably, when the surfactant molecule changes from the monomeric to the micellar form. As the interchange between both phases is very rapid, the chemical shift is an intermediate value weighted by the molar ratios of

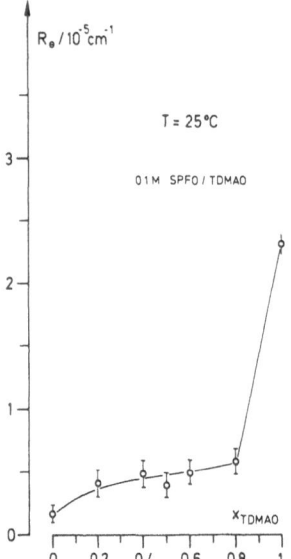

Fig. 7. Light scattering of SPFO/TDMAO in water-clycerol 50 : 50 w/w against molar ratio of TDMAO

both forms [21]. For pure SPFO in water and water-glycerol the typical behavior is found (Fig. 8). A plot of the chemical shift versus the reciprocal fluorosurfactant concentration shows that below the CMC the chemical shift is independent of concentration, whereas above the CMC it decreases linearly with the reciprocal concentration. For indefinitely high concentrations (abscissa value 0), the value in the micellar phase is reached. Figure 8 shows that in water-glycerol mixtures, as well as in water, micelles are formed. The CMC in water-glycerol is somewhat lower than in

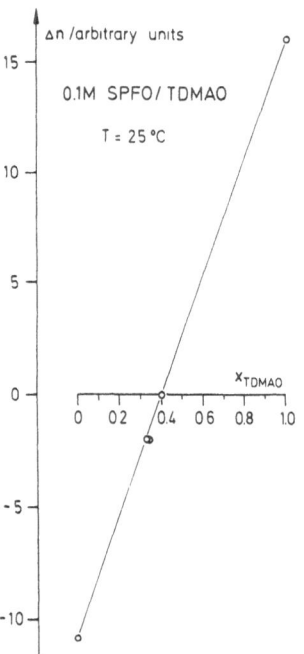

Fig. 6. Refractive index difference of mixtures of SPFO/TDMAO to the solvent water-glycerol 50 : 50 w/w against molar ratio of TDMAO

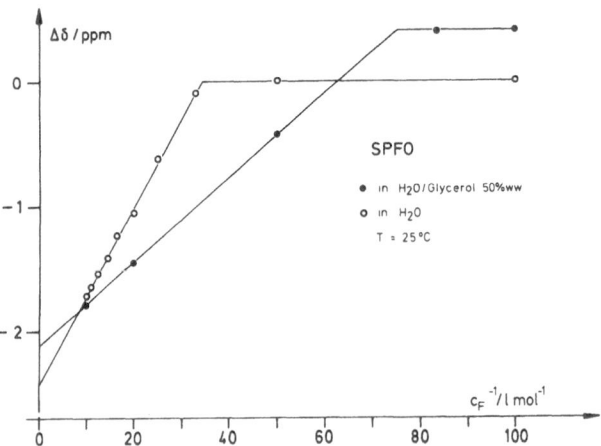

Fig. 8. Chemical shift of the CF_3-group of pure SPFO in water and water-glyerol 50 : 50 w/w against reciprocal fluorosurfactant concentration

water — 13 mM instead of 29 mM. The difference in the chemical shift is due to the difference in the magnetic susceptibility of the solvents. Figure 9 shows the chemical shift for a constant surfactant concentration of 100 mM and different mixing ratios of SPFO/TDMAO in water and water-glycerol. The curves are nearly identical, so as there is a difference in the susceptibilities, there is only a difference in the monomer concentrations, whereas the behavior is not changed.

Ultracentrifugation

Both kinds of coexistent micelles must have very different densities and should be easily separated by centrifugation. In order to detect the separation, the solutions were stained by dyes (1) and (2) (Fig. 10). It

was expected that the dye with the normal chain would be better solubilized by TDMAO-rich, and that with the fluorinated chain by SPFO-rich, micelles. For the system 4CFOTAI/TDMAO the sedimentation of only one kind of micelles was expected.

The concentration profiles were detected by the extinction values at the wavelength maxima of both dyes. For 4CFOTAI/TDMAO (Fig. 11) a monomodal distribution was found. For SPFO/TDMAO (Fig. 12) the distribution was bimodal. Whereas for 4CFOTAI/TDMAO the ratio of the extinction values of both dyes was constant, in the demixing system SPFO/TDMAO the fluorinated dye was enriched on the bottom of the sample tube (Fig. 13). This clearly shows that two kinds of micelles coexist which can be stained selectively by dyes with corresponding alkyl chains.

Stability

The investigated 100 mM solutions of SPFO/TDMAO are stable for several weeks or months. Then some clouds appear and the solutions split up into two phases within the demixing region given above. The lower phase also contains some clouds which are found to be formed by a lamellar liquid crystalline

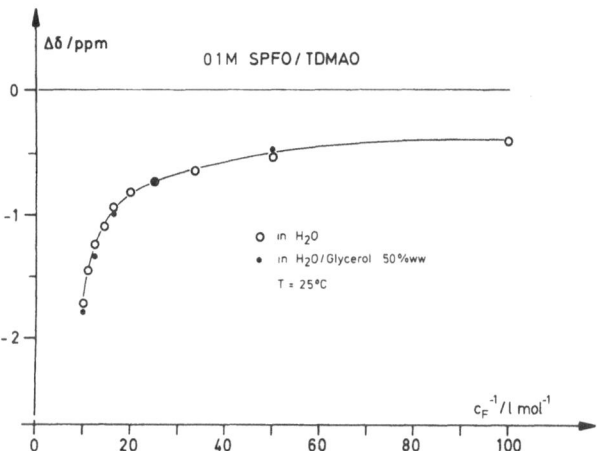

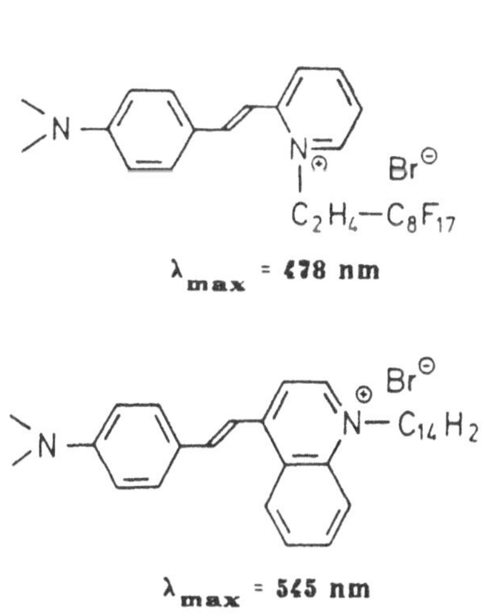

Fig. 9. Chemical shift of the CF$_3$-group of 0.1 M SPFO/TDMAO mixtures in water and water-clycerol 50 : 50 w/w against reciprocal fluorosurfactant concentration

Fig. 10. Dyes used for staining micelles

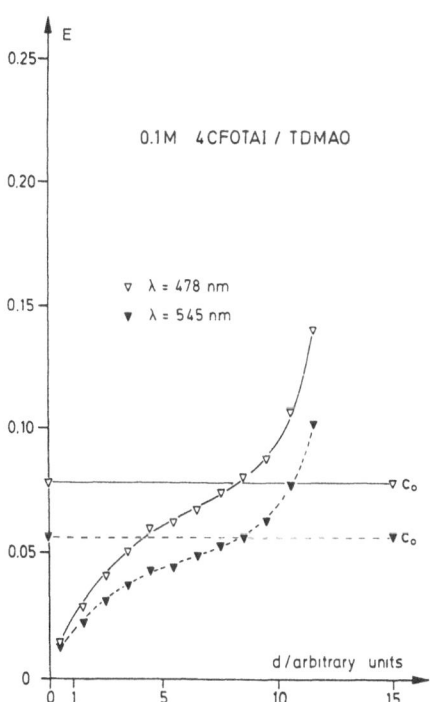

Fig. 11. Extinction of a stained 0.1 M mixture of 4CFOTAI/TDMAO 1 : 1 against distance of the rotor center after centrifugation

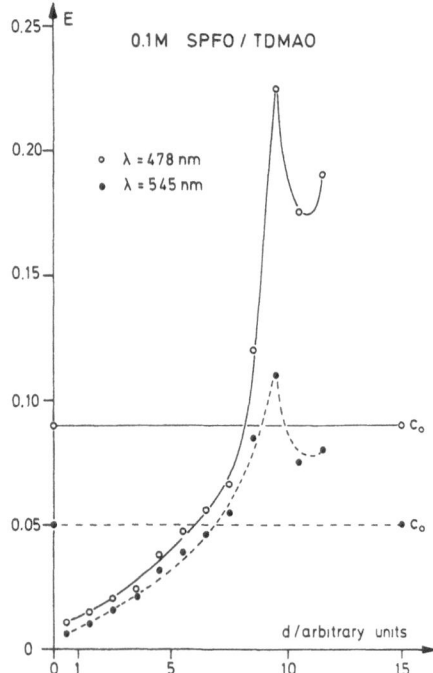

Fig. 12. Extinction of stained 0.1 M mixture of SPFO/TDMAO 1:1 against distance or the rotor center after centrifugation

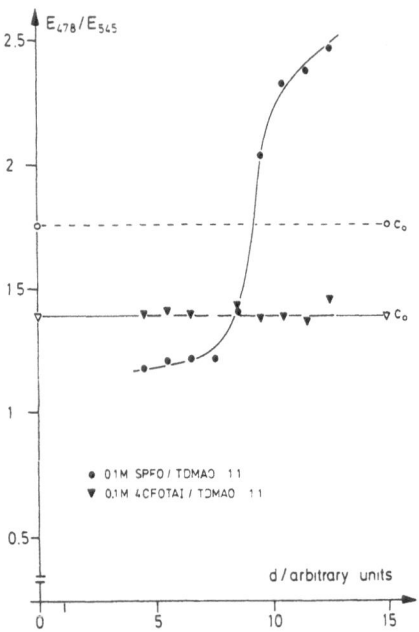

Fig. 13. Ratio of the extinctions of dyes (1) and (2) for 0.1 M SPFO/TDMAO 1:1 and 0.1 M 4CFOTAI/TDMAO 1:1 against distance of the rotor center after centrifugation

phase. This process can be inverted to some extent by shaking, but the solutions never become as clear as the freshly prepared ones.

Conclusions

The surfactant system SPFO/TDMAO has a complicated phase behavior. For low concentrations and high TDMAO contents, the system shows a strong synergistic effect, which is due to electrostatic effects on the micellar surface. A remarkable decrease of CMC and surface tension values is found. Both effects are connected with the increase of pH for mixtures of both surfactants. The preferred incorporation of SPFO into TDMAO increases the interaction and leads to high electric birefringence and viscosity values.

As shown by surface tension measurements, however, the system forms two kinds of micelles for TDMAO molar ratios between 0.2 and 0.85 at 100 mM total surfactant concentration. This is also clearly indicated by light scattering in water-glycerol mixtures and by centrifugation of the solutions. The behavior in water-glycerol mixtures is nearly identical to that in pure water. The centrifugation of the stained solutions shows also that it is possible to incorporate selectively dyes with fluorinated alkyl chains into fluorocarbon surfactant-rich, and dyes with hydrocarbon chains into hydrocarbon surfactant-rich micelles. This behavior can be used to build up two spatially segregated photosystems in micellar solution. This possibility is interesting for research in the field of photolytic water splitting. These systems could be used to mimic natural photosynthesis, a process which is driven by two photons in two coupled but spatially segregated photosystems.

Acknowledgement

Our thanks are due to Hoechst Gendorf company for the gift of some of the substances used.

References

1. Burkitt SJ, Ottewill RH, Hayter JB, Ingram BT (1987) Colloid Polym Sci 265:628–636
2. Funasaki N, Hada S (1980) J Phys Chem 84:736–744
3. Carlfors J, Stilbs P (1984) J Phys Chem 88:4410–4414
4. Asakawa T, Miyagishi S, Nishida M (1987) Langmuir 3(5)821:827
5. Mukerjee P, Mysels KJ (1975) Colloidal Dispersions and Micellar Behavior. The American Chemical Society Symposium 9:239–252
6. Mukerjee P, Yang AYS (1976) J Phys Chem 80:1388–1390
7. Bauernschmitt D, Hoffmann H (1980) Makromol Chem 181:2365–2384
8. Shinoda K, Nomura T (1980) J Phys Chem 84:365–369
9. Funasaki N, Hada S (1983) J Phys Chem 87:342–347

10. Kunitake T, Tawaki S, Nakashima N (1983) Bull Chem Soc Jpn 56:3235–3242
11. Asakawa T, Johten K, Miyagishi S. Nishida M (1985) Langmuir 1:347–351
12. Asakawa T, Miyagishi S, Nishida M (1985) J Colloid Interface Sci 104:279–281
13. Treiner C, Bocquet J-F, Pommier C (1986) J Phys Chem 90:3052–3054
14. Pletnev MY, Remizov YV, Frolov YG (1986) Kolloidnyi Zhurnal 48(1):103–107; (1986) Colloid J USSR Engl Transl 48(1):81–85
15. Hoffmann H, Oetter G, Schwander B (1987) Progr Colloid Polym Sci 73:95–106
16. Shelton RS, van Campen MG, Tilford CH, Lang HC, Nisonger L, Bandelin FJ, Rubenkoenig HL (1946) J Am Chem Soc 68:747–759
17. Clemo GR, Swan GA (1922) J Chem Soc 121:946–947
18. Schorr W, Hoffmann H (1981) J Phys Chem 85:3160–3167
19. Rubingh DN (1979) In: Mittal KL (ed) Solution Chemistry of Surfactants. Vol 1, Plenum Press, New York, pp 337–354
20. Rosen MJ, Friedman D, Gross M (1964) J Phys Chem 68(11):3219–3225
21. Muller M, Birkhahn RH (1968) J Phys Chem 72:583–588

Received December 10, 1987;
accepted December 18, 1987

Authors' address:

F.-H. Haegel
Lehrstuhl für Physikalische Chemie I
Universität Bayreuth
Postfach 10 12 51
D-8580 Bayreuth, F R G

Progress in Colloid & Polymer Science   Progr Colloid Polym Sci 76:140–143 (1988)

Microemulsions and other organized assemblies as media of fractal dimensions. A luminescence probe study

P. Lianos

University of Patras, School of Engineering, Patras, Greece

Abstract: Luminescence decay profiles in the presence of quenchers have been studied in water-in-oil microemulsions and lecithin vesicles. The application of various decay models has been critically examined. It was found that fractal modeling can be applied to describe decay in microemulsions when they allow quenching exchange between microdroplets. Preliminary results show that fractal modeling can also be applied to large lecithin vesicles.

Key words: Fractal, modeling, organized, assemblies, luminescence.

A novel but increasing interest has been shown recently in the modeling of organized assemblies as fractal objects [1]. Thus, electrical conductivity in water-in-oil microemulsions can be explained by treating them as percolation clusters [2], while energy transfer reactions in luminescence quenching experiments in organized assemblies are treated in the same manner as in disordered (or porous) solid structures [3]. In all cases the reaction medium is considered to possess a self similarity, i. e. a dilatation symmetry which allows its quantitative description by means of its fractal dimensions [4].

Meanwhile, microemulsions have been and are repeatedly studied by luminescence quenching methods, where association is made between the concentration of the reacting molecules and the concentration of the microemulsion droplets, by means of statistics of the molecular distribution among droplets. The decay of luminescence under pulsed excitation in the presence of quenchers is described by the following equations

$$I = A_1 \exp\left\{- A_2 t + A_3 \left[\exp\left(- A_4 t\right) - 1\right]\right\} \quad (1)$$

$$A_2 = k_0 + k_e[Q] \quad \text{and} \quad A_3 = [Q]/[M] \quad (2)$$

where k_0 is the decay rate constant in the absence of quencher, k_e the rate constant for the exchange of quenchers between micelles (or microdroplets), $[Q]$

the quencher and $[M]$ the micellar concentration, and A_4 (or k_q) is the quenching rate constant. These equations have been explained in several publications since literature is very rich on this subject [5]. When the exchange of quenchers is negligible, $A_2 = k_0$. On the contrary, improved but more complicated versions of Eq. (2) have been used in cases where the quencher also exchanges between the interface and the continuous phase [6]. In all cases, A_2 and A_3 depend linearly on $[Q]$, provided that $[M]$ stays constant with changing $[Q]$.

Equation (1) has been successfully employed to study a great variety of organized assemblies [5–7]. However, in extreme cases of multicomponent microemulsions and, in particular, water-in-oil microemulsions, departures from linearity between A_2 or A_3 and $[Q]$ are observed [8]. This problem has been directly or indirectly treated in some recent works. Thus, it has been found that large solubilizates affect the micellar size [9], while micellar polydispersity [10] or fragmentation-coagulation [11] complicate the kinetics of luminescence quenching in microemulsions. These questions can be circumvented if luminescence quenching in microemulsions is studied from a different point of view. Thus, in the present work, we will treat microemulsions as fractal objects and test the applicability of the recently developed fractal procedures, borrowed from solid state physics, to them. An organized assembly could be considered as a three-dimensional space embedding two domains of fractal dimensions, one hydrophobic and one hydrophilic. Each microdroplet of a microemulsion is a solubilization site for lumophore and quencher molecules, which are usually chosen to reside in only one of the two domains. In the time scale of the excited lumophore, a microemulsion allows or does not allow exchange of solubilizates between sites, depending on the nature of the water-oil interface [8,11]. A "rigid" microemulsion forces solubilizates to stay resident at sites, i. e. inside the microdroplets. In contrast, in a "non-rigid" microemulsion, the solubilizates can exchange between sites in a way that could be considered equivalent to a random walk of the lumophore from site to site. In all cases, the transient luminescence intensity is described by the following general equation [12,13]:

$$I = I_0 \exp\left(- k_0 t\right) \cdot \sum_{i=1}^{k} \exp\left\{- k_q(r_i)\, t\right\} \qquad (3)$$

where r_i is the center-to-center distance between interacting species. The main feature of this equation is the dependence of the quenching rate constant on space, and consequently on time. If an average k_q extending to all quenchers is considered instead of the time-dependent one, then k_q is proportional to the number of quenchers at a site. Assuming Poisson statistics for the distribution of quenchers, one obtains Eq. (1) from Eq. (3) [12]. It is obvious then that Eq. (1) does not take into account the inhomogeneity of the interaction medium.

The interaction between species diffusing in disordered media has been previously treated [14–17]. The inhomogeneity of space reflects on Eq. (3) as the dependence of k_q on r_i and subsequently on time. The interaction between an excited lumophore and a quencher of concentration $[L^*]$ and $[Q]$, respectively, obeys the differential equation $- d[L^*]/dt = k_q [L^*][Q]$, where k_q is a time-dependent value. If we consider the excited lomophore as a random walker visiting sites occupied by quenchers (traps) then, in analogy to Evesque [17a], $k_q = k'_q N(t)/t$, where $N(t)$ is the total number of sites visited; $N(t) \sim t^{\tilde{d}/2}$, $\tilde{d}$ being the so-called spectral dimension [14–17]. Substitution into Eq. (3) yields the following Eq. (4) which we propose in substitution of Eq. (1):

$$I = I_0 \exp\left\{- k_0 t - (k_q [Q]/f)\, t^f\right\} \qquad (4)$$

where $f = \tilde{d}/2$.

Equation (4) has been used to describe supertrapping processes in doped crystals [17a] where the concentration n_A of the excited donor is much smaller than the concentration n_B of the unexcited acceptors (corresponding to $[L^*] \ll [Q]$ in our case). It is also analogous to the reaction scheme employed in percolation clusters or porous membranes [15]. The value $f \to 1$ when the medium approaches homogeneity; f is indirectly related to the fractal dimension d of the diffusion medium and directly related to the so-called spectral dimension $\tilde{d}$ [14–17]. Thus, $f = \tilde{d}/2$ and $\tilde{d} = 2d/(2 + \delta)$, where δ is a scaling term related to the dependence of diffusion on distance [17]. It should be mentioned at this point that a very comprehensive treatment of intramicellar kinetic processes, where the role of dimensionality is taken into account has been published by Hatlee et al. [7e], but no consideration was given to the fractal behavior of the reaction medium.

Equations (1)–(4) have been tested in a series of water-in-oil microemulsions. Ruthenium tris (2,2'-bipyridine (Ru(bipy)$_3^{2+}$) was the lumophore and Fe(CN)$_6^{3-}$ was the quencher. Luminescence decay profiles were taken with the photon-counting technique and the data were analyzed with least-squares fit. Figure 1 shows a typical decay profile as well as the fitted curve. Notice that the values of the residuals and the autocorrelation function of the residuals have been inserted as visual tests of the quality of the fit [18]. We have found that when exchange of quenchers between droplets is allowed, i. e. when k_e in Eq. (2) has an appreciable value, then and only then is Eq. (4) simultaneously

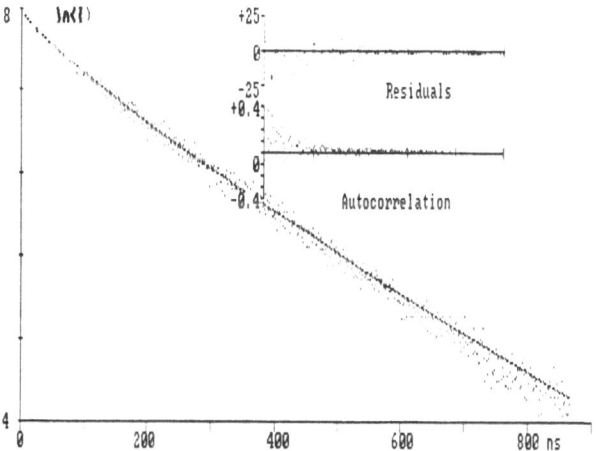

Fig. 1. Luminescence decay profiles of 20 μm Ru(bipy)$_3^{2+}$ in the presence of 1 mM Fe(CN)$_6^{3-}$ inside the microemulsion 2 (see Table 1)

Table 1. Quencher exchange rate constant k_e and fractal exponent f for various microemulsions

Microemulsion · Composition (percentage weight)	k_e $(s^{-1} M^{-1})$	f
(1) Cyclohexane 35, pentanol 40, water 18, SDS 7	1.2×10^9	0.69
(2) Toluene 62, butanol 16, water 14, SDS 8	5.1×10^9	0.77
(3) Toluene 45, butanol 21, water 23, SDS 11	$\sim 10^{10}$	0.93

applicable with Eq. (1). This means that if solubilizates can exchange between microdroplets, then, within a certain time scale, there is an effective communication between droplets establishing a domain of fractal dimensions. This, in turn, is equivalent to treating a microemulsion as a percolation cluster. Some values for the fractal exponent f and for the exchange rate constant k_e are given in the Table 1. Notice that f and k_e grow together. This is to be expected since very large values of k_e are consistent with a rather homogeneous environment where, of course, f is maximum, i.e. $f = 1$.

Luminescence decay profiles in organized assemblies and in the presence of quenchers possess the characteristic non-exponential form. In cases of "strong" quenching, this departure from monoexponentiality is very distinct, while in multicomponent microemulsions or large lipid vesicles it is only slightly observable [19–20]. Several authors have studied these cases, especially with vesicles, but because of the ambiguity of the experimental data they have proposed various, largely differing, fitting models [20]. One such model is the classical Einstein-Schmolukowski equation for diffusion limited reactions in viscous but homogeneous media

$$I = I_0 \exp \{- k_0 t - 4\pi NMRDt$$
$$- 8 (\pi D)^{1/2} NMR^2 t^{1/2}\} \qquad (5)$$

where N is Avogadro's number, M the quencher concentration, R is twice the reaction radius between lumophore and quencher, and D is the diffusion coefficient. We have tried this model with the quenching of $Ru(bipy)_3^{2+}$ by $Fe(CN)_6^{3-}$ in microemulsions, but did not obtain any fit at all. On the other hand, we have performed some preliminary experiments with pyrene excimer quenching in lecithin vesicles, where the model of Eq. (5) had been previously systematically applied [20]. We have obtained satisfactory fits with Eq. (4) as well as with Eq. (5), yielding fractal power f smaller than unity but different from 0.5. We

claim then that quenching in lecithin vesicles can also be modeled by fractals. At any rate, the study of this subject is being pursued further in our laboratory.

Acknowledgement

The author acknowledges financial aid from the Greek General Secretariat of Research and Technology.

References

1. Hilfiker R, Eicke H-F (1987) J Chem Soc Faraday Trans I 83:1621
2. Grest GS, Webman I, Safran SA, Bug ALR (1986) Phys Rev A 33:2842
3. (a) Takami A, Mataga N (1987) J Phys Chem 91:618; (b) Wagner GC, Colvin JT, Allen JP, Stapleton HJ (1985) J Am Chem Soc 107:5589; (c) Lin Y, Nelson MC, Hanson DM (1987) J Chem Phys 86:1586; (d) Sundheim BR (1987) J Colloid Interfac Sci 116:241; (e) Pines-Rojanski D, Huppert D, Avnir D (1987) Chem Phys Lett 139:109
4. Rammal R, Toulouse G (1983) J Phys Lett 44:L-13
5. (a) Atik SS, Thomas JK (1981) J Am Chem Soc 103:4367; (b) Atik SS, Nam M, Singer LA (1979) Chem Phys Lett 67:75; (c) Yekta A, Aikawa M, Turro NJ (1979) Chem Phys Lett 63:543; (d) Baxendale JH, Rodgers MA (1982) J Phys Chem 86:4906; (e) Lianos P, Dinh-cao M, Lang J, Zana R (1981) J Chim Phys 78:497
6. (a) Tachiya M (1975) Chem Phys Lett 33:289; (b) Infelta PP, Gratzel M, Thomas JK (1974) J Phys Chem 78:190; (c) Grieser F (1981) Chem Phys Lett 83:59; (d) Almgren M, Lofroth J-E, van Stam J (1986) J Phys Chem 90:4431; (e) Gelade E, De Schryver FG (1982) J Photochem 18:223; (f) Malliaris A, Lang J, Zana R (1986) J Phys Chem 90:655
7. (a) Lianos P, Lang J, Strazielle C, Zana R (1982) J Phys Chem 86:1019; (b) Zana R, Lianos P, Lang J (1985) J Phys Chem 89:41; (c) Almgren M, Swarup S, Lofroth J-E (1985) J Phys Chem 89:4621; (d) Gelade E, Boens N, De Schryver FC (1982) J Am Chem Soc 104:6288; (e) Hatlee M, Kozak JJ, Rothenberger G, Infelta PP, Gratzel M (1980) J Phys Chem 84:1508
8. Lianos P, Zana R, Lang J, Cazabat A-M (1986) In: Mittal KL, Bothorel P (eds) Surfactants in Solution, Plenum, New York, Vol 6, p 1365
9. Pileni MP, Brochette P, Hickel B, Lerebours B (1984) J Colloid Interface Sci 98:549
10. Almgren M, Lofroth J-E (1982) J Chem Phys 76:2734
11. (a) Malliaris A, Lang J, Sturm J, Zana R (1987) J Phys Chem 91:1475; (b) Lang J, Zana R (1986) J Phys Chem 90:5258

12. Habti A, Keravis D, Levitz P, van Damme H (1984) J Chem Soc Faraday Trans 2 80:67
13. Blumen A (1981) Nuovo Cimento 63:50
14. Blumen A, Klafter J, Zumofen G (1986) In: Pietronero L, Tosatti E (eds) Fractals in Physics. Elsevier Science Publishers, New York, p 399 and references therein
15. (a) Klymko PW, Kopelman R (1982) J Phys Chem 86:3686; (b) idem (1983) 87:4565; (c) Newhouse JS, Argyrakis P, Kopelman R (1984) Chem Phys Lett 107:48; (d) Kopelman R (1986) J Stat Phys 42:185; (e) Prasad J, Kopelman R (1987) J Phys Chem 91:265
16. Djordjevic Z (1986) In: Pietronero L, Tosatti E (eds) Fractals in Physics. Elsevier Science Publishers, New York, p 413
17. (a) Evesque P (1983) J Phys 44:1217; (b) Alexander S, Orbach R (1982) J Phys Lett 43:L-625; (c) Rammal R, Toulouse G (1983) J Phys Lett 44:L-13
18. (a) Grinvald A, Steinberg IZ (1974) Anal Biochem 59:583; (b) Grinvald A (1976) ibid 75:260; (c) Cundall RB, Dale RE (eds) (1983) Time-Resolved Fluorescence in Biochemistry and Biology. Plenum Press, New York
19. Duportail G, Lianos P, Chem Phys lett, in press
20. (a) Vanderkooi JM, Callis J (1974) Biochemistry 13:4000; (b) Kano K, Kawazumi H, Ogawa T, Sunamoto J (1981) J Phys Chem 85:2204; (c) Liu BM, Cheung HC, Chen K-H, Habercom MS (1980) Biophys Chem 12:341

Received December 10, 1987;
accepted December 23, 1987

Author's address:

P. Lianos
University of Patras
School of Engineering
26000 Patras, Greece

Three-phase behavior of brine/alkane/alcohol/alpha-olefinsulfonate mixtures

B. R. Voca[1]), J. P. Canselier[1]), C. Noik[2]) and M. Bavière[2])

[1]) Ecole Nationale Supérieure d'Ingénieurs de Génie Chimique, Chemin de la Loge, Toulouse, France
[2]) Institut Français du Pétrole, Rueil-Malmaison, France

Abstract: Phase diagrams of multicomponent systems containing α-olefinsulfonates are investigated within the framework of enhanced oil recovery. The relationship between the optimum salinity NaCl* and the alkane chain-length is logarithmic at low cosurfactant concentration and linear at higher concentration. At a given brine-to-oil ratio the variation of alcohol concentration alone is almost sufficient to account for the variations in the NaCl* values at the point common to the four Winsor regions. The change in the solubilization parameter SP* is a bit more sensitive to the effect of alcohol than to that of electrolyte. The effect of the overall alcohol-to-sulfonate ratio (A/S) on both NaCl* and SP* is explained with composition data. As the surfactant concentration decreases with constant alcohol content, the A/S ratio in the interfacial structures increases. Thus, the presence of alcohol molecules, reducing the density of surfactant at the interface, decreases surfactant efficiency. Besides, NaCl* decreases when the alcohol concentration increases in each phase. Alcohol concentrations can be calculated by means of the pseudophase model, as can the salt enrichment of the excess aqueous phase when NaCl brines are used.

Key words: Surfactant, microemulsion, phase diagram, α-olefinsulfonate, pseudophase model, surface efficiency, influence of alcohol concentration, Winsor regions.

Introduction

The phase behavior of brine/alkane/alcohol/surfactant mixtures has been studied by quite a number of authors, particularly for use in enhanced oil recovery methods based on chemical flooding. As a matter of fact, oil droplets trapped in porous rock can be displaced with surfactant solutions either by solubilization in a microemulsion or by strongly lowering the interfacial tension between oil and the micellar fluid injected. Interfacial tensions as low as 10^{-3} mN/m or less have been reached [27]. In the formulation adjustment, some parameters are practically fixed, such as the nature of the oil, as well as the temperature and the brine salinity. The adjustable parameters are then related to the water-to-oil ratio (WOR) and the nature and concentrations of amphiphilic compounds, i. e. surfactants and cosurfactants used to form the microemulsion.

The cosurfactant is usually a middle-chain aliphatic alcohol, for instance, one of the isomeric butanols or pentanols or a mixture of those. The main surfactant families, providing a good compromise between cost and performance and preferentially used for this reason in enhanced oil recovery, have been aromatic sulfonates synthetic alkylarylsulfonates and petroleum sulfonates. More recently, however, the behavior of other anionic compounds has been investigated and has shown promising results, especially in high salinity and hardness environments, namely aliphatic sulfonates, e.g. α-olefinsulfonates [1, 2, 4] vinylidenesulfonates [23], sulfonates from isomerized α-olefins [30] as well as primary and secondary alkanesulfonates [1, 31].

The present research was aimed at gaining better knowledge of the phase behavior of α-olefinsulfonates (AOS) in multicomponent systems, e. g. the influence of AOS composition. Indeed, being produced by the sulfonation of α-olefins, followed by the neutralization-saponification of the "acid mix", AOSs contain three main species: alkenesulfonates, hydroxysulfo-

Table 1. Main sulfonate characteristics

C18 AOS sample	Active matter (wt %)	Sulfonated species (wt % of sulfonated matter)			Na$_2$SO$_4$ (wt. %)	Unsulfonated matter (wt. %)
		Alkene[a]	Hydroxy[b]	Disulfonates[c]		
# 1 (Shell)	31	60	30	10	1.5	1
# 2 (Laboratory)	95	92	8	0	1.0	0
# 3 (Witco)	45	70	29	1	0.7	1

[a]) R–CH=CH–(CH$_2$)$_n$–SO$_3^-$Na$^+$; [b]) R–CHOH–(CH$_2$)$_n$–SO$_3^-$Na$^+$, n = 2,3; [c]) R–CH(SO$_3^-$Na$^+$)–(CH$_2$)$_n$–SO$_3^-$Na$^+$ (may include a C=C or an –OH group)

nates and disulfonates in various isomeric forms [3,12,17,24]. AOSs used in detergent products range from C$_{12}$ to C$_{18}$, whereas the longer C$_{18}$–C$_{24}$ compounds were found to better meet the requirements of oil recovery [2,4]. In this study, analytical results also provided the opportunity to check the applicability of the pseudophase model [19,22] to these mixtures.

Experimental

Materials

C$_{18}$ AOS samples were prepared by liquid-phase batch sulfonation in our laboratory or were obtained by pilot scale falling-film reactors. As described elsewhere in more detail [13], the laboratory sample was prepared by the dioxane-SO$_3$ sulfonation of 1-octadecene in 1,2-dichloroethane at – 20 °C; after a 3–4 day aging period the acid mix was titrated potentiometrically, then neutralized in a heterogeneous medium under vigorous stirring by a sub-stoichiometric amount of concentrated sodium hydroxyde solution, in order to avoid the saponification of sultones into hydroxysulfonates. The filtered precipitate was washed several times with an acetone-water mixture, with 95 % ethanol and finally with pure acetone before being dried under reduced pressure.

The surfactants were analyzed by various techniques, including conventional volumetric determinations (active matter, iodine and hydroxyl values, sulfate content), reversed-phase HPLC [14,15] and ^{1}H NMR [16]. The compositions of the samples studied are given in Table 1. Let us particularly notice that the laboratory sample is free of disulfonates and contains only a few percent hydroxysulfonates. Its Krafft point, obtained from conductivity measurements [25] was 40.8 °C, a much lower value than that found by Püschel with a specifically synthesized sodium 2-octadecenesulfonate (54 °C) [26], but much higher than that of a commercial C18 AOS (23 °C) [4]. Such a lowering of the Krafft point due to a mixture of isomers or similar compounds has already been noticed [32].

All the other chemicals were of commercial origin and were used without further purification. They include alkanes, alcohols, sodium and calcium chlorides (Prolabo or similar), and 1-octadecene (Shell, SHOP Process). Distilled water was used.

Phase diagrams and analyses

Phase boundaries were drawn from series of samples of known compositions, which were allowed to settle at constant temperature until phase equilibrium was reached, then analyzed [33]. Concentrations of volatile components (water, alcohol and hydrocarbons) were determined by GLC in a column filled with Porapak Q. Surface-active matter analyses were performed with Hyamine 1622 as a chemical reagent either by two-phase titration or by turbidimetry. The electrolyte content was either obtained by separate determination or simply calculated by difference.

Results and Discussion

1. Effect of alkane carbon number on optimum salinity

For fixed water-to-oil (WOR) and alcohol-to-surfactant (A/S) ratios, the phase diagrams of multicomponent systems containing brine, oil, surfactant and cosurfactant can be conveniently represented with the overall salinity of the brine used and any other parameter as coordinates. This kind of diagram shows the well-known phase transitions called I–III–II in Winsor nomenclature [34] and $\underline{2}$–3–$\bar{2}$ by other authors [19]. These transitions occur between a microemulsion surmounted by an excess oil phase (WI), a three-phase region with a microemulsion as the middlephase (WIII) and a microemulsion surmounting an aqueous phase (WII). The WIV notation refers to a totally miscible system (Fig. 1).

With a hydrophobic alcohol, increasing either the overall salinity at constant ratio of amphiphilic mixture, or the amount of amphiphilic mixture at constant salinity produces such transitions. On the other hand, the reverse sequence is found, at least with anionic surfactants when the temperature [27] or the pressure [20] is increased as well as while going from short-chain to long-chain hydrocarbons [28], or from long-

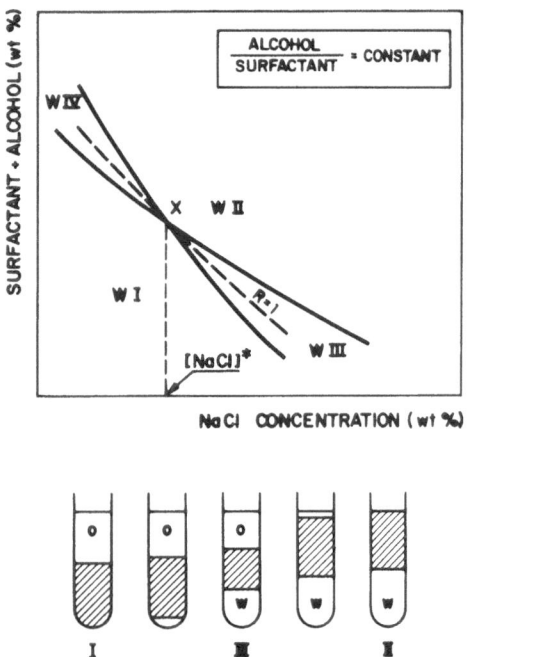

Fig. 1. Schematic phase behavior as a function of the amount of amphiphilic compounds and of brine salinity

chain to short-chain surfactants or alcohols, everything else being kept constant [27].

Several definitions of optimum salinity (NaCl*) have been considered [27] related to the minima of interfacial tensions in the three-phase regime or the equal solubilization of oil and water in the middle phase. By optimum salinity we mean the salinity of the initial brine at which equal weights of oil and water are solubilized in a middle-phase microemulsion. The corresponding points are found on the so-called optimum line whose origin is point X, common to the four Winsor regions (Fig. 1). On this line, Winsor's R ratio is equal to 1 [34].

The variations of optimum salinity (as part of the optimum line) with the alkane carbon number (ACN) are shown in Fig. 2 for fixed WOR (1.5 by weight) and surfactant content (1 wt % in the aqueous phase) and for two A/S ratios (3 and 5 by weight). Linear and logarithmic scales have been used for optimum salinity. The systems show higher optimum salinity, therefore better salt tolerance, with less alcohol. However, the addition of a minimum amount of cosurfactant is

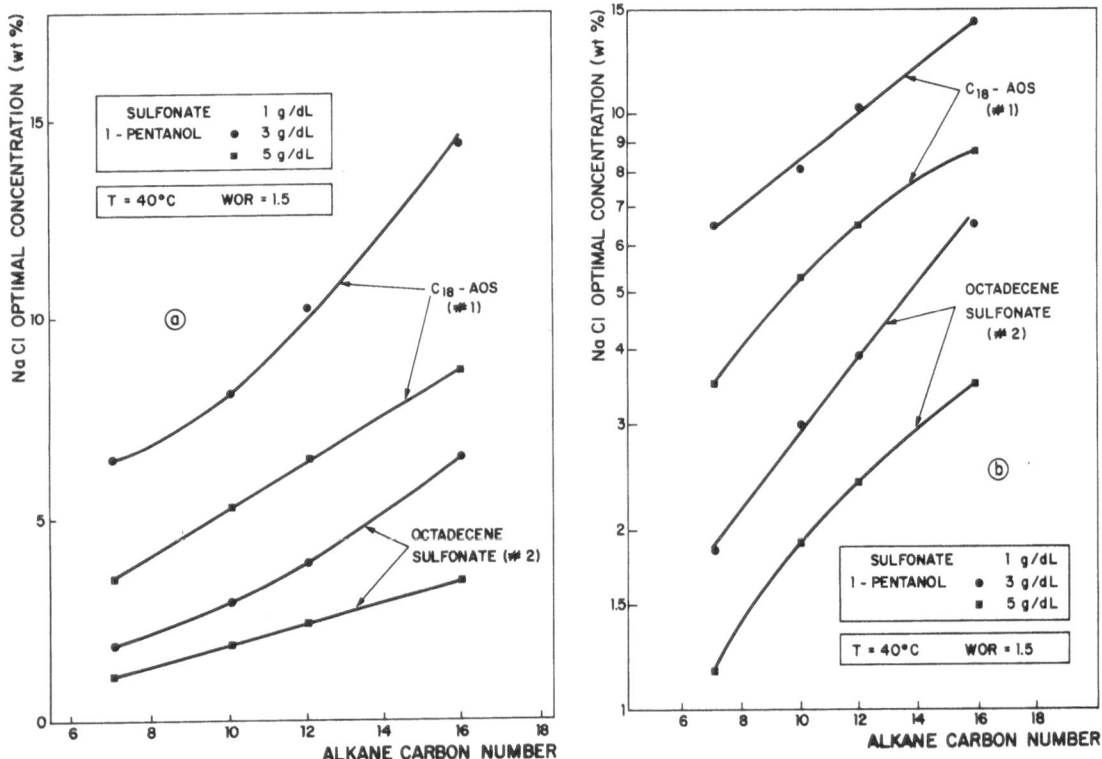

Fig. 2. Optimal salinity as a function of alkane carbon number

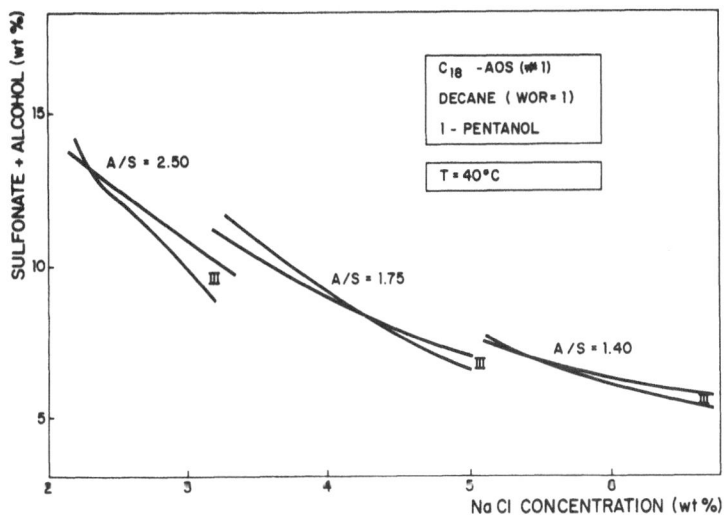

Fig. 3. Amphiphilic mixture concentration as a function of brine salinity (Commercial C18 AOS)

necessary. With 1 % alcohol or less, liquid crystals are observed.

The role played by hydroxysulfonates and disulfonates in improving the salt tolerance is evidenced when the behavior of C18 AOS # 1, which contains such species, is compared with that of almost pure alkenesulfonates (# 2). The effect of AOS composition is so important that for both alcohol concentrations the behavior of octadecenesulfonates (# 2) with hexadecane is practically the same as that of C18 AOS (# 1) with heptane, a difference of nine ACN units.

As expected, the optimum salinity increases regularly with ACN; but the relationship appears to be logarithmic at lower A/S ratio, as found with "conventional" surfactants such as alkylbenzenesulfonates [28], whereas it is simply linear with higher A/S ratio. Such a linear relationship has already been observed with the same type of compounds and high A/S ratios [2] and with nonionic surfactants [11].

2. Alcohol and salt effects on solubilization parameter

Figure 3 shows the behavior of a C18 AOS sample (# 1) in a system containing brine, decane and 1-pentanol at 40 °C. The boundary curves are drawn in the amphiphilic mixture (A + S)-salinity plane for various A/S ratios. These phase diagrams, with X-shaped curves bent towards the left, are characteristic of mixtures containing hydrophobic alcohols. On the contrary, hydrophilic alcohols yield X-shaped curves bent towards the right. With no alcohol, a true, standing X-shaped pattern is usually obtained [5, 6, 10, 29]. The WIII region observed is part of the „three-phase nonionic body" [19], but in fact this three-phase body does not show up with only brine, 1-pentanol and decane (at least when WOR = 1) [21], i. e. if no ionic amphiphile is present (S = 0 %), whatever the salinity may be. At point X the highest mutual solubility of oil and water is reached (the ordinate of X is inversely related to the solubilizing power of the amphiphilic mixture). Point X is shifted towards higher salinity values and lower concentrations of active mixture when A/S decreases. The solubilization properties of the amphiphilic mixture (A + S) are then improved in spite of a higher salinity. In other words II–III–I transitions are observed with decreasing A/S ratios.

The laboratory sample (# 2), consisting mainly of octadecenesulfonates, shows the same trend with lower salinities and a higher solubilizing power (Fig. 4).

The optimum solubilization parameter (SP*) was defined as the amount of oil present in a microemulsion per unit amount of surfactant along the optimum line. The separate contributions of surfactant and cosurfactant to the ordinate of X and the variations of SP* as a function of the A/S ratio are respectively given in Fig. 5 and 6; two rather striking results can be seen.

Firstly, the variation of the alcohol content nearly suffices to account for the change in salinity at X. But, in terms of solubilization with respect to surfactant alone, Fig. 6 shows that, at the point where the third phase emerges, an increase of the A/S ratio, reducing the optimum salinity, produces a decrease of SP*, which is more pronounced with the pilot plant AOS

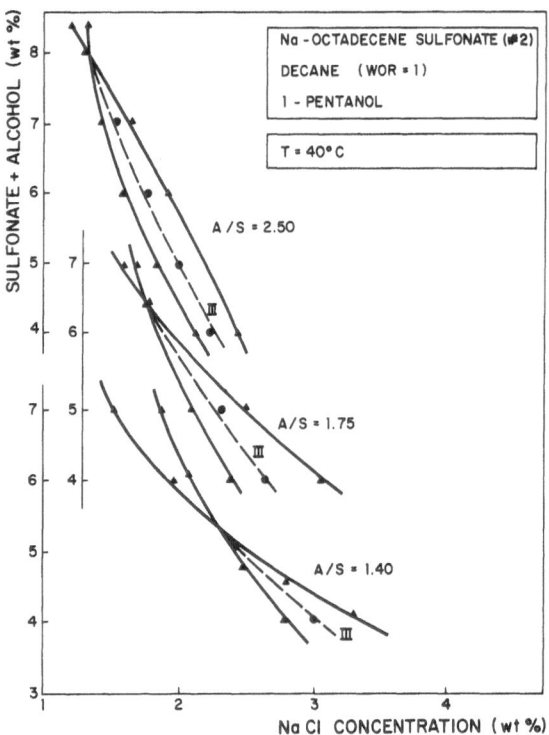

Fig. 4. Amphiphilic mixture concentration as a function of brine salinity (Laboratory octadecenesulfonates)

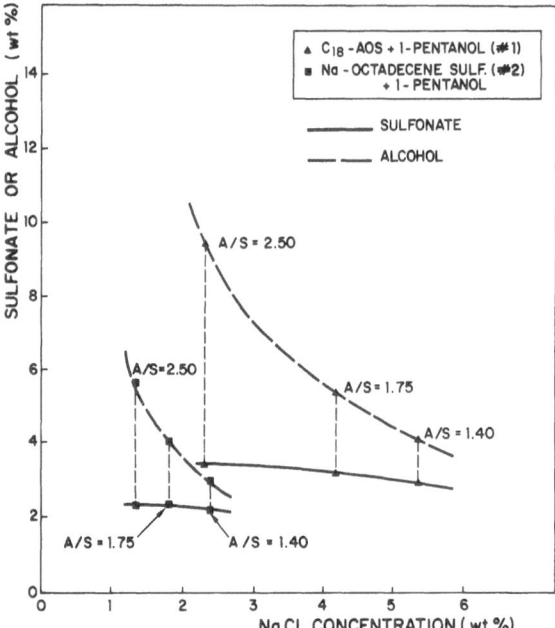

Fig. 5. Separate contributions of surfactant and cosurfactant to point X coordinates

(# 1) than with the laboratory sample (# 2). So, conversely, the effect of reducing the alcohol content is not exactly compensated for by an increase of salinity, usually reported as being detrimental to solubilization.

Secondly, the salinity difference observed between the two sulfonate samples increases when the alcohol-to-sulfonate ratio is reduced. In fact, the salinity range scanned with the octadecenesulfonate sample (# 2) is much narrower than with C18 AOS # 1. This can be explained by the higher content of hydroxysulfonates (and disulfonates) in the AOS sample (# 1). These two components favor higher optimum salinities (and thus lower solubilization parameters) [2].

The optimum salinity lines in Fig. 4 could be drawn through systematic chemical analysis of the middle phase along the salinity "window" (Fig. 7). This also allows precise determination of SP*. It makes sense that, as the microemulsion evolves from WI to WII, it gets richer in hydrocarbon and poorer in water and sodium chloride. All the surfactant is assumed to be present in the middle phase. Its concentration increases very slightly with the salinity of initial brine. Along the optimum path, A/S decreases from the overall value at X with increasing brine salinity.

The Chun Huh equation [18] for the solubilization parameter (SP) and interfacial tension (γ)

$$\gamma \cdot \frac{(SP)^2}{\cos \pi/4} = k$$

can be used to estimate interfacial tensions in the three-phase domain. For example, with a value of SP* taken

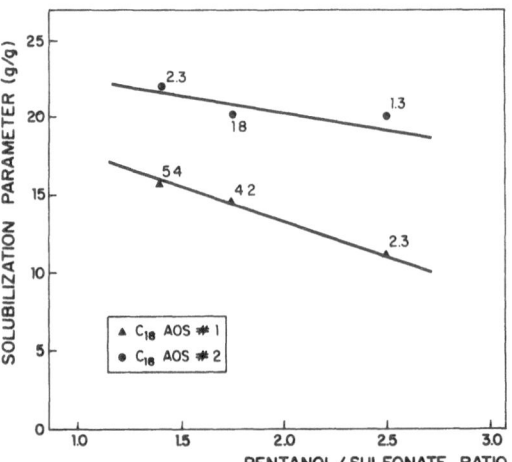

Fig. 6. Solubilization parameter as a function of A/S ratio at X. Numbers inside the graph indicate overall brine salinity (NaCl*) in wt. %

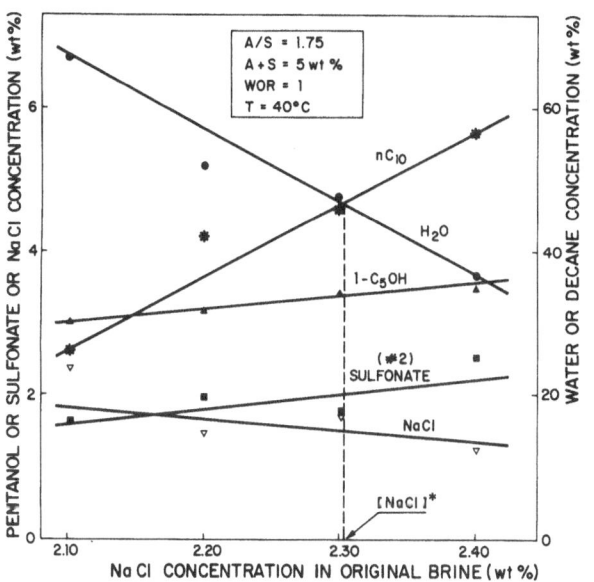

Fig. 7. Middle-phase composition along the salinity "window" with laboratory octadecenesulfonates

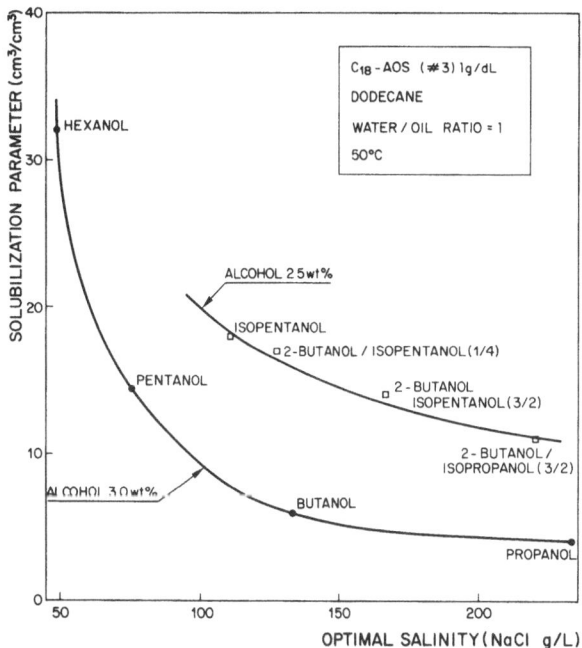

Fig. 8. Solubilization parameters as a function of optimal salinity for various alcohols

from Fig. 6 (SP* = 23.3 g/g for sample # 2) and a constant from the literature ($k = 0.4$) [2], the interfacial tension γ^* at the optimum salinity is found to be around 5×10^{-4} mN/m.

In the straight-chain alcohol series, the main trend observed is an increase in SP* and a concomitant decrease in the optimum salinity when the alcohol chain gets longer [4]. The replacement of 1-alkanols by 2-propanol, 2-butanol and/or isopentanol results in favorable effects on SP* and also on (NaCl)* in so far as replacing a linear alcohol by one of its branched isomers, e. g. 1-pentanol by isopentanol, is accompanied by an increase in optimum salinity (Fig. 8). Therefore, alone or as mixtures,, secondary or branched-chain alcohols appear to be much more advantageous than linear ones.

3. Alcohol and salt partitioning

According to the pseudophase model [9, 22] a microemulsion phase may be considered as consisting of three domains called pseudophases: an aqueous pseudophase W' containing all the brine and some alcohol, an organic one O' containing alcohol and all the oil, and a membrane formed by the amphiphilic com-

pounds: alcohol and all the surfactant. The main assumptions are as follows:

This membrane is two-dimensional, whereas the other two pseudophases are volumic ones.

The compositions of oil and water pseudophases are identical to those of the actual oil and water excess phases.

Equilibrium phenomena occur within these pseudophases and between each other, so that the chemical potential of each component is the same in different pseudophases. Four equilibrium constants are to be considered: K, self-association constant of the alcohol through hydrogen bonding into dimers, trimers, . . . polymers within the organic phase; k_W, partition constant of the alcohol between oil and water phases or pseudophases, which is a function of brine salinity; k_M, partition constant of the alcohol between oil and membrane, and ε partition constant of water between oil and water.

All these constants can be determined experimentally. Provided the overall composition of the starting mixture and the number of phases are known, this model allows determination of the composition of the phases in equilibrium.

Alcohol concentrations in WIII systems characterized by various amounts of surfactant and cosurfactant as well as different A/S ratios have been calculated by means of the pseudophase model. For instance, in the case of NaCl brine/dodecane/1-pentanol or 1-butanol and C18 AOS # 3 at 50 °C, it was possible to find a set of physically meaningful constants (Table 2) yielding calculated values which agree quite well with the experimental ones (Fig. 9).

Using pseudophase model hypotheses alone allows determination of the A/S ratio in the amphiphilic membrane. As a matter of fact, phase analyses show that alcohol concentration in the middle-phase is greater than would be calculated assuming alcohol to be distributed between brine and hydrocarbon pseudophases in the exact proportions found in excess phases. This excess of alcohol reveals its involvement in the interfacial structure (membrane pseudophase) of the surfactant – rich phase [7]. Calculation of the excess of alcohol A_e allows determination of the molecular alcohol-to-sulfonate ratio $(A_e/S)_M$ in the membrane pseudophase (about two alcohol molecules for one surfactant molecule with C18 AOS # 3) (Table 3). Adding alcohol with constant surfactant concentration results in a decrease of SP* from 18 to 13 g/g. This detrimental effect of alcohol on solubilization is clearly related to the increase in the $(A_e/S)_M$ ratio, which thus reduces the efficiency of the sulfonate by decreasing its interfacial density. On the other hand, the beneficial effect of branched-chain and/or secondary alcohols could perhaps be due to a lower A/S ratio in the membrane, with everything else being kept constant.

Let us now consider the sodium chloride partitioning between the middle phase and the excess aqueous

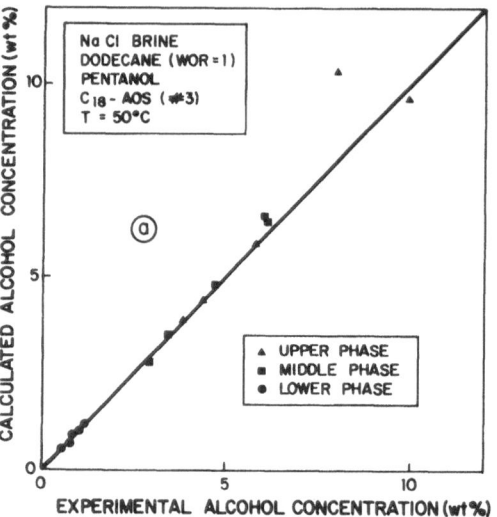

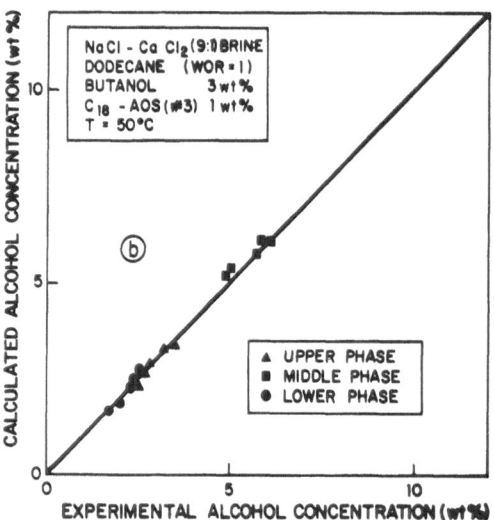

Fig. 9. Experimental and calculated alcohol concentrations with (a) NaCl and (b) NaCl–CaCl₂ brines

phase. In Fig. 10, the concentrations are in g NaCl/L water. It can be seen that the aqueous phase is always a bit richer in electrolyte than the microemulsion, and the higher the salinity, the larger the difference is. This is interpreted as an excluded volume effect towards the salt in the hydration shell of the surfactant polar heads [8].

Again within the framework of the pseudophase model, an extension involving the dissocation of the sulfonate in the micelle is used to compare calculated salt concentrations with the experimental ones shown

Table 2. Constants of the pseudophase model in systems brine/dodecane/alcohol/AOS # 3

Alcohol	NaCl (g/L H₂O)	CaCl₂, 2 H₂O (g/L H₂O)	K	k_W	k_M	ε
	39			1.9		
	49			1.8		
1-Pentanol	66.5		130[a]	1.7	94	0.025
	78			1.3		
	90			1.2		
	0	113		10.0		
	34	85		9.0		
1-Butanol	70	57	250	9.0	140	0.025
	101	28		7.0		
	135	0		6.0		

[a] Ref. [22]. The possible variation of K between room temperature and 50 °C was not taken into account

Table 3. Effect of overall alcohol concentration on phase composition, near optimum conditions, in mixtures of NaCl brine, dodecane (WOR = 1), C18 AOS # 3 (1.0 wt %) and 1-pentanol. $t = 50\,°C$

Overall pentanol conc. (wt. %)	Brine salinity (g/L)	Phase composition (wt %)					A_e (wt. %)	$(A_e/S)_M$ (Molar ratio)	Average SP* (g/g)
		Phase	Dodecane	Pentanol	Sulfonate	Brine			
3	78	Upper	95.54	4.46	—	—	1.15	1.84	18
		Middle	40.36	3.35	2.65	53.63			
		Lower	—	0.58	—	99.42			
6	39	Upper	90.35	9.65	—	—	2.05	2.39	13
		Middle	34.14	6.34	3.63	55.89			
		Lower	—	1.14	—	98.86			

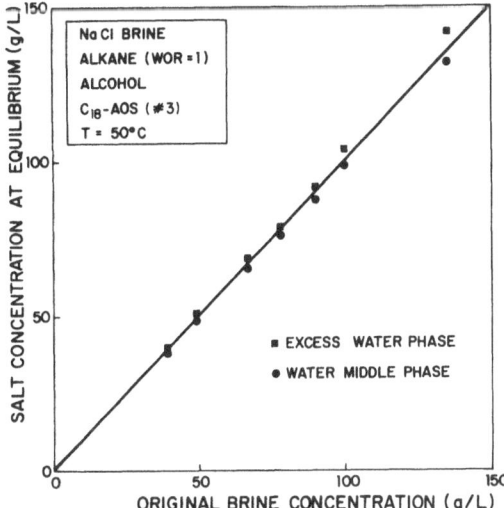

Fig. 10. Salt partitioning between excess water phase and water pseudophase

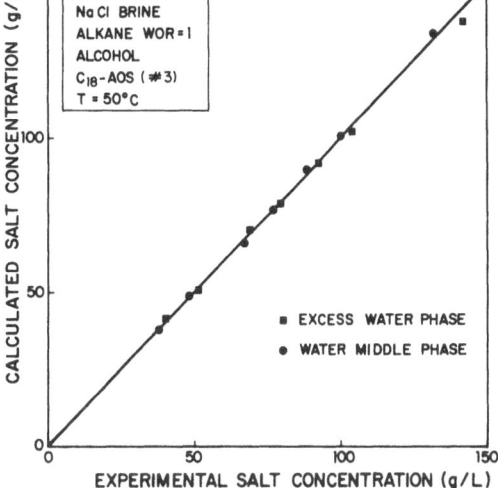

Fig. 11. Experimental and calculated salt concentrations in excess water phase and water pseudophase

in Fig. 10. With pure sodium chloride brines, the agreement between calculated and experimental values is quite good (Fig. 11).

Conclusions

The three-phase behavior of AOS in multicomponent systems of interest in enhanced oil recovery has been investigated, especially in conditions of high salinity. The following main features can be summarized:

At the point common to the four Winsor regions (emergence of the three-phase regime) the variations of the optimum salinity are accounted for by the alcohol concentration alone. With regard to solubilization parameter, as far as pentanol is concerned, the detri-

mental effect of increasing alcohol content can be incompletely compensated for by the beneficial decrease of optimum salinity.

The pseudophase model allows prediction of the partitioning of alcohol, hence the alcohol-to-surfactant ratio in the membrane pseudophase, as well as the salt enrichment of the excess aqueous phase.

The presence of alcohol in the membrane reduces the density of surfactant at the interface and decreases surfactant efficiency.

Acknowledgements

V. Castro for the preparation of the alkenesulfonate samples and HPLC analyses; Mrs. D. Deville and N. Berton for phase behavior studies and analyses; Shell (Amsterdam and London) and Witco Chemical S.A. (France) for providing some 1-alkene and AOS samples.

References

1. Barakat Y, Fortney LN, Lalanne-Cassou C, Schechter RS, Wade WH, Weerasooriya U, Yiv S (1983) Soc Pet Eng J Dec 913:918
2. Barakat Y, Fortney LN, Schechter RS, Wade WH, Yiv SH (1982) Proc of the 2nd European Symp on Enhanced Oil Recovery, Paris, Nov 8–10, Technip, pp 11–20
3. Baumann H, Stein W, Voss M (1970) Fette, Seifen Anstrichm 72(4):247–253
4. Bavière M, Bazin B, Noik C (1986) Paper SPE 14933, 5th SPE/DOE Symposium on Enhanced Oil Recovery, Tulsa, April 20–23, Vol II, pp 123–134, to be published in SPE Reservoir Engineering, May 1988
5. Bavière M, Wade WH, Schechter RS (1979) CR Acad Sci Série C 289:353–356
6. Bavière M, Wade WH, Schechter RS (1981) J Colloid Interface Sci 81(1):266–279
7. Bavière M, Wade WH, Schechter RS (1981) In: Shah DO (ed) Surface Phenomena in Enhanced Oil Recovery, Plenum Press, New York
8. Biais J, Barthe M, Bourrel M, Lalanne P (1986) J Colloid Interface Sci 109(2):576–585
9. Biais J, Bothorel P, Clin B, Lalanne P (1981) J Disp Sci Technol 2(1):67–95
10. Bourrel M, Chambu C, Verzaro F (1982) Proc of the 2nd European Symp on Enhanced Oil Recovery, Paris, Nov 8–10, Technip, Paris, pp 39–50
11. Bourrel M, Salager JL, Schechter RS, Wade WH (1980) J Colloid Interface Sci 75(2):451–461
12. Boyer JL, Canselier JP, Castro V (1985) Jorn Com Esp Deterg 16, Barcelone, March 13–15, pp 247–262
13. Castro V (1986) Thesis (Doctorat d'Etat), Inst Nat Polytech Toulouse, Jan 23, no 105
14. Castro V, Canselier JP (1985) J Chromatogr 325:43–51
15. Castro V, Canselier JP (1986) J Cromatogr 363:139–146
16. Castro V, Canselier JP, Boyer JL (1985) Jorn Com Esp Deterg 16, Barcelone, March 13–15, pp 373–390
17. Green HA (1976) In: Linfield WM (ed) Surfactant Science series 7(2) Anionic Surfactants. Marcel Dekker, New York, pp 345–380
18. Huh C (1979) J Colloid Interface Sci 71:408–426
19. Kahlweit M, Strey R (1987) J Phys Chem 91:1553–1557
20. Kim MW,, Gallagher W, Bock J (1987) 61st Colloid and Surface Science Symposium (ACS), Ann Arbor, June 21–24
21. Knickerbocker BM, Pesheck CV, Scriven LE, Davis HT (1979) J Phys Chem 83:1984–1990
22. Lalanne P, Clin B, Biais J (1984) Chem Eng Comm 27:193–208
23. Lalanne-Cassou C, Schechter RS, Wade WH (1986) J Disp Sci Technol 7(4):479–491
24. Mori A, Okumura O (1984) In: Cesio and Tegewa (eds) Proc of the 1st World Surfactants Congress, Münich, May 6–10, Kürle Druck und Verlag, Gelnhausen, Vol II, pp 93–104
25. Moroi Y, Oyama T, Matuura R (1977) J Colloid Interface Sci 60(1):103–111
26. Püschel F (1967) Tenside, 4(10):320–325
27. Reed RL, Healy RN (1977) In: Shah DO, Schechter RS (eds) Improved Oil Recovery by Polymer and Surfactant Flooding, Academic Press, New York, pp 383–437
28. Salager JL, Morgan JC, Schechter RS, Wade WH, Vasquez E (1979) Soc Pet Eng J, April 107–115
29. Salter SJ (1977) SPE paper 6843, 52nd Annual Fall Technical Conference and Exhibition, Denver, Colorado
30. Selliah RD, Schechter RS, Wade WH, Weerasooriya U (1987) J Disp Sci Technol 8(1):75–87
31. Schechter RS, Wade WH, Weerasooriya U, Weerasooriya V, Yiv SH (1985) J Disp Sci Technol 6(2):223–235
32. Tsuji K, Saito N, Takeuchi T (1980) J Phys Chem, 84:2287–2291
33. Voca BR (1985) Thesis (Docteur-Ingénieur), Inst Nat Polytech Toulouse, Nov 4, no 386
34. Winsor PA (1948) Trans Faraday Soc, pp 376–398

Received December 11, 1987;
accepted December 23, 1987

Authors' address:

J. P. Canselier
Ecole Nationale Supérieure d'Ingénieurs de Génie Chimique
Chemin de la Loge
F-31078 Toulouse, France

Progress in Colloid & Polymer Science

Progr Colloid Polym Sci 76:153–158 (1988)

Small angle X-ray scattering of microemulsions

U. Würz

Max-Planck-Institut für Biophysikalische Chemie, Göttingen, F.R.G.

Abstract: We have measured the small angle X-ray scattering of several water-n-octane-polyglycolether microemulsions covering a wide range of amphiphile concentrations, while keeping near to the top of the three-phase triangle in its symmetrical position. The resulting scattering curves are in excellent agreement with the formula derived by Teubner and Strey [13] in the context of a Landau-type expansion with *negative* gradient term. The correlation lengths involved increase by a factor of 25 when going from the highest to the lowest amphiphile concentrations used. The outer part of the curves are consistent with Porod's law, implying a black and white structure. At even larger angles, deviations from Porod's law occur, which can be attributed to a monolayer interface.

Key words: Small angle X-ray scattering, microemulsions, Porod's law, Teubner-Strey formula.

1. Introduction

Since the days of Schulman [5] many different models for the structure of microemulsions have been proposed: densely packed micelles or inverse micelles, random networks, bicontinuous structures of various kinds [2, 5, 14], etc. They all have in common that they were especially designed for strongly amphiphilic substances as the third component. On the other hand, Kahlweit et al. [6–8] have stressed the analogy in the phase behaviour of ternary systems with rather short-chained amphiphiles to that of the "classical" microemulsion systems. The models mentioned above seem much too detailed to discuss all these systems within the same framework. Our question then is: What are the minimal requirements needed to describe the structure (to be more precise: the scattering curves) of the middle phase in water-oil systems with amphiphiles as a third component?

Using $\eta = x_{water}/x_{oil} - 1$ as an order parameter, one can see from the work of Griffiths [4] that many features of the phase behaviour of ternary systems showing three phases can be modelled by a Landau-type expansion of the free energy. For the description of the fluctuations within these phases, however, this description is not sufficient. Following the lines of Orn-

stein and Zernike [11] and of Landau [9], one has to introduce a *local* order parameter $\eta(r)$ leading to gradient terms in the local free energy expansion. Keeping only the very first of the gradient terms — for stability reasons this is allowed only if its coefficient is positive — Ornstein and Zernike arrived at their famous scattering formula

$$I(q) = \frac{1}{a_2 + c_1 q^2} \tag{1}$$

where a_2 is the coefficient of the second order in η, and $q = 4\pi/\lambda \sin\theta$, and 2θ is the scattering angle. This function, however, monotonously decreases with the scattering angle, while the scattering curves of microemulsions typically show a definite hump.

A formal extension of the Ornstein-Zernike-Landau scheme, then, is to allow for a negative value of the coefficient of the first gradient term, $c_1 < 0$, and include a higher order term with positive coefficient, $c_2 > 0$, for stabilization [15]. Moreover, this formal procedure seems to be quite natural for the systems we are interested in: It is easy to conceive that by adding more and more amphiphilic molecules to a mixture of water and oil, it will be energetically favourable for the system, to form more and more interfaces and hence to form regions with gradients in the order parameter. With negative c_1 Teubner and Strey [13] deduced for the intensity

$$I(q) = \frac{\frac{8\pi}{\xi}\langle\eta^2\rangle c_2}{a_2 + c_1 q^2 + c_2 q^4} \equiv \frac{1}{a_2^* + c_1^* q^2 + c_2^* q^4} \quad (2)$$

and for the correlation function

$$\gamma(r) = \frac{d}{2\pi r} e^{-r/\xi} \sin\left(\frac{2\pi r}{d}\right) \quad (3)$$

with the *two* length scales

$$\xi = \left[\frac{1}{2}\left(\frac{a_2}{c_2}\right)^{1/2} + \frac{1}{4}\frac{c_1}{c_2}\right]^{-1/2} \quad (4)$$

and

$$d = 2\pi\left[\frac{1}{2}\left(\frac{a_2}{c_2}\right)^{1/2} - \frac{1}{4}\frac{c_1}{c_2}\right]^{-1/2}. \quad (5)$$

The "invariant" can be computed to be $Q = \pi\xi/(4 c_2^*)$ and $\langle\eta^2\rangle$ is equal to $\xi/(8\pi c_2^*)$.

We will restrict our interest to the structure of the middle phase near to the top of the three-phase triangle in the symmetric position, that is, at the very tail of the "fish" of Fig. 1. However, to test the range of applicability of Eq. (2), we must allow for a considerable variation in the amphiphilic character of the third component, and hence in its concentration.

2. Experimental

2.1 Systems and materials

Seven solutions were prepared from water, n-octane and several polyglycolehters n-$C_iH_{2i+1}O(CH_2CH_2O)_jH$ (abbreviated henceforth as C_iE_j). The concentrations of water and n-octane are equal (by weight), the respective concentrations of the amphiphile and the temperatures of homogenization and measurement are listed in Table 1. They correspond to the situation depicted in Fig. 1.

The water was twice distilled, n-octane (> 99%) was from Merck Schuchardt and the polyglycolethers (> 98%) were from Bachem. The solutions were homogenized at the appropriate temperature (compare Table 1 and Fig. 1), then quickly transferred with a syringe into thin-walled glass capillaries of about 1 mm in diameter and sealed immediately by fusion.

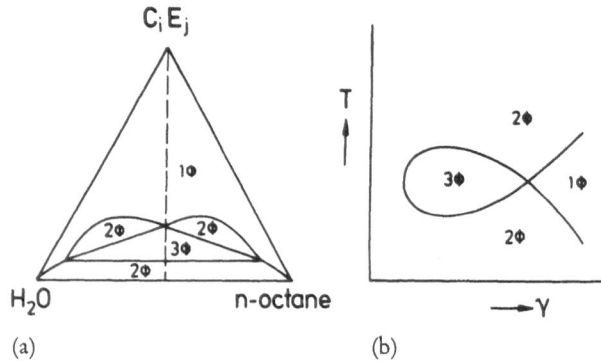

(a) (b)

Fig. 1. (a) Phase diagram (schematic, liquid crystalline phases ommitted). Samples were prepared and measured in the 1-phase region near to the top of the 3-phase triangle; (b) Temperature versus amphiphile concentration. Samples were prepared and measured in the 1-phase region near to the "fish" tail

Table 1. Compositions and temperatures of the samples

Sample No.	wt% C_iE_j	i	j	$T(°C)$
1	3.5	12	4	12.60
2	6.0	12	5	33.00
3	13.0	10	4	26.00
4	22.0	8	3	17.00
5	27.5	8	4	44.00
6	41.9	6	2	9.00
7	43.1	6	3	46.50

2.2 SAXS apparatus

The capillaries were mounted horizontally in a massive copper cylinder with appropriate apertures for the X-ray beam, making thermal contact with it over their entire length. This sample holder was introduced into the small, Peltier element-driven thermostat K-PR from A. Paar, Graz. Temporal and spatial variations of temperature were limited to less then 0.1 K. The thermostat was part of the Kratky compact small angle system KCLC/3 [3] used for the measurement. Cu radiation was provided by a Philips X-ray generator with an AEG fine-focus tube FK60-04 × 12Cu. The position-sensitive counter, OED 50 M from Braun, was mounted at the end of an extension tube at a distance of 98.35 cm from the sample. Monochromatization of the radiation was effected by an Ni-filter and by discriminating with a single channel analyzer. The intensities were normalized to the same units by measuring the intensity of the primary beam with the moving slit device (to obtain electron units, the units used in the graphs have to be multiplied by 1/0.6023).

The measured intensity curves $J(q)$ were affected by the slit collimation of the Kratky appartus [3]. The collimation conditions were such that the slit could be treated as being of infinite length. Figure 2 shows the relation between experimental intensity $J(q)$ (Fig. 2a) and the "desmeared" intensity $I(q)$, (Fig. 2b), for which the slit effect has been corrected.

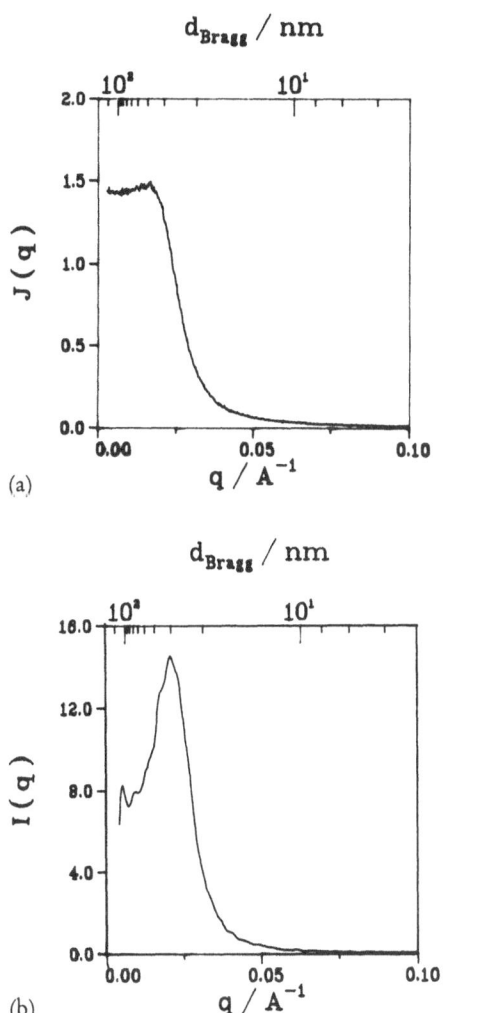

Fig. 2. (a) Experimental intensity $J(q)$ of sample 3; (b) Deconvoluted and smoothed intensity $I(q)$, same sample

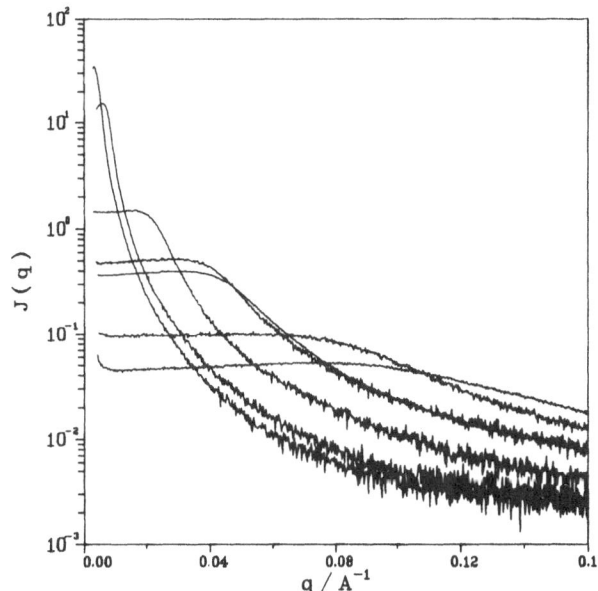

Fig. 3. Experimental scattering curves. The curves are ordered according to the number in Table 1 of the samples: Near to the origin, sample 1 has the highest intensity, sample 7, the lowest

Every deconvolution of noisy data has to be accompanied by a smoothing procedure, which introduces additional uncertainties into the comparison with the theoretical models (see e. g. the 'wiggles' in the computed intensity near to the origin in Fig. 2b). Since the convolution of Eq. (2) with an infinitely long slit can be computed exactly, it was decided to compare the experimental scattering curves directly with the convoluted version of Eq. (2).

3. Results and discussion

3.1 Inner part of the curves

Figure 3 gives an overview over the experimental scattering curves. They order according to the amphiphile concentration: the intensities at the smallest angles decrease by several orders of magnitude, when the "effectivity" of the amphiphile is decreased and the

concentration increased accordingly (compare Fig. 1 and Table 1). This is accompanied by a shift of the maximum to higher values of the scattering angle. On a linear scale, the curves look quite similar when the scale factore is chosen appropriately, as one can see from Fig. 4 for two extreme cases (samples 1 and 6).

They all show a definite hump, which, in the deconvoluted intensity, develops a marked maximum (see Fig. 2). There is no sign of critical or tricritical scattering, except perhaps in samples 6 and 7. The inner part of the curves (comprising all data up to values in q, were $J(q)$ has fallen to 10 % of its maximal value) was used to determine the parameters of Eq. (2) by a nonlinear least squares fit procedure (P3R from BMDP). Figure 4 gives an impression of the goodnes of fit for two extreme cases. It is similar in all cases, in spite of the strong variation in intensity and scattering angle region. The resulting parameters are listed in Table 2. Their estimated statistical errors, typically, are 1–2 % (one standard deviation).

Thus, we can state that all considered mixtures can be described with good precision with the same simple formula, using only three parameters. Moreover, relations exist between them, as one may conclude from the values of Table 2. Firstly, the ratio d/ξ is always near to 2, which is consistent with a similar shape of the intensity curves (Fig. 4), and hence the correlation

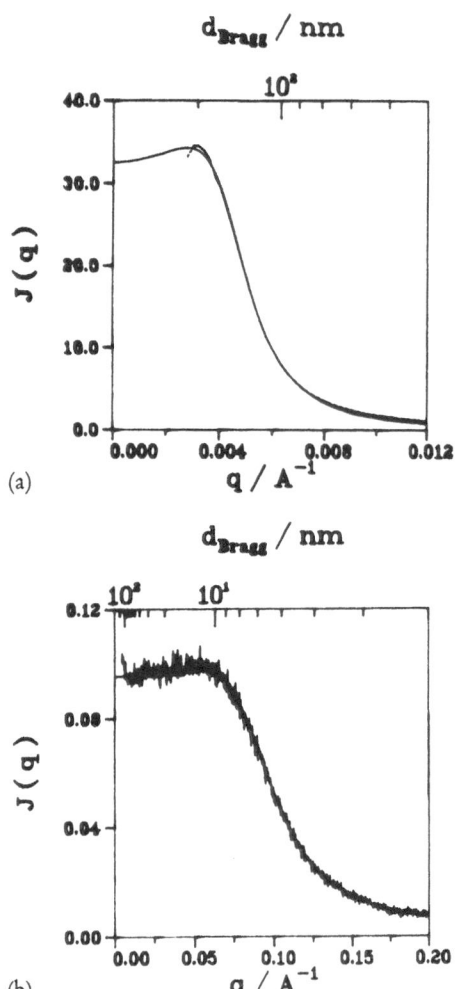

(a)

(b)

Fig. 4. (a) Sample 1. (b) Sample 6. Both the experimental intensity and the results of the fit are shown. Note the different scales of the plots.

Table 2. Results of the fit

Sample No.	ξ (Å)	d (Å)	$\varrho_1 - \varrho_2$ (g el/cm³)
1	598	1454	0.140
2	486	836	0.148
3	130	285	0.143
4	71.8	153	0.156
5	62.0	139	0.152
6	28.8	74.5	0.145
7	25.7	56.2	0.126

functions. Secondly, the mean square of the electron density $\langle \eta^2 \rangle$ is almost constant for all samples, which means that the degree of modulation in electron density $\varrho_1 - \varrho_2$ is near to the value of 0.150 deduced for pure water and n-octane domains [12].

This result is quite satisfying in so far as the same simple three-parameter formula applies perfectly for such a broad spectrum of microemulsions. On the other hand, it is rather disappointing, since when topologically well defined structures exist, as have been proposed recently [10], they cannot be discriminated from our crude model with its especially vague structure. To be more precise, the differences should show up in the residuals of the fit. For the inner part of the curves, these residuals do not appreciably exceed the experimental error (though some common pattern seems to exist). For the outer part of the curves, the fitted curve are systematically too low.

3.3 Outer part of the curves

Therefore, it seems worthwhile to take a closer look at the outer part of the curves, moreover, as theoretical predictions for $q \to \infty$ can be tested there. Figure 5 shows the intensities $J(q)$ of samples 1, 3 and 6 in a double logarithmic plot. For comparison, a line with slope -3 is added, which represents the limiting behaviour according to Porod [12] for black and white structures (because of the integration due to the slit collimation, this has to be expected, instead of the q^{-4}-dependence for the deconvoluted intensities). At even higher angles, the slope of curves in Fig. 5 decreases. This can be only partly due to a constant background,

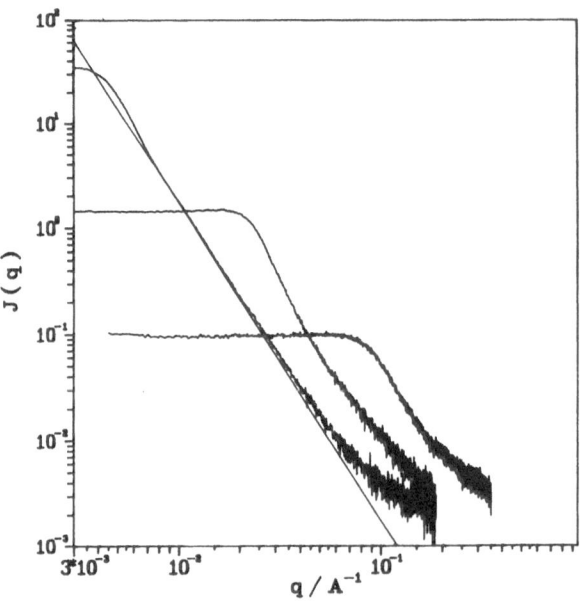

Fig. 5. Experimental scattering curves of samples 1, 3 and 6. A straigth line with slope -3 is drawn for comparison

but may be attributed to the scattering of the interface itself [1], which is expected to give a contribution to the experimental intensity $J(q)$ proportional to q^{-1}. We made an attempt to make nonlinear least squares fits to the outer part of the curves, too. In turned out, however, that for the first curves of the set, the fit was strongly influenced by slight 'wiggles' near to the right of the maximum, whereas, for the last curves, different contributions were strongly correlated. Thus, it seems to

be more appropriate to present a set of $q^3 J$ versus q^2 plots (Fig. 6a–f). Porod's law would result in a constant value in the high q limit, whereas the scattering of the interface shows up in a constant slope. For easy comparison, the length of the ordinate is chosen as proportional the molar concentration of the amphiphile. This would lead to the same intercept if the same area were occupied by the amphiphile molecule head groups. Moreover, the q-axis was dilated roughly according

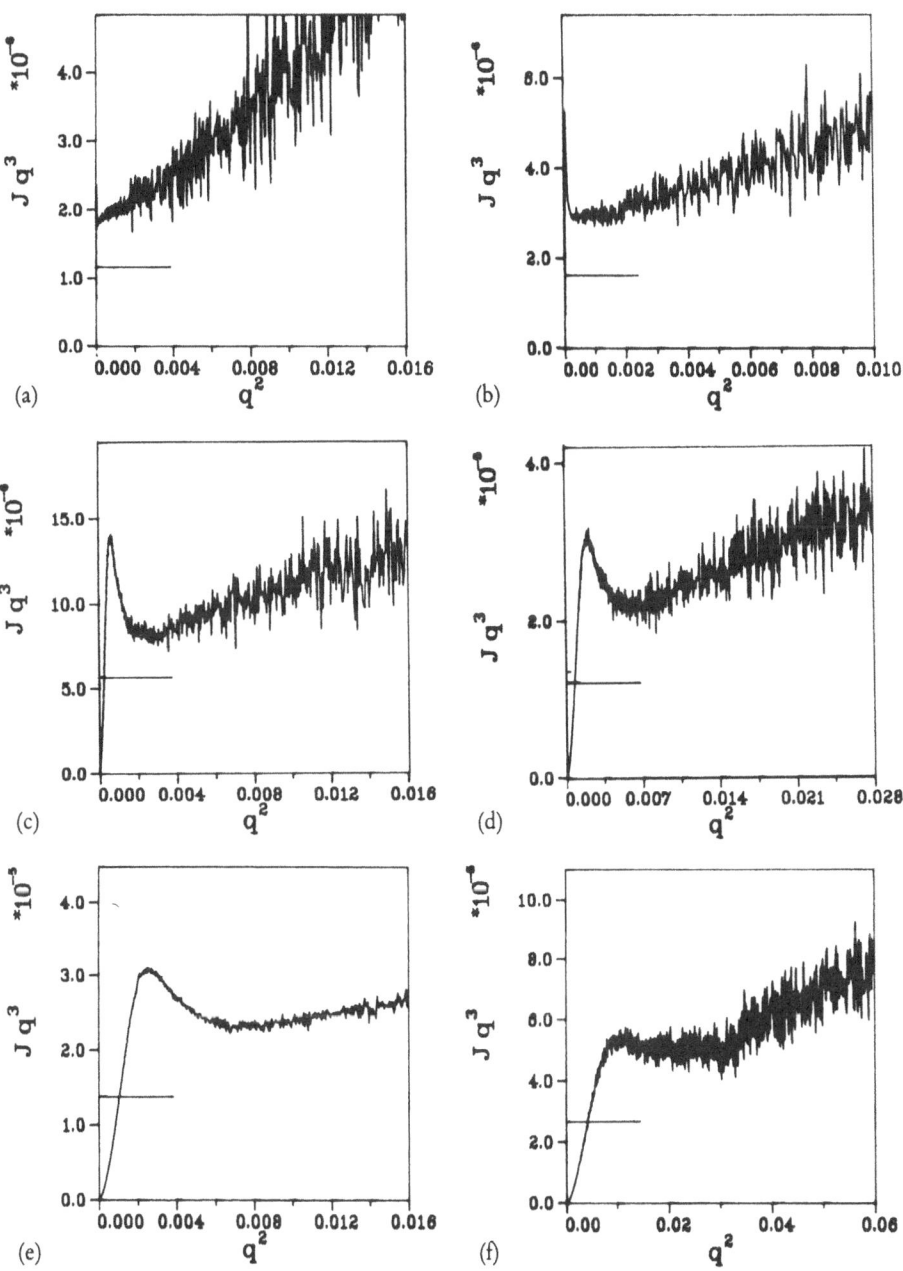

Fig. 6. (a)–(f) $J q^3$ versus q^2 curves for samples 1 to 6. The limiting values for the fitted curves are marked for comparison

to the number of ethoxy groups of the amphiphile, and hence would give the same slope, if the thickness of the interface is proportional to the length of the head group (the tail of the amphiphile will not give any contrast relative to the oil). It becomes evident from Fig. 6 that the expected behaviour is roughly fulfilled. However, the specific surfaces deduced from the outer part of the curves are larger by about a factor of 1.6 compared to the result $4 \varphi_1 \varphi_2 / \xi$ obtained from the fit, as one may see from Fig. 6a–f, where the latter is indicated by a line. This fact is not too surprising, regarding the different scales involved in "viewing" the surface.

4. Conclusions

In conclusion, we measured the small angle X-ray scattering of several water-n-octane-$C_i E_j$ microemulsions, covering a wide range of amphiphile concentrations while keeping near to the top of the three-phase triangle in its symmetric position. The resulting scattering curves were all consistent with the formula derived by Teubner and Strey [13] in the context of a Landau-type expansion with *negative* gradient term. The outer part of the curves were consistent with Porod's law, implying a black and white structure. At even larger angles, deviations from Porod's law occurred which can be attributed to the interface.

Acknowledgements

I thank Mr. Lieu for his help in performing the measurements and Dr. Teubner for many helpful discussions.

References

1. Auvray L, Cotton J-P, Ober R, Taupin C (1984) J Phys 45:913–928
2. de Gennes P, Taupin C (1982) J Phys Chem 86:2294
3. Glatter O, Kratky O (1982) Small angle X-ray Scattering. Academic Press, New York
4. Griffiths RB (1974) J Chem Phys 60:195
5. Hoar T, Schulmann JH (1943) Nature, London 152:102
6. Kahlweit M, Strey R (1985) Angew Chemie 24:654
7. Kahlweit M, Strey R, Firman P, Haase D (1985) Langmuir 1:281
8. Kahlweit M, Strey R, Firman P, Haase D (1986) J Phys Chem 90:671
9. Landau LD, Lifshitz EM (1979) Statistische Physik. Adademie Verlag, Berlin
10. Ninham BW, Barnes IS, Hyde ST, Derian PJ, Zemb TN (1987) Europhys Lett 4:561
11. Ornstein LS, Zernike F (1918) Phys Z 29:134
12. Porod G (1951) Kolloid Z 124:83; (1952) 125:51
13. Teubner M, Strey R (1987) J Chem Phys 87:3195
14. Widom B (1986) J Chem Phys 84:6943
15. Würz U, Chan SK, Döpkens J (1983) Ber Bunsenges Phys Chem 87:213

Received December 11, 1987;
accepted January 4, 1988

Author's address:

U. Würz
Max-Planck-Institut für Biophysikalische Chemie
D-3400 Göttingen, F.R.G.

Progress in Colloid & Polymer Science

Progr Colloid Polym Sci 76:159–164 (1988)

Solubilization of phosphorylase into microemulsion droplets. An ESR study

A. Xenakis and C. T. Cazianis

Institute of Biological Research, The National Hellenic Research Foundation, Athens, Greece

Abstract: The solubilization of glycogen phosphorylase in different microemulsion systems was studied by ESR spectroscopy. Spin-labelling technique was used to follow the behaviour of this enzyme when hosted in either ionic microemulsion systems formulated with AOT and isooctane or nonionic microemulsion systems with $C_{12}EO_4$ and decane. The differences between the two kinds of systems are discussed in terms of the polarity of the dispersed phase and of the mobilities of the labelled sites of the enzyme molecules. Depending on the composition of the microemulsion systems, the observed variations in both these parameters are shown to be related to structural changes which occur within the microemulsions.

Key words: Microemulsions, ESR spectroscopy, phosphorylase b, ionic systems, nonionic systems.

Introduction

Microemulsion systems comprise an increasing area of interest in the literature, due to their numerous potential applications in various domains of modern science and technology [1]. In the past few years, microemulsions have been used in biological studies extending from basic biological research [2] to biotechnology [3].

Especially, water-in-oil microemulsions can be regarded as model systems for studying various enzymatic reactions. It has been shown that the properties of the microdomain in the dispersed aqueous droplets of these systems are different from those of bulk water [4]. On the other hand, intracellular water has been proved to be quite structured, exhibiting different properties from bulk water [5], and resembling in many ways the dispersed aqueous phase of w/o microemulsions. Therefore, *in vivo* conditions can be better reproduced in experiments performed in microemulsion systems than in classic water solution systems [6, 7].

Glycogen phosphorylase, catalyzing the breakdown of glycogen, can be activated by some hydrophobic compounds and by certain amphiphiles, such as sodium cholate [8]. It seemed interesting to us to examine the behaviour of this enzyme in microemulsion systems. Spin-labelling techniques were used in order to examine particularly the solubilization of phosphorylase b (EC 2.4.1.1) in two different microemulsion systems, that is, an anionic system formulated with water/AOT/isooctane and an nonionic system consisting of water/tetraethyleneglycoldodecylether ($C_{12}EO_4$)/decane. The ESR technique can produce important information about the mobility of the probe studied and the polarity of the medium surrounding the spin label.

Phosphorylase b may be specifically and covalently spin-labelled at sites with different microenvironments and catalytic action. A spin-labelled 1-fluoro-2,4-dinitrobenzene derivative (FDNB) binds to a buried $-NH_2$ group critical for the activity of the enzyme, while an iodoacetamide spin label (IA) reacts with a surface $-SH$ group that is not essential for activity [9].

Experimental

Materials

Rabbit muscle phosphorylase b was isolated in our laboratory, according to the method of Fischer and Krebs [10] except that L-cysteine was replaced by 2-mercaptoethanol. The enzyme was recrystallized at least four times and bound AMP was removed, as described in Ref. [11]. AOT (bis-(2-ethylhexyl)sulfosuccinate sodium salt) was obtained from Serva and purified as described by Martin and Magid [12]. Moisture was removed by distillation in toluene and lyophilization, and checked periodically with Karl Fischer titrations. Tetra-ethyleneglycol dodecylether ($C_{12}EO_4$) was obtained from Nikko Chemicals Co. Ltd. Isooctane (2,2,4-trimethylpentane) was a product of Ferak, Berlin, n-decane was obtained from Merck, Darmstadt, and twice distilled water was used.

Spin labels: The iodoacetamide (IA) spin label: 4-(2-iodoacetamido) 2,2,6,6-tetramethylpiperidinyloxyl was a product of Synvar, while the spin-labelled fluoro dinitrobenzene derivative (FDNB): 4-(2,4,-dinitro-5-fluorophenoxy)-2,2,6,6-tetramethyl-1-piperidinyloxyl was prepared as described elsewhere [13].

Methods

Spin-labelling of phosphorylase b

Phosphorylase b was modified by the iodoacetamide spin label according to Campbell et al. [14]. The enzyme obtained had 0.7 labels per monomer while the excess label was removed by passage through a Sephadex G-25 column equilibrated with 1 mM EDTA, 20 mM glycerol-2-P buffer, pH 6.8. Spin-labelled phosphorylase b with the FDNB derivative (0.75 labels per monomer) was prepared as in Ref. [13].

Ionic microemulsions were prepared by adding the appropriate amounts of water in a 50 mM AOT solution in isooctane. The spin-labelled enzymes were solubilized in the reversed micelles by adding small aliquots (a few microliters) of a concentrated stock solution in buffer, to 1–2 ml of the AOT/isooctane solution. Solubilization was achieved within less than 1 min by gentle shaking. The total amount of water added was calculated to give the desired value of the molar ratio $[H_2O]/[AOT] = R$.

In the case of nonionic systems, a similar procedure was followed, using stock solutions of different concentrations of $C_{12}EO_4$ in decane. Special care was taken to keep the samples at constant temperature (22 °C).

ESR spectra were recorded on a Bruker ER 200 D spectrometer operating at X-band, equipped with a variable temperature unit. The samples were contained in an E-248 cell. The concentrations of the spin labels were about $4 \cdot 10^{-5}$ M. Typical settings were: field set, 3471 G; scan range, 100 G; time constant, 1 s; scan time, 16 min; microwave power, 31 mW; microwave frequency, 9.76 GHz; modulation amplitude, 2.5 G.

Results and discussion

A) Ionic systems

Figures 1a and 1b show the ESR spectra of the free FDNB and iodoacetamide spin-labels recorded in a series of w/o microemulsion systems of AOT in isooctane with various water contents, while the spectra of the corresponding spin-labelled phosphorylase b, recorded in a microemulsion, are shown in Fig. 1c and 1d.

Figure 2a presents the dependence of the hyperfine coupling constant a_N on the quantity of water, as

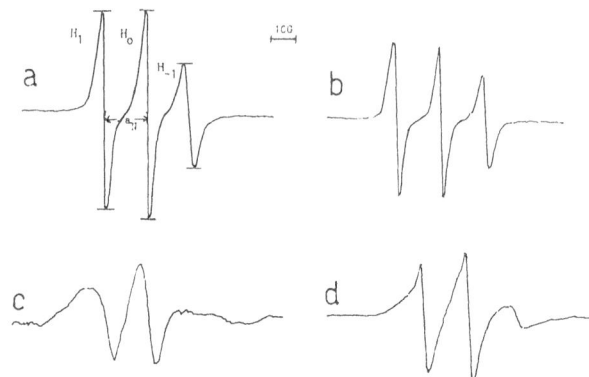

Fig. 1. ESR spectra of (a) FDNB spin label, (b) iodoacetamide spin label, (c) FDNB-phosphorylase b, (d) iodoacetamide-phosphorylase b, recorded in a microemulsion with $R = [H_2O]/[AOT] = 20$, $[AOT] = 5 \cdot 10^{-2}$ M. The concentration of the probe was $4 \cdot 10^{-5}$ M

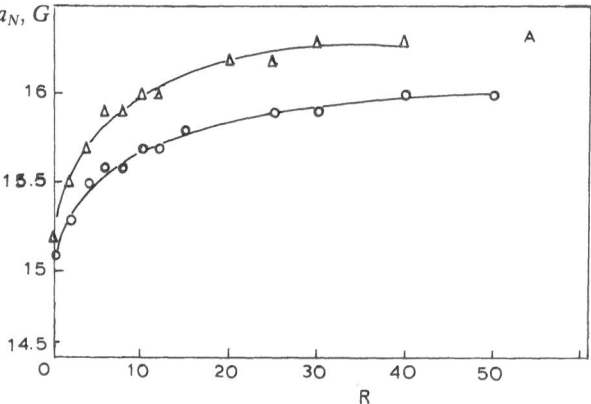

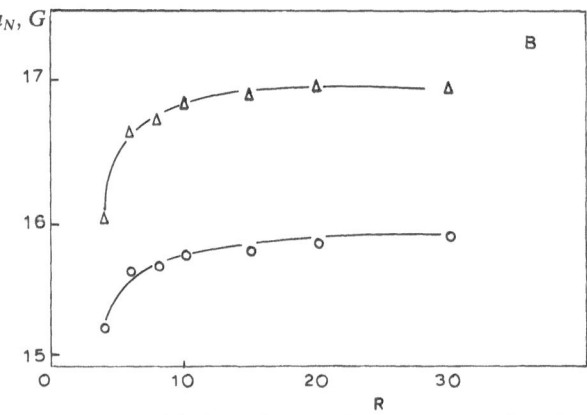

Fig. 2. Variation of the hyperfine coupling constant a_N of the free spin labels (A) and of the spin-labelled enzymes (B) as a function of R; ionic microemulsion systems. ($\triangle$) Iodoacetamide; ($\square$) FDNB

expressed by the hydration ratio of the reversed micelles, $R = [H_2O]/[AOT]$. The results show, as expected, that with increasing R the spin-label spends more time in the water core of the reversed micelles. In addition, the change of the slope of a_N observed when $R > 10$ can be attributed to the change of polarity of the dispersed water phase. Since, at low R values, the microdroplets have quite a small size (a radius of 31 Å has been reported for the same system when $R = 5$ [15]), most of the water molecules are bound at the interface as water of hydration. This bound water is motionally restricted; therefore, a lower polarity compared to that of bulk water is expected. With increasing R, the additional water molecules are located in the center of the water pools, the properties of which tend to those of normal water. The transition between bound and free water in these systems occurs at R-values around 10 [16].

Comparing the behaviour of the two probes in Fig. 2a, one can notice that Δa_N, corresponding to the iodoacetamide spin label (IA-SL) going from pure isooctane to $R = 40$, is greater than that of the FDNB spin label. This result is in accordance with the fact that the iodoacetamide spin label is more hydrophilic than the FDNB-SL. Also, the results of Fig. 3 show that the mobility of both free spin labels, expressed as the ratios of H_1/H_o, tend to the values of the mobilities of these labels in bulk water, respectively.

The increase of a_N and the decrease of mobility (H_1/H_o) can therefore be attributed to the capability of the free spin moieties to remain longer in the droplets than in the continuous phase.

After examining the behaviour of the free spin labels in these ionic microemulsion systems, we studied conformational and microenvironmental properties of the spin-labelled phosphorylase b in the same systems.

Figure 2b shows the results obtained with the spin-labelled enzymes under the same conditions as those shown in Fig. 2a for the free spin labels. The a_N variations as a function of R follow the same pattern: namely, the polarity of the microenvironment of both spin labels bound to the enzyme (expressed in terms of a_N) in increased with increasing R.

However, there is an important qualitative difference, i.e. the a_N values of the iodoacetamide spin-labelled phosphorylase b are much higher than those of the free iodoacetamide spin label obtained for the same R values. This observation suggests that the part of the enzyme to which the label is attached, is merged in the core of the microdroplets without any partition into the continuous phase. This could not occur in the

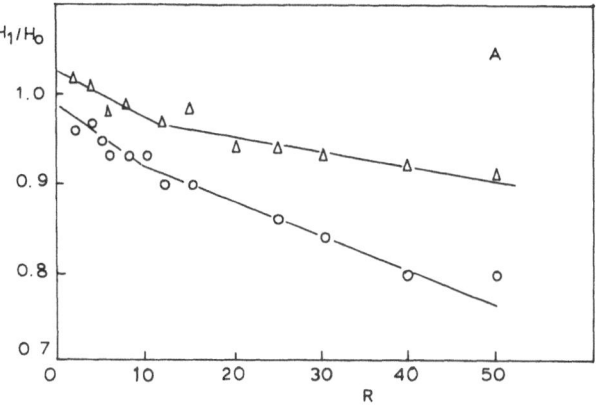

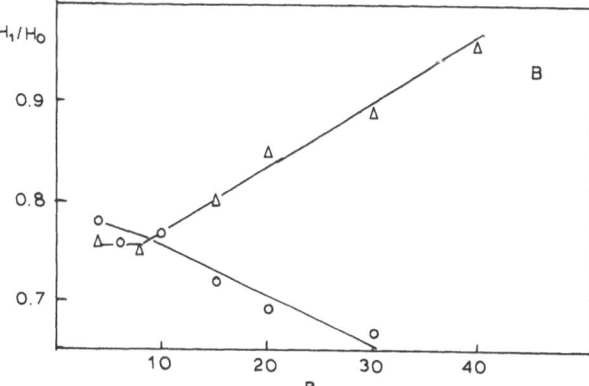

Fig. 3. Variation of the ratios H_1/H_o of the free spin labels (A) and of the spin-labelled enzymes (B) versus R; ionic microemulsion systems. (△) Iodoacetamide; (○) FDNB

case of the free spin label. This is more pronounced for higher R values, where the size of the droplet is large enough to accomodate the entire macromolecule without any structural distortions [7]. However, for the very low R values, the small size of the droplets could preferentially host the most hydrophilic part of phosphorylase b, which seems to contain the site at which the iodoacetamide spin label is located.

A similar comparison for the free FDNB spin label and FDNB spin-labelled phosphorylase b did not show any significant differences, which means that the polarity of the microenvironment surrounding the protein part where the label is bound, is more hydrophobic than that of iodoacetamide spin-labelled phosphorylase b.

The variation of the ratios H_1/H_o of the spin-labelled enzymes versus R is presented in Fig. 3b. In small droplets ($R < 10$), the mobilities of the spin labels are quite similar. However, for $R > 10$ the two enzymes exhibit completely different behaviour: namely, increaseing R increases the mobility of the iodoacetamide spin-labelled phosphorylase b, while the mobility of the FDNB spin-labelled phosphorylase b is slightly decreased. This observation could indicate that increasing the size of the water core gives the iodoacetamide spin label, being on the surface of the enzyme, more 'room to move', while the FDNB spin label, attached at a rather hydrophobic buried site, is additionally restrained by water molecules.

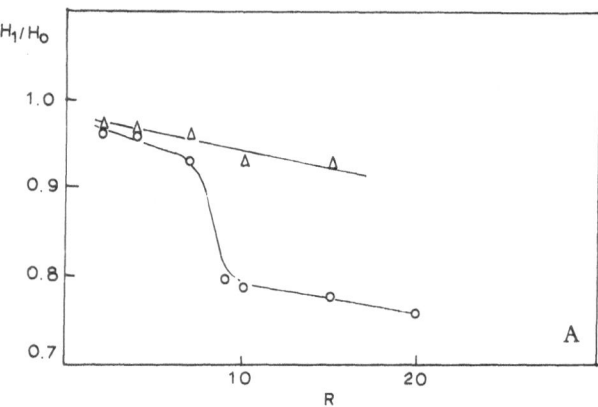

B) Nonionic systems

The behaviour of the free spin labels and the corresponding spin-labelled phosphorylase b was also examined in nonionic microemulsion systems, consisting of water/n-decane/tetraethylene glycoldodecylether ($C_{12}EO_4$). The samples were prepared by adding the appropriate amounts of either the free spin label or the corresponding spin-labelled enzyme to a stock solution of 20% $C_{12}EO_4$ in decane. The samples were properly adjusted to contain the same final spin concentration with different total amounts of water.

Examining the polarity of the microenvironment in these systems (expressed in terms of a_N) as a function of water content, the free iodoacetamide label exhibits the same tendency as that in the ionic systems described above. The same tendency is observed for the free FDNB label in the nonionic systems but for low water content (up to $R = 8$). Further increase of water content results in ESR spectra of the shape shown in Fig. 4. This type of spectrum, with the characteristic two high field peaks, has been observed in a biological membrane system [17] and interpreted as the sum of two spectra: one arises from radicals in aqueous solution (B) and the other from radicals in a low viscosity fluid-like hydrophobic region (A). This preliminary observation could have significant implications concerning the structural changes occuring at a certain water content in these nonionic systems [18].

Considering the H_1/H_o of the free probes as a function of R (Fig. 4), one notices that the mobility of both spin labels decreases with increasing R. However, an abrupt falling of the mobility is observed for a water content of $R = 8$. Most likely, this observation can be related to the structural changes mentioned above.

Concerning the iodoacetamide and FDNB labelled phosphorylase b in these nonionic systems, the ESR spectra of Fig. 5 are characteristic of a rather strong immobilization which was not observed under similar

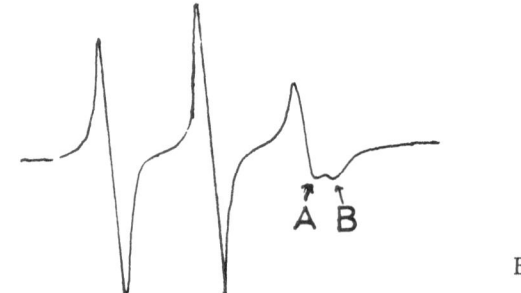

Fig. 4. (A) Variation of the ratios H_1/H_o of the free spin labels versus R, obtained in the nonionic microemulsion systems of $C_{12}EO_4$ in decane. ($\triangle$) Iodoacetamide; ($\bigcirc$) FDNB. (B) spectrum of the FDNB spin label recorded in a system with $R = 15$

conditions (low R values) in the ionic microemulsions described in the previous section.

In cases of strong immobilization, a more sensitive parameter to express mobility is the ratio H_1/H_1' (increase in this ratio reflects a decrease in mobility) or the ratio H_1'/H_o (increase in this ratio means increasing mobility) [19]. Figure 5, showing the plot of these two parameters versus R, indicates that: (i) the mobility of the iodoacetamide labelled phosphorylase b increases with increasing R; (ii) the FDNB labelled enzyme follows the opposite pattern, that is, mobility decreases with increasing R.

Concerning the dependence of mobility on water content, a similar explanation is proposed to that used for ionic microemulsion systems. More specifically, the different behaviour of the two enzymes is due to the different sites at which the labels are attached (IA, a hydrophilic label bound to a surface $-$ SH group and, FDNB, a hydrophobic label attached to a buried $-NH_2$ group).

The higher immobilization of the two labelled enzymes at low R values observed in the nonionic

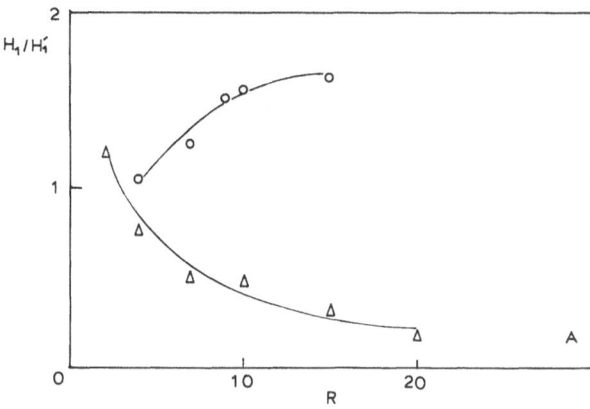

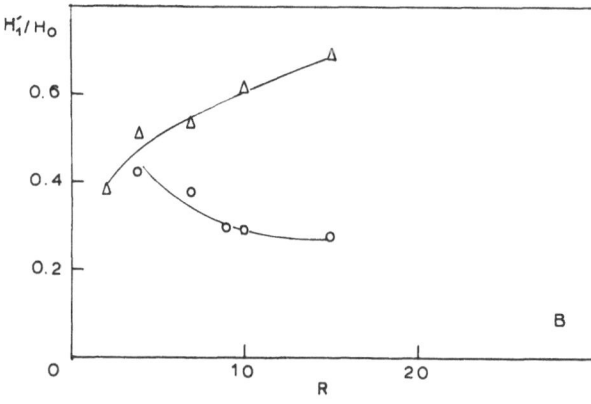

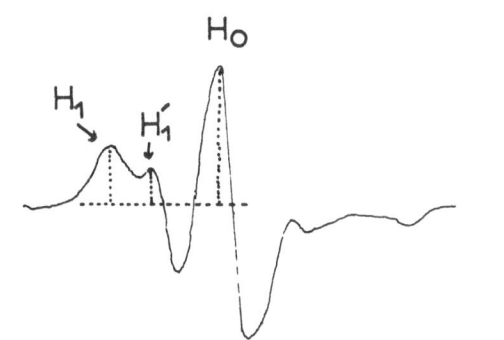

Fig. 5. Variation of the ratios H_1/H_1' (A) and H_1'/H_o (B) of the spin-labelled enzymes as a function of R; nonionic microemulsion systems. (△) Iodoacetamide; (○) FDNB. (C) typical strong immobilization spectrum

microemulsions might be attributed to the particular structure of the dispersed phase in these systems. In the ionic microemulsion systems formulated with AOT as surfactant, the globular structure of the dispersed phase has been proved by many research teams over the last 10 years [14, 20]. This is not the case when microemulsions based on $C_{12}EO_4$ are considered. Recent results of Ravey et al. [21], obtained by small angle

neutron scattering and small angle X-ray scattering, showed two distinct kinds of structures: (i) small "lamellar" aggregates for low water contents and (ii) water-in-oil oblate globules for higher water contents.

According to these structures, a rather high immobilization of the two spin-labelled enzymes is expected when working with low water content systems. The lamellar aggregates, being quite small, cannot include the enzyme molecule without any structural changes. Thus, it is more probable that the enzyme promotes the formation of greater aggregates where the whole molecule is restricted.

For higher water contents, where globular structures exist, the increase of iodoacetamide-labelled phosphorylase b mobility is due to the contact of the hydrophilic labelled site with an increasing size of the water core. In contrast, the FDNB spin label is bound to a hydrophobic region of phosphorylase b; thus, higher water contents cause tightening of this site of the enzyme.

In conclusion, our results show that the use of specific spin labels attached to different parts of the phosphorylase b molecule makes it possible to examine the properties of the microenvironments of these sites, related to the nature and size of the microemulsion droplets. On the other hand, use of hydrophobic and hydrophilic spin-probes in such microheterogeneous systems can be very useful for studying these systems by ESR spectroscopy, since spectral variations of such spin-probes are related to structural changes occuring when the composition is varied.

Acknowledgements

We thank Mr. A. Papageorgiou, V. Zevgolis and S. Kyriakidis for the isolation and purification of the enzyme.

References

1. Gillberg G (1984) Surfactant Sci Ser 6:1
2. Martinek K, Levashov AV, Klyachko N, Pantin VI, Berezin IV (1981) Biochem Biophys Acta 657:277
3. Kadam KL (1986) Enzyme Microb Technol 8:266
4. Maitra A (1984) J Phys Chem 88:5122
5. Drost-Hansen W, Clegg JS (eds) (1979) Cell Associated water. Academic Press, New York
6. Martinek K, Levashov AV, Klyachko N, Khmelnitski Y, Berezin IV (1986) Eur J Biochem 155:453
7. Luisi PL, Magid L (1986) CRC Critical Reviews in Biochemistry 20:409
8. Sotiroudis TG, Cazianis CT, Oikonomakos NG, Evangelopoulos AE (1983) Eur J Biochem 131:625
9. Avramovic-Zicic O, Smillie LB, Madsen NB (1968) J Biol Chem 243:6202
10. Fisher EH, Krebs EG (1962) Methods Enzymol 5:369

11. Melpidou AE, Oikonomakos NG (1983) FEBS Lett 154:105
12. Martin CA, Magid LJ (1981) J Phys Chem 85:3985
13. Cazianis CT, Sotiroudis TG, Evangelopoulos AE (1980) Biochem Biophys Acta 621:117
14. Campbell ID, Dwek RA, Price NC, Randall RJ (1972) Eur J Biochem 30:339
15. Zulauf M, Eicke HF (1979) J Phys Chem 83:480
16. Eicke HF, Rehak J (1976) Helv Chim Acta 59:2883
17. Hubbell WL, McConnell HM (1968) Proc Natl Acad Sci 61:12
18. Ravey JC, Buzier M (1984) In: Mittal KL, Lindman B (eds) Surfactants in Solution. Vol 3, Plenum Press, New York, p 1759
19. Swartz HM, Bolton JR, Borg DC (eds) (1972) Biological Applications of Electron Spin Resonance. Wiley Interscience, New York
20. Kotlarchyk M, Huang JS, Chen SH (1985) J Phys Chem 89:4382
21. Ravey JC, Buzier M, Oberthur R (1987) Progr Colloid Polym Sci 73:113

Received December 11, 1987;
accepted January 4, 1988

Authors' address:

A. Xenakis
Institute of Biological Research
The National Hellenic Research Foundation
48, Vas. Constantinou Avenue
GR-11635 Athens, Greece

Progress in Colloid & Polymer Science　　　Progr Colloid Polym Sci 266:165–168 (1988)

Nonequilibrium phenomena in critical microemulsions

M. A. López Quintela, A. Fernández Nóvoa, J. Quibén, D. Losada, and C. Ferreiro

Grupo Biodinámica Física, Departamento de Química Física, Universidad de Santiago, Santiago de Compostela, Spain

Abstract: Different nonequilibrium phenomena in critical microemulsions are reported. Hydrodynamic instabilities and quenching experiments show, among others, the influence of fluctuations on the behaviour of these systems. Pressure-jump relaxation experiments on conducting microemulsions clearly show the existence of both percolation and non-linear effects which may be responsible for the nonequilibrium phenomena observed.

Key words: Critical microemulsions, hydrodynamic instabilities, nonlinear relaxation, nonequilibrium phenomena, fluctuations.

Introduction

When a small amount of surfactant is dissolved in water, molecular aggregates are formed which can solubilize oils giving rise to clear solutions called microemulsions. It has long been known that most microemulsions exhibit critical phenomena which are, in certain respects, similar to those of multicomponent fluids. The similarities are that there exist scaling laws when these systems approach the critical point or cloud point, and that their dynamics exhibit critical slowing down [4–8, 11, 14–18, 20, 21]. However, the physics of these systems differs appreciably from the general properties of multicomponent fluids in that the critical exponents found for micelles and microemulsions seem to be non-universal [6, 7, 14, 15, 34] and in the existence of critical phenomena at temperatures which differ appreciably from the critical temperature, T_c [34]. These facts, especially the latter, are the reasons for the special significance of microemulsions for the study of non-equilibrium and/or non-linear phenomena [13].

In this report we will refer to various non-equilibrium phenomena which can be observed in critical microemulsions.

Experimental

Two microemulsions have been employed in this study with the following compositions by weight: M1 composed of water (0.6587), 2-butoxyethanol (C_4E_1) (0.3147), n-dodecane (0.0050) and KCl (0.0210); M2 composed of Aerosol-OT (AOT) (0.0750), n-decane (0.4625), water (0.4602) and NaCl (0.0023).

Figures 1 and 2 represent the phase diagrams for M1 and M2. Experiments were carried out in the indicated 1-phase regions.

Hydrodynamic instabilities

In order to study the nature of the observed hydrodynamic instabilities, we carried out experiments in a Rayleigh-Bénard cell with sapphire windows made in our laboratory [22].

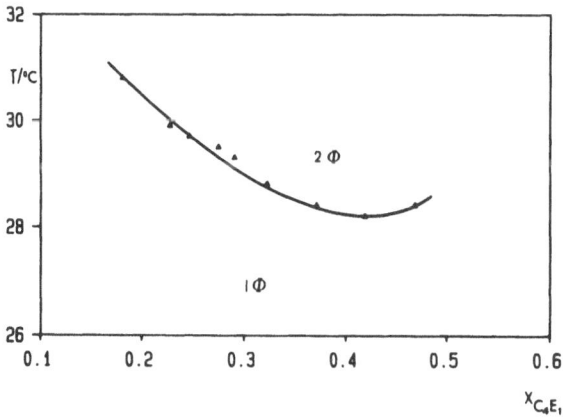

Fig. 1. Approximate phase diagram of the M1 microemulsion employed in this study. 1 Φ: one-phase region. 2 Φ: two-phase region. $X_{C_4E_1}$: composition by weight of C_4E_1

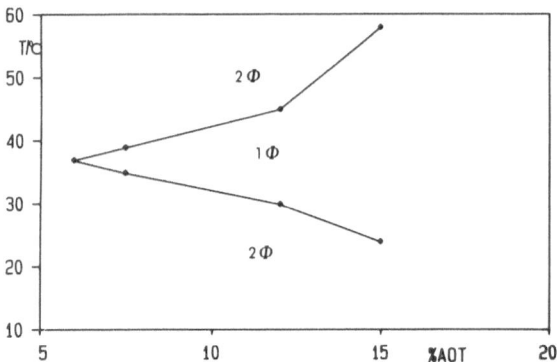

Fig. 2. Approximate phase diagram of the M2 microemulsion employed in this study. 1 Φ: one-phase region. 2 Φ: two-phase region

When microemulsions (M1 or M2) are introduced into the cell and heated very slowly from beneath, the formation of spatial dissipative structures is observed, as can be seen in Fig. 3.

Patterns are observed as convection lines with stronger scattering than the rest of the fluid. It can also be seen that the fluid has a dextrorsum motion, i.e., the fluid moves upwards mainly at the corners of the polygons and downwards at the center.

At 1° below the critical point, we investigated the influence of the temperature difference ΔT between the top and the bottom of the cell. It was then observed that the structures did not disappear when this difference was reduced, although the Rayleigh number was sub-critical [29, 35]. This implies that forces other than

buoyancy are significant for the establishment of the observed hydrodynamic instabilities. One factor which might contribute to the instability of the microemulsions is interfacial tension. If we consider a microemulsion as a microheterogeneous solution, it is possible to define an interfacial tension between the droplets of the microemulsion and the bulk phase. Such interfacial tensions may contribute to the formation of dissipative structures in the same way as surface tension contributes to Marangoni instabilities [19, 29, 30, 32, 33, 35, 36]. Besides a temperature gradient, variation of the interfacial tension with temperature may also force the self-organization of the fluid. At this stage, it is not easy to quantify the influence of this force, because it is difficult to relate it to the measured interfacial tensions between microemulsions and their excess phases. However, it is possible to deduce, by means of a theoretical model [23], that the Marangoni number for such an interfacial tension at 1° below the critical temperature should be supra-critical for $\Delta T \approx 0.1$ K, due to the interfacial tension tending to zero on approaching the critical point [11, 21].

Quenching

We have carried out some experiments in which the single phase microemulsion in the cell was changed to a two-phase state by a rapid change in temperature. It was observed that the phase separation takes place via the formation of spatial dissipative structures (Fig. 4). These spatial structures are transient patterns which

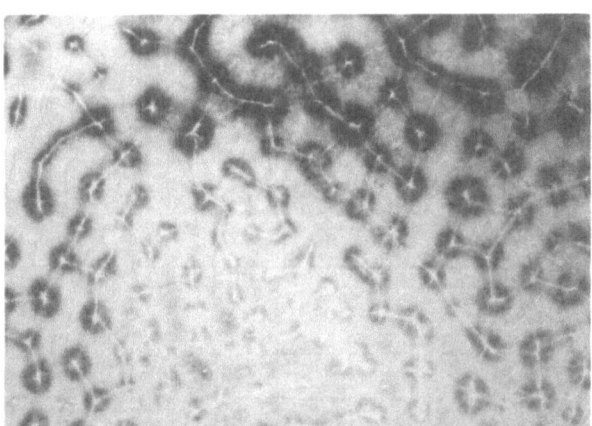

Fig. 3. Spatial patterns observed with microemulsion M1 in a rounded ($\emptyset$: 6 cm) Rayleigh-Bénard cell of 1 cm depth at $|T - T_c| = 1$ K and $\Delta T = 0.1$ K

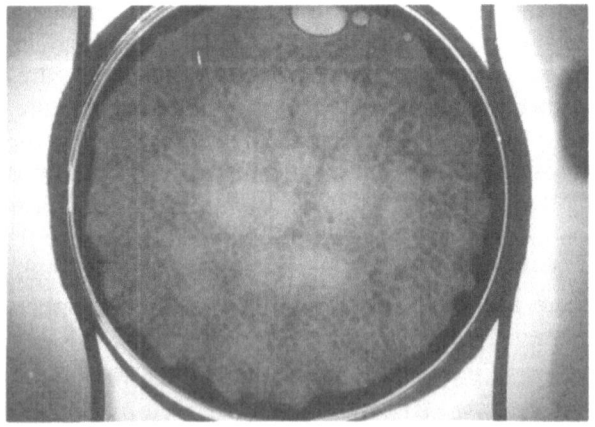

Fig. 4. Pattern formation in quenching experiments with microemulsion M2

become more and more irregular until a chaotic structure is observed and two phases appear. Experiments are planned to see whether this occurs by well-defined routes (deterministic chaos [1, 2, 37]).

Nonlinear relaxations of fluctuations

The great fluctuations which appear when the critical point is approached are able to cause important changes in the behaviour of the systems, as compared to the same system far away from the critical temperature. The importance of stochastic fluctuations for the formation of dissipative structures has recently been analysed by various authors [10, 27, 28] who predict the amplification of the fluctuations in the neighbourhood of a bifurcation point to bring the system to a new far-from-equilibrium state. As far as we know, the influence of internal fluctuations on the onset of instabilities has not been observed experimentally, because of their small size: only the influence of external fluctuations has been investigated [9, 12, 16].

The experiments reported above show that interesting non-linear phenomena should be associated with the large, slow, non-Poisson fluctuations present in critical microemulsions. In order to obtain more information about these non-linear phenomena, we have carried out some experiments involving the relaxation of these fluctuations by the pressure-jump method with conductimetric detection [24, 26]. It was observed that the relaxations were not purely exponential but fitted very well to the Kohlrausch-Williams-Watts (KWW) relaxation law [38]: $y = A \exp (- (t/\tau)^\beta)$. Table 1 shows the results obtained with microemulsion M2 at different temperatures. Firstly, it can be observed that the relaxation time, τ, decreases as the temperature increases, which clearly indicates the presence of a critical slowing down. Secondly, the parameter β tends to the value $\beta \approx 0.5$ when the percolation temperature ($Tp = 307.2$ K) [26] is approached. This value agrees with the theoretical value ($\beta = 0.6 \pm 0.1$) predicted for the relaxation in a percolating system at the percolation threshold ($p = p_c$) [3, 31]. When the temperature of the microemulsion is increased ($p > p_c$) then the value of β should increase until $p \to 1$, in which case $\beta \to 1$ [24]. It can be seen in Table 1 that β increases with the temperature attaining values greater than 1. This can only be explained assuming a non-linear behaviour for the system [24, 25], which agrees with the observation of spatial dissipative structures reported above.

Table 1. Variation of parameters of the KWW-relaxation equation $[y = A \cdot \exp (- (t/\tau)^\beta)]$ with temperature, for microemulsion M2

T/K	β	τ/ms
311.86	1.65 ± 0.03	34.3 ± 0.2
311.46	1.34 ± 0.04	28.2 ± 0.3
310.76	1.01 ± 0.05	19.1 ± 0.3
310.16	0.82 ± 0.05	11.9 ± 0.3
309.66	0.82 ± 0.06	5.18 ± 0.11
309.16	0.64 ± 0.07	1.92 ± 0.06
308.76	0.60 ± 0.07	1.28 ± 0.04

The findings described above clearly show the existence of very interesting non-linear phenomena in critical microemulsions. At present we are studying these phenomena in critical microemulsions by means of different techniques, such as relaxation methods, dynamic light scattering, etc. in order to obtain more information for a quantitative interpretation of the various results reported here, a more detailed description of which will be published elsewhere.

Acknowledgements

We thank the Secretaría de Estado de Universidades e Investigación (SEUI), the Consellería de Educación de la Xunta de Galicia and the Stiftung Volkswagenwerk for their financial support of different parts of this work.

References

1. Ahlers G, Behringer RP (1978) Phys Rev Lett 40:712–716
2. Bergé P, Pomeau Y, Vidal C (1984) Hermann, L'Ordre dans le chaos: Vers une approche déterministe de la turbulence. Paris
3. Blumen A, Klafter J, Zumofen G (1986) In: Pietronero L, Tosatti E (eds) Fractals in Physics. Elsevier, Amsterdam, pp 399–408
4. Cazabat AM, Chatenay D, Langevin D, Meunier J (1982) Faraday Disc Chem Soc 76:291–303
5. Chen SH, Kotlarchyk M (1985) In: Degiorgio V, Corti M (eds) Physics of Amphiphiles: Micelles, Vesicles and Microemulsions. North Holland, Amsterdam 768–792
6. Corti M, Degiorgio V, Zulauf M (1982) Phys Rev Lett 48:1617–1620
7. Dorshow R, de Buzzaccarini F, Bunton CA, Nicoli DF (1981) Phys Rev Lett 47:1336–1339
8. Fourché G, Bellocq AM, Brunetti S (1982) J Colloid Interface Sci 88:302–307
9. Gollub JP, Steinman JF (1980) Phys Rev Lett 45:5511–5514
10. Graham R (1975) In: Riste T (ed) Fluctuations, Instabilities and Phase Transitions. Plenum Press, New York, pp 215–279
11. Guest D, Langevin D (1986) J Colloid Interface Sci 112:208–220
12. Horsthemke W, Lefever R (1983) Noise-Induced Transitions. Springer-Verlag, Berlin

13. Haken H (1983) Advanced Synergetics: Instability Hierachies of Selforganizing Systems and Devices. Springer-Verlag, Berlin
14. Huang JS, Kim MW (1981) Phys Rev Lett 47:1462–1465
15. Huang JS, Kim MW (1985) In: Degiorgio V, Corti M (eds) Physics of Amphiphiles, Micelles, Vesicles and Microemulsions. North-Holland, Amsterdam, pp 864–875
16. Kabashima S, Kawakubo T (1980) In: Garrido L (ed) Systems far from Equilibrium. Springer-Verlag, Berlin, pp 395–402
17. Kim MW, Huang JS (1982) Phys Rev B 26:2703–2706
18. Kim MW, Bock J, Huang JS (1985) Phys Rev Lett 54:46–48
19. Koschmieder EL (1974) Adv Chem Phys 26:177–212
20. Kotlarchyk M, Chen SH, Huang JS (1985) Phys Rev A 28:508–510
21. Langevin D (1985) In: Degiorgio V, Corti M (eds) Physics of Amphiphiles: Micelles, Vesicles and Microemulsions. North-Holland, Amsterdam, pp 181–201
22. López Quintela MA, Fernández Novoa A, to be published
23. López Quintela MA (1987) J Nonequil Thermod, submitted for publication
24. López Quintela MA, Losada D (1988) Phys Rev Lett, accepted for publication
25. López Quintela MA (1988) In: Velarde MG (ed) Synergetics, Order and Chaos, World Scientific, in print
26. Losada D (1987) Tesina de Licenciatura, Universidad de Santiago
27. Nicolis G, Van den Broeck C (1984) In: Velarde MG (ed) Nonequilibrium Cooperative Phenomena in Physics and Related Fields, Plenum Press. New York, pp 473–504
28. Nicolis G (1980) In: Garrido C (ed) Systems far from Equilibrium. Springer, Berlin, pp 91–124
29. Normad C, Pomeau Y, Velarde MG (1977) Rev Modern Phys 49:581–624
30. Pearson JRA (1958) J Fluid Mech 4:489–500
31. Rammal R, Toulouse G (1983) J Phys Lett 44:L13–L22
32. Scriven LE, Sternling CV (1964) J Fluid Mech 19:321–340
33. Smith KA (1966) J Fluid Mech 24:401–414
34. Toprackcioglu C, Dore JC, Robinson BH, Howe A (1984) J Chem Soc Faraday Trans I, 80:413–422
35. Velarde MG, Castillo J (1984) In: Velarde MG (ed) Nonequilibrium Cooperative Phenomena in Physics and Related Fields. Plenum Press, New York, pp 179–196
36. Velarde MG (1979) In: Sorensen TS (ed) Dynamics and Instabilities of Fluid Interfaces. Springer, Berlin, pp 260–275
37. Velarde MG (1981) In: Riste T (ed) Nonlinear Phenomena at Phase Transitions and Instabilities. Plenum Press, New York, pp 205–247
38. Williams G, Watts DC (1970) Trans Faraday Soc 66:80–85

Received December 14, 1987;
accepted January 4, 1988

Authors' address:

M. A. López Quintela
Grupo Biodinámica Física
Departamento de Química Física
Universidad de Santiago
E-15706 Santiago de Compostela, Spain

Progress in Colloid & Polymer Science

Progr Colloid Polym Sci 76:169–175 (1988)

The relation between the cubic phase and the neighbouring solution phase in systems of di(alkyl)dimethylammonium bromide-hydrocarbon-water

K. Fontell and M. Jansson

Chemical Center, University of Lund, Lund, Schweden
Department of Physical Chemistry, University of Uppsala, Uppsala, Sweden

Abstract: A liquid crystalline cubic phase occurs in systems of di(alkyl)dimethylammonium bromide, liquid hydrocarbon and water. Its location in the ternary isothermal phase diagram is about the same irrespective of the hydrocarbon. Its microstructure is not yet known but must be continuous with respect both to hydrocarbon and water. The structure is probably closely related to that in neighbouring parts of the optically isotropic, transparent fluid phase occurring in the same systems, usually called the "microemulsion".

Key words: Liquid crystalline, cubic phases, double chain quaternary ammonium halides, X-ray diffraction, molecular diffusivity, spin-echo pulsed-field gradient NMR.

Introduction

The ubiquity of cubic liquid crystalline phases in lipid-water systems

Liquid crystalline cubic phases are common in lipid-water systems [1–5]. The lipids encompass such a wide variety as anionic, cationic, zwitterionic and non-ionic surfactants (single chain and double chain), monoglycerides, lecithins, lysolecithins, galactolipids etc. The cubic phases are found in pure compounds [6], in binary [1–28, 36–39] and in multicomponent systems [13, 15, 20–22, 25–27, 30–39] where one component is water. The additional components may vary in polarity, being long-chain alcohols [20, 21, 25, 33], fatty acids [2, 21, 33], hydrocarbons [21, 26, 34–37] and so on. The phases have been described by various designations, such as "viscous isotropic phase", "ringing gel", "microemulsion gel", "transparent gel", and "Brummgele". The existence of two, three or multiphase zones in the isothermal phase diagrams has not always been recognised and the existence of cubic phases has not been noticed and accepted.

It is not always easy to show that these phases posses a liquid crystalline cubic structure. They are dark in a polarizing microscope and X-ray studies are seldom conclusive with respect to the microstructure. The X-ray diffractograms are often spotty and irregular, which means that a proper indexing of the reflections is difficult. When the existence of cubic phases was first recognised it was considered that the structures were face-centered arrays of closed aggregates of the type oil-in-water and water-in-oil [8, 10]. The phases occur in widely different parts of the phase diagrams in equilibrium with other phases with varying structure and therefore one cannot expect that all have the same basic structure. At least six different structures have been considered. Both bicontinuous structures and structures composed of closed aggregates have been proposed [1, 4, 5, 40, 45, 46].

Observations in the didodecyldimethylammonium bromide system

The isotropic solution region

Mixtures of didodecyldimethylammonium bromide and water form optically isotropic transparent liquids (usually called microemulsions) together with long-chain liquid alifatic hydrocarbons. In an isothermal ternary phase diagram the solution region (the L_2-phase) has large extension and has the shape of a pear-shaped drop hanging from the hydrocarbon corner, as

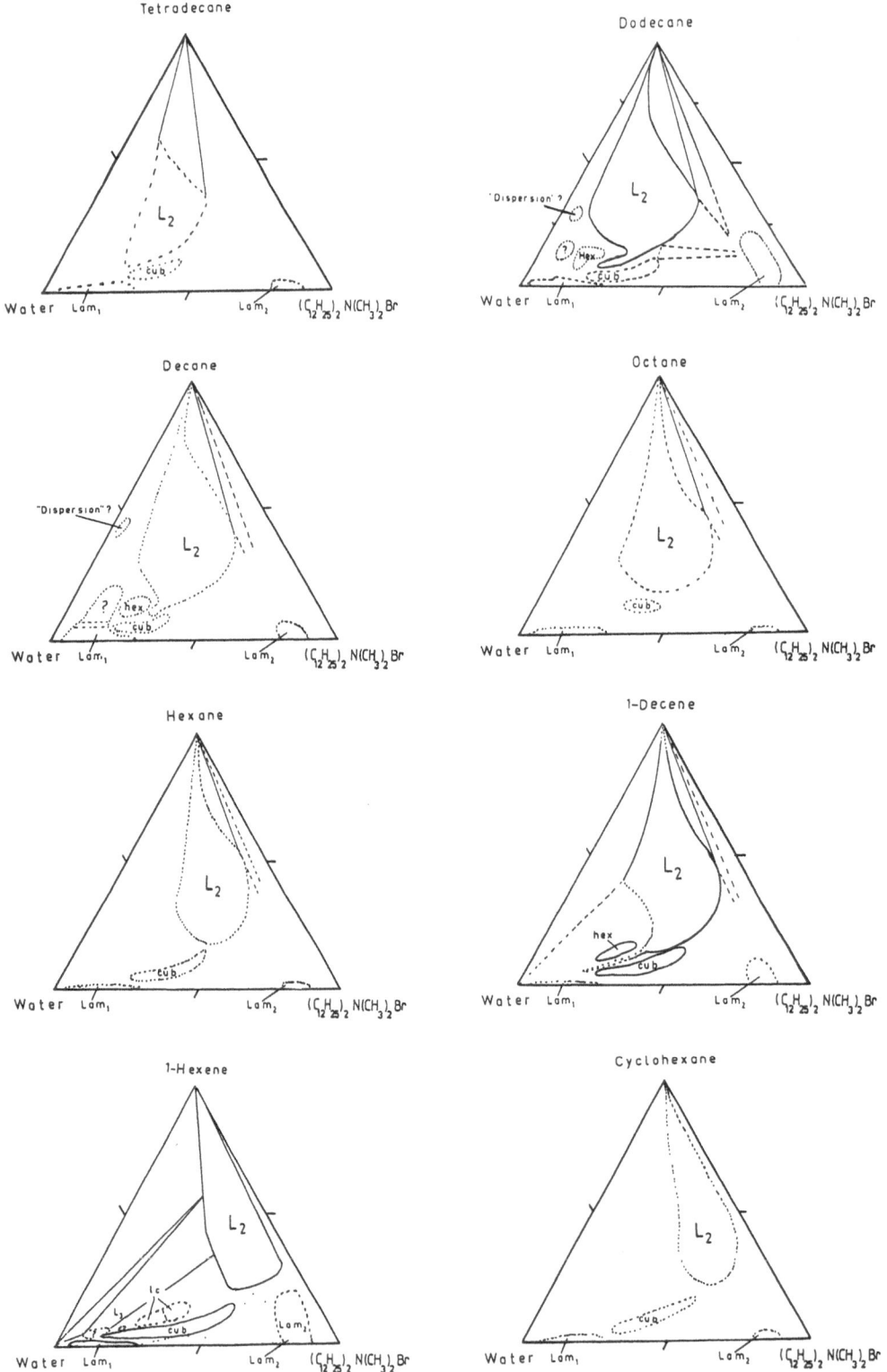

Fig. 1. Isothermal phase diagrams for systems of didodecyldimethylammonium bromide/hydrocarbon/water (20 °C). The dodecane diagram is taken from Ref. [36], the others are composite diagrams based upon work by Evans-Ninham groups [43] and our own experimental observations

long as the chain-length of the hydrocarbon is below or equal to that of the surfactant. When the alifatic hydrocarbon is unsaturated or the hydrocarbon is slightly polar the location of the pendant drop is shifted towards the surfactant/hydrocarbon axis. When the chain-length of the hydrocarbon is longer than that of the surfactant, the L_2-phase region is an isolated island in the center of the triangle diagram. The conditions for the occurrence of this solution phase and its properties have been extensively studied by Ninham and Evans, and their coworkers [43]. The reason for the interest in the structure of this group of association colloid solutions is that they have been suggested to be model systems for some of the systems termed "microemulsions".

Liquid crystalline lamellar and cubic phases

Two lamellar liquid crystalline phases occur in the binary system of didodecyldimethylammonium bromide and water (Fig. 1, surfactant-water axis), while there is only one large region with lamellar liquid crystalline structure in the corresponding chloride system [29]. The phases are found in the bromide system between about 4% and 30% and between about 80% and 90% of surfactant, respectively, and are separated by a wide two-phase region (Fig. 2). The phases have a restricted capability for incorporating hydrocarbon without losing the structure. When about 3 – 15% of hydrocarbon is added to the system there is, at intermediate surfactant contents, a cubic phase which in

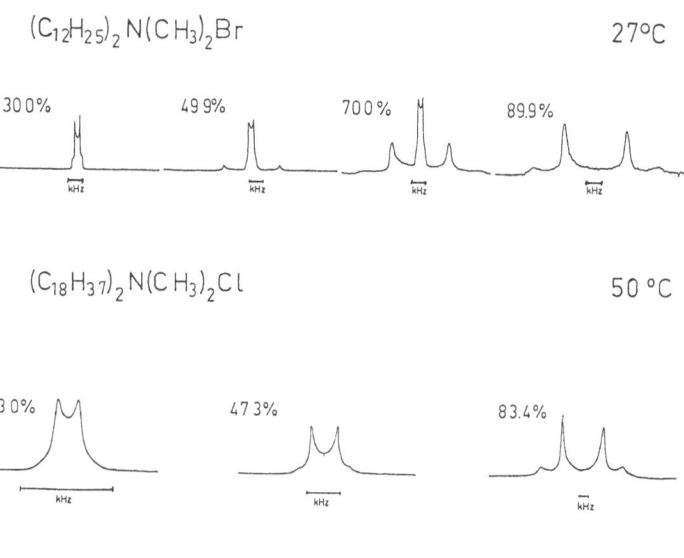

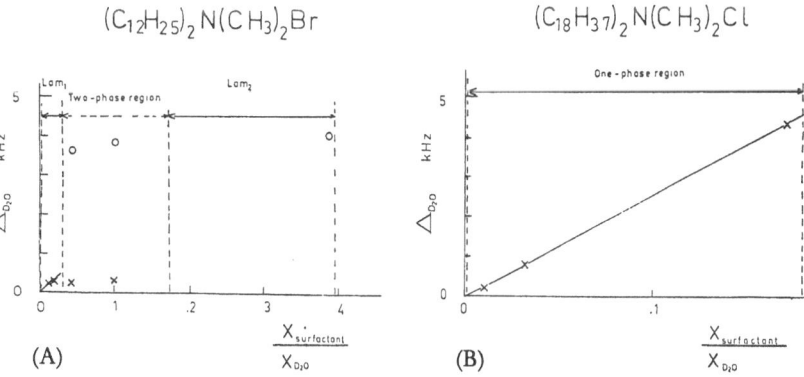

Fig. 2. Quadrupole splittings in binary systems of didodecyldimethylammonium bromide and dioctadecyldimethylammonium chloride, respectively, in heavy water. (A) Water deuteron NMR spectra, two of the spectra in the didodecyldimethylammonium bromide system are from liquid crystalline lamellar one-phase regions, 30.0 and 89.9%, while the two other are superimposed spectra from the intermediate two-phase zone, 49.9 and 70.0% (courtesy of Marianne Svärd). The spectra from the dioctadecyldimethylammonium chloride system indicate the existence of a large liquid crystalline lamellar region (courtesy of Ali Khan). (B) Plots of water deuteron quadrupole splittings *versus* molar ratio of surfactant to heavy water. In the didodecyldimethylammonium bromide system X denotes the splitting due to the Lam_1-phase and o to the Lam_2-phase

general is situated at about the same location of the phase map, irrespective of the nature of the hydrocarbon; only the extension is affected. The cubic phase is separated from the neighbouring phases by two- and three-phase regions. Figure 1 shows the location and extension of the cubic phase in eight systems with different hydrocarbons. A rather detailed phase diagram has been published for the dodecane system [36]. With decanol as the third component, a cubic phase has also been found in systems of dialkyldimethylammonium chloride and bromide, respectively ([29, 37], unpublished results K.F.).

Macroscopically cubic phases are clear, isotropic, rather soft gels. In this respect they differ from the cubic phases in binary and ternary soap systems which are rather stiff. The cubic liquid crystalline structure is evident from the appearance of "crystalline" X-ray diffraction peaks in the 30 to 150 Å range, together with the broad diffuse reflection at 4.5 Å demonstrating the liquid state of the hydrocarbon chains. However, the diffraction patterns in the low-angle region are often of poor quality, so it is futile to attempt an indexing of the peaks. That has been accomplished only in a few cases. The structure seems to be primitive or body-centered. In the cubic phases studied by us the indexing has seemed to favour a primitive structure (Table 1), while

the cubic phase in the decane system, according to Anderson, is body-centered [37].

Self-diffusion in the cubic phase

A simple way of establishing whether a structure is bicontinuous or not is to study the molecular self-diffusion coefficients of the components [41]. That can be done by using spin-echo pulsed-field gradient NMR [42]. By this technique one measures the diffusivity of each component of the system. Molecules of components forming domains which extend over macroscopic distances diffuse very rapidly; except for obstruction and solvation effects, they will diffuse as in pure liquid.

The diffusion coefficients have been determined for the molecular components of the cubic phase in the 1-hexene system. A few measurements have also been performed in the cubic phases of the hexane and dodecane systems. The diffusion coefficients of both hydrocarbon and water are rather high throughout, in the range $10^{-9} - 10^{-10}$ m^2/s, while the values for the surfactant are about one order of magnitude lower (Table 2). The water diffusion increases and the hydrocarbon diffusion decreases somewhat with the water content in the 1-hexene system. The dependence of the coeffi-

Table 1. X-ray diffraction findings in cubic samples in systems of didodecyldimethylammonium bromide-hydrocarbon-water (20°C)

Composition (%) Surfactant	Hydrocarbon	Water	Spacings (Å)	$(h^2 + k^2 + l^2)$ m	Unit cell a_0 (Å)
Decane					
30.0	6.0	64.0	128	3	~ 210–220
			105	4	
			90	6	
			77	8	
Hexane					
35.0	5.0	60.0	66	2	~ 93
			54	3	
			38	6	
			33	8	
			31	9	
1-Hexene					
55.0	15.0	30.0	47	2	~ 67
			39	3	
			33	4	
			28	6	
			51	2	~ 70– 72
			41	3	
			35	4	
			29	6	

Comments: The diffractograms were of poor quality, therefore the spacing values and the indexing are uncertain.

Table 2. Self-diffusion coefficients in cubic samples of the systems didodecyldimethylammonium bromide-hydrocarbon-water (26 °C)

Composition (%) Surfactant	Hydrocarbon	Water	D_w	$D_{hydrocarbon}$ 10^{-10} m²/s	$D_{surfactant}$
Dodecane					
34.95	5.05	60.60	5.52 ± 0.37	1.46 ± 0.05	[a]
23.85	4.78	71.34	8.31 ± 0.43	2.15 ± 0.12	[a]
22.83	3.76	73.41	9.14 ± 0.41	2.25 ± 0.27	[a]
Hexane					
31.78	5.05	63.17	5.99 ± 0.28	1.68 ± 0.20	[b]
25.05	4.95	70.00	5.09 ± 0.36	1.12 ± 0.15	[b]
1-Hexene					
55.04	14.04	29.92	4.25 ± 0.01	7.20 ± 0.2	0.19 ± 0.02
53.27	14.55	32.18	4.57 ± 0.01	7.62 ± 0.2	0.21 ± 0.02
45.1	10.0	45.0	5.97 ± 0.02	4.83 ± 0.2	0.21 ± 0.03
			6.1/ ± 0.05	4.59 ± 0.4	0.21 ± 0.03
44.95	8.12	46.93	5.66 ± 0.02	5.43 ± 0.2	0.18 ± 0.02
34.97	5.57	59.46	9.23 ± 0.04	5.45 ± 0.3	0.17 ± 0.03
			7.38 ± 0.03	5.43 ± 0.4	
26.68	4.23	69.09	8.75 ± 0.4	4.74 ± 1.4	
			8.08 ± 0.4	4.69 ± 1.4	0.63 ± 0.09
19.98	6.06	73.96	10.77 ± 0.5	5.89 ± 1.8	1.54 ± 0.23
20.01	3.97	76.02	11.92 ± 0.6		0.9 ± 0.1
15.87	2.91	81.23	11.60 ± 0.6	5.70 ± 1.7	1.24 ± 0.19

Pure compounds: Water 22.72 · 10^{-10} m²/s; Dodecane 8.8 · 10^{-10}; Hexane 44.2 · 10^{-10}; 1-Hexene 47.10 ± 0.85, experimentally determined; [a]) courtesy of Andrea Ceglie; [b]) courtesy of Kali Das

cients for surfactant, water and hydrocarbon on the water content is presented in Fig. 3.

Discussion

The structure of the cubic phases in the systems didodecyldimethylammonium bromide-hydrocarbon-water is probably related to that in the neighbouring parts of the L_2-solution phases. The two-phase zones between the two phases are narrow and the self-diffusion coefficients (of hydrocarbon, water and surfactant) in the cubic phases are of the same order of magnitude as in the L_2-phase. (In the dodecane system, the self-diffusion coefficients for water, dodecane and surfactant have values of about 5–9, 2 and 0.5, and about 5, 2.5 and 0.7 10^{-10} m²/s, respectively, in the cubic phase and the neighbouring part of the L_2-phase) ([36, 43] unpublished results, K.F.). The values of the diffusion coefficients show, furthermore, that the structures must be continuous with respect to both hydrocarbon and water. However, as has been stressed above, the X-ray diffractograms have so far failed to

provide any useful information about the structure, apart from the fact that it must be liquid crystalline and that it seems to be primitive or body-centered cubic.

The solubility of the surfactant is low in both hydrocarbon and water and the surfactant therefore resides

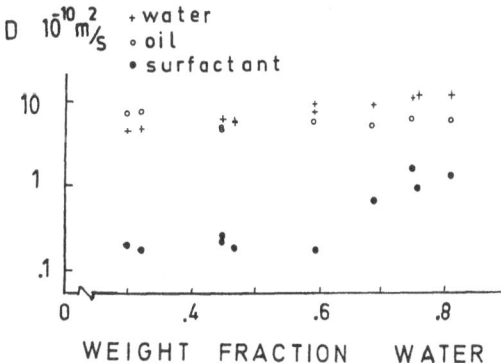

Fig. 3. Plots of water, hydrocarbon and surfactant self-diffusion coefficients, respectively, *versus* water content for cubic samples in the system didodecyldimethylammonium bromide/1-hexane/water (27 °C)

both in the cubic phase and in the L_2-phase at the interfaces between polar and nonpolar regions of the structure. A further notable feature is that the phase occurs in about the same location of the phase diagram, irrespective of the hydrocarbon. Geometric constraints are obviously important for the formation of the cubic phase. For the L_2-phase, it has been suggested that the structure consists of two interconnected continuous networks of disordered worm-like aggregates of hydrocarbon and water, respectively, with a monomolecular layer of surfactant at the polar/nonpolar interface which layer has a constant curvature [43, 44]. The structure of the cubic phase can be visualised to be essentially similar, but somewhat more ordered.

A thought-provoking feature is the narrow outgrowth of the L_2-phase which penetrates between the cubic and hexagonal phase regions in some of the phase diagrams. The feature is especially prominent in the 1-decene system but is also noticeable in the decane and dodecane systems (Fig. 1). A more detailed study of the phase behaviour may also reveal the same phenomenon in other systems.

The cubic phase in the 1-hexene system is especially interesting from various points of view. It has a large extension with respect to the content of water, from about 30% to about 80%. In the preparation of the samples in this system, there is the practical observation that it is often more difficult to obtain equilibrium in samples with high water content (it takes weeks) although the consistency is very soft, than at low contents of water. There is also a change in the self-diffusion of the surfactant: below 60% the coefficients are constant at about the same level as in the L_2-phase [36, unpublished results, K.F.], but above there is an almost tenfold increase (Fig. 3). One may speculate about a change of the structure inside the cubic phase.

Such a change in structure has been shown to occur inside the cubic phase in the system monoolein-water. Monoolein is practically insoluble in water, but forms a cubic phase between about 25% and 40% of water. The structure is body-centered at low water contents and primitive at high contents [45]. The monoolein bilayers form an infinite periodic minimal surface (IPMS) of the gyroid type in the body-centered structure, but as the water content is increased the structure is transformed, without any change in the curvature of the interface, into a diamond type one [46]. Actually, the gap between the methyl end groups of the infinite lipid bilayer constitutes the IPMS-interface. In this manner, the lipid bilayer separates two equal continuous water networks without connection. The transition from the gyroid-type to the diamond-type structure involves very small energy differences, and could occur by a Bonnet transformation. Any two-phase zone will be very difficult to observe.

There may be a similar transition in the cubic phase of the 1-hexene system. In any case, it is quite probable that the structure of the cubic phase in the di(alkyl)dimethylammonium halide systems is related to that in the neighbouring part of the L_2-phase.

References

1. Luzzati V (1969) In: Chapman D (ed) Biological Membranes. Academic Press, New York, p 71
2. Ekwall P (1975) In: Brown GH (ed) Advances in Liquid Crystals. Vol 1, Academic Press, New York, p 1
3. Tiddy GT (1980) Phys Rep 57:1
4. Fontell K (1981) Mol Cryst Liquid Cryst 63:59
5. Luzzati V, Mariani P, Gulik-Krzywicki T (1987) Springer Proc Phys 21:131; Charvolin J, Sadoc JF (1987) Springer Proc Phys 21:126
6. Spegt PA, Skoulios AE (1966) Acta Cryst 22:892; Luzzati V, Spegt PA (1967) Nature 215:701
7. Luzzati V, Mustacchi H, Skoulios AE (1958) Disc Faraday Soc 25:53
8. Luzzati V, Mustacchi H, Skoulios A, Husson F (1960) Acta Cryst 13:660; Husson F, Mustacchi H, Luzzati V (1960) Acta Cryst 13:668
9. Lutton ES (1965) J Am Oil Chem Soc 42:1068; (1966) J Am Oil Chem Soc 43:28
10. Luzzati V, Reiss-Husson F (1966) Nature 210:1351
11. Larsson K (1967) Z Phys Chem, Frankfurt 56:173
12. Reiss-Husson F, Luzzati V (1967) Adv Biol Med Phys 11:87
13. Small DM (1967) J Lipid Res 8:551
14. Clunie JS, Corkhill JM, Goodman JF (1968) Proc Roy Soc A285:520; Balmbra RR, Clunie JS, Goodman JF (1968) Proc R Soc A285:528
15. Fontell K, Mandell L, Ekwall P (1968) Acta Chem Scand 22:3209
16. Luzzati V, Tardieu A, Gulik-Krzywicki T (1968) Nature 217:1028
17. Luzzati V, Gulik-Krzywicki T, Tardieu A (1968) Nature 218:1031
18. Luzzati V, Tardieu A, Gulik-Krzywicki T, Rivas E, Reiss-Husson F (1968) Nature 220:485
19. Lawson KD, Mabis AJ, Flautt TJ (1968) J Phys Chem 72:2958
20. Ekwall P, Mandell L, Fontell K (1969) J Colloid Interface Sci 31:508
21. Ekwall P, Mandell L, Fontell K (1969) Mol Cryst Liquid Cryst 8:157
22. Friberg S, Mandell L, Fontell K (1969) Acta Chem Scand 23:1055
23. Rivas E, Luzzati V (1969) J Mol Biol 41:261
24. Rogers J, Winsor PA (1969) J Colloid Interface Sci 30:247
25. Ekwall P, Mandell L, Fontell K (1970) J Colloid Interface Sci 33:215
26. Friberg S, Mandell L (1970) J Am Oil Chem Soc 47:149
27. Arvidson G, Brentel I, Khan A, Lindblom G, Fontell K (1983) Eur J Biochem 159:753

28. Gulik A, Luzzati V, de Rosa M, Gambacorta P (1985) J Mol Biol 182:131
29. Kuneida H, Shinoda K (1978) J Phys Chem 82:1710
30. Spegt PA, Skoulios AE, Luzzati V (1961) Acta Cryst 14:866
31. Luzzati V, Husson F (1962) J Cell Biol 12:207
32. Small D, Bourges M (1966) Mol Cryst 1:541
33. Fontell K (1973) J Colloid Interface Sci 43:156
34. Müller-Goymann C (1984) Pharmaceutical Research 141; Usselman B, Müller-Goymann CC (1984) Progress Colloid Polym Sci 69:56; Müller-Goymann CC (1984) Seifen-Fette-Öle-Wachse 110:395; Müller-Goymann CC (1985) Pharm Z 130:682
35. Nürnberg E, Pohler W (1984) Progr Colloid Polym Sci 69:48
36. Fontell K, Ceglie A, Lindman B, Ninham B (1986) Acta Chem Scand A40:247
37. Anderson DM, Thesis Ph (1986) University of Minnesota, Minneapolis; Anderson DM, US Pat Appl Series 052713, May 20, 1987
38. Tabony J (1986) Nature 319:400; de Geyer A, Tabony J (1987) Springer Proc Phys 21:372
39. Fontell K (1974) In: Gray GV, Winsor PA (eds) Liquid Crystals and Plastic Crystals, Ellis Horwood, Vol 2, p 80
40. Fontell K, Fox KK, Hansson E (1985) Mol Cryst Liq Cryst Letters 1:9
41. Bull T, Lindman B (1974) Mol Cryst Liquid Cryst 28:155
42. Lindman B (1987) Springer Proc Phys 21:357; Lindman B, Stilbs P (1986) In: Friberg S, Bothorel P (eds) Microemulsions, CRC Press, Cleveland, Ohio, p 119; Lindman B, Stilbs P (1987) In: Rosano H, Clausse M (eds) Microemulsion systems, Marcel Dekker, New York, p 129
43. Ninham BW, Evans FD, Wei T (1983) J Phys Chem 84:5029; Angel LR, Evans FD, Ninham BW (1983) J Phys Chem 87:538; Hashimuoto S, Thomas JK, Evans DF, Ninham BW (1983) J Colloid Interface Sci 95:594; Talmon Y, Evans FD, Ninham BW (1983) Science 221:1047; Chen SJ, Evans DF, Ninham BW (1984) J Phys Chem 88:1631; Ninham BW, Chen SJ, Evans FD (1984) J Phys Chem 88:5855; Brady JE, Evans DF, Kachar B, Ninham BW (1984) J Am Chem Soc 106:4279; Blum FD, Pickup S, Ninham BW, Chen SJ, Evans FD (1985) J Phys Chem 89:711; Chen SJ, Evans DF, Ninham BW, Mitchell DJ, Blum FD, Pickup S (1986) J Phys Chem 90:482
44. Zemb TN, Hyde ST, Derian P-J, Barnes IS, Ninham BW (1987) J Phys Chem 91:4814
45. Lindblom G, Larsson K, Johansson L, Fontell K, Forsén S (1979) J Am Chem Soc 101:5465; Longley W, McIntosh J (1983) Nature 303:612; Larsson K (1983) Nature 304:664
46. Hyde ST, Anderson S, Ericsson B, Larsson K (1984) Z Kristallogr 168:213

Received December 15, 1987;
accepted December 30, 1987

Authors' address:

K. Fontell
Chemical Center
University of Lund
Box 124
S-22100 Lund, Sweden

Progress in Colloid & Polymer Science
Progr Colloid Polym Sci 76:176–182 (1988)

Thermotropic phase transitions in totally synthetic surfactant vesicles and bilayers

A. Malliaris

N.R.C. Demokritos, Athens, Greece

Abstract: The thermotropic phase transitions observed in vesicles and bilayers of totally synthetic surfactants are discussed from the physicochemical point of view. The formation, characterization and morphology of these aggregates are briefly presented. Some general trends and the behaviour of these temperature-induced phase transitions are examined.

Key words: Amphiphile, surfactant, vesicle, bilayer, phase transition.

Introduction

Most biological membranes exhibit characteristic thermotropic phase transitions, which have been termed "solid-to liquid crystalline" or "gel-to liquid crystalline" transitions [1]. The temperature (T_c) of this reversible transition depends primarily on the nature of the phospholipids which make up the membrane bilayer [2]. This has been established by a variety of physicochemical techniques, such as differential scanning calorimetry (DSC), electron paramagnetic resonance (EPR), nuclear magnetic resonance (NMR), fluorescence, dilatometry X-ray diffraction, etc. It has been found that several changes in structural parameters occur at T_c. Thus the thickness of the lipid membrane [3], the volume of the bilayer [4], the proton NMR line width [3–5], the order parameter of EPR probes [6], the fluorescence polarization [7–9], etc., are affected by the appearance of the phase transition. Also, biologically relevent functions of the membrane, such as cell growth, transport, immunological response, etc., are known to undergo important changes as well at the onset of the phase transition [10, 11]. Strong experimental evidence has attributed the thermotropic phase transition in biomembranes exclusively to their phospholipid bilayer [12]. In fact, naturally occuring lipids present in biological membranes often form in water bilayer structures which

demonstrate thermotropic polymorphism. Thus, several bilayers of natural phospholipids exhibit two first-order thermal phase transitions, the so-called "main" transition at higher temperature, and "pretransition" at lower temperature.

On the other hand, synthetic surfactant vesicles, which are easily formed upon dispersal of single-chain, double-chain and triple-chain amphiphiles having cationic, anionic, non-ionic and zwitterionic head groups, have been extensively studied as membrane bilayer models [13–20]. The reason is that natural membranes are quite complicated structures [21] and therefore, synthetic surfactant vesicles offer good alternatives for the study of natural lipid bilayers. The purpose of the present paper is to discuss the physicochemical aspects of thermotropic phase transitions occuring in synthetic surfactant vesicles and bilayers.

Vesicle-forming amphiphiles

The first reported case of a totally synthetic bilayer membrane was that of didodecyldimethylammonium bromide [22] which, when dispersed in water by extended sonication at 50 °C, produces single bilayer vesicles with a diameter of ca. 300–500 Å. Latter it was shown that several salts with two long aliphatic chains (I–IV) could produce lamellar structures, as well as large multi-walled and single compartment bilayer vesicles [23–25]. Moreover, it was shown that stable bilayers are formed even with single chain surfactants, provided the hydrophobic part contains a rigid segment, e.g., the diphenylazomethine moity (V–VI) [26, 27, 30]. Note that diphenylazomethine derivatives belong to a major class of liquid crystalline materials, which emphasizes the close relationship between bilayers, micelles and liquid crystals [28, 29]. Also, synthetic dialkyl amphiphiles with zwitterionic as well as nonionic head groups (VII–X) [31], when dispersed in water, either by sonication or by the injection method [32], can produce vesicles, bilayers, or other structures, depending on the experimental conditions [31]. A number of other special bilayer and vesicle-forming synthetic surfactants have been reported (XI–XV) [33–36]. On the other hand, in an attempt to obtain long-term stability of the vesicles for use in practical applications, e.g. drug carriers, a large number of polymerizable surfactants have also been synthesized, but they will not be discussed in the present account.

Sonic dispersion of surfactants at temperatures above T_c is the most efficient method for vesicle and bilayer formation [17]. Once the organized aggregates have been formed, they remain stable, even if the temperature is lowered below T_c [15]. Apart from sonication, other procedures have also been developed for the preparation of vesicles, most notably the injection method [32]. Increasing the sonication time, at a given power, results in a decrease of the viscosity and turbidity of the solution down to a point beyond which further sonication has no additional effect [17, 23]. From the preceding discussion, becomes quite evident that bilayer formation is not a specific property of biolipids, but instead it is a rather general physicochemical property common to many classes of amphiphiles.

Synthetic surfactant vesicles and bilayers can be characterized by many physicochemical methods such as surface tension [26, 31], fluorescence [37], electrical conductivity [13, 38, 39], NMR [14], transmission electron microscopy (freeze-fracture, negative staining) [17, 26, 31, 34, 37, 40, 41], small angle X-ray scattering [37], EPR [42], low angle laser light scattering [43], photon correlation spectroscopy [43], positron annihilation [15], osmotic shock techniques [15], fluorescence quenching [13], etc. Finally, it is worth mentioning the use of the very efficient self-quenching of the fluorescence of carboxyfluorescein for proving the formation of closed vesicles [43]. In conclusion, the essential structural characteristics of

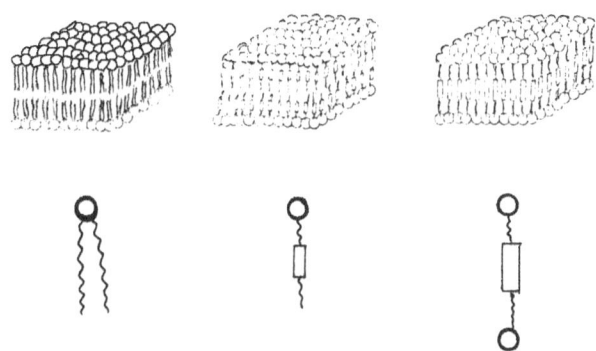

Fig. 1. Monolayers and bilayers formed from the various types of synthetic surfactants shown in Scheme A. (O) polar head; (∿∿) aliphatic chain; (▭) rigid segment

synthetic surfactant membranes are very similar to those of biomembranes, i. e. the bilayer is formed only when the alkyl chain length exceeds 10 carbon atoms, while its thickness is 30–50 Å depending on the length of aliphatic chain. Figure 1 shows a schematic illustration of bilayer membranes formed from the various types of synthetic surfactants listed in Scheme A.

Thermotropic phase transitions

Temperature-induced phase changes in pure phospholipids were originally observed by i-r and DSC methods in the lipid dimyristoylphosphatidyl ethanolamine which contains two saturated aliphatic chains [44, 45]. It was subsequently found that when increasing amounts of water were added to phospholipids, the temperature of the phase transition was reduced, eventually reaching a limiting value independent of water content [45]. This thermotropic phase transition of water-rich phospholipid dispersions has been extensively studied in several natural lipid bilayers, for both its biological interest and also the important relationship between cell membranes and liquid crystalline structures [46, 47]. Similar studies have been extended to vesicles and bilayers made of synthetic surfactants, and it is now well established that identical behaviour is also exhibited by most of these aggregates. The most important technique for the study of thermotropic phase transitions in biomembranes and synthetic bilayers is differential scanning calorimetry, and therefore the subsequent discussion will be based mainly upon results of thermal analysis.

An accurate treatment of the thermotropic phase transition would require evaluation of the partition

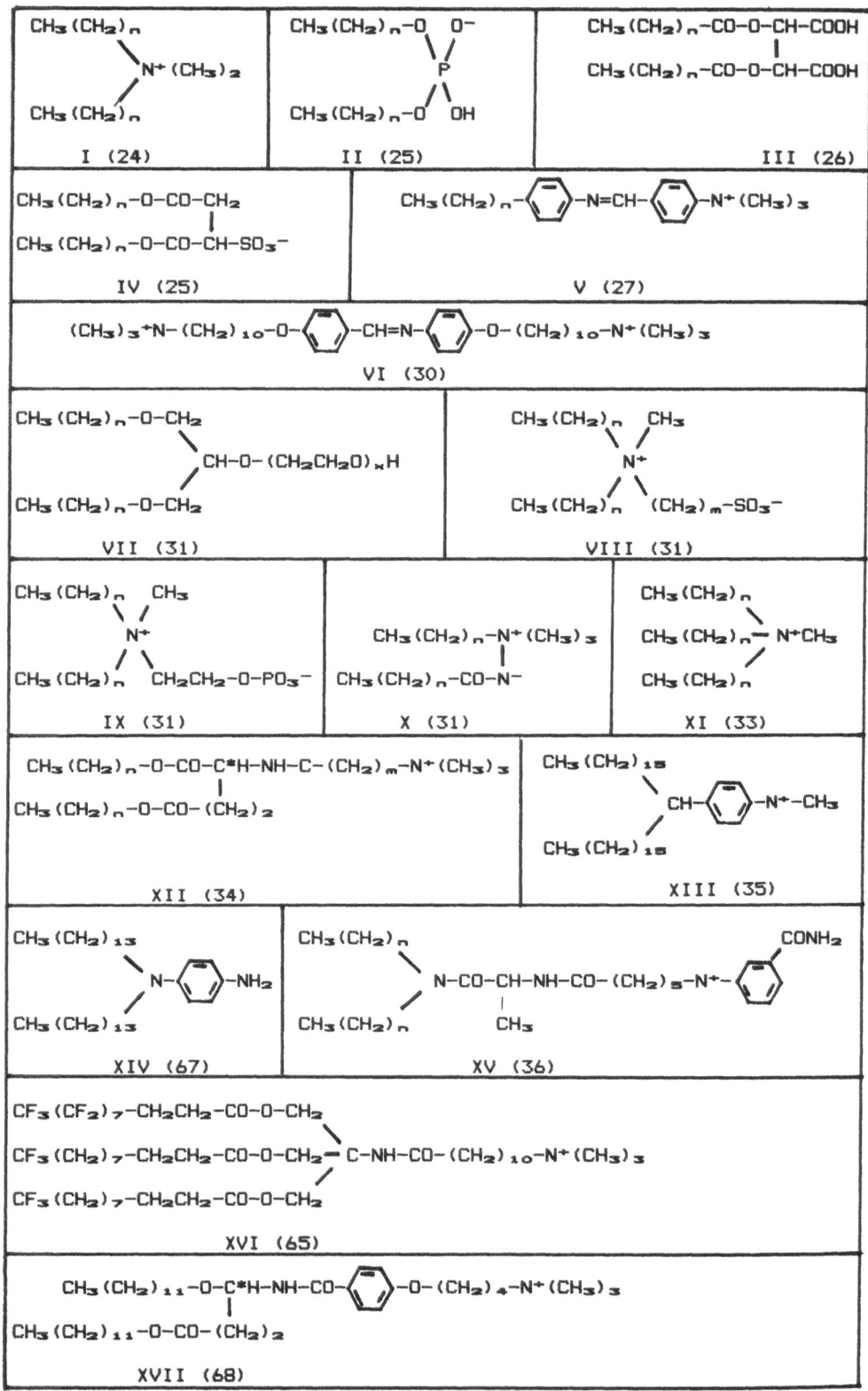

Scheme A: Structural formulae of some vesicle- and bilayer-forming synthetic surfactants. Reference numbers in parentheses

function which involves all the possible configurations of the system. Moreover, the total energy of a given configuration is the sum of a number of contributions, the most important of which, in the case of a bilayer, include the repulsive excluded-volume interactions, the attractive van der Waals, rotamer, and head group interactions. A rigorous statistical mechanical solution to this problem is extremely complicated, mainly because of the excluded volume and van der Waals interactions. Consequently, only qualitative aspects of the phase transition are discussed here from a purely phenomenological point of view. Thus, from the transition enthalpy, ΔH (measured from the area under the endothermic peak in the DSC thermograms) is ca. 4–14 Kcal/mol, depending on the particular surfactant [48], the associated entropy increase is estimated ($\Delta S = \Delta H/T$) to be approximately 13–45 cal/deg. mol. If, on the other hand, the additional degrees of freedom liberated during the transition are assumed to be of a simple two-state nature, then it is evident ($\Delta S = R \ln 2 = 1.4$) that a rather large number of degrees of freedom (10–30) must be activated as a result of the phase transition. Clearly, only the hydrocarbon chains can be associated with so many degrees of freedom through rotations about their numerous carbon-carbon bonds, i.e. through chain conformations (rotational isomerism). This conclusion is in agreement with the common characterization of the phase transition in lipid and surfactant bilayers as chain "melting" [1, 2, 10]. The importance of the gauche conformers above T_c is also evident from the changes in dimensions occuring in membranes at the phase transition. Thus, in bilayers of dipalmitoyl phosphatidyl choline, the membrane surface area per molecule increases from 0.403 nm² below T_c, to 0.445 nm² above T_c, while the bilayer thickness changes from 4.6 nm at 20 °C to 3.5 nm at 50 °C [45, 50]. These changes are necessary to accommodate the increased number of gauche conformations present above the phase transition [51]. Note that the pretransition is also associated with limited trans-gauche isomerization, especially towards the central part of the acyl chain [51, 52].

Ordinarily, phase transitions occur sharply at a single temperature T_c, e.g. 0.08 ° in N-p-cyanobenzylideno-p-octyloxyaniline [53], and the two phases coexist in equilibrium only at this particular temperature. In the case of biomembranes and surfactant vesicles and bilayers, however, the transition from the gel state of high rigidity and conformational order to the more fluid liquid crystalline state takes place over an appreciable range of temperatures, e.g. the transi-

tion in dipalmitoyl phosphatidyl choline occurs within ca. 1 °C [54]. Normally, a first-order transition is characterized by non-zero latent heat ΔH, as well as discontinuous changes of entropy and volume, whereas second order transitions have zero ΔH. Order phase transitions have zero latent heat of transformation. Experiments have shown that as far as the change of latent heat is concerned, the phase transition in bilayers is of the first order [53], but volume changes are not discontinuous [54]. It is, of course, possible that the lack of discontinuity in volume changes is due to experimental artifacts, such as impurities, temperature inhomogeneities in the sample, etc. Moreover, it should be mentioned that for a first-order transition, the two phases involved must be well separated and distinct from each other. Clearly, this is not quite the case in bilayers, where the one phase is developed within the matrix of the other [55]. Therefore, from the experimental point of view, a transition in these systems is considered to be of the first order if simply the latent heat of the transformation is not zero. Another important physical parameter of the phase transition, i.e. the size of the average cooperative unit, can be estimated from the DSC graph. Indeed, the number of surfactants N in the cooperative unit is given approximately by the expression $N = 4RT_c^2 C_{max}/\Delta H^2$ where C_{max} is the maximum value of the excess heat capacity and ΔH is the latent heat of the transition [56, 57].

Figure 2 shows the proposed structural model associated with the two first-order phase transitions, the main transition and the pretransition, drawn on the basis of DSC and X-ray diffraction data [17, 51, 58]. According to this figure, at low temperature, a one-dimensional lamellar structure is formed, where the hydrocarbon chains are fully extended and tilted with respect to the plane of the bilayer. At the pretransition temperature, the structure is transformed to lamellae distorted by periodic ripples. Finally, at the main transition temperature, the chains "melt", assuming a nearly liquid-like conformation, as shown at the right side of Fig. 2. Note that the chain melting in bilayers is not as extensive as it is in hydrocarbons, and this is evidenced by the fact that the latent heat of transition and the volume change per methylene group is about threefold higher in hydrocarbons than in bilayers [12]. This may be understood in terms of the additional constraints existing in bilayers, i.e. the attachment of the surfactant chains to the water/bilayer interface. Note that in some cases an additional third transition, the "subtransition" has been reported at lower temperature than the pretransition [57]. In the following we

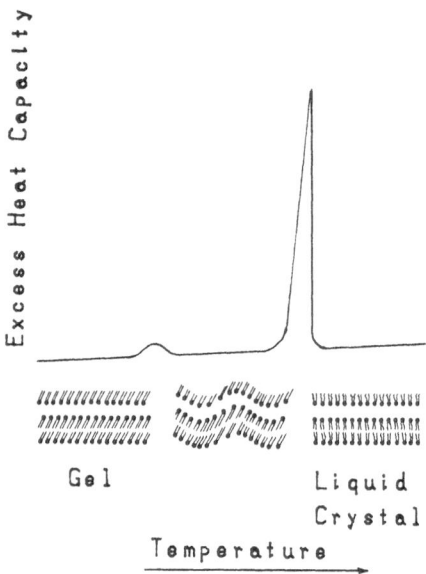

Fig. 2. Thermogram and membrane structure at various temperatures

will examine the effect of various parameters on the phase transition.

Effect of experimental conditions

An important factor which affects the thermotropic behaviour of synthetic surfactant aggregates is the method of dispersion. It has been observed that samples treated by simple sonication give different thermograms than samples in which sonication was followed by slow freezing to −50 °C (freezing method). Thus, frozen samples of I, II, III and IV have a higher T_c than the corresponding sonicated samples, and the difference increases with increasing chain length. Also, the endothermic peak in sonicated samples is broad with shoulders and provides evidence for pretransitions, whereas frozen samples produce rather sharp DSC graphs. Electron microscopy has shown that these differences reflect differences in the structure of the aggregates, which, upon repeated heating cycles, eventually disappear [48]. In contrast, aggregates of nonionic and zwitterionic (phosphocholines) amphiphiles produce DSC curves independent of the method of dispersion [59, 60]. Likewise, in tri-alkylammonium aggregates, T_c is independent of the method of preparation [33]. Also, dispersions of single as well as double chain synthetic phosphocholines

seem to produce DSC graphs independent of the method of dispersion [61]. It is worth mentioning here that bilayers of di- and tri-dodecyl quaternary ammonium salts show a phase transition only when frozen samples are examined, while sonicated samples do not demonstrate thermotropism [33, 48]. The starting temperature of a thermogram with respect to T_c is another experimental factor which usually affects the thermotropic behaviour of bilayers. In general, DSC thermograms are not reproducible, in either sonicated or frozen samples, if the starting temperature is close to T_c. Thus, the latent heat of the transition increases progressively as the starting temperature is lowered, until it reaches its regular value when the starting temperature is 10°−20 °C below T_c [48]. Finally, the scan rate (degrees/min) at which the thermograph is recorded has been shown to influence transition parameters such as ΔH, width of the endothermic transition peak, temperature of pretransition, etc. [56]. A possible interpretation of this effect is that although the main transition is a fairly rapid process [62], the pretransition ocurs at quite a slow rate [63].

Effect of hydrophobic chain

Among the synthetic surfactant vesicles and bilayers of Scheme A, the thermotropic behaviour of the double chain amphiphiles has been studied most extensively [14, 31, 48]. Figure 3 shows plots of T_c and ΔH vs. the length of the aliphatic chain for symmetric double chain quaternary ammonium surfactants [48]. It is generally observed that T_c increases regularly with increasing length of the alkyl or acyl chain [61], while, in most cases, amphiphiles with hydrophobic chain length less than 12 carbon atoms do not exhibit phase transition. Changes in the structure of the long chain influence the thermotropic behaviour of vesicles and bilayers, e.g. introduction of a double bond lowers T_c by several degrees [64]. Replacing H by fluorine atoms results in a decrease of the enthalpy change ΔH, whereas the effect on T_c is rather complicated [65]. Fluorocarbons behave differently from their respective hydrocarbons, primarily because of the shorter C–F bond (1.317 Å) and the large van der Waals radius of F (1.35 Å). The variation observed in the values of ΔH, which range from less than 1 to more than 10 kcal/mol in the different synthetic surfactant aggregates, reflects differences in the extent of the change of molecular motion in the bilayer during phase transition [16]. The anomalous dependence of ΔH on chain length in the

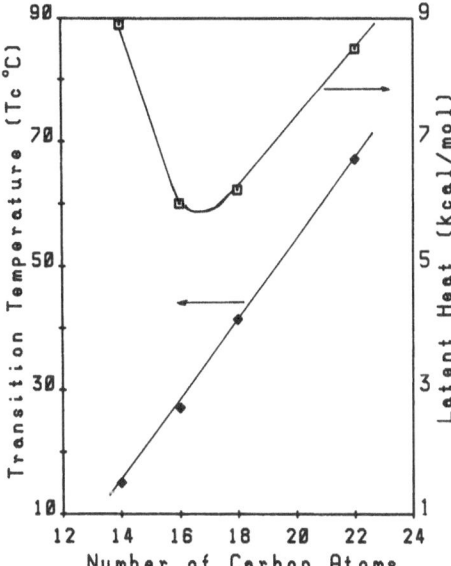

Fig. 3. Plots of T_c and ΔH vs. chain length for dialkyldimethylammonium surfactants (I). From Ref. [48]

same homologous series (Fig. 3), probably indicates nonuniform alignment of the alkyl chains, because otherwise ΔH should be proportional to the chain length.

Effect of head group

The nature of the hydrophilic head group also affects the thermotropic transition in the bilayer membrane mainly in two ways, i.e. by stabilizing the bilayer structure, and by affecting the alignment of the hydrophobic chains. One example of the stabilizing effect is the case of membranes made of the zwitterionic lipids dimyristoylphosphatidylcholine (di-my-ph-O-CH_2-CH_2-$N(CH_3)_3^+$) and dimyristoylphosphatidyl ethanolamine (di-my-ph-O-CH_2-CH_2-NH_3^+). Although the only difference between these two phospholipids is in the substituents on the nitrogen in the head group, the two corresponding membranes have T_c values which differ by about 24 °C, the membrane of the ethanolamine derivative being the more stable. The explanation for this difference is that the bilayer surface is differently stabilized by the electrostatic attractions among neighbouring ionic groups of opposite charge [66]. Other evidence, however, disputes this role attributed to the zwitterionic head group. Indeed, bilayers of synthetic dialkyl amphiphiles have

higher T_c values when the polar head is anionic than when it is zwitterionic (and also cationic or nonionic) as the identical hydrophobic part of the surfactants is the same [48]. As for the alignment effect of the head group, it becomes evident from the dependence of the phase transition of bilayers of some anionic and cationic amphiphiles on the method of sample preparation, as discussed earlier. This indicates that different molecular arrangements can correspond to a certain bilayer. Note, however, that the introduction of interacting groups to the polar head tends to eliminate the dependence of T_c on the method of sample preparation [48]. A different effect of the head group is demonstrated by the nonionic polyoxoethylene dialkyl ethers, where the length of the polyoxoethylene chain attached to the polar head has a crucial influence on T_c. Thus, although the phase transition occurs at the same temperature in the homologues with 10 or 15 oxoethylene units, the phase transition disappears when the polyoxoethylene chain contains 30 units. This behaviour has been attributed to the high mobility of the long polyoxoethylene chain, which evidently overcomes the stabilizing effect of the two hydrophobic alkyl chains [48].

References

1. Sackmann E (1978) Ber Bunsenges Phys Chem 82:891
2. Oldfield E, Chapman D (1972) FEBS = Federation of European Biochemical Societies, Lett 23:285
3. Chapman D, Williams RM, Ladbrooke BD (1967) Chem Phys Lipids 1:445
4. Trauble H, Haynes DH (1971) Chem Phys Lipds 7:324
5. Lee AG, Birdsall NJM, Levine YK, Metcalfe JK (1972) Biochim Biophys Acta 255:43
6. Hubbell WL, McConell HM (1971) J Am Chem Soc 93:314
7. Lussan C, Daucon JF (1971) FEBS Lett 19:186
8. VanderKooi JM, Chance B (1972) FEBS Lett 22:23
9. Cogan U, Shinitzky M, Weber G, Nishida T (1973) Biochemistry 12:521
10. Chapman D (1976) Quart Rev Biophys 8:185
11. McMurchie EJ, Raison JK (1979) Biochim Biophys Acta 554:364
12. Nagel JF, Scott HL (1978) Phys Today Febr:38
13. McNeil R, Thomas JK (1980) J Colloid Interface Sci 73:522
14. Nagamura T, Mihara S, Okahata Y, Kunitake T, Matsuo T (1978) Ber Bunsenges Phys Chem 82:1093
15. Kano K, Romero A, Djermouni B, Ache HJ, Fendler JH (1979) J Am Chem Soc 101:4030
16. Kunitake T, Okahata Y (1980) J Am Chem Soc 102:549
17. Fendler JH (1980) Ac Chem Res 13:7
18. Nomura T, Escabi-Perez JR, Sunamoto J, Fendler JH (1980) J Am Chem Soc 102:1484
19. Fendler JH, Tundo P (1984) Ac Chem Res 17:3
20. Okahata Y (1986) Ac Chem Res 19:57

21. Singer SL, Nicolson GL (1972) Science 175:720
22. Kunitake T, Okahata Y (1977) J Am Chem Soc 99:3860
23. Deguchi K, Mino J (1978) J Colloid Interface Sci 65:155
24. Kunitake T, Okahata Y, Tamaki K, Kumamaru F, Takayanagi M (1977) Chem Lett 387
25. Kunitake T, Okahata Y (1978) Bull Chem Soc Japan 51:1877
26. Kunitake T, Okahata Y (1980) J Am Chem Soc 102:549
27. Okahata Y, Kunitake T (1980) Ber Bunsenges Phys Chem 84:550
28. Malliaris A, Christias C, Margomenou-Leonidopoulou G, Paleos CM (1982) Mol Cryst Liq Cryst 82:161
29. Margomenou-Leonidopoulou G, Malliaris A, Paleos CM (1985) Thermochim Acta 85:147
30. a) Okahata Y, Kunitake T (1979) J Am Chem Soc 101:5231; b) Kunitake T, Okahata Y, Shimomura M, Yasunami S-I, Takarabe K (1981) J Am Chem Soc 103:5401
31. Okahata Y, Tanamachi S, Nagai M, Kunitake T (1981) J Colloid Interface Sci 82:401
32. Deamer D, Bangham AD (1976) Biochim Biophys Acta 443:629
33. Kunitake T, Kimizuka N, Higashi N, Nakashima N (1984) J Am Chem Soc 106:1978
34. a) Nakashima N, Asakuma S, Kunitake T (1985) J Am Chem Soc 107:509; b) Nakashima N, Asakuma S, Kim J-M, Kunitake T (1984) Chem Lett 1709
35. Sudholter EJ, Engberts JBFN, Hoekstra D (1980) J Am Chem Soc 102:2467
36. Murakami Y, Aoyama Y, Kikuchi J, Nishida K, Nakano A (1982) J Am Chem Soc 104:2937
37. Kumano A, Kajiyama T, Takayanagi M, Kunitake T, Okahata Y (1984) Ber Bunsenges Phys Chem 88:1216
38. Malliaris A, Babilis D, Paleos CM, unpublished results
39. Israelachvili JN, Mitchel DJ, Ninham BW (1977) J Chem Soc Faraday II 10:133
40. Sheets MP, Chan SI (1972) Biochemistry 11:4573
41. Bangham AD, Horne RW (1964) J Mol Biol 8:660
42. Lim YY, Fendler JH (1979) J Am Chem Soc 101:4023
43. Weinstein JN, Yoshikami S, Henkart P, Blumenthal R, Hagins WA (1977) Science 195:489
44. Byrne P, Chapman D (1964) Nature 202:987
45. Chapman D, Collin DT (1965) Nature 206:189
46. Papahadjopoulos D, Jacobson K, Nir S, Isac T (1973) Biochim Biophys Acta 311:330

47. Soutar AK, Pownali HJ, Hu AS, Smith LC (1974) Biochemistry 13:2828
48. Okahata Y, Ando R, Kunitake T (1981) Ber Bunsenges Phys Chem 85:789
49. Levine YK (1973) Prog Surf Sci 3:279
50. Janiak MJ, Small DM, Shipley GG (1976) Biochemistry 15:4575
51. Cameron DG, Casel HL, Mantsch HH, Boulanger Y, Smith ICP (1981) Biophys J 35:1
52. Levin IR, Bush SF (1981) Biochim Biophys Acta 640:760
53. Torza S, Cladis PE (1974) Phys Rev Lett 32:1406
54. Nagle JF (1973) Proc Natl Acad Sci USA 70:3443
55. Lee AG (1977) Biochim Biophys Acta 472:237
56. Chen SC, Sturtevant JM, Gaffney BJ (1980) Proc Natl Acad Sci USA 77:5060
57. Mabrey S, Sturtevant JM (1978) Methods in Membrane Biology 9:237
58. Kajiyama T, Kumano A, Takayanagi M, Okahata Y, Kunitake T (1979) Chem Lett 645
59. Hinz H-J, Sturtevant JM (1972) J Biol Chem 247:6071
60. Mabrey S, Sturtevant JM (1976) Proc Natl Acad Sci USA 73:3862
61. Kunitake T, Okahata Y, Tawaki S-I (1985) J Colloid Interface Sci 103:190
62. Tsong TY (1974) Proc Natl Acad Sci USA 71:2684
63. Lentz BR, Freire E, Biltonen RL (1978) Biochemistry 17:4475
64. Mc Garrity JT, Armstrong JB (1981) Biochim Biophys Acta 640:544
65. Kunitake T, Higashi N (1985) J Am Chem Soc 107:692
66. Yeagle PL (1978) Ac Chem Res 11:321
67. Fuhrhop J-H, Bartch H, Fritsch D (1981) Angew Chem Int Ed Engl 20:804
68. a) Kunitake T, Nakashima N, Shimomura M, Okahata Y, Kano K, Ogawa T (1980) J Am Chem Soc 102:6642; b) Nakashima N, Kunitake T (1982) J Am Chem Soc 104:4261

Received December 15, 1987;
accepted January 4, 1988

Author's address:

A. Malliaris
N.R.C. "Demokritos"
G-153 10 Athens, Greece

Progress in Colloid & Polymer Science   Progr Colloid Polym Sci 76:183–187 (1988)

Electron spin resonance and electron spin echo modulation studies of N,N,N',N'-tetramethylbenzidine photoionization in sodium dodecylsulfate micelles: structural effects of alcohol and crown ether addition

P. Baglioni and L. Kevan[1])

Department of Chemistry, University of Florence, Florence, Italy
[1]) Department of Chemistry, University of Houston, Houston, Texas, U.S.A.

Abstract: Electron spin echo modulation (ESEM) and electron spin resonance (ESR) spectra of the photogenerated N,N,N',N'-tetramethylbenzidine cation radical (TMB$^+$) in frozen micellar solutions of sodium dodecylsulfate containing 2-propanol, 1-propanol, 1-pentanol, 1-octanol, 2-propanol-d$_7$, 1-octanol-d$_{17}$ in D$_2$O and H$_2$O or 15-crown-5, 18-crown-6 ethers in D$_2$O have been studied as a function of the alcohol or crown ether concentration from 0 to 200 mM. Modulation effects due to the TMB$^+$ interactions with deuteriums in D$_2$O and in H$_2$O in TMB/SDS/alcohol systems give direct evidence that 2-propanol is mainly located at the micellar interface, whereas the alkyl chain of 1-octanol is located deeper in the micelle. Alcohol addition leads to an increase of water penetration into the micellar interface in the order 1-propanol < 2-propanol ≤ 1-pentanol < 1-octanol. Modulation effects due to TMB$^+$ in TMB/SDS/crown systems indicate that the TMB molecule moves toward the interfacial region when sodium cation is complexed by the crown ether. The efficiency of charge separation upon TMB photoionization increases with alcohols or crown ethers additon and correlates with the increased TMB$^+$- water interactions.

Key words: Electron spin resonance, electron spin echo modulation, N,N,N',N'-tetramethylbenzidine, photoionization, influence of alcohols and crown ethers, water penetration, micellar solutions of Na-dodecylsulfate.

Introduction

During the past decade, aqueous micellar solutions have been extensively used as media to enhance photochemical charge separation reaction of electron donor solutes. Electron escape from the micelle and inhibition of the charge neutralization backreaction can be assisted by modification of the electrostatic barrier at the micelle-water interface.

N,N,N',N'-tetrameteyhlbenzidine (TMB) has been used as a photoionizable solute since it has a rather low ionization potential and can be photoionized in solution by 370 nm light. It was shown that the photoionization efficiency of TMB to produce TMB$^+$ in micellar solutions depends on various factors including the micelle size and charge [2, 3, 14] the nature of counterion [17] and ionic salt addition [12, 13]. Similar results

have been obtained on the photoionization of perylene [8, 9, 11].

In this study we report the effects of added alcohols (1-propanol, 2-propanol, 1-pentanol, 1-octanol, 2-propanol-d$_7$ and 1-octanol-d$_{17}$) and crown ethers (15-crown-5 and 18-crown-6) to TMB in sodium dodecylsulfate (SDS) micellar solutions, with the double aim of clarifying the solubilization process, in terms of solubilization site, of alcohols in SDS micelles and of determining how the "perturbation" of the micellar interface due to the alcohols or crown ethers addition affects the TMB photoionization yield. In fact, alcohol molecules intercalate between the surfactant ionic headgroups to decrease the micellar surface charge density, decrease the dielectric constant at the micellar surface and to change the molecular order of the interface region [1, 18]. Crown ethers complex the micellar counterion (in this case sodium) and thus change the micellar surface charge density.

Experimental

SDS and TMB were purchased from Eastman Kodak Co. SDS was recrystallized three times from ethanol, washed with diethyl ether and dried under moderate vacuum. The crown ethers (purity > 99%), 1-octanol, 1-pentanol, 1-propanol and 2-propanol (HPLC products, purity > 99%) were obtained from Aldrich Chemical and used without further purification. 2-propanol-d_7 and 1-octanol-d_{17} were obtained from Merck Isotopes and used as received. A stock solution of 0.1 M of TMB was prepared in chloroform. A film of the amount of TMB needed to obtain a 0.1 mM solution was generated in a flask by evaporating the chloroform. The solution of surfactant was added to this film which was solubilized by sonicating for 30 min and by stirring the solution for 3 h at 50 °C in a nitrogen atmosphere. TMB is sensitive to oxygen which favours formation of dimers that react in an unknown way in the photoionization process and thus can introduce an uncontrolled error in the TMB$^+$ yield. The concentrations of the samples studied were 0.1 M SDS, 0.1 mM TMB, and 0–200 mM alcohols or crown ethers. Alcohols and 15-crown-5 were added directly to the samples. A solution of 18-crown-6 was prepared in chloroform. A film of 18-crown-6 was generated on the walls of the sample tubes by evaporating the solvent under nitrogen atmosphere at the beginning of the sample preparation.

All the samples were sealed in 2 × 3 mm quartz tubes and frozen rapidly in liquid nitrogen. Photoirradiation of TMB was carried out at 77 K with a Cermax 300 W Xenon lamp filtered with 10 cm^3 of water and a Corning filter 7–51, which passes radiation centered at 370 nm with 80% transmittance. Each sample was irradiated for 12 min. ESR spectra were recorded on a Bruker ESP 300 system. Two-pulse electron spin echo signals were recorded at 4.2 K on a home-built spectrometer by using 40 ns exciting pulses.

The normalized deuterium modulation depths were computed as described previously [13]. The echo decay curves show detectable modulations with periods of 0.5 μs and 0.08 μs, corresponding respectively to electron-deuterium and electron-proton interactions. No other modulations (e. g. sodium modulation) were detected. The concentration of TMB cation was measured from the ESR signal intensity.

Discussion

The electron spin echo technique (ESE) enables the detection of very weak electron-nuclear dipolar hyperfine interactions. In solids, the echo amplitude is often modulated by frequencies associated with nearly nuclear spin coupled weakly to the electron spin. This modulation can be detected up to about 6 Å and can be analyzed in terms of a structural model for the arrangement of surrounding nuclear spin and is extremely useful for the determination of solvation structures and the number, distance and orientation of molecules in the local environment around paramagnetic species, particularly in disordered systems.

The results reported in this study were analyzed in terms of relative modulation depth, which is related to the strength of the electron-nuclear dipolar interaction distance. Note that since the ESE modulation arises from a dipolar interaction, it is averaged out by rapid molecular tumbling in liquid solution and consequently can only be observed in rapidly frozen solutions.

In micellar systems, several specific experiments indicate that the micellar structure is retained. In homogeneous solutions, TMB$^+$ has a lifetime of only a few microseconds, while in anionic micellar solutions it is stable for tens of minutes. It is possible to generate TMB$^+$ in frozen micellar solution, that the solution, and still observe TMB$^+$ by electron spin resonance, thus indicating that the micellar "structure" was retained upon freezing [14]. Hashimoto and Thomas [10] have recently used luminescence quenching to measure micellar aggregation numbers and have found similar aggregation numbers in liquid and frozen SDS micellar solutions. Baglioni et al. [5, 7] have measured the partition coefficients of several alcohols and crown ethers in frozen SDS micellar solutions and have found values in good agreement with those obtained by NMR at room temperature [16]. Further evidence that the micellar structure was retained in rapidly frozen micellar solutions comes from freeze fracture electron microscopy [4].

The results obtained with alcohols and crown ethers are reported separately.

a) Alcohols/SDS/TMB

Figures 1 and 2 show the variation of normalized deuterium modulation depths as a function of alcohol concentration. The alcohol addition causes an increase in the TMB$^+$-water deuterium interactions, indicating that the micellar surface of SDS becomes more hydrated in the presence of the alcohol. Furthermore, this effect is dependent on the alcohol nature. In fact, the plateau values of the normalized deuterium modulation depth versus alcohol concentration increase in the order 1-propanol < 2-propanol ≤ 1-pentanol ≪ 1-octanol. This order, then, reflects the order of increasing water penetration at the micellar interface and seems to be correlated with a different alcohol location at the micellar surface. A direct proof of different locations for alcohols with different chain length is obtained from analysis of the trend of the deuterium

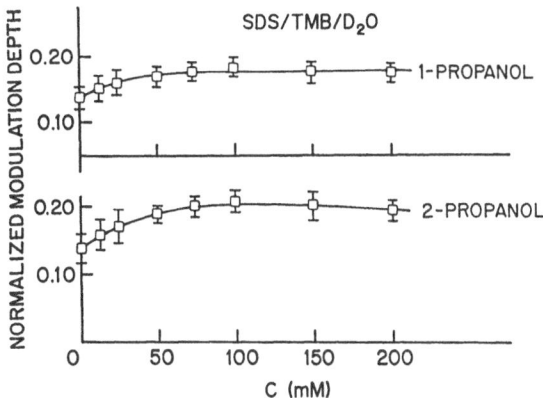

Fig. 1. Dependence of normalized deuterium modulation depths as a function of 1-propanol and 2-propanol concentrations in frozen TMB$^+$/SDS/D$_2$O micellar solutions

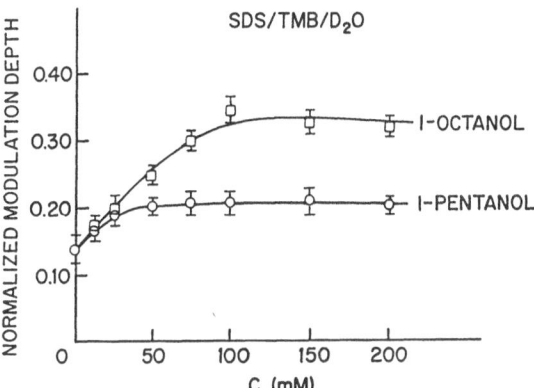

Fig. 2. Dependence of normalized deuterium modulation depths as a function of 1-pentanol and 1-octanol concentrations in frozen TMB$^+$/SDS/D$_2$O micellar solutions

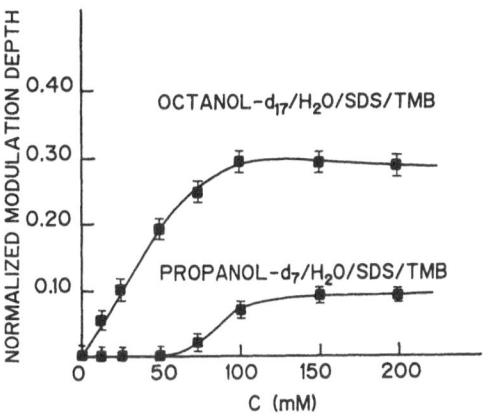

Fig. 3. Dependence of normalized deuterium modulation depths as a function of 2-propanol-d$_7$ and 1-octanol-d$_{17}$ concentrations in frozen TMB$^+$/SDS/H$_2$O micellar solutions

same alcohols but using the 5-doxylstearic acid probe [5].

The above results are relevant to the analysis of TMB$^+$ yield versus alcohol concentration show in Fig. 4. Szajdzinska-Pietek et al. [17] suggested that the increase in the photoionization yield was primarily due to enhancement of the TMB$^+$-water interactions. Since the alcohols change the extent of TMB$^+$-water interaction, they are expected to differently affect the photoionization efficiency of TMB.

The initial slopes of the photoionization yield reported in Fig. 4 increase in the order 1-propanol $\leq$ 2-propanol $<$ 1-pentanol $<$ 1-octanol which correlate with the order of increasing TMB$^+$-water interactions. However, the maximum yield shows exactly the

modulation in H$_2$O for two perdeuterated alcohols, 2-propanol-d$_7$ and 1-octanol-d$_{17}$ (see Fig. 3). With 2-propanol-d$_7$ there is an initial concentration range of about 0 to 0.70 alcohols/SDS mole ratio over which TMB$^+$ shows no interaction with alcohol deuteriums while TMB$^+$ does show such interaction with 1-octanol-d$_{17}$.

The TMB$^+$ cation radical has been found by ESEM to interact weakly with water deuteriums in SDS micellar solutions, indicating that the photoproduced TMB$^+$ cation must be located asymmetrically within the micelle close to the micellar interface, below the micellar headgroups [6]. Since TMB$^+$ is located below the micellar headgroups [6,14], the previous results shows that 1-octanol-d$_{17}$ penetrates below the micellar surface while 2-propanol-d$_7$ is located mainly at the micellar surface at low alcohol concentrations. This is also in agreement with the results obtained with the

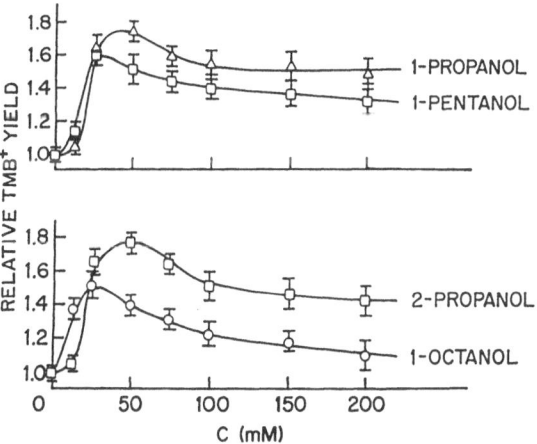

Fig. 4. TMB$^+$ yields measured by ESR as a function of the indicated alcohol concentrations in frozen SDS micellar solutions

opposite trend, with 1-octanol < 1-pentanol < 2-propanol ≤ 1-propanol. This indicates that another factor affects the photoionization efficiency. One factor is probably the disorder of the surfactant molecules at the micelle surface, which has been suggested to correlate with enhanced TMB+-water interactions. The other factor is likely associated with the degree of water disorder at the micellar interface. This disorder should be maximized by 1-propanol and 2-propanol which concentrate at the micellar surface and such water disorder should facilitate electron solvation in the same way electron solvation is facilitated in glassy versus polycrystalline matrices [15].

b) Crown ethers/SDS/TMB

Crown ethers are known to complex alkali ions in solution. The formation constants for the metal cation-crown complex are dependent on the nature of the alkali ion, the crown structure and the solvent. In previous studies [7] the results obtained from the analysis of ESE and EPR spectra of n-doxylstearic acid probes in SDS and lithium dodecylsulfate (LDS) micellar solutions containing 15- or 18-crown ethers showed that the crown ethers are mainly located at the micellar interface and that their interaction with the surfactant headgroups causes a decrease of the hydration of the micellar interface and the complexation of the micellar counterion, which in turn leads to a decrease of the local ionic strength at the micellar surface. Furthermore, the complexation constants depend on the nature of the micellar counterion and of the crown structure and are in the order 18-crown-6 > 15-crown-5 for SDS, 15-crown-5 ≤ 18-crown-6 for LDS micellar solutions [7].

In Figs. 5 and 6 show the dependence of the normalized deuterium modulation depth as a function of crown ether concentration, and the TMB+ yield measured by ESR as a function of crown concentration, respectively. The dotted lines refer to the crown ethers concentrations where a complex lineshape, which is difficult to unambiguously interpret, is present in the ESR spectra.

As shown in Fig. 5, the deuterium modulation depth increases to a plateau over a small concentration range of about 15 mM 18-crown-6 or 15-crown-5 ether. This correlates with a small increase in the TMB+ photoionization yield over the same crown ether concentration range. This increase in deuterium modulation would normally be interpreted as increased TMB+-water interaction from increased wa-

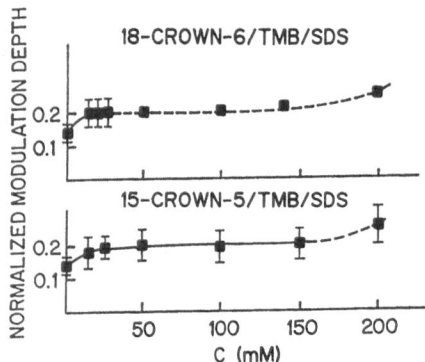

Fig. 5. The dependence of the normalized deuterium modulation depths as a function of crown ether concentration in frozen TMB+/SDS/D$_2$O

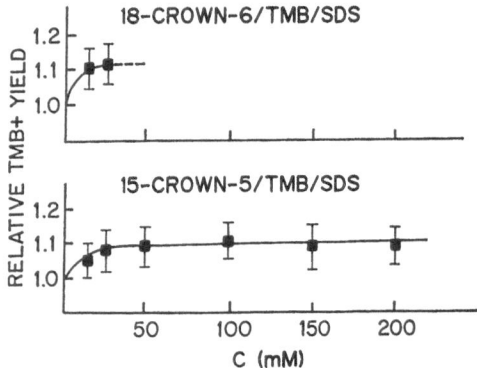

Fig. 6. TMB+ yield measured by ESR as a function of crown ether concentration in frozen micellar solutions

ter penetration at the micellar surface associated with crown ether complexation of Na+. However, with n-doxylstearic acid probes [5] the deuterium modulation from D$_2$O decreases with crown ether addition. So TMB must move toward the micellar surface with Na+ complexation by crown ether. The driving force could be a decrease in local ionic strength at micellar surface. This makes the interfacial region more energetically favourable for TMB.

This behavior is complementary to the movement of TMB away from the interfacial region when its local ionic strength is decreased by high added salt concentration [13].

Conclusion

The results obtained from the analysis of electron spin echo and electron spin resonance spectra lead to the following principal conclusions:

1. Alcohol addition to a SDS micellar solution leads to an increase of water penetration into the micellar interface in the order 1-propanol < 2-propanol ≤ 1-pentanol < 1-octanol.

2. Modulation effects due to TMB^+ interactions with deuterium in SDS/D_2O or $/H_2O$ with 2-propanol-d_7 or 1-octanol-d_{17} give direct evidence that 2-propanol is mainly located at the micellar interface whereas the alkyl chain of 1-octanol is located deeper into the micelle.

3. The initial efficiency of charge separation upon photoionization of TMB as a function of alcohol concentration correlates with the degree of water penetration into the micelle, but the maximum photoionization efficiency seems more related to the degree of water organization at the micellar surface due to specific perturbing effects on the micellar structure dependent on the alcohol nature.

4. Modulation effects due to TMB^+ radical indicate that the TMB molecules move toward the micelle interfacial region with the addition of crown ethers, suggesting that the relative TMB^+ yield increase of about 10 % does correlate with an increase in deuterium modulation depth, as found for alcohol.

5. The sodium cation complexation by the crown ethers may be considered to decrease the local ionic strength in the interfacial region which promotes a change in the TMB locus toward the micellar interface. This enhances the TMB^+-water interactions and may also reflect the local ordering of the interfacial water.

Acknowledgment

This research was supported by the Division of Chemical Sciences, Office of Basic Energy Sciences, Office of Energy Research, U.S. Department of Energy and by the Italian Ministry for Public Instruction (MPI).

References

1. Almgren M, Swarup S (1983) J Colloid Interface Sci 91:256; Almgren M, Lofroth JE (1981) J Colloid Interface Sci 81:486
2. Arce R, Kevan L (1985) J Chem Soc Faraday Trans 1 81:1025
3. Arce R, Kevan L (1985) J Chem Soc Faraday Trans 1 81:1669
4. Bachmann L, Dasch W, Kutter P (1981) Ber Bunsenges Phys Chem 85:883
5. Baglioni P, Kevan L (1987) J Phys Chem 91:1516
6. Baglioni P, Kevan L (1987) J Phys Chem 91:2106
7. Baglioni P, Rivara-Minten E, Kevan L (1988) J Phys Chem, 92:2613
8. Bernas A, Grand D, Hautecloque S, Chambaudet A (1981) J Phys Chem 85:5236
9. Grand D, Hautecloque S, Bernas A, Petit A (1983) J Pyhs Chem 87:5236
10. Hashimoto S, Thomas JK (1983) J Am Chem Soc 105:5230
11. Hautecloque S, Grand D, Bernas A (1985) J Phys Chem 89:2705
12. Hiromitsu I, Kevan L (1986) J Phys Chem 90:3088
13. Maldonado R, Kevan L, Szajdzinska-Pietek E, Jones RRM (1984) J Phys Chem 81:3958
14. Narayana PA, Li ASW, Kevan L (1982) J Am Chem Soc 104:6502
15. Narayana M, Suryanarayana D, Kevan L (1982) J Am Chem Soc 104:3552
16. Stilbs P (1982) J Colloid Interface Sci 87:547
17. Szajdzinska-Pietek E, Maldonado R, Kevan L, Jones RRM (1984) J Am Chem Soc 106:4675; (1985) J Am Chem Soc 107:6467
18. Zana R, Yiv S, Strazielle C, Lianos P (1981) J Colloid Interface Sci 80:208; Lianos P, Lang J, Strazielle C, Zana R (1982) J Phys Chem 86:1019

Received December 16, 1987;
accepted January 4, 1988

Authors' address:

P. Baglioni
Department of Chemistry
University of Florence
Florence, Italy

Progress in Colloid & Polymer Science

Progr Colloid Polym Sci 76:188–202 (1988)

Molecular calculations for interacting surfactant monolayers

J. C. Eriksson and S. Ljunggren

Department of Physical Chemistry, The Royal Institute of Technology, Stockholm, Sweden

Abstract: An outline of a new, rather general molecular calculation scheme for interacting, ionic surfactant/long-chain alcohol monolayers is presented. These surfactant layers are physically adsorbed either at a surfactant solution/air interface or at a surfactant solution/ (solid) hydrophobic substrate interface, or, are present in a lamellar liquid crystal. They may be molecularly dispersed or of the condensed, liquid-expanded type.

The most important free energy contributions included are caused by: (i) the hydrophobic effect associated with the hydrocarbon chains, (ii) the electrostatic interactions, (iii) the constraints imposed on the hydrocarbon chain conformations, (iv) the interfacial contacts, (v) the mixing of surfactant and alcohol in the monolayers, (vi) the hydrophobic attraction between hydrocarbon-covered surfaces, (vii) the van der Waals attraction.

Good agreement is demonstrated with surface tension and surface force measurements.

Key words: Ionic surfactant, long-chain alcohol monolayers, hydrocarbon chains, electrostatic interactions, van der Waals attraction.

Introduction

Condensed monolayers composed of water-soluble surfactant as a rule form in the concentration range $c > 0.1$ CMC, at air/solution interfaces and likewise at hydrophobic substrate/solution interfaces. The driving "force" behind this adsorption process is evidently the hydrophobic effect, which is mainly a consequence of the free energy increase, relatively speaking, for the water molecules in the vicinity of hydrocarbon chains.

In a series of papers [1–4] Motomura and coworkers have presented very precise surface tension data for various surfactant solutions, either in contact with air or with a liquid hydrocarbon phase. Through these experimental studies it has been demonstrated that for surfactant solution/air interfaces, there is generally a Henry's law region at low surfactant concentrations ($c < 0.1$ CMC) where the surface tension drops rather slowly in a linear fashion with increasing surfactant concentration. This is consistent with the notion of a "gaseous" surface state of molecularly dispersed surfactant molecules. For ionic surfactants this behaviour is readily accounted for by employing the conventional Gouy-Chapman theory of the repulsive electrostatic interactions at a planar interface [5].

At about $c = 0.1$ CMC a first-order surface phase transition occurs, resulting in the formation of a condensed, coherent monolayer of the liquid-expanded type [2]. Our previous analysis [5] shows that a major contribution to the surface pressure in this condensed regime arises below about 50 $Å^2$/molecule in surface area because of the progressively tighter packing of the hydrocarbon chains upon raising the surfactant concentration.

It is well-known that micelle formation is also a process driven by the hydrophobic "attraction" and counteracted by the electrostatic and hydrocarbon chain packing "repulsions". Earlier, in a series of papers [6–9], we developed quantitative models for the formation of micelles of different geometries (spherical, rod- and disc-shaped). These calculations involved the following free energy contributions:

i. The free energy of transferring a hydrocarbon chain from water to hydrocarbon as quantified by Tanford [10].

ii. The electrostatic free energy computed within the extended Gouy-Chapman framework developed by Evans and Ninham [11].

iii. The configurational free energy of the hydrocarbon chains computed by Gruen and Lacey [12] which depends upon the hydrocarbon chain packing density.

iv. A hydrocarbon/water contact free energy which was set equal to the macroscopic value of 50 mJm^{-2}.

In this paper we present similar calculations for interacting film surfaces containing surfactant in a gaseous or condensed state. Hence we have added interaction free energies caused by the electrostatic double layer repulsion and by the hydrophobic attraction [13,14]. The two approaching surfaces can either be soap film surfaces, i.e. two interacting solution/air interfaces, or two (solid) hydrophobic surfaces with adsorbed surfactant. This latter experimental situation can be realized in the surface force apparatus designed by Israelachvili.

As our theoretical scheme embraces the formation of, as well as the interaction between, the surfactant-loaded interfaces at *constant chemical potentials*, there is no need to assume constant surface potential or constant surface charge conditions, i.e. we are dealing with fully *regulating* surfaces. As a matter of fact, values of the surface potential and surface charge density are generated at each calculation carried out for a given surface separation, H.

Finally, we have allowed for the possibility of having mixed surfactant layers composed of an ionic surfactant and the corresponding alcohol with the same hydrocarbon chain length. In subsequent papers we shall present more extensive calculation results obtained using the program described below.

Thermodynamics

The thin film primarily studied here is depicted in Fig. 1. It has an overall thickness equal to H, where H is the separation between the two (irreversibly) hydrophobed solid surfaces. The thickness of the water core is denoted by h. The (smeared-out) surface charges are assumed to be located at the hydrocarbon/water interfaces which are positioned at $z = 0$ and $z = h$. This thin film is open towards, and in full equilibirium with, a bulk solution containing a cationic surfactant at the concentration c_2 and possibly also a long-chain alcohol at the concentration c_3 with the same (straight) hydrocarbon chain length as the surfactant. For reasons of simplicity, in this context we will assume that the bulk solution does not contain an additional salt compo-

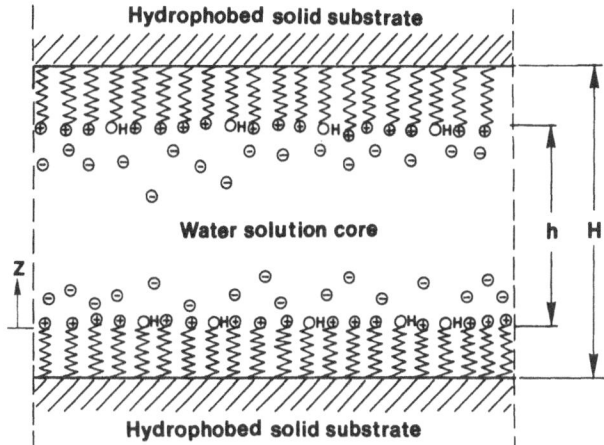

Fig. 1. Sketch of the thin film between hydrophobic surfaces primarily studied, which is composed of a water solution core surrounded by condensed surfactant/long-chain alcohol monolayers in liquid-expanded state. The film is *open* in the thermodynamic sense towards an adjacent bulk phase of surfactant/long-chain alcohol solution. In lateral directions there acts a resultant film tension γ and in the normal direction an interaction pressure π. For a soap film, the hydrophobed surfaces are replaced by air contacts and for a lamellar liquid crystal, by the vicinal monolayers of the hydrocarbon bilayer parts. Note that the drawing is highly schematic as to the state of the hydrocarbon chains

nent, although our calculation scheme actually pertains to this case, too.

The number of surfactant ions in the two monolayer is N_2 and the corresponding number of alcohol molecules is N_3. Thus, the total number of hydrocarbon chains in the monolayers is $N_2 + N_3$.

The Helmholtz free energy of the monolayers is

$$F = F^f - hA\bar{F}^b \tag{1}$$

where F^f is the Helmholtz free energy of the entire film and $\bar{F}^b$ the Helmholtz free energy per volume unit in the bulk solution. Let us now introduce the excess free energy per hydrocarbon chain, g, by means of the relation

$$(N_2 + N_3)\,g = F + p_e V - n_1\mu_1 - n_2^+\mu_2^+$$
$$- n_2^-\mu_2^- - n_3\mu_3 \tag{2}$$

where n_1 is the excess number of water molecules in the core volume hA, n_2^+ the excess number of coions, n_2^- the excess number of counterions and n_3 the excess number of alcohol molecules. p_e denotes the external, bulk solution, pressure whereas V is the monolayer volume $A(H - h)$. As the next step, we define the corresponding Ω-potential:

$$\Omega = (N_2 + N_3)\,g - N_2\mu_2^+ - N_3\mu_3 = (N_2 + N_3)\,\varepsilon \,. \quad (3)$$

Thus, ε is the Ω-potential per hydrocarbon chain in the monolayers.

The free energy differential for the whole film becomes:

$$dF^f = - S^f dT - p_e dV^f + \gamma dA - \pi A dH$$
$$+ \mu_1 dn_1^f + \mu_2^+ dn_2^{f+} + \mu_2^- dn_2^{f-} + \mu_3 dn_3^f \,. \quad (4)$$

Upon subtracting the corresponding bulk expression for the water core volume hA, viz.

$$dF^b = - S^b dT - p_e d(hA) + \mu_1 dn_1^b$$
$$+ \mu_2^+ dn_2^{b+} + \mu_2^- dn_2^{b-} + \mu_3 dn_3^b \quad (5)$$

we get

$$dF = - SdT - p_e dV + \gamma dA - \pi A dH + \mu_1 dn_1$$
$$+ \mu_2^+ d(N_2 + n_2^+) + \mu_2^- dn_2^- + \mu_3 d(N_3 + n_3) \,. \quad (6)$$

In the above expressions γ denotes the *film tension* and π the *interaction (disjoining) pressure*, i. e.

$$\pi = p_N - p_e \quad (7)$$

where p_N is the normal component of the pressure tensor. Employing the Legendre transformation (2), Eq. (6) can be rewritten as follows.

$$d[(N_2 + N_3)g] = - SdT + Vdp_e + \gamma dA - \pi A dH$$
$$- n_1 d\mu_1 - n_2^+ d\mu_2^+ - n_2^- d\mu_2^-$$
$$- n_3 d\mu_3 + \mu_2^+ dN_2 + \mu_3 dN_3 \,. \quad (8)$$

Hence the chemical potentials in the monolayers can be computed by means of differentiating $(N_2 + N_3)g$:

$$\left(\frac{\partial[(N_2 + N_3)g]}{\partial N_2}\right)_{T,p_e,A,H,N_3,c_2,c_3} = \mu_2^+ \quad (9)$$

$$\left(\frac{\partial[(N_2 + N_3)g]}{\partial N_3}\right)_{T,p_e,A,H,N_2,c_2,c_3} = \mu_3 \,. \quad (10)$$

The differential of the Ω-potential differs from Eq. (8) only with respect to the last two terms: it is

$$d\Omega = - SdT + Vdp_e + \gamma dA - \pi A dH + n_1 d\mu_1$$
$$- (N_2 + n_2^+)\,d\mu_2^+ - n_2^- d\mu_2^- - (N_3 + n_3)\,d\mu_3 \,. \quad (11)$$

Integration at constant T, p_e, H and all μ yields

$$\Omega/A = \gamma = 2\varepsilon/a \quad (12)$$

where the last equality follows from the self-evident relation

$$(N_2 + N_3)\,a = 2A \quad (13)$$

a denoting the surface area per chain. The factor 2 arises because of the two film faces. Thus, knowing the Ω-potential per chain, ε, we can easily obtain the film tension γ by dividing by the chain surface area. The film tension, γ, can also be calculated as a derivative of g (cf. Eq. (8)):

$$\left(\frac{\partial[(N_2 + N_3)g]}{\partial A}\right)_{T,p_e,H,N_2,N_3,\text{all}\,\mu} = \gamma \,. \quad (14)$$

By inserting $dA = (1/2)(N_2 + N_3)\,da$ this yields

$$\left(\frac{\partial g}{\partial a}\right)_{T,p_e,H,x_S,\text{all}\,\mu} = \gamma/2 \quad (15)$$

where x_S denotes the surfactant mole fraction in the monolayers.

By introducing $\Omega = \gamma A$ into Eq. (11) we can readily put this equation in the same form as the Gibbs surface tension equation:

$$- d\gamma = (S/A)\,dT - (H - h)\,dp_e + \pi dH$$
$$+ \Gamma_1 d\mu_1 + \Gamma_2 d\mu_2 + \Gamma_3 d\mu_3 \,. \quad (16)$$

Here the dependent differential $d\mu_1$ may be eliminated by making the film excess of water, $\Gamma_1 = n_1/A$ exactly equal to zero. Equation (16) includes the important thermodynamic relation

$$\left(\frac{\partial\gamma}{\partial H}\right)_{T,p_e,\text{all}\,\mu} = - \pi \quad (17)$$

which must be fulfilled for any thin equilibrium film which is composed of constituents that are present also in the adjacent bulk solution.

Returning to Eqs (9) and (10) and making use of the volume condition $(N_2 + N_3)v = A(H - h)$ (where v

denotes the hydrocarbon chain volume) it is an easy matter to demonstrate that

$$\left(\frac{\partial[(N_2 + N_3)g]}{\partial N_2}\right)_{T, p_e, A, h, N_3, c_2, c_3} + \pi v = \mu_2^+ \quad (18)$$

$$\left(\frac{\partial[(N_2 + N_3)g]}{\partial N_3}\right)_{T, p_e, A, h, N_2, c_2, c_3} + \pi v = \mu_3 \quad (19)$$

where the partial derivatives refer to constant h rather than H. Hence, the circumstance that the monolayer hydrocarbon chains are subjected to a pressure component in the z- direction different from p_e will affect the values of the chemical potentials μ_2^+ and μ_3. We have assumed this to be of importance only for surfactant monolayers adsorbed on solid surfaces (surface force apparatus case) but not for a soap film, where we have put $H = h$.

The central thermodynamic property of a thin (liquid) equilibrium film is the film tension y. Our calculation scheme is aimed at obtaining y as a function of the independent state variables T, H, c_2 and c_3. To achieve this goal we need to employ the results of various molecular calculations and experimental studies which enable the quantification of the main free energy contributions involved.

The electrostatic free energy

The electrostatic free energy contribution can be computed in the same way in all the different cases considered. As discussed by Wennerström et al. [15, 16], it can be calculated to a sufficient degree of accuracy using the Poisson-Boltzmann (PB) approximation, However, at large surface separations relative to the Debye length, $l/\varkappa$, the full PB scheme results in quite time-consuming computations and, for that reason, it is helpful to have access also to the generally less satisfactory weak *overlap approximation* described in the following.

For comparatively thick films the electrostatic free energy may be thought of as consisting of two parts, (i) twice the free energy of a charged plate immersed in an electrolyte solution which can be calculated according to the Gouy-Chapman theory, and (ii) the additional free energy caused by the overlap of the ion-distributions at the two charged surfaces.

The contribution (i) for two charged surfaces at a large distance from one another equals

$$g_{el}^{(i)} = 2kTx_S\left[\ln\left(S + \sqrt{S^2 + 1}\right) + \frac{1}{S} + \frac{\sqrt{S^2 + 1}}{S}\right] \quad (20)$$

per hydrocarbon chain [17]. The reduced charge parameter S is defined by

$$S = (\sigma^2/8RT\varepsilon_0\varepsilon_r c_2)^{1/2} \quad (21)$$

where σ is the surface charge density, e the charge of the monovalent head-group, ε_r, the relative dielectric number for water $(= 306.7 \exp(-T/218.6))$ and c_2 the electrolyte (i. e. surfactant) concentration.

In order to calculate the second contribution, (ii), we observe that, according to the weak overlap approach, the interaction pressure between two charged surfaces immersed in an electrolyte solution of concentration c_2 equals

$$\pi_{el}(h) = 4kTc_2[e\phi(h/2)/kT]^2 \quad (22)$$

per unit surface area. The two charged plates are assumed to be situated at $z = 0$ and $z = h$, respectively, i. e. their separation equals h, and $\phi(h/2)$ is the potential at mid-distance due to the (single) surface at $z = 0$. Now, according to the Gouy-Chapman theory,

$$\phi(z) = (2kT/e)\ln\left[(1 + ye^{-\varkappa z})/(1 - ye^{-\varkappa z})\right] \quad (23)$$

where, as usual, the ratio y (not to be confused with the film tension) is defined by

$$y = (e^{e\phi(0)/2kT} - 1)/(e^{e\phi(0)/2kT} + 1). \quad (24)$$

Here, $\phi(0) = \phi^0$ is the surface potential of the charged plates.

Generally, it is not permissible, for reasons of accuracy, to replace $\phi(h/2)$ in Eq. (22) by using the Debye-Hückel approximation, $\phi(z) = \phi^0 \exp(-\varkappa z)$, where $\phi^0 = \sigma/\varepsilon_0\varepsilon_r\varkappa$ and

$$\varkappa = (2N_A^2 c_2 e^2/\varepsilon_0\varepsilon_r RT)^{1/2}. \quad (25)$$

On the other hand, the linearized Poisson-Boltzmann approximation should be valid to a good approximation if $z > h/2$ because the overlap approximation yielding Eq. (22) presupposes that $e\phi/kT \ll 1$. Thus $e\phi/kT \ll 1$ when $z > h/2$ and, in addition, $\exp(-\varkappa z) \ll 1$. Hence, for $z > h/2$ we have

$$\phi(z) = \phi(h/2)\frac{\ln\left(\frac{1 + ye^{-\varkappa z}}{1 - ye^{-\varkappa z}}\right)}{\ln\left(\frac{1 + ye^{-\varkappa h/2}}{1 - ye^{-\varkappa h/2}}\right)} \approx \phi(h/2)e^{-\varkappa(z - h/2)} \quad (26)$$

and it follows form Eq. (22) that

$$\pi_{el}(h') = \pi_{el}(h) e^{-\varkappa(h'-h)} . \tag{27}$$

The overlap contribution, (ii), to the electrostatic free energy per unit surface can, therefore, be approximated as

$$\frac{(N_2 + N_3) g_{el}^{(ii)}}{A} = \int_h^\infty \pi_{el}(h') \, dh' = \pi_{el}(h) \int_h^\infty e^{-\varkappa(h'-h)} dh'$$

$$= \pi_{el}(h)/\varkappa . \tag{28}$$

The overall electrostatic free energy thus becomes

$$g_{el} = 2kTx_s \left[\ln \left(S + \sqrt{S^2 + 1} \right) + \frac{1}{S} - \frac{\sqrt{S^2 + 1}}{S} \right]$$

$$+ (2kTac_2/\varkappa)[e\phi(h/2)/kT]^2 \tag{29}$$

where, as before, a equals the surface area occupied by each hydrocarbon chain.

It turns out, however, that the overlap approximation as described above *is generally not sufficiently accurate* when the purpose is to account for interaction forces of an electrostatic origin. Then one has to resort to (exact) numerical solutions of the Poisson-Boltzmann (PB) equation [18]. It may be verified that within the range of the PB approximation we have [19]

$$g_{el} = e\phi^0 + a_{charge} \gamma_{el}/2 \tag{30}$$

i. e.

$$g_{el} = e\phi^0 + a\gamma_{el}/2x_S \tag{31}$$

where γ_{el} is the electrostatic contribution to the film tension:

$$\gamma_{el} = \int_0^h (p_e - p_T) \, dz \tag{32}$$

and p_T is the tangential component of the Maxwell tensor, i. e.

$$p_T = p_N + \varepsilon_0 \varepsilon_r E_z^2 \tag{33}$$

E_z denoting the electric field strength (cf. Ref. [19]). Inserting the electrostatic interaction pressure, π_{el}, Eq. (32) can be recast as

$$-\gamma_{el} = \varepsilon_0 \varepsilon_r \int_0^h E_z^2 dz + \pi_{el} h \tag{34}$$

where

$$\pi_{el} = c_2(c^{e\phi_m/kT} + e^{-e\phi_m/kT} - 2) \tag{35}$$

ϕ_m denoting the mid-plane potential. From the numerical solution of the PB equation we can readily obtain ϕ^0, ϕ_m, $E_z^2(z)$ and, hence, also g_{el}.

The hydrophobic free energy

Following Tanford [10], the free energy gained upon transferring a C 12 hydrocarbon chain from the standard monomer state in water solution to a hydrocarbon bulk phase has a value equal to 19.960 kT at 25 °C. Hence,

$$\varepsilon_{\text{Tan}} = - [19.960 + x_S \ln x_2 + x_{OH} \ln x_3] kT \tag{36}$$

is the corresponding contribution to $\varepsilon = g - x_S \mu_2^+ - x_{OH} \mu_3$ where x_2 and x_3 are the bulk monomer mole fractions of surfactant and alcohol, respectively. This is, of course, an average value for the surfactant and alcohol molecules.

It should be noted that in our theory the following two free energy contributions are treated as being effectively additive:

i. The free energy change associated with bringing the hydrocarbon chains from the monomer state in water solution first to a hydrocarbon bulk phase and then further on to the adsorbed state at the interface, and

ii. The free energy change of bringing the charged head groups from the monomer state in solution to the interface. The predominant part of this second contribution is the electrostatic free energy g_{el}.

At temperatures deviating from 25 °C we have used $\Delta \bar{C}_p = - 772$ JK^{-1} mol^{-1} for the transfer reaction $C_{12}(aq) \rightarrow C_{12}(hydrocarbon)$ [20]. Assuming that $\Delta H^0 = 0$ at 25 °C we thus obtain

$$\varepsilon_{\text{Tan}} = - [19.960 + x_S \ln x_2 + x_{OH} \ln x_3] kT$$

$$+ 92.840 \, k \left[T \ln \left(\frac{T}{298.15} \right) - (T - 298.15) \right] . \tag{37}$$

The configurational free energy

The configurational free energy of the hydrocarbon chains is not the same in the condensed surface monolayer as in a corresponding hydrocarbon bulk phase because in the monolayer one end of each chain is fixed to the head group and because of excluded volume and packing effects. This free energy has been calculated by Gruen and Lacey [12] for a C12 chain and different geometries by employing a statistical-mechanical single-chain mean field approach. Thus, for a planar bilayer:

$$g_{conf}/kT = 3.126301 - 1.155336\,L + 0.1827336\,L^2$$
$$- 0.01407631\,L^3 + 0.0004774232\,L^4$$
(38)

per chain, where L is the thickness of a hydrocarbon monolayer, i.e. $L = v/a$, v denoting the C12 hydrocarbon chain volume $= 351$ Å. The above g_{conf}-function has a minimum value for $L = 7.3$ Å, corresponding to $a = 48.1$ Å^2. This means that for a-values < 48.1 Å^2 a chain *surface pressure* arises whereas for a-values > 48.1 Å^2 the hydrocarbon chain part gives a positive contribution to the *surface tension*.

For those cases in which the hydrocarbon chain ends are in contact with air rather than with a hydrophobic substrate, the configurational free energy might well differ from Eq. (38). However, as no calculations are available for this case we shall assume that the difference in g_{conf} can be included in the head group constants ε_{pg} and ε_{OH}, to be introduced below.

The free energy of mixing surfactant and alcohol

Assuming ideal mixing, the free energy of mixing surfactant and alcohol molecules in the monolayer is simply

$$g_{mix}/kT = x_S \ln x_S + x_{OH} \ln x_{OH}\,.$$
(39)

Contact free energies

In line with our previous calculations for micelles, we have put the hydrocarbon/water contact free energy equal to

$$g_{hc/w} = a\gamma_{hc/w}$$
(40)

with $\gamma_{hc/w}$ equal to its macroscopic value [1]

$$\gamma_{hc/w} = 50.0 - 0.94\,(T - 305.0)$$
$$+ 0.0011\,(T - 305.0)^2 \; (\text{mNm}^{-1})\,.$$
(41)

In those applications of our theory where the condensed hydrocarbon chains are in direct contact with air, there is also an additional contribution,

$$g_{hc/air} = a\gamma_{hc/air}$$
(42)

and we have inserted the following numerical expression [21]

$$\gamma_{hc/air} = 20.0 - 0.1\,(T - 293.15) \; (\text{mNm}^{-1})$$
(43)

which holds true for a dodencane/air interface. The case of the surface concentration of surfactant and alcohol molecules being so low that the interface is not fully covered requires special treatment, as below.

The hydrophobic attraction free energy

For thin films between hydrophobic surfaces, there is also a rather strong and long-range hydrophobic attraction, seemingly resulting from the overlap of interfacial zones where the water molecules are orientationally ordered to a slight extent [13]. In order to introduce this sizeable physical effect into our calculation scheme we have made use here of the experimentally based, two-exponential expression generated by Claesson et al. [14] yielding

$$g_{hyph} = - (1/4\pi)\,(a - a_{pg})\,[0.36\,C_1 \exp\,(- h/12.0)$$
$$+ 0.0066\,C_2 \exp\,(- h/55.0)]$$
(44)

where h should be inserted in Å units and where a_{pg} is the surface area actually covered by the polar group. To be more precise, this expression was only employed for molecularly dispersed surfactant monolayers with $a < 200$ Å/chain and putting $h = H$. The coefficients C_1 and C_2 with values between 0 and 1, were included in order to be able to enable fits to experimental surface force data. In fact, a recent experimental study shows that the more long-range part of g_{hyph} is quite sensitive to the presence of ionic groups in the hydrophobic surface [22].

The van der Waals attraction

The van der Waals (non-retarded) interaction free energy equals

$$g_{vdW} = - (1/2)\,aA_H/12\pi h^2$$
(45)

per chain, where A_H is the Hamaker constant and h, as usual, refers to the thickness of the water core. The factor (1/2) arises because g denotes the free energy *per chain*. For a thin film there are, of course, two chains (one on each side) corresponding to the surface area a.

The van der Waals attraction is rather weak, particularly so for thin water films between hydrophobic mica surfaces. For these films we have used the experimental Hamaker constant value $A_H = 0.5 \times 10^{-20}$ J whereas for the soap film case we have inserted $A_H = 3.7 \times 10^{-20}$ J.

The total free energy excess of the film

The total free energy cost per chain of forming the thin film from dispersed surfactant and alcohol molecules in solution at the mole fraction x_2 and x_3, respectively, becomes (Eq. (3)):

$$\varepsilon = g - x_s \mu_2^+ - x_{OH} \mu_3$$
$$= g - \mu^0 - kT(x_s \ln x_2 + x_{OH} \ln x_3) \quad (46)$$

where

$$\mu^0 = \mu_{Tan}^0 + x_s \mu_{pg}^0 + x_{OH} \mu_{OH}^0 \quad (47)$$

is the standard chemical potential per chain in solution. μ_{Tan}^0 denotes the hydrophobic free energy part of this standard chemical potential whereas μ_{pg}^0 and μ_{OH}^0 are the corresponding head group parts. Let us now introduce the free energy excess quantity

$$\tilde{g} = g - \mu^0. \quad (48)$$

In this definition of $\tilde{g}$ the solution standard states are employed in order to introduce a suitable, fixed, free energy reference level.

Adding all the different contributions to $\tilde{g}$ we get

$$\tilde{g} = g_{el} - 19.960 \, kT$$
$$+ 92.840 \, k \left[T \ln \left(\frac{T}{298.15} \right) - (T - 298.15) \right]$$
$$+ g_{conf} + a(\gamma_{hc/w} + \gamma_{hc/air})$$
$$+ kT(x_S \ln x_S + x_{OH} \ln x_{OH})$$
$$+ x_S \varepsilon_{pg} + x_{OH} \varepsilon_{OH} - (1/2) \, aA_H/12\pi h^2$$
$$- (1/4\pi) \, (a - a_{pg}) \, [0.36 \, C_1 \exp (- h/12.0)$$
$$+ 0.0066 \, C_2 \exp (- h/55.0)] \quad (49)$$

where ε_{pg} and ε_{OH} are the free energy changes of the head groups upon transfer from the bulk to the inter-

face apart from the purely electrostatic contribution, cf. below.

Equilibrium conditions

It follows from Eqs. (9) and (10) that the relations (throughout T, p_e, H, c_2, and c_3 are kept constant here)

$$(\partial[(N_2 + N_3) \tilde{g}]/\partial N_2)_{A, N_3} = kT \ln x_2 \quad (50)$$

$$(\partial[(N_2 + N_3) \tilde{g}]/\partial N_3)_{A, N_2} = kT \ln x_3 \quad (51)$$

must hold at equilibrium between the thin film and the surrounding water solution. The left-hand sides of Eqs. (50) and (51) can easily be transformed:

$$(\partial[(N_2 + N_3) \tilde{g}]/\partial N_2)_{A, N_3}$$
$$= (\partial[(N_2 + N_3) \tilde{g}]/\partial N_2)_{a, N_3}$$
$$+ (\partial[(N_2 + N_3) \tilde{g}]/\partial a)_{N_2, N_3} (\partial a/\partial N_2)_{A, N_3}$$
$$= \tilde{g} + x_{OH}(\partial \tilde{g}/\partial x_S)_a - a(\partial \tilde{g}/\partial a)_{x_S} \quad (52)$$

and

$$(\partial[(N_2 + N_3) \tilde{g}]/\partial N_3)_{A, N_2}$$
$$= (\partial[(N_2 + N_3) \tilde{g}]/\partial N_3)_{a, N_2}$$
$$+ (\partial[(N_2 + N_3) \tilde{g}]/\partial a)_{N_2, N_3} (\partial a/\partial N_3)_{A, N_2}$$
$$= \tilde{g} + x_S(\partial \tilde{g}/\partial x_{OH})_a - a(\partial \tilde{g}/\partial a)_{x_S} \quad (53)$$

where, in accordance with Eq. (15):

$$\left(\frac{\partial \tilde{g}}{\partial a} \right)_{x_S} = \gamma/2. \quad (54)$$

Hence, the equilibrium conditions (50) and (51) can be written in the form

$$\tilde{g} + x_{OH}(\partial \tilde{g}/\partial x_S)_a - a\gamma/2 = kT \ln x_2 \quad (55)$$

$$\tilde{g} + x_S(\partial \tilde{g}/\partial x_{OH})_a - a\gamma/2 = kT \ln x_3. \quad (56)$$

These equations are consistent with Eqs. (3) and (12) as is easily verified by multiplying Eq. (55) with N_2 and Eq. (56) with N_3 and then adding:

$$\Omega = (N_2 + N_3)\varepsilon = (N_2 + N_3)\tilde{g} - N_2 kT \ln x_2$$
$$+ N_3 kT \ln x_3 = (a\gamma/2) \, (N_2 + N_3) = A\gamma. \quad (57)$$

In our calculation program, the film tension γ was computed by means of Eq. (54), i. e. through *differentiation* of $\tilde{g}$, whereas ε was computed from the equilibrium value of $\tilde{g}$:

$$\varepsilon = \tilde{g} - kT \left[x_S \ln x_2 + x_{OH} \ln x_3 \right]. \tag{58}$$

Finally, we note that since our calculation scheme in principle involves the free energy differential Eq. (8) as well as the integrated relation (12), i. e.

$$(N_2 + N_3) g = A\gamma + N_2 \mu_2^+ + N_3 \mu_3, \tag{59}$$

the generalized Gibbs surface tension equation, Eq. (16), will automatically be satisfied.

The head group constants ε_{pg} and ε_{OH}

The head group constants, ε_{pg} for surfactant molecule and ε_{OH} for an alcohol molecule account for several, less well known effects, e. g.:

i. The reduction of the overall free energy of the hydrocarbon/water contact because of the presence of polar head groups. This contribution has a negative sign and might be regarded as a kind of effective head group area.

ii. The unfavourable asymmetric hydration of the ionic species in close proximity of the hydrocarbon layer and image charge effects.

iii. The repulsive interactions between the hydrated head groups that are not already included in the electrostatic free energy calculated according to the PB approximation assuming a smeared-out surface charge.

No quantitative theories of these effects are as yet available. Throughout the calculations to be presented below we have used the values

$$\varepsilon_{pg} = 1.445 \, kT$$

$$\varepsilon_{OH} = 0 \tag{60}$$

at 25 °C. This value of ε_{pg} was determined by fitting to the surface tension data for $C_{12}NH_3^+Cl^-$ solutions of Motomura et al. [2]. Some support for putting ε_{OH} equal to zero was obtained from our previous study of mixed micelles [9].

Gaseous interfacial monolyers

In the pure surfactant case, i. e. without the $C_{12}OH$ component, and at sufficiently low bulk phase concentrations, one has to anticipate the formation of dispersed ("gaseous") monolayers with a larger than about 200 Å^2/molecule. In order to deal with this situation a few modifications of our theory are needed.

First of all, the volume constraints giving rise to changes of g_{conf} upon varying the hydrocarbon chain area a are removed and the residual, constant chain configurational free energy may be included in the head group constant, ε_{pg}. Secondly, it is evident that $\tilde{g}_{Tan}$ might deviate considerably from the standard value $-19.960 \, kT$ at 25 °C because the chains are only partially being removed from water exposure at transfer to the interface and they do not form a coherent monolayer. For the dispersed solution/air interface we have deduced

$$\tilde{g}_{Tan} = -12.873 \, kT \quad (25\,°C). \tag{61}$$

For the hydrophobic/solution interface we have estimated that

$$\tilde{g}_{Tan} = -18.498 \, kT \quad (25\,°C). \tag{62}$$

In addition, there are some rather obvious changes of the contact free energies which must also be taken care of. In particular, we may note for the solution/air case that the hydrocarbon/water contact free energy was estimated by assuming the contact area to be 124 Å^2, corresponding with the surface area occupied by a $C12$ hydrocarbon chain lying down in the surface, i. e. as $\gamma_{hc/w} \times 124 \times 10^{-20}$ J/chain.

Programming considerations and numerical procedures

In Eq. (49) the free energy quantity $\tilde{g}$ is given as a function of T, a and x_S (or x_{OH}) at some fixed values of the water core thickness h and the surfactant concentration c_2. In order to compute c_2 and c_3 for certain given values of a and x_S an iteration procedure was invoked since g_{el} depends upon the electrolyte concentration, i. e. on c_2. Damping of the strong numerical oscillations was achieved by using the mean value of the old and new c_2 values in the subsequent iteration step.

When c_2 and c_3 were fixed at the outset, a and x_S were determined by an iterative solution of the resulting non-linear equations.

The solutions of the PB equation was made by means of separate subroutines using, in essence, the numerical method devised by Jönsson [18]. Our PB program was checked against the corresponding program developed by Chan, Pashley and White [23] and exact agreement was established in both the constant charge and constant potential modes.

In so far as the electrostatic weak overlap approximation could be employed, the computing times were rather short (of the order of minutes) when making use of a conventional PC computer like IBM XT. However, calculations on interacting monolayers as a rule have to be performed by exact numerical solution of the PB equation. With about 25 c_2, c_3 iteration steps and 12 PB iterations in each step the computing time to get one set of equilibrium data for given T, h, c_2, c_3 values was in the range of 1 h on a Microvax 2000 computer.

The surface tension of $C_{12}NH_3^+Cl^-$ solutions

Figure 2 shows the surface tension curve generated at 25 °C for dodecylammonium chloride ($C_{12}NH_3^+Cl^-$) solutions of various concentrations upon making the film thickness h equal to 10 000 Å. This surface separation is well beyond the range of the electrostatic interaction. The head group constant ε_{pg} was adjusted to 1.445 kT so as to reach agreement between the calculated $\gamma/2$ value and the surface tension value actually measured by Motomura et al. [2] at 25 °C and $c_2 =$ 5.09 mM. In general, the agreement between our calculated curve and the experimental points is quite satisfactory, although not perfect. Hence, at higher concentrations there appears to be some additional surface pressure contribution besides the electrostatic and lateral chain pressure components. The break point at

1.34 mM indicates a (first order) transition from a gaseous to a condensed, liquid-expanded monolayer.

The corresponding step-wise isotherm is plotted in Fig. 3. As expected, it is very similar to the isotherm derived thermodynamically by Motomura et al. (loc. cit.) from their experimental γ/c_2 function by means of the Gibbs surface tension equation. At c_2 concentrations just above the surface phase transition at 1.34 mM, the hydrocarbon layer is in a *stretched* state thus contributing positively to the surface *tension*. At higher monolayer packing densities (a less than about 48 Å) the hydrocarbon chain pressure turns positive and together with the electrostatic surface pressure component it contributes to the surface tension lowering.

We may note that it would be considerably more difficult to account for the observed surface phase transition had we assumed a significantly lower value of the hydrocarbon/water contact free energy than $\gamma_{hc/w}$ = 50.70 mNm^{-1} at 25 °C. Thus, we have not been able to gather support for Jönsson's much lower value $\gamma_{hc/w}$ = 18.2 mNm^{-1} [18]. However, his value was deduced on the basis of purely electrostatic calculations for lamellar liquid crystals, i. e. without explicitly invoking the chain configurational free energy.

Interfacial tension of $C_{12}NH_3^+Cl^-$ solutions in contact with a hydrophobic surface

In Fig. 4 we present plots of $\gamma/2$ calculated for 10 000 Å thick films of $C_{12}NH_3^+Cl^-$ solution between hydrophobic solid surfaces. Without added dodecylalcohol ($C_{12}OH$, upper curve) there is a striking similarity in

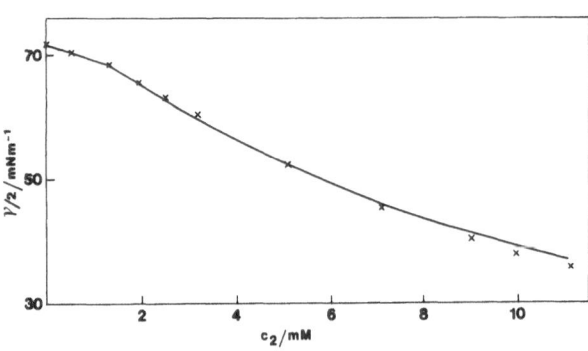

Fig. 2. The surface tension at 25 °C of $C_{12}NH_3^+Cl^-$ solutions below the CMC. The line represents the calculated $\gamma/2$ vs. c_2 function with $\varepsilon_{pg} = 1.445$ kT. The experimental points (×) of Motomura et al. [2] are also shown. The break point at $c_2 = 1.34$ mM corresponds to a first order phase transition from a gaseous to a liquid-expanded monolayer

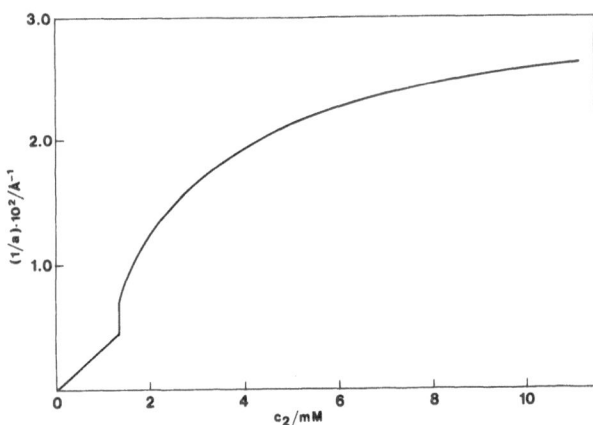

Fig. 3. Stepwise surfactant adsorption isotherm at 25 °C of $C_{12}NH_3^+Cl^-$ at the air/solution interface corresponding to the $\gamma/2$ vs. c_2 function in Fig. 2. The isotherm shown would follow directly from the $\gamma/2$ function of Fig. 2 by applying the Gibbs surface tension equation that is implicit in our calculation procedure

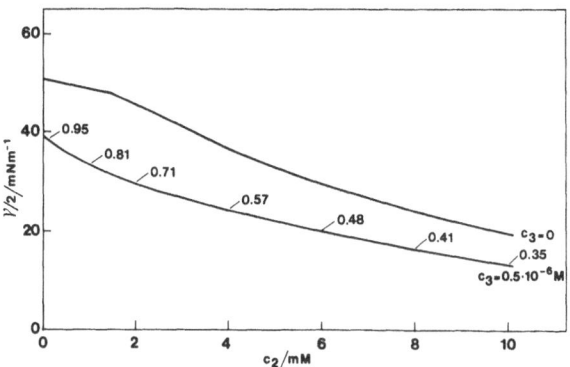

Fig. 4. Calculated film tensions at 25 °C of a 10 000 Å thick $C_{12}NH_3^+Cl^-/C_{12}OH$ solution film between impenetrable hydrophobic surfaces chosing $\varepsilon_{pg} = 1.445 \, kT$. The upper curve was obtained without the $C_{12}OH$ component whereas the lower curve was generated for $c_3 = 0.5 \cdot 10^{-6}$ M. The mole fraction of $C_{12}OH$ in the monolayer, x_{OH}, is shown at various points along the lower curve

shape with the curve in Fig. 2 for the solution/air interface. In particular, there is a phase transition at about the same c_2 concentration, 1.5 mM. The whole curve is shifted downwards by about 20 mNm^{-1} on the ordinate axis as compared with the solution/air case. This feature is not unexpected, however, as the major difference is that the contact free energy contribution $y_{hc/air}$ × A has been annihilated upon replacing the air phase with an impenetrable hydrophobic wall (cf. Eq. (9)). Hence, for a methylated glass substrate we would predict an adsorption isotherm practically identical to the one in Fig. 3. The adhesion force measurements by means of methylated glass spheres due to Shchukin, Amelina and Yaminsky [24] which yielded $y(H = \infty)$ + constant (molecular contact) term for a series of sodium dodecyl sulphate and cetylpyridinium bromide solutions lend support to this conclusion. This may seem counter-intuitive at first, since one might well have anticipated a larger extent of surfactant adsorption on a hydrophobic surface than at the solution/air interface.

Adding $C_{12}OH$ to a fixed, rather low concentration c_3 (Fig. 4, lower curve) has a profound effect on the course of the $y/2$ curve. The gaseous Henry's law range has disappeared entirely. At all c_2 concentrations there is a condensed liquid-expanded monolayer with the hydrocarbon chain area a in the range 35–38 Å^2. Note that the composition of the monolayer varies substantially with the surfactant concentration.

Towards this background we may argue that one major stumbling-block associated with providing firm

experimental evidence for the phase transition between a gaseous and a condensed, liquid-expanded monolayer is probably that small amounts of sparingly soluble, surface-active impurities like $C_{12}OH$ must not be present in the solution. If, however, this can not be avoided, the surface tension measurements should be carried out rapidly enough to limit the transport of such impurities to the interface.

The soap film case

Figure 5 shows the film tension y as a function of the surface separation h for a thin soap film formed from a 5 mM $C_{12}NH_3^+Cl^-$ solution. For comparison, some points resulting from purely electrostatic calculations at constant charge density (i. e. a) and constant surface potential ϕ^0, respectively, are also marked. To be more precise, these latter y-values were obtained by adding $\Delta y_{el}(h)$ to the equilibrium y-value at $h = 10\,000$ Å.

The rather rapid change of y for small surface separations as computed by means of our complete calculation program is surprisingly well reproduced by the purely electrostatic calculations at constant a unchanged in spite of the fact that the hydrocarbon chain area does vary somewhat (cf. Fig. 6). The constant potential condition yields an equally good agreement in this case.

For this soap film with a about 47 Å^2 and $h > 25$ Å we have assumed that there is no substantial hydrophobic attraction left associated with the hydrocarbon/water contacts. Inserting Eq. (44) with $C_1 = 1$ and $C_2 = 0$ would have resulted in a y-function passing through a maximum value in the range of 50 Å. Whether or not there is still some residual hydrophobic attraction between the two soap film faces at small

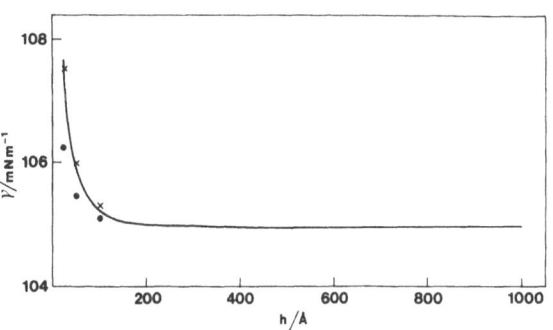

Fig. 5. The film tension y of a soap film formed at 25 °C from a 5 mM $NH_3^+Cl^-$ solution as a function of the surface separation h. The crosses mark the results of purely electrostatic calculations at a constant surface charge density or at a constant surface potential

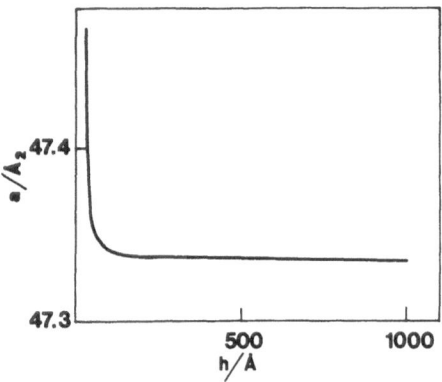

Fig. 6. The hydrocarbon chain surface area, a, as a function of the surface separation h in the same soap film as in Fig. 5

separations is so far an unsettled question, although the stability of Newton black films might indicate that this is indeed the case.

Concerning the electrostatic double layer interaction in soap films, it is interesting to note that Exerowa, Kolarov and Khristov have treated this problem recently [25] on the basis of the PB approximation which, however, yielded clear discrepancies with disjoining pressure measurements. Additional work in this area is being planned by us in order to investigate whether by employing our full equilibrium calculation scheme we can account for this lack of agreement between experiments and the DLVO theory.

In Table 1 we have listed the various contributions to the change in film tension, Δy, for a 50 Å thin film as compared with a 10 000 Å thick film. These contributions were computed in accordance with Eq. (12), $y = 2\varepsilon/a$, i. e. as free energy changes rather than as lateral pressure/tension changes. It appears that a predominant contribution to Δy in this case is caused by the electrostatic double layer interaction. However, the somewhat lower monolayer packing density at 50 Å as

Table 1. The various contributions to $\Delta y = y(h) - y(\infty)$ for a soap film with thickness $h = 50$ Å formed from a 5 mM $C_{12}NH_3^+Cl^-$ solution

Free energy change per surface area unit	μJm^{-2}
$2\Delta(g_{el}/a)$	889.4
$2\Delta[(\varepsilon_{Tan} + \varepsilon_{pg})/a]$	129.4
$2\Delta(g_{conf}/a)$	−5.6
$2\Delta(g_{vdW}/a)$	−39.3
Total $\Delta y = 974.1 \mu Jm^2$	

compared with 10 000 Å (Fig. 6) also gives a significant Tanford term contribution to Δy. The configurational free energy term is small here because a is very close to 48.1 Å² where g_{conf} has a rather broad minimum, for the 50 Å film as well as for the 10 000 Å film.

Surface force measurements

Recently, Herder [26] has carried out a series of surface force measurements employing hydorphobed mica surfaces immersed in $C_{12}NH_3^+Cl^-$ solutions of various concentrations. The hydrophobation was achieved by Langmuir-Blodgett deposition of dimethyldioctadecylammonium bromide (DDOAB). Figures 7 and 8 display normalized surface force data recorded for $c_2 = 1$ mM and $c_2 = 10^{-2}$ mM. The theoretical F/R curves were obtained by means of the Derjaguin approximation, according to which

$$F/R = 2\pi[y(h) - y(\infty)] \tag{63}$$

where R is the local radius of curvature of the cylindrical mica surfaces [27].

In order to achieve acceptable agreement between theory and experiments we found it necessary to invoke an additional free energy gain associated with the surfactant adsorption on the hydrophobic mica

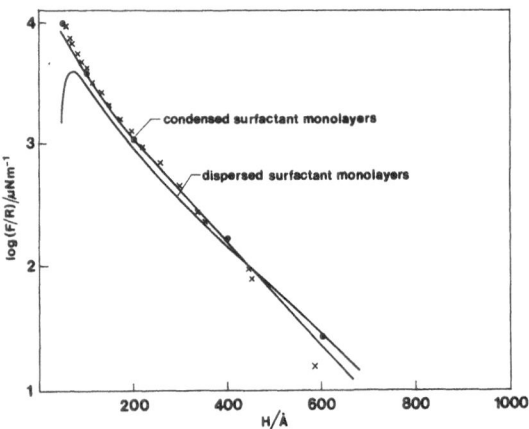

Fig. 7. Normalized surface forces, F/R, of thin films between hydrophobed mica surfaces formed from a 1 mM $C_{12}NH_3^+Cl^-$ solution. The crosses denote experimental results [26]. The filled points (●) indicate the results of purely electrostatic calculations at constant a as well as at constant ϕ^0. Assuming a gaseous monolayer and choosing $C_1 = 1$, $C_2 = 0$ in Eq. (44) for the hydrophobic attraction, the lower curve with a maximum was obtained. The upper curve refers to condensed, liquid-expanded monolayers. In both cases g_{mix}^{surf} was put equal to -0.5×10^{-20} J/chain and ε_{pg} to 1.445 kT

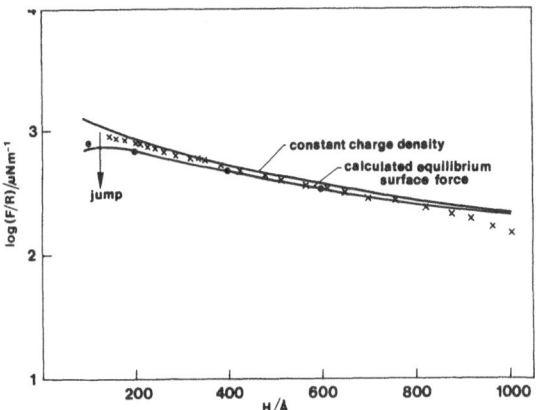

Fig. 8. Normalized surface forces, F/R, of thin films between hydrophobed mica surfaces formed from a 0.01 mM $C_{17}NH_3^+Cl^-$ solution. The crosses denote the experimental points [26] and the lower curve the calculated surface forces chosing $g_{mix}^{surf} = -1.25 \cdot 10^{-20}$ J/chain and $\varepsilon_{pg} = 1.445\ kT$. The upper curve shows the results of purely electrostatic calculations at constant a and the filled points (●) the corresponding results obtained at constant ϕ^0. At about $H = 150$ Å the hydrophobed surfaces jump to molecular contact ($H = 0$)

surfaces. We have supposed this free energy gain to be caused by a partial mixing of the C12 surfactant hydrocarbon chains with the C18 chains of DDOAB on the mica. In the two cases, this extra g_{mix}^{surf} quantity (per chain) was made equal to

$$g_{mix}^{surf}\ [1\ mM] = -0.5 \times 10^{-20}\ J$$

$$g_{mix}^{surf}\ [0.01\ mM] = -1.25 \times 10^{-20}\ J. \qquad (64)$$

For the 1 mM $C_{12}NH_3^+Cl^-$ solution, a somewhat better fit to the experimental points was obtained assuming condensed surfactant monolayers with a about 90 Å^2 and y about 93 mNm^{-1} rather than assuming a gaseous monolayer with a in the range of 220 Å^2 and y about 94 mNm^{-1} (Fig. 7). In particular, at small separations, there is a rather marked difference in shape between the two curves. At least in principle, the crossing of the curves at 450 Å would imply a first order monolayer transition occurring at the shortest distance between the curved mica sheets. However, this question is very much of an academic character because the relative experimental accuracy is worse in this H-range where the surface forces are weak.

It is noteworthy that the constant potential and constant charge density modes of carrying out the electrostatic calculations both yield fair agreement with

our fully equilibrium calculation as well as with the experimental measurements. In the next section we shall discuss this matter further in general terms.

In Table 2 the various contributions to the surface force F/R have been listed for a thin film with $H = 100$ Å. The total surface force can be evaluated in two different ways, from (i) the *free energy* of formation as $F/R = 2\pi\Delta(2\varepsilon/a)$, (which is based on a comparison of the film and bulk states, the monolayer being regarded as a *thermodynamically open* system) and from (ii) the infinitesimal work of stretching the surface (being regarded as a *thermodynamically closed* system) as $F/R = 2\pi[(\partial\tilde{g}/\partial a)(H) - (\partial\tilde{g}/\partial a)(H = \infty)]$. As shown above, the two expressions for F/R must yield identical results for the film tension but the individual contributions (e. g. electrostatic, configurational, etc.) to the film tension need not be the same in the two calculation schemes as shown in Table 2. Thus, the relationship $2\varepsilon/a = y$, Eq. (12), only demands that the *sum* of the different contributions to the Ω-potential per unit area and to the mechanical surface tension, respectively, must be the same.

It is rather obvious that the constant terms of $\tilde{g}$, like $\tilde{g}_{Tan}$, ε_{pg}, ε_{OH}, g_{mix}^{surf} etc., do not affect the mechanical Δy-value. However, they do contribute to $\Delta(2\varepsilon/a)$ in so far as the hydrocarbon chain area depends upon the surface seaparation H. This is normally the case as soon as there is some sensible surface-surface interaction.

In the much more dilute case, $c_2 = 0.01$ mM, the Debye length is 960 Å and one may anticipate that the surface concentration of surfactant is quite low. A reasonably good fit to the experimental $F/R, H$ points was obtained by assuming a gaseous monolayer state and

Table 2. The different components of the surface force F/R for a 100 Å thin film between hydrophobed mica surface formed from 1 mM $C_{12}NH_3^+Cl^-$ solution. The values in the left column were obtained as ε-contributions and those in the right column as $\partial\tilde{g}/\partial a$-contributions (see text)

Component of surface force	F/R free-energy-wise μJm^{-2}	F/R tension-wise μJm^{-2}
Electrostatic	640.9	4246.2
Tanford term + ε_{pg}	3061.2	0.0
Configurational	116.9	− 399.6
Hydrophobic attraction	− 138.2	− 165.9
van der Waals	− 10.1	− 10.1
Total $F/R = 3670.7$		Total $F/R = 3670.6$

setting $g_{mix}^{surf} = -1.25 \times 10^{-20}$ J resulting in $a(H = \infty) =$ 8925 Å^2 (Fig. 8). Furthermore, the full g_{hyph} expression (44) with $C_1 = C_2 = 1$ was inserted in order to account for the hydrophobic attraction which presumably causes the hydrophobed mica surfaces to jump to molecular contact ($H = 0$) from a surface separation of about 150 Å. For pure water this occurs already from about $H = 550$ Å.

Note that the constant potential condition yields a rather satisfactory approximation in this case to our full equilibrium calculation but the constant charge condition does not. This is hardly surprising in view of the circumstance that a changes drastically upon diminishing H whereas the surface potential ϕ^0 stays rather constant (Fig. 9).

The constant surface charge density and constant surface potential constraints

When discussing the results obtained for interacting surfaces we have pointed out that the purely electrostatic, PB-calculated surface forces as a rule approximate the equilibrium surface forces rather well, in spite of the coupling which occurs between different kinds of surface forces. Hence, it is relevant to inquire about possible reasons for this behaviour.

Beginning with the constant charge density case we may analyze this question as follows. Let us first of all assume that T and all μ are kept constant during the variations considered below. Then we may write

$$d\gamma = \left(\frac{\partial \gamma}{\partial h}\right)_a dh + \left(\frac{\partial \gamma}{\partial a}\right)_h da \qquad (65)$$

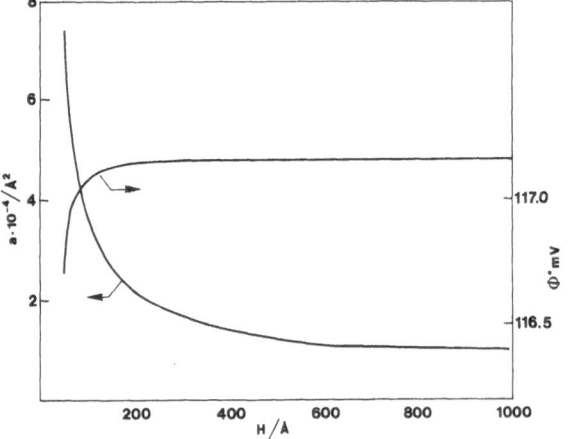

Fig. 9. Hydrocarbon chain surface area, a, and surface potential, ϕ^0, corresponding with the calculated F/R vs. H function displayed in Fig. 8

or

$$\frac{d\gamma}{dh} = \left(\frac{\partial \gamma}{\partial h}\right)_a + \left(\frac{\partial \gamma}{\partial a}\right)_h \frac{da}{dh}. \qquad (66)$$

Generally, in our calculations, we have found that $da/dh \neq 0$. However, supposing the initial state to be an equilibrium state with minimal Ω-potential we must have

$$\left(\frac{\partial \gamma}{\partial a}\right)_h = 0. \qquad (67)$$

Accordingly, $\Delta \gamma$ can be obtained by the integration

$$\Delta \gamma = \gamma(h) - \gamma(\infty) = \int_{\infty}^{h} (\partial \gamma / \partial h')_a dh' \qquad (68)$$

where $(\partial \gamma / \partial h')_a$ should be evaluated at the equilibrium value of $a(h')$. But it is reasonable to assume that the partial derivative $(\partial \gamma / \partial h')_a$ is only weakly dependent on a because of the close balance between repulsive and attractive forces normally found at least for aggregated surfactant systems (Refs. [6, 7]). If so, one can choose to keep $a = a(\infty) = $ constant during the integration of Eq. (68) without introducing appreciable errors, thus leaving aside all a-dependent surface forces. Evidently, this reasoning would not be valid when large changes of a result from varying the surface separation h. We should add that the constant charge condition assumes a hypothetical monolayer that is equilibrated neither with the adjacent film part nor with the bulk solution.

It is different with the constant surface potential condition which actually serves the purpose of mimicking the (correct) constant chemical potential condition. Writing, in analogy with Eq. (66),

$$\frac{d\gamma}{dh} = \left(\frac{\partial \gamma}{\partial h}\right)_{\phi^0} + \left(\frac{\partial \gamma}{\partial \phi^0}\right)_h \frac{d\phi^0}{dh} \qquad (69)$$

we note that we would actually have that

$$d\phi^0 / dh = 0 \qquad (70)$$

provided that $\phi^0 = $ constant were always a consequence of the general, constant μ condition. Hence, the success of this *ansatz* will clearly depend upon what molecular factors determine the chemical potentials

μ_2^+ and μ_3 in the thin film. These questions are of a general importance and we intend to return to them utilizing more extensively the calculation methods presented in this paper.

Concluding remarks

The calculation scheme for surfactant monolayers presented in this paper embraces a number of different experimental systems, viz.

i. A surfactant solution/air interface. Here we have demonstrated satisfactory agreement with experimental results for one particular surfactant, $C_{12}NH_3^+Cl^-$, at 25 °C. Additional calculations on reliable surface tension data are needed to explore the effects of varying the chain length and head group, changing the temperature, adding salt or a long-chain nonionic amphiphile etc., before we might claim to have provided an adequate theoretical framework for discussing the molecular mechanisms of the chief functional property of a surfactant: its ability to lower the surface tension of water.

ii. Interacting soap film faces composed of monolayers formed from a surfactant solution. Hitherto, investigations of the interactions in soap films have often been hampered by not knowing precisely enough the monolayer surfactant density and composition. On this point our theory may prove helpful. But there are still some major issues which have to be studied further: is there some additional repulsive force, besides the electrostatic one, operating in the 100 Å thickness range [25]; does the hydrophobic attraction play some role in the stability of the very thin black films? We are planning experimental and theoretical work on mixed films together with the Bulgarian surface science group in Sofia to approach these matters.

iii. A surfactant solution film between hydrophobed solid surfaces. Here again, the present investigation has been restricted to one type of surfactant, one temperature, one kind of hydrophobic solid surface, etc. Hence, much work remains to fill the gaps with respect to studying surface forces as well as adhesion forces employing penetrable as well as impenetrable hydrophobic surfaces.

As yet our theory does not apply to a surfactant solution/*liquid* hydrocarbon interface because the statistical thermodynamics of the mixing of hydrocarbon chains and solvent needs a more satisfactory treatment than is currently available.

iv. The lamellar liquid crystal case. This is a very important special case of (iii) where the film tension γ

and the interaction pressure π are both equal to zero because of mechanical reasons. Jönsson and Wennerström [28] have devoted an extensive theoretical study to the phase diagrams of water/surfactant/long-chain alcohol systems employing PB electrostatics and calculation methods similar to ours. However, it would be valuable to check the chain packing constraints which they have introduced and their choices of parameters in the lamellar liquid crystal region. This can easily be done using our calculation scheme as a starting-point.

v. Surfactant penetration of insoluble monolayers. Recently, this surface balance experiment has been investigated thermodynamically by Ter-Minassian-Saraga [29] and Hall [30] but a quantitative molecular treatment is lacking. By adjusting the $C_{12}OH$ bulk concentration c_3 so as to keep Γ_3 constant while varying the surfactant concentration we may elucidate the competition between the lateral chain surface pressure and the entropy of mixing effects.

In forthcoming papers we shall report on our ensuing works concerning the experimental situations (i)–(v) described above which are based on the theory and calculations presented in this article.

Acknowledgements

We wish to thank Peter Herder for putting his surface force data at our disposal before publication. We also thank him and Per Claesson for assistence with checking the electrostatic subroutines and for helpful comments.

References

1. Motomura K, Aranato M, Matubayasi N, Matuura R (1978) J Colloid Interface Sci 67:247
2. Motomura K, Iwanaga S-I, Hayami Y, Uryu S, Matuura R (1981) J Colloid Interface Sci 80:32
3. Aranato M, Yamanaka M, Matubayasi N, Motomura K, Matuura R (1980) J Colloid Interface Sci 74:489
4. Motomura K, Iwanaga S-I, Yamanaka M, Aranato M, Matuura R (1982) J Colloid Interface Sci 86:151
5. Eriksson JC (1982) Finn Chem Lett p 105
6. Eriksson JC, Ljunggren S, Henriksson U (1985) J Chem Soc Faraday Trans 2 81:833
7. Eriksson JC, Ljunggren S (1985) Ibid, p 1209
8. Ljunggren S, Eriksson JC (1986) J Chem Soc Faraday Trans 2 82:913
9. Ljunggren S, Eriksson JC (1987) Progr Colloid Polym Sci 74:38
10. Tanford C (1980) The Hydrophobic Effect. Wiley, New York, 2nd ed
11. Evans DF, Ninham B (1983) J Phys Chem 87:5025
12. Gruen DWR, Lacey EHB (1984) In: Mittal KL, Lindman B (eds) Surfactants in Solution. Vol I, Plenum Press, New York, p 279

13. Eriksson JC, Ljunggren S, Claesson PM (1988) J Chem Soc Faraday Trans 2, accepted for publication
14. Claesson PM, Blom CE, Herder PC, Ninham BW (1986) J Colloid Interface Sci 114:234
15. Jönsson B, Wennerström H (1981) J Colloid Interface Sci 80:482
16. Gunnarsson G, Jönsson B, Wennerström H (1980) J Phys Chem 84:3114
17. Missel DJ, Mazer NA, Benedek GB, Young CY, Carey MC (1980) J Phys Chem 84:1044
18. Jönsson B (1981) The Thermodynamics of Ionic Amphiphile-Water Systems, A Theoretical Analysis. Thesis, Lund
19. Ljunggren S, Eriksson JC (1988) J Chem Soc Faraday Trans 2 84:329
20. Mazer NA, Olofsson G (1982) J Phys Chem 86:4584
21. (1982) Handbook of Physics and Chemistry. 63rd Ed, CRC Press Inc, Bota Raton, Fa
22. Claesson PM, Christenson HK, Herder PC, Berg J (1988) Langmuir, submitted
23. Chan DY, Pashley RM, White LR (1980) J Colloid Interface Sci 77:283
24. Shchukin ED, Amelina EA, Yaminsky VV (1981) Colloids Surf 2:221
25. Exerowa D, Kolarov T, Khristov K (1987) Colloids Surf 22:171
26. Herder PC (1988) Surfactant Adsorption on Mica Studied by Surface Force and ESCA Measurements. Thesis, Stockholm
27. Derjaguin BV (1934) Kolloid Z 69:155
28. Jönsson B, Wennerström H (1987) J Phys Chem 91:338
29. Ter-Minassian-Saraga L (1985) Langmuir 1:391
30. Hall DG (1986) Langmuir 2:809

Received December 23, 1987;
accepted January 7, 1988

Authors' address:

J. C. Eriksson
Department of Physical Chemistry
The Royal Institute of Technology
S-100 44 Stockholm, Sweden

Progress in Colloid & Polymer Science   Progr Colloid Polym Sci 76:203–210 (1988)

Phase behavior of the quinary mixture:
H₂O-NaCl-dodecane-pentanol-SDS. Origin of the Winsor III equilibria

A. M. Bellocq and D. Gazeau

Centre de Recherche Paul Pascal (CNRS) and GRECO "Microemulsions", Domaine Universitaire, Talence Cedex, France

Abstract: The phase behavior of the quinary mixture H₂O-NaCl-dodecane(D)-pentanol(P)-SDS is represented at fixed P and T in a serie of pseudoquaternary diagrams, each corresponding to a constant brine salinity. In practice, each diagram is itself constructed by examining a series of cuts X in which the brine over surfactant ratio is kept constant. This procedure allows one to describe completely the extent of the three-phase region Winsor III (W III) and to determine its mode of appearance. Our results show that this region, which starts to be observed at 4 g/l NaCl in water is generated by an indifferent state. In such a state, the compositions of the three coexisting phases are located along a straight line in the four-dimensional space of representation. As salinity goes above 5.5 g/l NaCl, the volume W III is bounded in the oil-rich region by a line of critical end points, LC_1. A second line of such points, LC_2, occurs in the water-rich region at salinities higher than 8.1 g/l NaCl. Then, the three-phase region W III appears by two critical end points only when salinity is higher than 8.1 g/l NaCl. Our experiments show that the lines LC_1 and LC_2, respectively, originate in the quaternary mixtures H₂O-D-P-SDS and H₂O-NaCl-P-SDS.

Key words: Phase behavior, quinary mixtures, W, NaCl, Dodecane, pentanole sodium dodecyl sulfate, dodecylsulfate, diagrams, pseudoquaternary diagrams.

1. Introduction

The quaternary mixtures water-oil-surfactant-alcohol may form several phases and complex multiphase equilibria. In certain conditions, the so-called Winsor III three-phase equilibria, where a microemulsion is in equilibrium with an aqueous phase and an organic phase are obtained. For anionic surfactants, these equilibria are only obtained in presence of salt. Up to now, little has been known about the mechanism involved in the generation of this three-phase region. It is commonly known that this equilibrium appears and disappears by a critical end point [1–3]. Recently, Kahlweit et al. [4] have proposed that in systems containing a nonionic amphiphile the three-phase region evolves from a tricritical point. Other possibilities may also be considered to account for the formation of a three-phase equilibrium. In a recent work, we have obtained clear evidence that three-phase regions may also arise from an indifferent state [5].

In this paper we report results on the origin of the Winsor III three-phase region of a quinary mixture at fixed T. We present an experimental study of the phase diagram of the quinary system: water-sodium chloride-dodecane-pentanol-sodium dodecylsulfate (W-NaCl-D-P-S). We have investigated the progression of phase behavior as one adds salt to the quaternary mixture water-dodecane-pentanol SDS (W-D-P-S). In previous papers we have reported the complete three-dimensional phase diagrams at constant pressure and temperature of the two quaternary systems: water-dodecane-pentanol-SDS (W-D-P-S) (I) and water-sodium chloride-pentanol-SDS (W-NaCl-P-S) (II) [5, 6].

The outline of the paper is as follows. Part 2 is a brief review of the modes of appearance of a three-phase region for four and five components mixtures. In part 3 we present the experimental procedure used to determine the region of existence of the three-phase

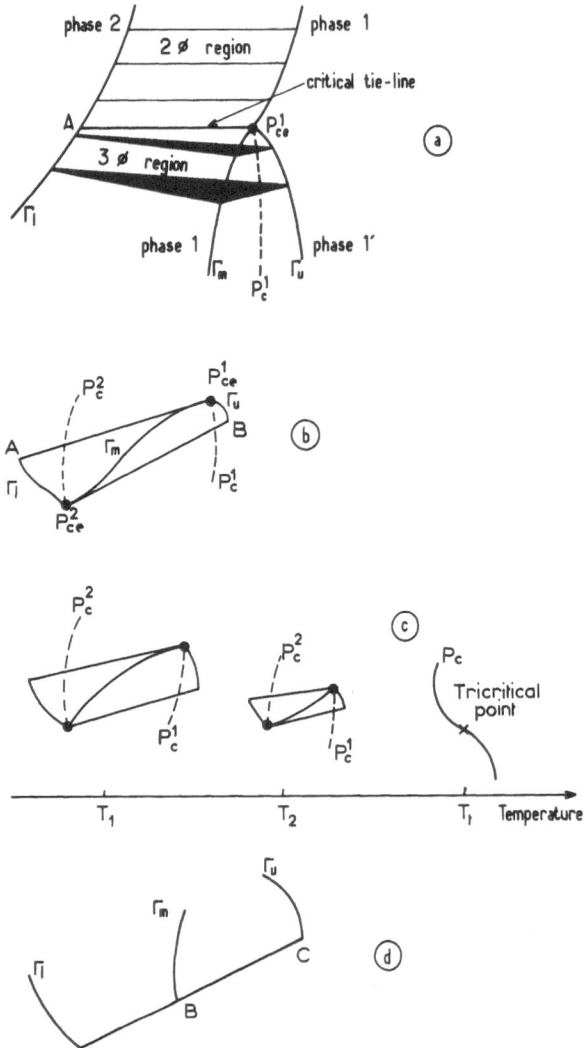

Fig. 1. Modes of appearance of a three-phase region for a quaternary mixture. (a) Critical end point; (b) two critical end points; (c) one tricritical point; (d) one indifferent state

equilibria in a five components mixture. After a brief description of the main features of the phase diagrams of the systems I and II in part 4, we present, in part 5, the phase diagram of the quinary system. Finally, part 6 is devoted to the discussion of the results.

2. Modes of Appearance of three-phase regions in quaternary and quinary mixtures

2.1 Quaternary mixtures

At constant pressure and temperature, a mixture of n components has $n-1$ independent variables; there-

fore, the representation of the phase diagram of a quaternary mixture requires a three-dimensional space. A three-phase region is a volume made by a stack of tie-triangles. The phase rule indicates that the variance of a three-phase state is one. This means that the points representative of the three phases, i. e., the vertices of the tie-triangles, are located along three curves Γ_u, Γ_m and Γ_l corresponding respectively to the upper, middle and lower phases. The topology of the three-phase volume is then described by these curves. Depending upon the relative positions of these curves in the three-dimensional space, the three-phase region may be generated in several different ways. Some of them are illustrated in Fig. 1. The first possibility is the existence of one critical end point P_{ce}^1. In this case (Fig. 1a) the three-phase volume is bounded by a critical tie-line P_{ce}^1-A. At the point P_{ce}^1, the phases 1 and 1' become identical. In certain systems, the three-phase volume is bounded by two critical tie-lines; then the three curves Γ_u, Γ_m and Γ_l form a single twisted curve (Fig. 1b). The critical end points P_{ce}^1 and P_{ce}^2 are the ends of two critical lines P_c^1 and P_c^2. The location and size of the three-phase region change with changing temperature. As the tricritical point temperature T_t is approached the critical end points are closer and the three-phase region is smaller. At the tricritical temperature, the three phases become identical; the three-phase region shrinks and the critical lines P_c^1 and P_c^2 form a connected critical line P_c. For a quaternary mixture, at fixed pressure, the variance of a tricritical state is zero (Fig. 1c).

The three-phase volume may also be generated by an indifferent state. In this case as shown in Fig. 1d, the points representative of the compositions of the three phases in equilibrium are along a straight line A B C. In an indifferent state, the compositions of all the phases are unchanged although the masses of the phases are changed. According to Prigogine and Defay [7] for a system made of c components, a state consisting of ψ phases is indifferent when all the determinants of order ψ composed with the lines of the following table are equal to zero.

$$\begin{bmatrix} x_1^1 & \cdots & x_1^\psi \\ x_2^1 & \cdots & x_2^\psi \\ \vdots & & \\ x_c^1 & \cdots & x_c^\psi \end{bmatrix}$$

In this table, the coefficients $x_1^1, \ldots x_c^1, \ldots x_1^\psi, \ldots x_c^\psi$ represent the concentrations of the c components in the ψ phases. The $c + 1 - \psi$ conditions of indifference are only relative to the compositions of the phases; they are independent of pressure and temperature. The geometric significance of the $1 + c - \psi$ conditions of indifference is that the ψ phases are along a straight line in the $c - 1$ dimensional space.

2.2 Quinary mixtures

At fixed P and T, a quinary mixture is described by four independent variables, namely four compositions. Three-phase equilibria are divariant. This means that the vertices of the tie-triangles are located only three surfaces S_u, S_m and S_l corresponding respectively to the upper, middle and lower phases. The modes of appearance of a three-phase region for a quinary mixture are the same that those for a quaternary one. But the variance of the corresponding states is increased by one. At fixed P and T a three-phase region in a quinary mixture may result from one tricritical point and develops between two lines of critical end points L_{c1} and L_{c2}. As well as for quaternary mixtures, three-phase regions in five component mixtures may originate in an indifferent state.

3. Experimental determination of a three-phase region in the phase diagram of a quinary mixture

Since at fixed P and T the representation of the phase diagram of a quinary mixture is impossible, we have chosen to represent it by a series of tetrahedrons, each corresponding to a fixed salinity. In practice, each tetrahedron is itself built by examination of a series of pseudoternary sections X^s where the ratio of the brine and surfactant concentrations is kept constant (this ratio is denoted X^s).

As already mentioned, the vertices of the tie-triangles are on three surfaces S_u, S_m and S_l. In the tetrahedron corresponding to the salinity s, (T_s), the three-phase region, if it exists, is bounded by the curves Γ_u^s, Γ_m^s, Γ_l^s which are the intercepts of the surfaces S_u, S_m and S_l with the tetrahedron T_s. In order to localize the boundaries of the Winsor III three-phase region we have searched in the sections X^s, the points R of connection between the one-phase and three-phase regions (Fig. 2). These points obviously belong to one of the curves Γ_u^s, Γ_m^s or Γ_l^s. These latter can be easily identified from the examination of the relative phase

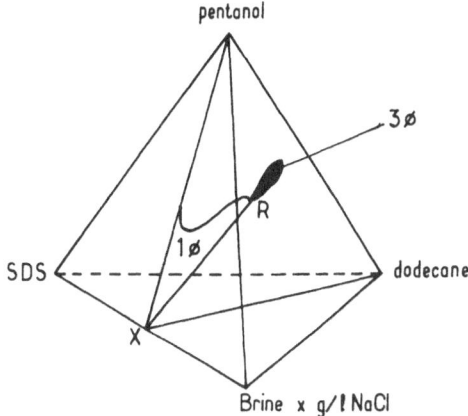

Fig. 2. Tetrahedron T_s for representing the phase diagram of the quinary mixture water-NaCl-dodecane-pentanol-SDS at fixed salinity(s). Section at constant brine over surfactant ratio X^s. R is the point at which the three-phase region touches the one-phase region

volumes for three-phase samples close in composition to the point R. An illustration of the procedure is given in Fig. 3 for a three-phase volume generated by two critical end points P_{ce}^1 and P_{ce}^2. The phase which takes up the largest volume is of the same type as that represented by the point R. When the point R is close to one of the two critical end points, the volumes of the nearly critical phases are almost equal. On Fig. 3 are shown the volumes for four three-phase samples close in composition to the points R_A to R_D in the four sections X_A^s to X_D^s.

4. Phase diagrams of the quaternary mixtures I and II

Prior to presenting the experimental phase diagrams obtained for the quinary system water-NaCl-dodecane-pentanol-SDS, let us briefly describe the main features of the phase behavior of the ternary mixture water-pentanol-SDS (W-P-S) and that of the two quaternary mixtures I and II.

The phase diagram of the ternary system W-P-S presents three liquid crystal phases (lamellar, hexagonal and rectangular phases) and one isotropic phase. This latter is continuous from the water rich region to the alcohol rich region [5, 6]. The main consequence of this continuity is the existence of a critical point P_c^A. At 25 °C, this point occurs for a water over SDS ratio $X = 6.6$. By addition of dodecane to the ternary mixture, a line of critical points P_c^1 develops between P_c^A

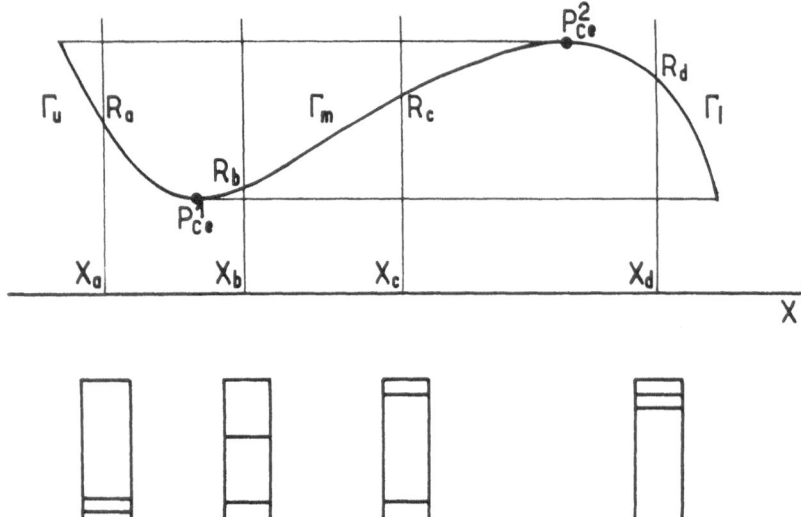

Fig. 3. Three-phase volume issued from two critical end points P^1_{ce} and P^2_{ce}. Phase volumes of three-phase equilibria observed close to the connection points R_a, R_b, R_c and R_d in the sections X_a to X_d

and a critical end point P^B_c. P^B_c is found in the oil rich region at $X = 0.95$ [6]. In addition to the microemulsion phase L, two mesomorphic phases (hexagonal and lamellar phases) are identified. All these phases appear as an extension of those found for the ternary mixture W-P-S. In certain regions of the phase diagram the microemulsions present special characteristics; for example, some oil rich mixtures are flow birefringent. The one-phase regions are separated by complex multiphase regions where two, three and even four phases coexist. All the equilibria observed are different from the so-called Winsor I, II or III equilibria. A complete description of the phase diagram of the system I has been proposed. In this description the considerable complexity of the multiphase region appears to be generated by the occurrence of several lines of critical points and one indifferent state. One of the critical line, the line P^1_c as well as two critical end points have been

located. P^1_c is located, as previously mentioned, along the coexistence surface of the microemulsion phase and extends from $X = 0.95$ to $X = 6.6$ (Fig. 4).

Addition of salt to the ternary system W-P-S modifies the extent of the phases and produces several multiphase regions [5]. By the addition of salt, the critical point P^A_c shifts towards the brine corner. These critical points form a line P^{1-s}_c which terminates at the salinity of 3.5 g/l NaCl at a critical end point P^c_{ce}. At this point a three-phase region emerges. At higher salinity, it disappears by an indifferent state. A second line of critical points P^2_c exists in the water rich region for salinities above 8.1 g/l NaCl. This second line gives rise to a second three-phase region. The critical lines observed

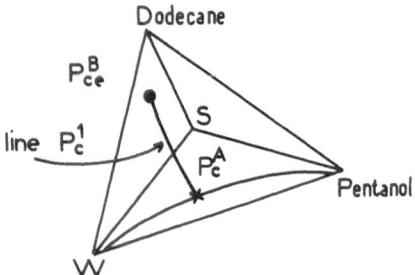

Fig. 4. Line of critical point P^1_c for the quaternary mixture: water-dodecane-pentanol-SDS

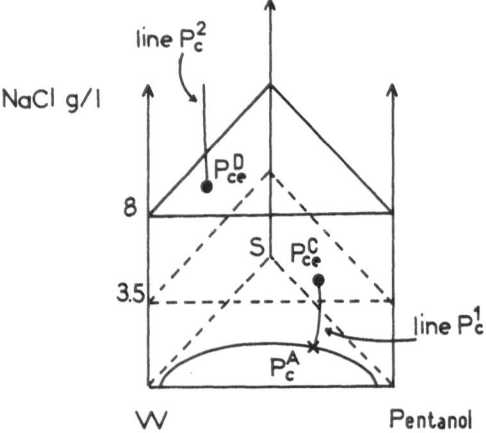

Fig. 5. Representation in a prism of the lines of critical points P^{1s}_c and P^2_c for the quaternary mixture water-NaCl-pentanol-SDS

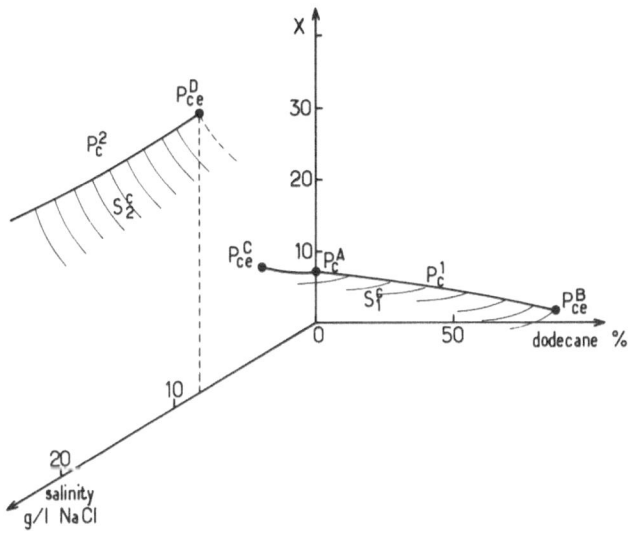

Fig. 6. Representation in the space X-dodecane-salinity of the critical lines P_c^1 and P_c^2 observed for the quaternary mixtures I and II. P_{ce}^B, P_{ce}^C and P_{ce}^D are critical end points

in the phase diagram of the W-NaCl-P-S system are schematically represented in Fig. 5 in a prism with salinity as ordinate and the ternary system W-P-S as base.

The phase diagrams of the two other quaternary mixtures W-D-S-NaCl and W-D-P-NaCl have not been determined. For these mixtures, one does not expect critical lines at 25 °C.

In the phase diagram of the quinary mixture W-NaCl-D-P-S one thus expects at least two surfaces of critical points, one S_c^1 issued from the critical line P_c^1 and a second S_c^2 issued from the critical line P_c^2 (Fig. 6).

5. Phase diagram of the quinary mixture W-NaCl-D-P-S

As shown in Fig. 7, addition of salt to the quaternary mixture I strongly changes the phase diagram. Salt reduces the extent of the microemulsion region and produces the appearance of a Winsor III three-phase region. This region exists as salinity is higher than 4 g/l NaCl. We have explored in detail the tetrahedrons corresponding to the salinities 5; 6; 7; 8.5 and 20 g/l NaCl. Using the procedure presented above we have searched for the points of connection between the one-and three-phase regions. For a given salinity s, as X increases, one expects it to successively describe the curves Γ_u^s, Γ_m^s, Γ_l^s. This is not always the case and depends on the tetrahedron T_s i. e. on the salinity. Three ranges of salinity can be distinguished.

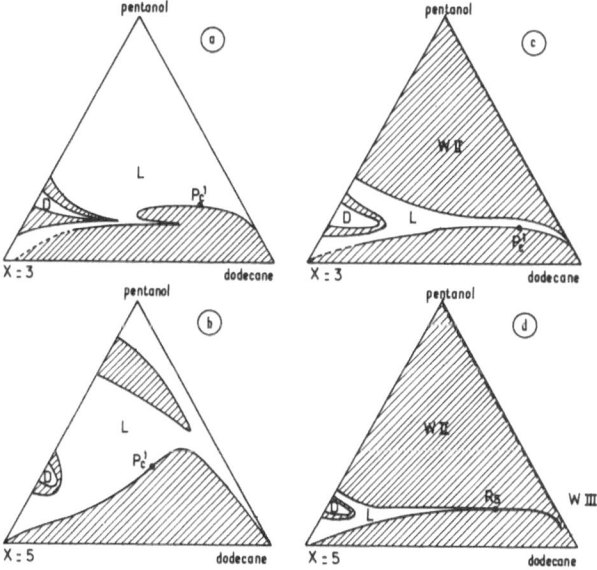

Fig. 7. (a, b) Sections X = 3 and X = 5 of the phase diagram of the water-dodecane-pentanol-SDS system. (c, d) Same sections through the tetrahedron corresponding to the salinity 20 g/l NaCl. (L) an isotropic phase; (D) a lamellar phase. (W II) and (W III) represent Winsor II and Winsor III equilibria

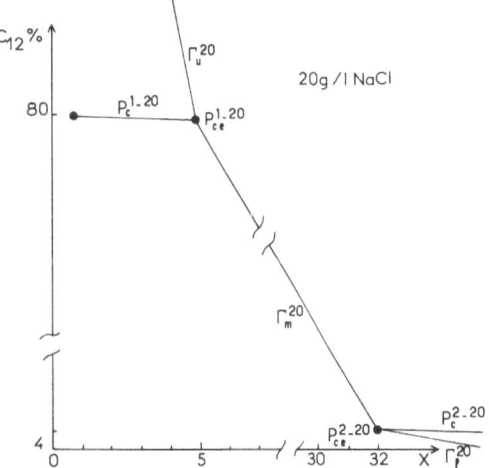

Fig. 8. Projections on the plane X-pentanol-dodecane of the three curves Γ_u^{20}, Γ_m^{20} and Γ_l^{20} obtained at the salinity 20 g/l NaCl. The points P_{ce}^{1-20} and P_{ce}^{2-20} are critical end points. At P_{ce}^{1-20} the middle and upper phases become identical; at P_{ce}^{2-20} the middle and lower phases become identical

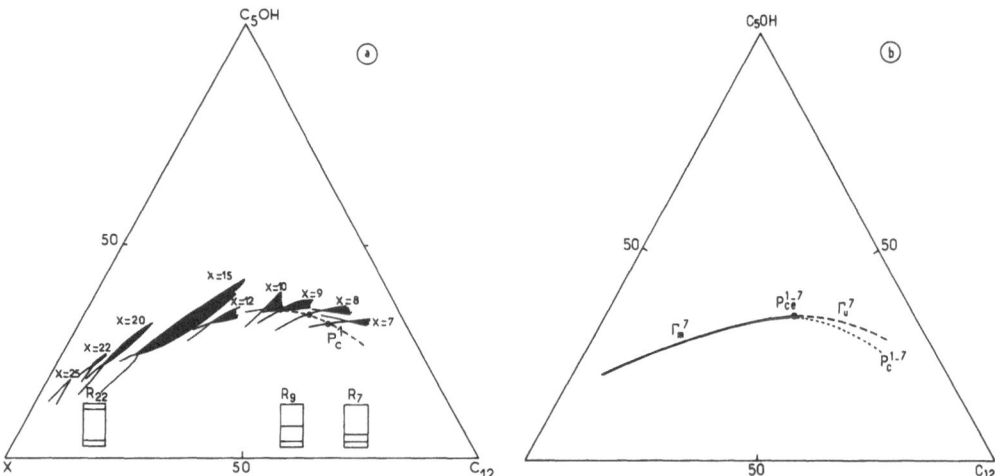

Fig. 9. (a) Sections X-pentanol-dodecane through the tetrahedron defined by the salinity 7 g/l NaCl. Positions of the three-phase regions (black) and one-phase regions (clear) are shown in these sections. Phase volumes near the points R_7, R_9 and R_{22} where the one-phase region touches the three-phase region in the sections $X = 7$, 9 and 22. (b) Curves Γ_u^7 and Γ_m^7 deduced from the diagrams shown in Fig. 9a

5.1 Salinities above 8 g/l NaCl

At the salinities 8.5 and 20 g/l NaCl as X increases one observes successively the three curves Γ_u^s, Γ_m^s and Γ_l^s. They form a single twisted curve. A projection of these curves on the plane alcohol dodecane-X is shown in Fig. 8. The curve Γ_m^s is connected at each of its ends to the Γ_u^s and Γ_l^s curves. The upper and middle phases become identical at the critical end point P_{ce}^{1-s} while the middle and lower phases become identical at the critical end point P_{ce}^{2-s}. P_{ce}^{1-s} and P_{ce}^{2-s} are respectively the ends of the two lines of critical points P_c^{1-s} and P_c^{2-s}. These two lines belong to the critical surfaces S_c^1 and S_c^2. Then, at the salinities above 8 g/l NaCl, the region Winsor III evolves between two lines of critical end points L_{ce}^1 and L_{ce}^2. L_{ce}^1 is located in the oil rich region and L_{ce}^2 in the brine region.

5.2 Salinities comprised between 6 and 8 g/l NaCl

In the tetrahedrons corresponding to the salinities 7 and 6 g/l NaCl, only the curves Γ_u^s and Γ_m^s are found as X increases (Fig. 9–10). Likewise, the higher salinities, the curves Γ_u^s and Γ_m^s are connected. The middle and upper phases become identical at the critical end points P_{ce}^{1-6} and P_{ce}^{1-7}.

5.3 Salinities below 6 g/l NaCl

In the tetrahedron 5 g/l NaCl, only the Γ_u^5 curve exists (Fig. 11). The three-phase region disappears before having reached the critical end point P_{ce}^{1-5}. At the salinity 5 g/l NaCl, the line of critical points P_c^{1-5} goes continuously from the oil-rich part of the diagram to the water-rich one.

6. Discussion

At 25 °C, the three-phase region Winsor III of the system water-NaCl-dodecane-pentanol-SDS appears by two lines of critical end points L_{ce}^1 and L_{ce}^2, only as salinity goes above 8 g/l NaCl. At lower salinity, the mode of appearance of the three-phase region is different. In the range of salinity between 5 and 8 g/l NaCl, the extent of the three-phase body evolves similarly as X increases. The three-phase region becomes smaller and thinner and finally disappears for a given value X_d of X; this value X_d depends on salinity. In the sections $X > X_d$, the one-phase region is surrounded by two-phase regions.

Analysis of phase compositions of several three-phase equilibria located in the tetrahedrons corresponding to the salinities 5, 6 and 7 g/l NaCl confirm the above results [8]. Indeed, for all the equilibria investigated, one founds that the salinity in the lower

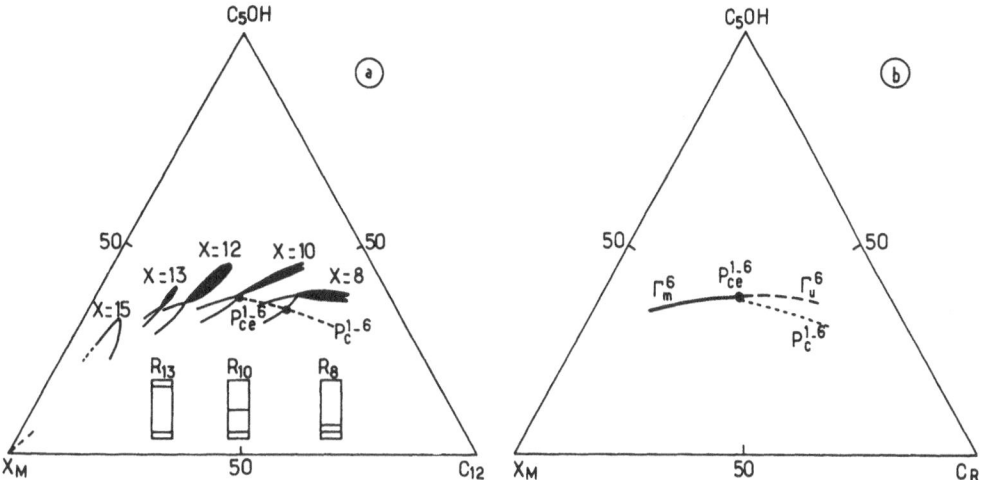

Fig. 10. (a) Sections X-pentanol-dodecane through the tetrahedron, defined by the salinity 6 g/l NaCl. Positions of the three-phase regions (black) and one-phase regions (clear) are shown in these sections. Phase volumes near the points R_8, R_{10} and R_{13} where the one-phase region touches the three-phase regions in the sections X = 8, 10 and 13. (b) Curves Γ_u^6 and Γ_m^6 deduced from the diagrams in Fig. 10a

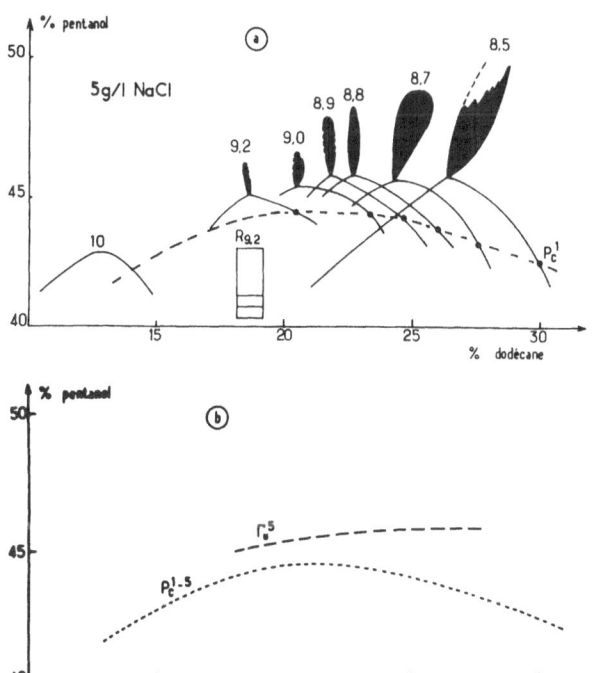

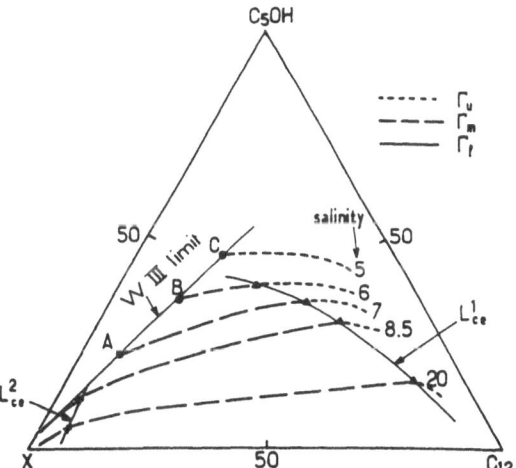

phase is always higher than 8 g/l NaCl. This result is in agreement with the fact that we do not observe the Γ_l curve in the tetrahedron corresponding to the salinities 7, 6 and 5 g/l NaCl.

In Fig. 12, we have plotted the projections on the plane X-pentanol-dodecane of the curves obtained at the salinities 5, 6, 7, 8,5 and 20 g/l NaCl. This figure,

Fig. 11. (a) Sections X-pentanol-dodecane through the tetrahedron defined by the salinity 5 g/l NaCl. Positions and extensions of the three-phase regions (black) and one-phase regions (clear) in the sections indicated on the figure. Phase volumes near the point of connection R = 9.2. (b) Curve Γ_u^5 deduced from the diagrams in (a)

Fig. 12. Projections on the plane X-pentanol-dodecane of the curves Γ_u^s, Γ_m^s and Γ_l^s observed at the salinities 5, 6, 7, 8.5 and 20 g/l NaCl. The symbols (▲, +) represent respectively the critical end points P_{ce}^1 and P_{ce}^2, and (●) corresponds to the limit of existence of the three-phase equilibria Winsor III

which represents the projections of the three surfaces S_u, S_m and S_l discussed in IIb, gives the limits of existence of the three-phase volume Winsor III. It clearly appears that at low salinity, the three surfaces S_u, S_m and S_l are limited by a straight line. The points where the three-phase region disappears are all along a straight line A B C. This result is an indication that at low salinity the three-phase region Winsor III of the quinary mixture water-NaCl-dodecane-pentanol-SDS arises from an indifferent state.

Let us mention that for all the three-phase equilibria close to the straight line A B C, the three phases strongly scatter light. This could indicate that the system is close to a tricritical point. Such point would result from the connection of both two lines of critical end points L_{ce}^1 and L_{ce}^2.

Acknowledgements

We are indebted to Mrs Maugey and M. Babagbeto for their assistance with the experiments. This work was supported by Total — Compagnie Française des Pétroles.

References

1. Bellocq AM, Bourbon B, Lemanceau B, Fourche G (1982) J Colloid Interface Sci 89:427
2. Kahlweit M (1982) J Colloid Interface Sci 90:197
3. Kilpatrick PK, Scriven LE, Davis HT (1982) SPEJ 11209
4. Kahlweit M, Strey R (1986) J Phys Chem 90:5239
5. Guérin G, Bellocq AM (1988) J Phys Chem, 92:2550
6. Bellocq AM,, Roux D (1987) In: Friberg SE, Bothorel P (eds) Microemulsions: structure and dynamics. CRC Press, pp 33–78
7. Prigogine I, Defay R (1950) In: Thermodynamique chimique. Liège
8. Gazeau D (1988) PhD thesis, University Bordeaux I

Received January 7, 1988;
accepted January 13, 1988

Authors' address:

A. M. Bellocq
Centre de Recherche Paul Pascal (CNRS) and
GRECO "Microemulsions"
Domaine Universitaire
F-33405 Talence Cedex, France

Progress in Colloid & Polymer Science
Progr Colloid Polym Sci 266:211–215 (1988)

PCPS 035

Structure and dynamics of microemulsion-based gels

A. M. Howe[1]), A. Katsikides[2]), B. H. Robinson[2]), A. V. Chadwick[3], and A. Al-Mudaris[3])

[1]) Institute of Food Research, Norwich Laboratory, Colney Lane, Norwich, U.K.
[2]) School of Chemical Sciences, University of East Anglia, Norwich, U.K.
[3]) University Chemical Laboratory, University of Kent, Canterbury, U.K.

Abstract: The structure and dynamic properties of microemulsion based gels formed by water/AOT/heptane in the presence of gelatin have been investigated. The gels are optically transparent, single phase systems. Measurements of electrical conductivity and tracer diffusion (T_2O and $^{22}NaCl$) have been made. Transmission electron micrographs of the freeze-fractured gels are also presented. The results indicate that the dynamic properties of the system are a sensitive function of both temperature and composition. With suitable formulations the conductivity can change over several orders of magnitude for a narrow temperature range. Results from the tracer diffusion studies parallel those from the conductivity data. In a stiff gel (high gelatin content), electron microscope studies indicate a very different structure from that in weakly-gelled systems.

Key words: Dynamics, structure, microemulsion-based gels, tracer diffusion, conductivity.

Introduction

Water-in-oil (w/o) microemulsions have been extensively studied over recent years using a range of physical techniques [4]. Of particular interest are systems stabilised by the anionic surfactant Aerosol-OT (AOT), since the L_2 domain formed in water-oil mixtures is extensive and is characterised by essentially droplet-like structures of water dispersed in a continuous oil phase. It has been fashionable to incorporate macromolecular solubilisates e.g. enzymes, metal particles, polymers in these dispersions [4]. In general hydrophilic solubilisates studied are incorporated inside the water droplets with little change in the overall nature of the dispersion. However, over the last year Haering and Luisi [3] and Quellet and Eicke [6] have described the formation of microemulsion-based gels (MBGs) on addition of gelatin to a w/o microemulsion. Such gel-based systems may well have novel properties e.g. as membranes in separation processes. It has also already been suggested that enzymes may be immobilised in such media with retention of activity and stability [3]. Such a system offers exciting possibi-

lities for biosynthesis, especially for enzyme reversal procedures [5, 7].

In this paper we describe some of our results from a preliminary investigation into structural and dynamic aspects of these MBGs. This study forms part of a systematic investigation now underway in our laboratories. The following techniques have been used in an attempt to characterise these rather complex and unusual systems: (i) electrical conductivity, (ii) tracer diffusion and (iii) transmission electron microscopy.

Preparation of MBGs

We follow closely the methods described in the pioneering papers of Luisi [3] and Eicke [6]. However, variations in the experimental protocol did not significantly affect the final properties of the gels. Gelatin was obtained from Fluka (Bloom 300) and was used without further purification. For the work described in this paper the following procedure was adopted: a dispersion of gelatin in water was first prepared by adding the required amounts of each substituent. This was then heated to approximately 50 °C and thermostatted for one hour at that temperature. A clear solution of AOT in heptane (also at 50 °C) was then added and the mixture was vigorously shaken. The viscous, optically clear dispersion thus pro-

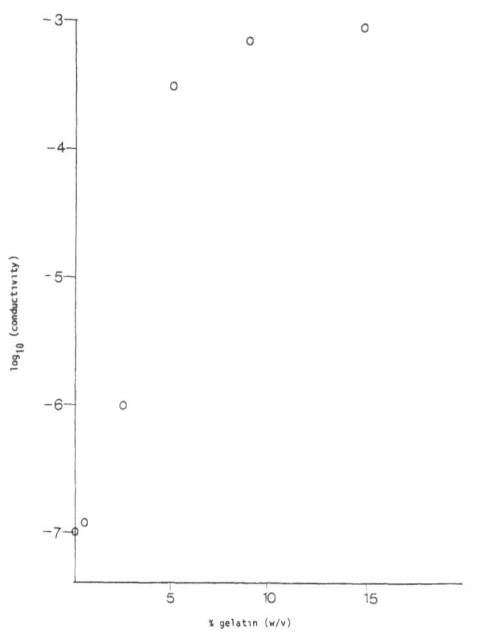

Fig. 1. Plot of $\log_{10}$ (conductivity) vs. gelatin concentration at $[AOT] = 0.2$ mol dm^{-3} and 24% v/v water

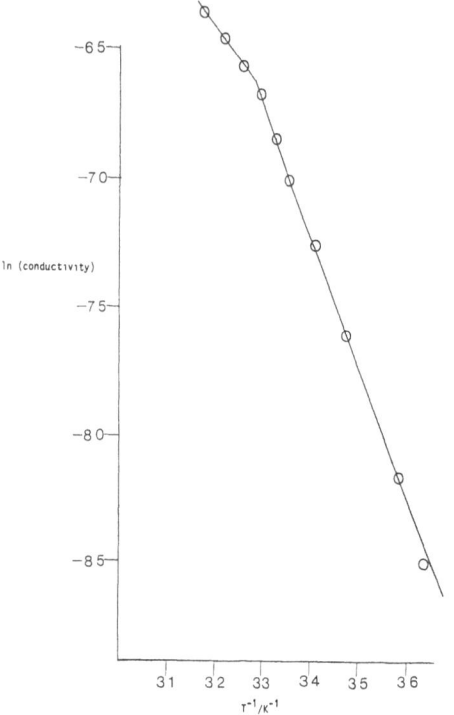

Fig. 2. Plot of ln (conductivity vs T^{-1} for the system $[AOT] = 0.4$ mol dm^{-3}. $[H_2O] = 40$% v/v, gelatin = 14% w/v

duced was allowed to cool slowly to room temperature in air. (The rate of cooling does not seem to significantly affect the properties of the final gel). Measurements of electrical conductivity were taken after allowing a few hours for ageing of the sample. In general the conductivity was reproducible to better than ± 10%, and this was used as the prime criterion for consistency of our sample preparation. No significant long-time ageing effect could be detected by this method. The gels obtained under our concentration conditions are invariably one-phase optically transparent systems.

Results and discussion

1. Electrical conductivity

Figure 1 shows that the electrical conductivity increases dramatically to a limiting value as the gelatin concentration is increased to 15% w/v at fixed concentrations of AOT (0.2 mol dm^{-3} overall concentration) and water (24% v/v). This is a predictable result and is consistent with the formation of extensive gelatin-water-AOT networks in the system.

The effect of temperature on the conductivity is also of interest. The conductivity increases with temperature and for temperatures less than 32 °C the results

can be adequately expressed by means of an Arrhenius plot with a measured E_a of approximately 60 kJ mol^{-1}. However, at 32 °C there is a discontinuity in the Arrhenius slope, such that the E_a associated with conductivity is reduced. This phenomenon appears to be general, as it is observed for many of our AOT/water/oil systems in the presence of gelatin and is thought to be associated with a structural change in the gelatin itself, as has been found previously in aqueous gelatin solutions at around the same temperature. A typical plot is shown in Fig. 2.

For some compositions there is abrupt change in conductivity over a relatively narrow temperature range (high activation energy) which suggests a possible application of these systems in the form of thermal switches.

Figure 3 shows that the activation energy associated with low-temperature conductivity decreases sharply as the gelatin concentration increases. A particularly abrupt change is observed at ~ 5% gelatin for the system 0.2 mol dm^{-3} AT, 24% v/v water. At the same gelatin concentration there is an abrupt increase in the conductivity as expected. It would appear from these

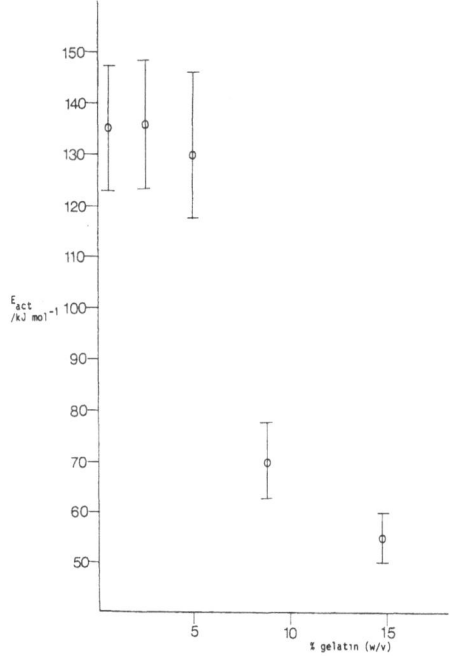

Fig. 3. Plot of E_{act} (conductivity) vs [gelatin] for the system [AOT] = 0.2 mol dm^{-3}, 24% v/v H$_2$O

results that a critical amount of gelatin is required in order that conducting networks may form.

2. Tracer diffusion

The method of Barr and El Messiery [1], subsequently modified by Freer and Sherwood [2], was used to determine the diffusion coefficient of tritium in

the form of T$_2$O in our rigid gels. The gels were degassed and allowed to set in a glass tube of 1 cm diameter. A filter paper (1 cm in diameter) saturated with tritiated water was placed on the gel; the apparatus was then sealed and thermostatted at 25 $\pm$ 0.5 °C. After 72 h penetration of T$_2$O through the gel to a depth of a few cm had occurred. The gel was then cut into sections of thickness $\sim$ 2 mm and placed in a solution of toluene scintillant. The samples were then sonicated until the gel section had dissolved in the scintillation medium. Samples were then placed on a scintillation counter and the count rate determined. The diffusion coefficient (D) of T$_2$O through the gel network was determined from the slope of plots of log (count rate) against the square of the penetration distance.

$$\text{Slope} = - (2.303 \times 4 \times D \times t)^{-1}$$

The value of D in a MBG of 40% water and 14% gelatin increases proportionally as the concentration of AOT is decreased (Fig. 4) from 2 $\times$ 10^{-6} cm^2 s^{-1} at an AOT concentration of 0.4 mol dm^{-3} AOT to 8 $\times$ 10^{-6} cm^2 s^{-1} at 0.1 mol dm^{-3} AOT. At low AOT concentrations, D approaches 2 $\times$ 10^{-5} cm^2 s^{-1}, the value obtained in bulk aqueous solution. Similarly D increases (at constant AOT and gelatin concentrations) on addition of water to the MBG. Similar diffusion studies were carried out with ^{22}Na in the form of ^{22}NaCl. The diffusion coefficient of ^{22}Na was found to be approximately a factor of three lower than that of ^{3}H under the same range of conditions.

It is interesting that the general conductivity behaviour of the MBG systems (Fig. 5) parallels that of the

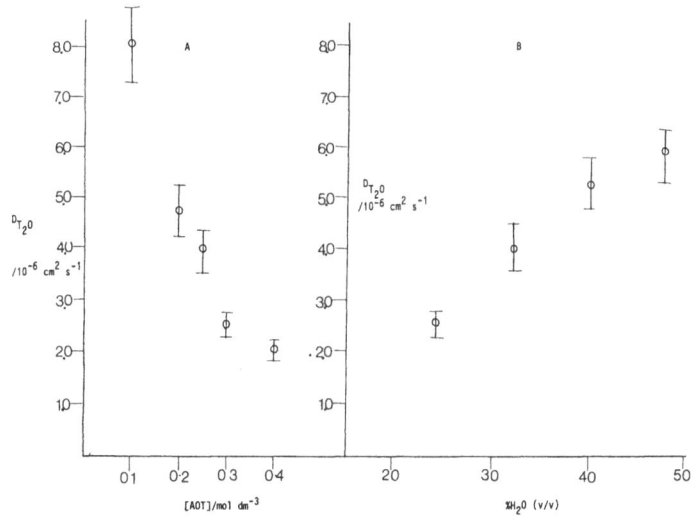

Fig. 4. (A) Plot of DT$_2$O vs (AOT) for a MBG of 40% water and 14% gelatin. Temp = 20 °C. (B) Plot of DT$_2$O vs % H$_2$O v/v [AOT] = 0.2 mol dm^{-3}, [gelatin] = 14% w/v. Temp = 25 °C

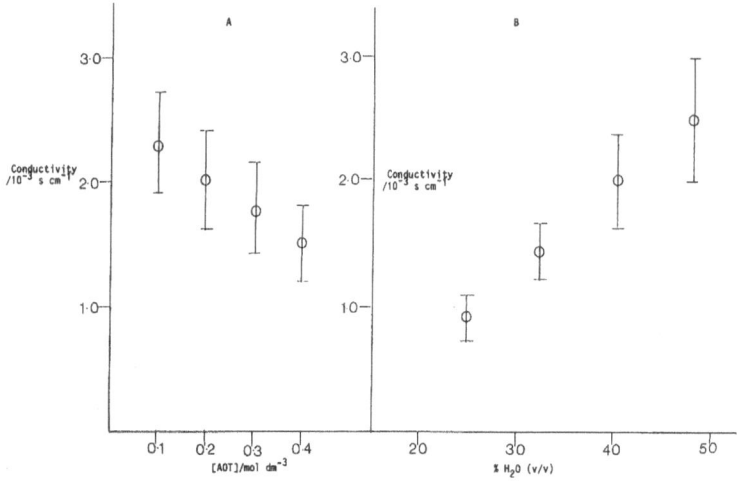

Fig. 5. (A) Plot of conductivity vs [AOT] at 40 % H_2O, 14 % gelatin, Temperature = 25 °C. (B) Plot of conductivity vs % H_2O (v/v) at [AOT] = 0.2 mol dm^{-3}. Temp = 25 °C

water and sodium diffusion. The sodium ions are likely to be the main charge carriers in the conductivity process, and they presumably travel through the water-continuous networks in the MBGs as this correlation is to be expected.

3. Electron microscopy

Transmission electron microscopy studies of the MBGs were carried out in collaboration with Brian Wells of the John Innes Institute, Norwich. Samples were frozen on contact with a copper block cooled in liquid helium, then fractured, etched and coated (with platinum at 26°, for contrast, and carbon at 80°, as a support) in a Ballzers 400D freeze etching unit. The replica was removed with acid and examined in a JEOL 1200EX transmission electron microscope at 80 k eV.

Micrographs of a weakly gelled gelatin-containing microemulsion composition (Fig. 6A) and a strongly-gelled microemulsion (0.2 mol dm^{-3} AOT, 40 % v/v H_2O, 14 % w/v gelatin) (Fig. 6B) are shown. The structures in Figs. 6A and 6B are very different, with no "droplet like" structure evident for the strongly-gelled system.

Conclusions

It is of particular interest that these MBGs have reproducible physical properties. From our measurements and those reported previously [3, 6], it is clear that the onset of characteristic gel behaviour occurs

when > 5 % gelatin is present (dependent on the water and surfactant content of the system). In our systems we believe that gelatin is strongly interacting with itself

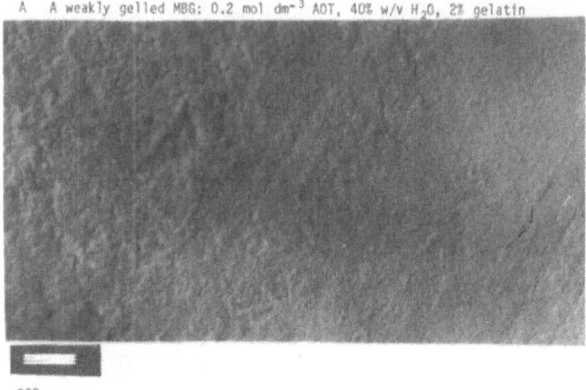

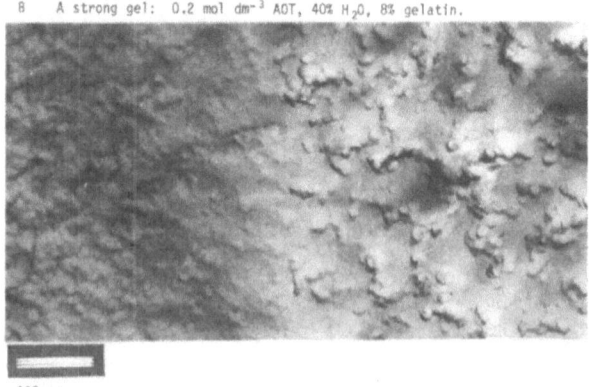

Fig. 6. Freeze fracture electron micrographs of (a) a weakly gelled MBG: 0.2 mol dm^{-3} AOT, 40 % w/v H_2O, 2 % gelatin, and (b) a strong gel: 0.2 mol dm^{-3} AOT, 40 % H_2O, 8 % gelatin

to form networks with attendant hydration of the chains and stabilisation by adsorption of AOT at the oil-water interface. At this stage the sizes of the water channels are not known. It should however be noted that any residual water and AOT not involved in the formation of the networks could coexist in the form of reversed micelles or water droplets, the balance of water involvement in the types of superaggregate (networks vs droplets) will be sensitively dependent on the composition of the system chosen. If this hypothesis were valid, we would expect a rapid dynamic exchange of AOT and water between the two structural forms which is consistent with the reproducibility of preparation. Evidence for the two coexisting structures should be forthcoming from further investigations using scattering methods (e.g. neutron, X-ray and light).

Acknowledgements

We thank Brian Wells of the John Innes Institute, Norwich, for performing the electron microscopy experiments. We also thank the Science and Engineering Research Council and the Agricultural and Food Research Council for financial support for this work. The work at Norwich was carried out within the auspices of the Biocolloid Centre.

References

1. Barr LW, El Messiery MAMI (1979) Nature 281:553–561
2. Freer R, Sherwood JN (1980) J Chem Soc Faraday Trans 1, 76:1021–1037
3. Haering G, Luisi PL (1986) J Phys Chem 90:5892–5895
4. Luisi PL, Magid L-J (1986) CRC Crit Rev Biochem 20:409–474
5. Luisi PL (1985) Ang Chemie, Int Ed Engl 24:4339–450
6. Quellet C, Eicke HF (1986) Chimia 40:233–238
7. Martinek K, Berezin IV, Khmeinski YL, Klyachko NL, Levashov AV (1987) Biocatalysis 1, 1:9–16

Received January 8, 1988;
accepted January 15, 1988

Authors' address:

A. M. Howe
Institute of Food Research
Norwich Laboratory
Colney Lane
Norwich NR4 7UA, U.K.

Neutron scattering from ganglioside micelles

L. Cantú[2]), M. Corti[1]), V. Degiorgio[1]), R. Piazza[1]), and A. Rennie[3])

[1]) Dipartimento di Elettronica-Sezione di Fisica Applicata, Università di Pavia, Pavia, Italy
[2]) Dipartimento di Chimica e Biochimica Medica, Università di Milano, Milano, Italy
[3]) Institut Laue-Langevin, Grenoble, France

Abstract: Dilute solutions of ionic micelles formed by a biological glycolipid (the ganglio-side GM1) have been investigated at various ionic strengths by small angle neutron scattering. The size and shape of the GM1 micelle is not appreciably affected by added salt concentration in the range 0–100 mM NaCl. Despite the fact that the solute concentration is below 0.5 %, with no added salt the structure factor shows a pronounced peak at a value k_m which roughly corresponds to $2\pi/d$, where d is the interparticle distance. From the height of peak we derive the electric charge Z of the micelle by fitting the data with a theoretical calculation which uses a screened Coulomb potential for the intermicellar interaction, and the hypernetted chain approximation for the calculation of the radial distribution function. The experimental k_m is somewhat larger than the theoretical one. This effect could be connected with the nonsphericity of GM1 micelles.

Key words: Ganglioside micelles in water, neutron scattering, influence of NaCl, interparticle distance, electric charge of micelles, micellar interactions, sphericity of micelles.

Introduction

Gangliosides are anionic glycolipids of biological origin which form micelles in aqueous solution. The aggregation properties of various gangliosides have been investigated in the last few years by static and dynamic light scattering [1, 2]. It is interesting to note that ganglioside micelles are rather large: molecular weights are in the range 3×10^5 to 5×10^5 daltons, and hydrodynamic radii are between 5.5 and 6 nm. Recently, we have performed light scattering measurements on dilute solutions of the ganglioside GM1 (see Ref. [3] for the nomenclature) at various ionic strengths ranging from 0 to 100 mM of added NaCl [4,5]. When 100 mM NaCl is added to the solution, the electric charge of the micelle is well screened by the small ions, so that the solution behaves essentially as a system of noninteracting micelles. As a consequence the scattered intensity is proportional to the aggregation number m of the micelle, and the decay time of the intensity correlation function is proportional to the hydrodynamic radius of the micelle. When no salt is added to the solution, the effect of electrostatic intermicellar interactions is so strong that the scattered intensity is almost one order of magnitude lower than the value found with high ionic strength. A fit of the data with theory allows us to calculate the fractional electric charge of the ganglioside micelle, $\alpha = Z/m$, which was found to be 0.16.

We present in this paper a small-angle neutron scattering study [6] of the system previously studied by light scattering. The first aim of the experiment was to check in a direct way whether the micelle aggregation number m depends on the ionic strength I of the solution: usually, m depends on I for ionic micelles, whereas the light scattering data suggested that m is independent of I in the case of gangliosides. The second aim was to measure the full angular dependence of the scattered intensity, and to derive the structure factor. When the repulsion is sufficiently long-range, we find that the micelles tend to be localized around the maximum possible mutual distance. As a consequence, the structure factor shows a peak at a position which is, approximately, $k_m \approx 2\pi/d$, where d is the average intermicellar distance.

Experimental results

Gangliosides are anionic glycolipids occurring in neuronal plasma membranes. The material used in the present investigation was prepared as a sodium salt by Tettamanti and coworkers. Detailed information about nomenclature, properties and preparation procedure can be found in Ref. [3]. Gangliosides are known to form micelles above a critical micelle concentration which is smaller than 10^{-8} M. The results reported in this paper concern the ganglioside GM1 which contains one sialic-acid residue. The solvent was D_2O with added NaCl in the concentration range from 0 to 100 mM. The temperature was 25 °C.

The neutron scattering measurements were performed at the Institut Laue-Langevin (Grenoble, France) on the instrument D17. The data were taken at an incident wavelength of 1.0 nm (wavelength spread $\Delta\lambda/\lambda$ about 10 %) with a detector to sample distance of 2.88 m. In order to increase the range of scattering vectors k, the multidetector was placed off-axis at an angle of 5°. The quartz sample cells had a pathlength of 2 mm.

We report in Figs. 1–3 the coherently scattered intensity $I(k)$ at various GM1 and salt concentrations. The background scattering from the solvent and sample cells was subtracted and the data normalized to absolute units with data from a 1 mm H_2O sample. The results of Figs. 1 and 2 refer to a 2.7 mM solution of GM1 at 4 different salt concentrations. The region $k \leq 6 \times 10^{-3}$ Å^{-1} is not accessible because it is masked by the beam-stop. When no salt is added to the solution, a marked peak appears in the angular dependence of the scattered intensity. The peak position is at $k_m = 1.5 \times 10^{-2}$ Å^{-1}. When salt is added, the peak becomes progressively smaller and broader until it disappears at salt concentrations above 10 mM. The peak position is weakly dependent on the salt concentration, drifting from $k = 1.5 \times 10^{-2}$ to 1.7×10^{-2} Å^{-1} as the salt concentration increases from 0 to 5.3 mM. In the region $k \geq 3 \times 10^{-2}$ Å^{-1}, the data for different salt concentrations are all perfectly superposable, that is, $I(k)$ becomes independent of the ionic strength of the solution. We have also investigated a 1.35 mM GM1 solution at many concentrations of added salt. Since the trend is identical to that observed at the higher GM1 concentration, we present in Fig. 3 only the two extreme cases: no salt and high salt. The peak is slightly less pronounced and appears at a lower value of k, $k_m = 1.3 \times 10^{-2}$ Å^{-1}. For $k \geq 3 \times 10^{-2}$ Å^{-1}, $I(k)$ is independent of salt concentration, and has a value which is exactly half that obtained at 2.7 mM GM1.

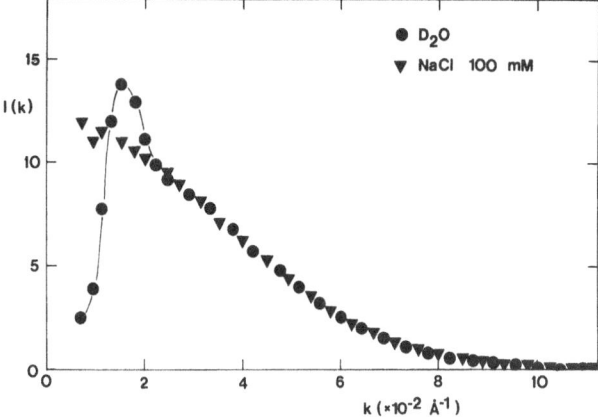

Fig. 1. Intensity of scattered neutrons as a function of k for a 2.7 mM GM1 ganglioside solution in pure D_2O and in 100 mM NaCl D_2O

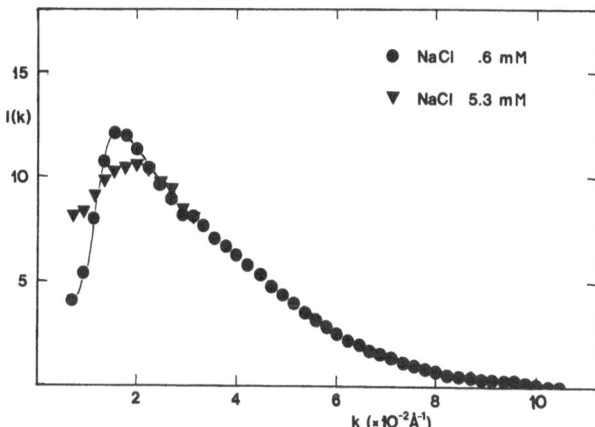

Fig. 2. The same as Fig. 1 for two different concentrations of added salt

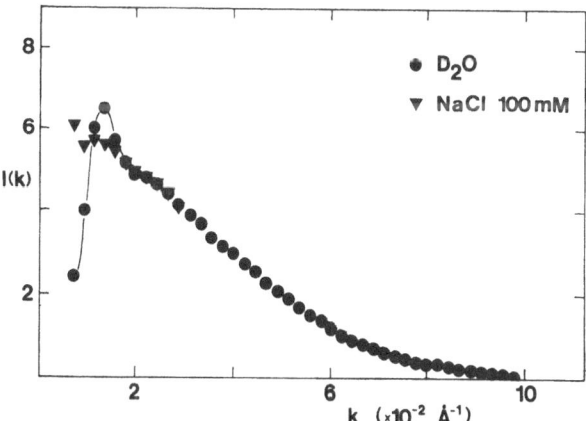

Fig. 3. The same as Fig. 1 for a 1.35 mM GM1 solution

Discussion

The intensity scattered by the micellar solution can be expressed as

$$I(k) = AcMP(k)\, S(k) \qquad (1)$$

where A is an instrumental constant, c is the amphiphile molar concentration, M the micellar molecular weight, $P(k)$ the micelle form factor, and $S(k)$ the structure factor which takes into account the effect of intermicellar interactions. Eq. (1) is valid under a number of assumptions: $I(k)$ should represent the excess scattering over the solvent scattering, c is the total concentration minus the critical micellar concentration, the particle orientation should be decoupled from interparticle interactions (this applies strictly only to spherical particles). The structure factor is related to the radial distribution function $g(r)$ as follows:

$$S(k) = 1 + 4\pi\, c_p k^{-1} \int [g(r) - 1]\, \sin kr\, dr \qquad (2)$$

where $c_p = cN_A/(1000\, m)$ is the micelle concentration (micelles/cm^3), and m is the micelle aggregation number. The function $g(r)$ can be calculated once the pair interaction potential $V(r)$ is known.

In the case of the 2.7 mM GM1 solution, taking $m = 300$ [1, 2], we find $c_p = 5.5 \times 10^{15}$ micelles/cm^3. This gives an average intermicellar distance of $d = 57$ nm. Since d is much larger than the hydrodynamic radius of the micelle, $R_H = 5.9$ nm, we can assume that the only relevant contribution to $V(r)$ comes from electrostatic interactions. Treating the micelles as spheres of radius a and electric charge Z, the screened Coulomb potential is:

$$V(r) = (Ze^2 \exp[-\varkappa(r - 2a)]/[\varepsilon r(1 + \varkappa a)^2] \qquad (3)$$

where ε is the dielectric constant of water, r is the distance between the centers of the two micelles, and $\varkappa$ is the inverse Debye-Hückel screening length expressed as [4]

$$\varkappa = \left[(2e^2/\varepsilon k_B T) \left(\frac{1}{2} Zc/m + c_s \right) \right]^{1/2} \qquad (4)$$

where c_s is the concentration of added NaCl.

Our data can be interpreted in the following way. At $c_s = 100$ mM, the number of small ions is sufficiently large to screen very effectively the Coulomb repulsion, so that $S(k) \approx 1$ over the entire range of scattering vectors and $I(k)$ reflects essentially the form factor of the micelle. When the salt concentration is reduced, repul-

sive interactions become progressively more important and the micelles tend to stay at the maximum possible mutual distance. In this state, $S(k)$ becomes very small at low k (the osmotic compressibility is small) and develops a peak at $k_m \approx 2\pi/d$. At zero added salt the situation is reminiscent of dispersions of charged latex spheres which form the so-called colloidal crystals [7]. In our case, the system of particles is still in the liquid state, probably because the particles are smaller and consequently the "freezing" temperature of the system is lower than room temperature [8].

At large values of k, $S(k)$ becomes ≈ 1, independent of the ionic strength of the solution. Therefore, the experimental $I(k)$ reflects, at high k, essentially the behavior of the form factor $P(k)$. The fact that $I(k)$ does not depend on the ionic strength in the range $k \geq 3 \times 10^{-2}$ Å^{-1} clearly indicates that the size and shape of the GM1 micelle are not influenced by the NaCl concentration. This extends and confirms previous light scattering results. Such a finding may at first seem surprising, since it is known for many ionic amphiphiles (for instance, the alkyl sulfates) that the aggregation number depends considerably on the concentration of added salt. In the case of gangliosides, the head group is very bulky. As a consequence, the head-group interactions are probably determined by steric rather than electrostatic contributions. This would explain the lack of sensitivity of m to the NaCl concentration.

Before discussing a quantitative interpretation of the data, it is useful to recall first the light scattering results [2–4]. The GM1 micelle has an aggregation number of 300 ($\pm$ 10%), a hydrodynamic radius of 5.9 ($\pm$ 2%) nm and a fractional electric charge $\alpha = 0.16$ ($\pm$ 10%). As discussed in detail in Ref. [2], the GM1 micelle cannot be spherical because the measured hydrodynamic radius is much larger than the length of the monomer. We simply recall here that for the purpose of interpreting scattering data the micelle can be modelled as an oblate ellipsoid with an axial ratio ≈ 2.

In order to calculate the form factor seen in the neutron scattering experiment, the micellar volume should be divided into two parts, the hydrophobic core which is filled by the hydrophobic tails, and the hydrophilic layer which contains all the hydrophilic headgroups and solvent [9]. Whereas the core has a uniform scattering-amplitude density, the hydrophilic layer presents a density which depends on the radius, going from the value typical of saccharide groups at the inner surface to the value typical of heavy water at the outer surface. We have, however, simplified the calculation by considering a uniform "dry" micelle having

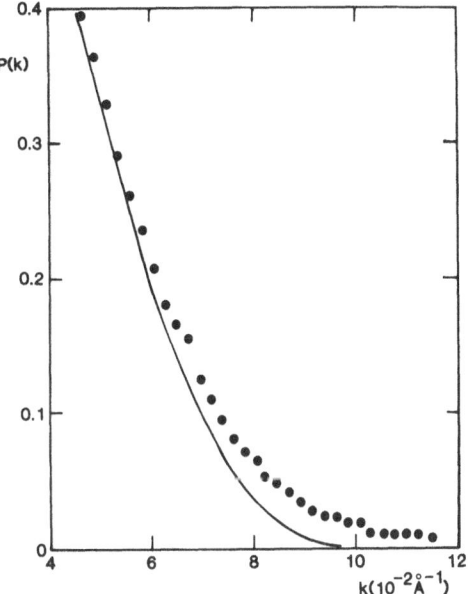

Fig. 4. Fit of the experimental $P(k)$ to the spherical form factor

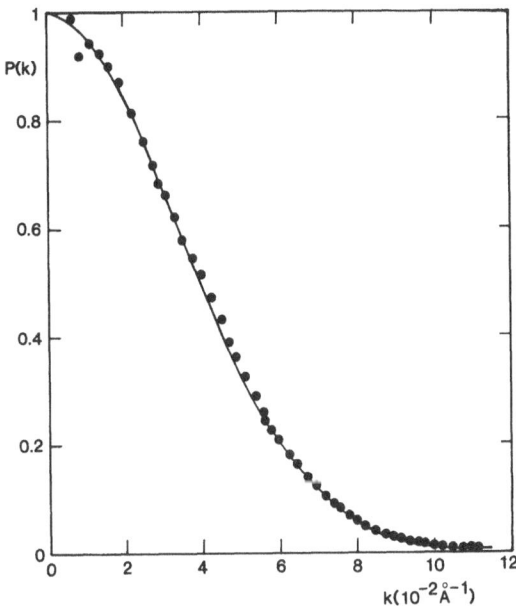

Fig. 5. Fit of the experimental $P(k)$ to the form factor of an oblate ellipsoid with axial ratio 0.6

the scattering-amplitude density of the GM1 monomer. We assume that the scattered intensity $I(k)$ measured with 100 mM NaCl has the same k-dependence as the micelle form factor $P(k)$. We have fitted the experimental $P(k)$ with the form factor calculated for a sphere. The fit is good in the range $0 < k < 5 \times 10^{-2}$ Å^{-1} but a systematic deviation is found at larger values of k, as shown in Fig. 4. The best-fit radius of the sphere is 4.5 nm which corresponds, considering a monomer-volume of 2.05 nm^3, to an aggregation number of 185. The fit is considerably improved if we consider an ellipsoid characterized by a semiaxis ratio $z = a/b$, where a, b, b are the axes of the ellipsoid. Clearly, $z = 1$ corresponds to a sphere, $z \leq 1$ to an oblate ellipsoid and $z \geq 1$ to a prolate ellipsoid. The fit shown in Fig. 5 corresponds to $z = 0.6$. The aggregation number derived from the fit with the oblate ellipsoidal shape is 272 ± 30, and the radius of the sphere having the equivalent volume is $R_e = 5.1$ nm. It should be noted that a good fit could also be obtained with a value of $z \approx 2$ (prolate ellipsoid), but the corresponding aggregation number and radius of the equivalent sphere would be too small to be compatible with the light scattering data.

By taking the ratio $I(k)/P(k)$ we have derived from our data the structure factor $S(k)$ as a function of the NaCl concentration. As an example, we report in Fig. 6 the structure factor obtained with no added salt. We

have calculated the theoretical structure factor $S(k)$ by taking a pair interaction potential consisting of a hard-core repulsion plus a screened Coulomb potential and by using the hypernetted chain approximation for $g(r)$ [5, 10]. We have taken as radius of the particle the hydrodynamic radius measured by light scattering. We have checked that the calculated $S(k)$ does not change if, at fixed particle concentration, we take R_e instead of R_H. The particle density is calculated from the known concentration in grams of amphiphile per liter of solution and from the known aggregation number of the GM1 micelle ($m = 272$). The only free parameter in the fit is the fractional electric charge of the micelle α. The calculated $S(k)$ is smoothed to take into account the wavelength spread of the incident neutron beam. We find that the height of the first peak is well reproduced by choosing $\alpha = 0.15$, which is close to that found in the fit of the light scattering data [3]. However the peak of the theoretical $S(k)$ appears at a value of k which is systematically lower than that of the experimental $S(k)$, as shown in Fig. 6. The peak position is weakly dependent on the particle concentration and essentially independent of the particle radius and charge, so that the uncertainties in the fit parameters do not seem sufficient to explain the discrepancy between theory and experiment. A possible explanation is suggested by a very recent theoretical paper [11] which presents the calculation of $S(k)$ for rod-like particles.

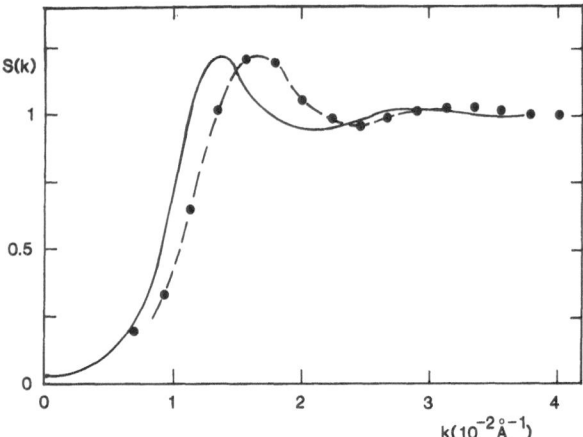

Fig. 6. The experimental and the theoretical $S(k)$. The dashed line is drawn only to guide the eye through the experimental points

The theoretical results show that, at fixed particle concentration, the peak position is displaced to a higher value of k_m when going from spherical to rod-like shape, whereas the height of the peak is only modestly influenced by the particle shape. For an axial ratio around 2, the shift of the peak is comparable to that which we have observed experimentally.

Acknowledgements

We are very grateful to L. Belloni for kindly giving us the computer program used to calculate $S(k)$. We thank G. Tettamanti and coworkers for the preparation of the GM1 ganglioside. This work was supported by grants of the Italian Ministry of Public Education. L. C. thanks Fidiz S.p.A., Abano Terme, Italy, for financial support.

References

1. Corti M, Degiorgio V, Ghidoni R, Sonnino S, Tettamanti G (1980) Chem Phys Lipids 26:225
2. Cantú L, Corti M, Sonnino S, Tettamanti G (1986) Chem Phys Lipids 41:315
3. Tettamanti G, Sonnino S, Ghidoni R, Masserini M, Venerando B (1985) In: Degiorgio V, Corti M (eds) Physics of Amphiphiles, Micelles, Vesicles and Microemulsions. North-Holland, Amsterdam, p 607
4. Cantú L, Corti M, Degiorgio V (1986) Europhys Lett 2:673
5. Cantú L, Corti M, Degiorgio V (1987) Faraday Discuss Chem Soc 83:287
6. Hayter JB (1985) In: Degiorgio V and Corti M (eds) Physics of Amphiphiles, Micelles, Vesicles and Microemulsions. North-Holland, Amsterdam, p 59; Chen S-H (1985) ibid, p 281. (These articles provide a general discussion about the application of neutron scattering to micellar solutions)
7. Pusey PN, Tough RJA (1985) In: Pecora R (ed) Dynamic Light Scattering, Applications of Photon Correlation Spectroscopy. Plenum Press, New York, p 85
8. Kremer K, Robbins MO, Grest GS (1986) Phys Rev Lett 57:2694
9. Zulauf M, Weckström K, Hayter JB, Degiorgio V, Corti M (1985) J Phys Chem 89:3411
10. Goodstein D (1975) States of Matter. Prentice Hall, New York, p 260
11. Schneider J, Karrer D, Dhont JKG, Klein R (1987) J Chem Phys 87:3008

Received December 11, 1987;
accepted December 15, 1987

Authors' address:

V. Degiorgio
Dipartimento di Elettronica-Sezione di Fisica Appplicata
Università di Pavia
I-27100 Pavia, Italy

Progress in Colloid & Polymer Science   Progr Colloid Polym Sci 76:221–223 (1988)

Comparison of sodium dodecylsulfate apparent molar volumes and micellar aggregation numbers*)

Sz. Vass, T. Török[1]), Gy. Jákli, and E. Berecz[1])

Central Research Institute for Physics of the Hungarian Academy of Sciences, Budapest, Hungary
[1]) Technical University of Heavy Industry, Miskolc-Egyetemváros, Hungary

Abstract: The apparent molar volume of sodium dodecylsulfate (SDS) was determined vs. surfactant concentration in normal and 99.85 % heavy water solutions at 25 °C up to 0.8 mol/dm³. The apparent molar volume data in normal water were found to be systematically greater than those in heavy water. The apparent volume of aggregated SDS molecules v was found to be constant; in normal water for SDS concentrations greater than 0.04 mol/dm³, $v = (4.1472 \pm 0.0049) \cdot 10^{-22}$ cm³, indicating no change in the intermolecular forces inside SDS micelles of increasing size.

Key words: Sodium dodecylsulfate, apparent molar volume, micellar aggregation numbers, normal and heavy water, aggregate molecules.

Introduction

The evaluation of scattering experiments carried out in micellar systems usually requires knowledge of the volume fraction of the micellar pseudophase, sometimes in heavy water solutions. This volume fraction is calculated from apparent molar volume data, which, in contrast with the concentration range applied in the scattering experiments, are published around the region of the critical micelle concentration (cmc) and in normal water solutions.

In the present short communication, apparent molar volumes of SDS in normal and in heavy water are reported up to approximately 0.8 mol/dm³ surfactant concentration at 25 °C. Qualitative conclusions for SDS micellar structure are drawn by comparing the apparent volume of aggregated SDS molecules with SDS micellar aggregation numbers obtained from small angle neutron scattering (SANS) experiments by different groups.

Methods and materials

The calculation of apparent molar volumes in isotopic waters is based on the general definition of apparent molar properties [1]: "if

X is the value of a particular property for mixture n_1 moles of one component and n_2 moles of the other, and X_1^0 is the value of the property per mole of pure component (1), then the apparent molar property Φ_2 for component (2) is given by $\Phi_2 = (X - n_1 X_1^0)/n_2$". Let Q_s be the mass of the surfactant of molar mass M_s dissolved in isotopic water of mass Q_w, and let M_w and V_w^0 be the mean molar mass and molar volume of the solvent, respectively. Let ϱ be the density of the solution; then $X = (Q_s + Q_w)/\varrho$, $n_1 = Q_w/M_w$, $X_1^0 = V_w^0$ and $n_2 = Q_s/M_s$. Denoting now the solvent to solute mass ratio Q_w/Q_s by β and the solvent density M_w/V_w^0 by ϱ_w, for the apparent molar volume Φ_s of the surfactant we have the following equation:

$$\Phi_s = M_s \left(\frac{1 + \beta}{\varrho} - \frac{\beta}{\varrho_w} \right). \tag{1}$$

Molar concentration c and mole fraction x are also determined by the solvent to solute mass ratio β, as given in Eqs. (2) and (3):

$$c = \frac{1000}{M_s} \frac{\varrho}{1 + \beta} \tag{2}$$

and

$$x = \frac{M_w}{M_w + \beta M_s}. \tag{3}$$

Surfactants in micellar solutions exist in monomer and in aggregate forms. Due to the different hydrational behavior of monomers and of aggregates, one can assume that their contribu-

---
*) This work has been supported by the OTKA foundation of the Hungarian Academy of Sciences.

tions to the apparent molar volume will be different. An attempt is made to separate these contributions from each other as follows. Let p_m and p_a be the probabilities of finding a surfactant molecule in monomer form and inside one of the aggregates, respectively; let Φ_m and Φ_a be the corresponding contributions per mole to the apparent molar volume Φ_s. Provided that the monomer surfactant concentration remains constant and equals the critical micelle concentration c_0 for any surfactant concentration c, probabilities p_m and p_a are expressed by the concentrations as

$$p_m = c_0/c$$

and

$$p_a = 1 - p_m = 1 - c_0/c. \qquad (4)$$

By assuming that, similarly to the partial molar volumes, the apparent molar volumes of the monomers and those of the aggregates are, at least in relatively dilute solutions, additive, the apparent molar volume Φ_s can be written as a sum of two terms:

$$\Phi_s = p_m \Phi_m + p_a \Phi_a. \qquad (5)$$

In Eq. (5) Φ_m is taken to be equal to the constant value of surfactant molar volumes obtained below cmc [2,3]. By combining Eqs. (4) and (5), the apparent molar volume of aggregates Φ_a is

$$\Phi_a = \frac{c}{c - c_0} \Phi_s - \frac{c_0}{c - c_0} \Phi_m. \qquad (6)$$

Densities necessary to evaluate Eq. (1) were determined in two different ways. Solution densities ϱ were measured by an Anton Paar 602 type vibrating tube densitometer equipped with a Haake F3 ultrathermostate. Solution temperature was measured by a high precision Knauer thermometer with sensitivity of 0.001 K and with a calibration precision of 0.05 K. The mean value of solution temperatures was found to be 24.993 °C and the mean value of the overall temperature fluctuations was less than 0.003 K; however, temperature was kept constant within ± 0.001 K during each density measurement. Solvent densities ϱ_w were calculated from Kell's formula [4] and were corrected for temperature and pressure fluctuations.

The densitometer was calibrated with triple-distilled normal water and with double-distilled heavy water; the heavy water was assumed to consist of naturally distributed oxygen isotopes [4], 0.0015 atomic fraction ^{1}H and 0.9985 atomic fraction ^{2}H. The densities of the calibrating samples were calculated in the same way as those of the solvents.

Solutions were prepared from the pure solvents used for calibration and from > 99 weight percent SDS obtained from Merck. The water content was removed by heating the surfactant at 50 °C under vacuum for 48 h; a liquid chromatograph/mass spectroscopic analysis found 0.3 % decyl-, 0.2 % tetradecyl- and 0.1 % hexadecyl sulfate impurity in the dried sample.

Results and discussion

Apparent molar volumes of SDS in normal and 99.85 % heavy water solutions calculated from Eq. (1)

are drawn vs. surfactant concentration in Fig. 1. Most of data points are averages taken form independent results; the reproducibility of a particular molar volume was guessed from the standard deviation of the averages and is about 0.1–0.3 cm³. The normal water data in the low concentration region are in good agreement with those available in the literature [2,3]. In order to compare thermodynamic quantities arising from different composition isotopic waters, abscissas in each figure are given in mole fractions. Since molar volumes of ^{1}H$_2$O and of ^{2}H$_2$O at 25 °C and 0.101325 MPa atmospheric pressure equal within 0.12–0.15 %, the abscissas, for demonstration purposes, are given in molar concentrations as well.

Apparent molar volumes of the aggregates are also calculated from Eqs. (4)–(6) and $v = \Phi_a/6.023 \cdot 10^{23}$, the apparent volume of aggregated SDS molecules is drawn vs. surfactant concentration in Fig. 2. Critical micelle concentration and monomer apparent molar volume of SDS necessary to calculate Φ_a are taken from the literature [2,3,5]: in normal water $c_0 = (8.184 \pm 0.1) \cdot 10^{-3}$ mol/dm³, $x_0 = (1.478 \pm 0.018) \cdot 10^{-4}$ mole fraction, $V_m = 237.5 \pm 0.6$ cm³. The monomer apparent molar volume of SDS in heavy water was not available and thus v is drawn only for normal water solutions.

In both solvents, apparent molar volumes increase very slowly and show a slight, but definite isotope effect $\Phi_s(^1$H$_2$O) - \Phi_s(^2H_2$O) \sim 0.5$–1.0 cm³ for surfactant concentrations greater than 0.1 mol/dm³. Apparent volumes of aggregated SDS molecules (i. e.

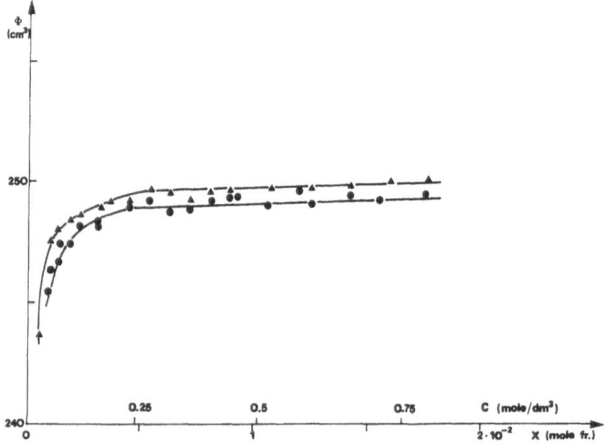

Fig. 1. Apparent molar volume of SDS in normal (▲) and in 99.85 % heavy water (●) vs. surfactant mole fractions. For demonstration purposes, approximate surfactant molar concentrations are also marked

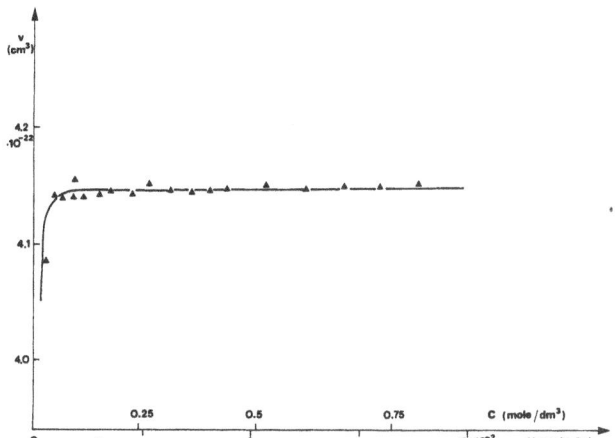

Fig. 2. Apparent volume of aggregated SDS molecules vs. surfactant mole fractions in normal water

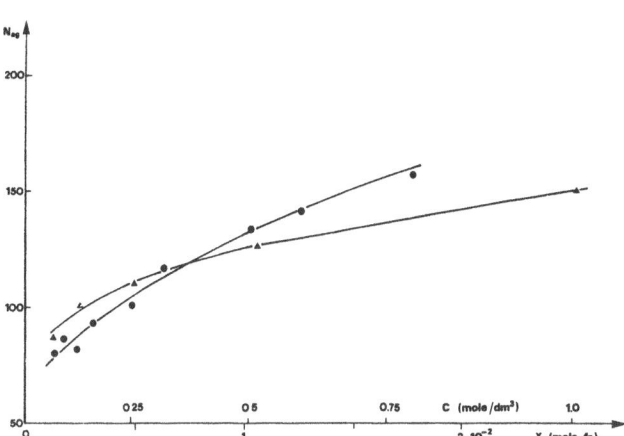

Fig. 3. Aggregation number of SDS micelles vs. surfactant mole fractions (in heavy water). Aggregation numbers were obtained from SANS experiments carried out in two different laboratories; data points are taken from Refs. [7] (▲) and [8] (●)

apparent molar volumes of the aggregates) show even less change: by averaging the normal water data in the concentration range of 0.04–0.8 mol/dm³, $v = (4.1472 \pm 0.0049) \cdot 10^{-22}$ cm³ is obtained. The relative change in the apparent volume of aggregated SDS molecules remains less than $1.2 \cdot 10^{-3}$, while the surfactant concentration changes more than one order of magnitude. (The same calculations were carried out for the heavy water apparent molar volumes by using [5] $c_0 = 8.050 \cdot 10^{-3}$ mol/dm³ and the monomer apparent molar volume obtained for normal water; these hypothetic calculations lead to similar results, as in the case of normal water.)

According to SANS experiments [6–8], the aggregation number of SDS micelles strongly depends on the surfactant concentration. At 25 °C, SDS aggregation numbers vs. surfactant concentration are available from two different research groups [7, 8] and the results are drawn in Fig. 3. High resolution SANS experiments [9] indicate that labelled segments of the hydrocarbon chain (labelled CH_2- and CH_3-groups) are evenly distributed in a dry core and that the depth of the interface between the dry core of SDS micelles and the solvent phase is well below 0.5 nm, this upper limit set by the resolution of the SANS spectrometer. In agreement with the conclusions drawn in Ref. [9], from all these experimental findings one may conclude that SDS micelles do not contain regions in which

the packing of the monomer chains and of the solvent molecules would depend on geometric arrangements.

References

1. Glasstone S (1946) Textbook of Physical Chemistry. 2nd ed, Van Nostrand, New York, p 239
2. Musbally GM, Perron G, Desnoyers JE (1974) J Colloid Interface Sci 48:494
3. Doughty NA (1979) J Phys Chem 83:2621
4. Kell GS (1977) J Phys Chem Ref Data 6:1109
5. Mukerjee P, Mysels KJ (1971) Critical Micelle Concentrations of Aqueous Surfactant Systems. Nat Stand Ref Data Ser, Nat Bur Stand 36 US Government Printing Office, Washington DC
6. Hayter JB, Penfold J (1983) Colloid Polym Sci 261:1022
7. Sheu EY, Wu CF, Chen SH, Blum L (1983) Phys Rev A32:3807
8. Cser L, Jákli Gy, Kajcsos Zs, Vass Sz, Borbély S, Bezzabotnov VYu, Ostanievich YuM, Juhász E, Lelkes M (1986) Proc 6th Int Symposium on Surfactants in Solution. New Dehli, India, August
9. Cabane B, Duplessix R, Zemb T (1985) J Phys 46:2161

Received January 20, 1988;
accepted January 25, 1988

Authors' address:

Sz. Vass
Central Research Institute for Physics
Hungarian Academy of Sciences
P. O. Box 49
H-1525 Budapest, Hungary

Progress in Colloid & Polymer Science

Progr Colloid Polym Sci 76:224–227 (1988)

High-internal-phase-volume emulsions in water/nonionic surfactant/hydrocarbon systems

C. Solans[1]), N. Azemar[2]) and J. L. Parra[1])

[1]) Instituto de Tecnología Química y Textil (C.S.I.C.)
[2]) Asociación de Investigación de Detergentes (A.I.D.)
Jorge Girona Salgado, Barcelona, Spain

Abstract: Emulsions with gel properties which form at the water-rich region of ternary water/nonionic surfactant/hydrocarbon systems have been studied. The investigations of gel emulsions with a constant water content of 99 % by weight showed that they are high-internal-phase-volume W/O emulsions (the water being the disperse phase), and that their formation is closely related to the HLB temperature of the corresponding system. The stability of these gel emulsions is highly dependent on temperature, hydrocarbon/surfactant ratio and the presence of additives.

Key words: Gel, emulsion, HLB temperature, nonionic surfactant, viscoelasticity, influence of addition.

Introduction

Gels, as well as emulsions, are very important in formulations widely used in pharmaceutical, cosmetic, food and other technologies [1].

Emulsions with gel properties that form in the water-rich region of water/nonionic surfactant/hydrocarbon systems are the subject of the research reported. With only three components they are rather simple as compared to gels normally used in practical applications or found in living systems [1, 2]. In addition, they show two advantages of special interest from a practical viewpoint: they form with a very low surfactant content (i.e. 0.5 % by weight) and with a water concentration as high as 99 % by weight. This kind of gel has been recently described for the first time [3]; in contrast, gels formed in the oil-rich region of similar systems were known and described earlier [4–6].

Gels in surfactant-containing systems have been the object of intense research for a long time [4–10]. Although there is still little systematic information regarding some important aspects, such as formation, structure, stability etc., their overall properties can be related to those of the systems to which they belong. The most important property of nonionic surfactant systems is the pronounced sensitivity of their association capacity to temperature. Shinoda [11] rationalized the complex behaviour of this type of system by introducing the HLB temperature of phase inversion temperature (PIT) concept, as the temperature at which nonionic surfactants change their preferential solubility from water to oil as temperature increases.

In an attempt to elucidate the mechanisms of formation of gel-type emulsions produced at the water-rich region of water/nonionic surfactant/hydrocarbon systems, a characterization of emulsion type, their transport and rheological properties and the relation between HLB temperature and gel formation has been undertaken.

Experimental

Nonionic surfactants are commercially available as ethoxylated lineal alcohols. They are designated as $C_m (EO)_n$ where m is the carbon number of the major component of the hydrophobic part and n is the approximate number of ethylene oxide groups per molecule of surfactant. They were supplied by Tenneco. Aliphatic hydrocarbons (heptane, decane, hexadecane) were of puris. Grade, from Fluka, and water was distilled. NaCl was from Merck.

Electrical conductivities were determined with a Crison conductimeter model CDTM 525 using a dip cell containing platinized platinum electrodes. A Ferranti-Shirley Model MK III viscometer was used to determine rheological properties. Optical microscopy was performed with a Wild M-20 polarizing microscope attached to an automatic exposure camera for photomicrography.

Results and discussion

Gel formation of water/nonionic surfactant/hydrocarbon mixtures with a water content of 99 % by weight was determined by weighing all the components followed by vigorous shaking under thermostated conditions. Glass balls were added to facilitate their formation. The common characteristic of gel emulsions formed was the high viscosity they showed. Their visual aspect varied from clear transparent to bluish to white depending on the particular system, concentration and temperature.

Table 1. T_{HLB}-values for different water/nonionic surfactant/hydrocarbon systems

$T_{HLB} = K_{OIL} (N_{HLB} - N_{OIL})$ [12]		
K_{OIL} 17 °C/unity of HLB	N_{OIL} Heptane: 8.90 Decane: 8.50 Hexadecane: 7.70	N_{HLB} $\frac{E}{5}$ (Griffin Ec)
T_{HLB} $\begin{cases} C_{12}(EO)_4 - \text{Heptane} - H_2O \\ C_{12}(EO)_4 - \text{Decane} - H_2O \\ C_{12}(EO)_4 - \text{Hexadecane} - H_2O \end{cases}$	14 °C 20 °C 34 °C	

Figure 1 shows gel regions (shaded areas) as a function of temperature and hydrocarbon/(hydrocarbon + surfactant) weight ratio for several $C_{12}(EO)_4$ surfactant systems. Formation of gel could not be induced at any temperature in those mixtures with lower or higher hydrocarbon contents than the ones corresponding to gel regions. For a given system, the hydrocarbon to surfactant weight ratio and temperature were found to be the determining factors that induced formation of a gel. By comparing the minimum temperature needed to form gels (Fig. 1) and the HLB temperature of the corresponding system (Table 1) it was evident that gel emulsions form at temperatures above the HLB temperature, at which the surfactant shows lipophilic properties and acts as a W/O emulsifier. The HLB temperature values were calculated according to the equation derived by Kunieda et al. [12] who established a relation between HLB temperature (T_{HLB}), HLB of the surfactant (N_{HLB}) and the parameter, N_{oil}, a characteristic of a given hydrocarbon. Other systems, such as water/$C_{14}(EO)_4$/decane, water/$C_{18}(EO)_2$/ decane, water/$C_{18}(EO)_2$/hexadecane, etc., not shown graphically, were checked for gel formation and the same trend was found. Gel emulsions could only be induced at higher temperatures than the HLB temperature of the corresponding system.

In order to determine electrical conductivity, the samples were prepared by using 10^{-1} M, sodium chloride solution instead of water. The presence of electrolyte at such concentration was found not to change the temperature interval of the gel region. However, the hydrocarbon to sufactant weight ratios at which the gel could be formed was in increased and properties such as viscosity and stability were also improved.

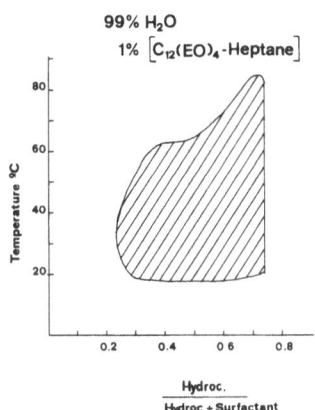

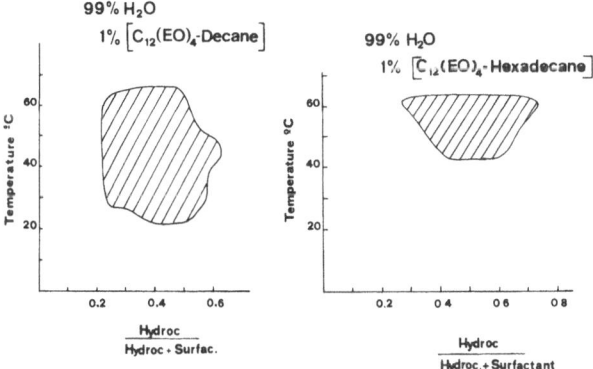

Fig. 1. Gel regions (shaded areas) as a function of temperature and hydrocarbon to surfactant weight ratio for systems containing 99 % water and 1 % surfactant plus hydrocarbon (on a weight basis)

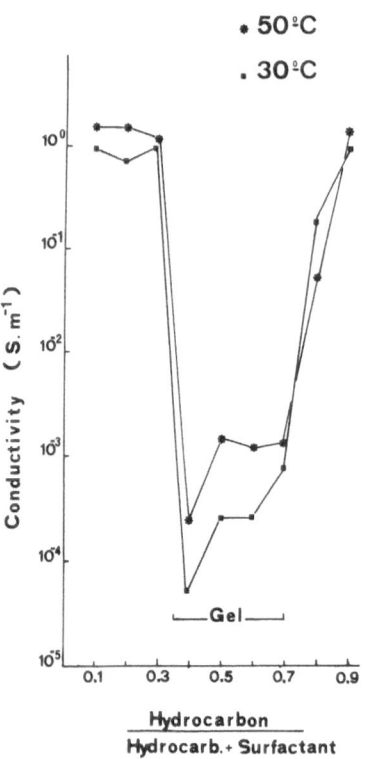

Fig. 2. Electrical conductivity of compositions with 99% aqueous 10^{-1} M NaCl and 1% C_{12} (OE)$_4$ and decane at 30 °C and 50 °C

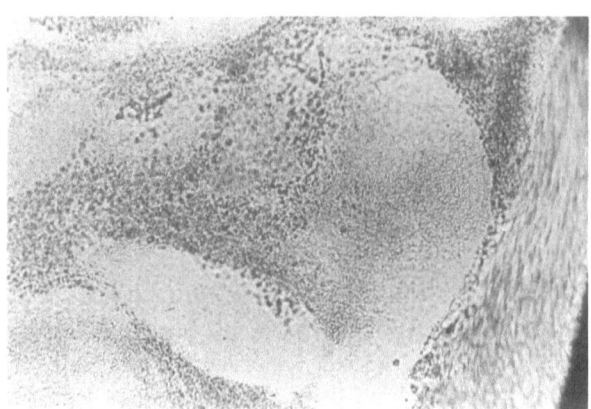

Fig. 3. Photomicrographs of a gel-emulsion viewed under normal light Magnification, × 100

Figure 2 shows the electrical conductivity for the system with $C_{12}(OE)_4$-decane, as a function of decane/ $(C_{12}(OE)_4 + $ decane) weigth ratio. The values of electrical conductivity of gel emulsions (mixtures with hydrocarbon weight ratios between 0.35 and 0.75) were four orders of magnitude lower than those mixtures with no gel properties. It is worth nothing that when electrical conductivities of gels formed in the hydrocarbon-rich region of similar ternary systems were determined, the lowest values obtained were of the order of 4.5×10^{-3} Sm^{-1}.

The low values of electrical conductivity and the fact that gel emulsions formed only with surfactants showing lipophilic properties were clear indications that they consisted of high-internal-phase-volume emulsions where water was the disperse phase. Optical microscopy revealed the existence of close-packed spherical droplets ranging from 2 μm to submicrometer order (Fig. 3). The results were confirmed by electron microscopy [13].

When samples were viewed through polarized light, no liquid crystalline phase was observed. Ultra- centrifugation of gels gave two layers, a lower or water phase (90 – 96% of the total volume) and an upper phase containing liquid crystal of the lamellar type. A detailed study of the phase behaviour close to the water corner was not attempted since commercial surfactants were employed. However, studies of phase properties of gels formed with pure nonionic surfactants showed the gel emulsions to consist of two isotropic liquid phases: a water phase and a hydrocarbon phase with solubilized water [14]:

Gel emulsions showed a typical viscoelastic behaviour. The viscosity values of ternary mixtures before acquiring gel properties were of the order of 1 to 2 mPas while this value could be 10^3 times higher after emulsions acquired those properties. Full characterization of their rheological properties is rather complex and beyond the scope of this paper. Continuous shear experiments only will be reported. They were carried out by measuring shear stress (τ) versus shear rate (D) at constant sweep time (t), (Fig. 4), and shear stress (τ) versus time (t) at constant rate of shear (D), (Fig. 5).

In the shear stress (τ) versus time (t) plots (Fig. 5), a pronounced increase of shear stress during the first 30 s was observed and then it levelled off. The value of shear stress at the stationary state increased with the constant shear rate applied. For each system, hydrocarbon/surfactant ratio and temperature there was a critical shear rate value at which the shear stress, after a pronounced increase, instead of keeping to its stationary state, reached a maximum and decreased rapidly towards zero. That was interpreted as due to the breakdown of the emulsion structure.

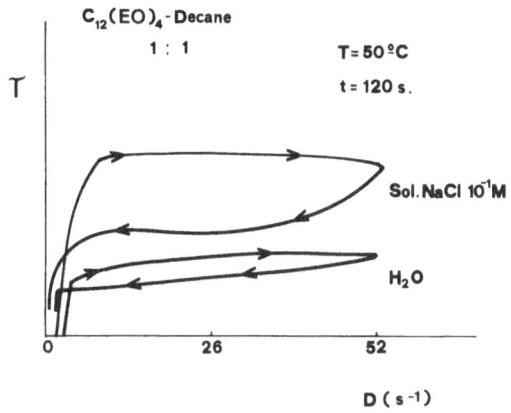

Fig. 4. Shear stress (τ) versus shear rate (D) at constant sweep time (t) of gels with 99% water or a 10^{-1} M NaCl aqueous solution and 1% equal amounts of C_{12} (EO)$_4$ and decane.

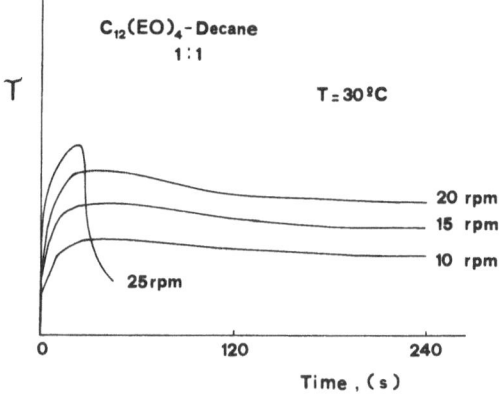

Fig. 5. Shear stress (τ) versus time at constant shear rates of a gel with 99% water and 1% equal amounts of C_{17} (EO)$_4$ and decane

A similar viscoelastic behaviour as that showed in Fig. 4 and 5 was obtained with gels formed with other systems or with other hydrocarbon surfactant ratios.

Concluding Remarks

Gel emulsions in systems of water/nonionic surfactant/hydrocarbon with a water concentration of 99% by weight were formed at temperatures higher than the HLB temperature of the corresponding system, at which nonionic surfactants show lipophilic properties. These gel emulsions consist of high-internal-phase — volume emulsions. The water at concentrations as high as 99% by weight is the disperse phase.

Acknowledgements

The authors would like to acknowledge Mrs I. Carrera for her valuable technical assistance in this work.

References

1. Krog NJ, Riisom TH, Larsson K (1985) In: Becher P (ed) Encyclopedia of Emulsion Technology. Vol 2, Marcel Dekker, New York, pp 321–365
2. Tanaka T, Sun S, Nishio I (1981) NATO Adanced Study Institute Series, Series B Physics. Vol 73, Plenum Press, N 4
3. Solans C, Comelles F, Azemar N, Sánchez Leal J, Parra JL (1986) J Com Esp Deterg 17:109–122
4. Ali AA, Mulley BA (1978) J Pharm Pharmac 30:205–213
5. Sagitani H, Hattori T, Nabeta K, Nagai M (1983) Nippon Kagaku Kaishi 10:1399–1404
6. Sagitani H, Hirai Y, Nabeta K, Nagai M (1986) J Jpn Oil Chem Soc 35:102–107
7. Laing ME, McBain JW (1920) J Chem Soc 11:1506
8. Courtney DL (1972) Am Cosmet Perfum 87:31
9. Ekwall P (1975) In: Brown EG (ed) Advances in Liquid Crystals. Vol 1, Academic Press, New York, p 1
10. Barry BW (1975) Adv Colloid Interface Sci 5:37
11. Shinoda K, Saito H (1968) J Colloid Interface Sci 26:70
12. Kunieda H, Shinoda K (1985) J Colloid Interface Sci 107:107
13. Solans C, García-Domínguez JJ, Parra JLO, Heuser J, Friberg SE (1987) submitted to Colloid Polymer Sci 5
14. Kunieda H, Solans C, Shida N, Parra JL (1987) Colloids Surfaces 24:225–237

Received January 26, 1988;
accepted January 29, 1988

Authors' address:

C. Solans
Instituto de Tecnologia Quimica y Textil (C.S.I.C.)
Jorge Girona Salgado, 18–26
E-08034 Barcelona, Spain

Progress in Colloid & Polymer Science   Progr Colloid Polym Sci 76:228–233 (1988)

Rheological behaviour of microemulsions

C. Blom and J. Mellema

Rheology group, Department of Applied Physics, Twente University of Technology, Enschede, The Netherlands

Abstract: The steady state viscosity and the complex dynamic shear viscosity of micro-emulsions in the I, II and III phase regions have been measured. The rheological behaviour of the microemulsions studied could be partly analysed with the help of a model based on a system with almost spherical droplets, resulting in values for the interfacial tension of the droplet interface. In the III-phase region of an ionic and a nonionic micro-emulsion, the structure deduced is one in which rapidly fluctuating deformed droplets exist. Outside the III-phase region, droplet-like microemulsions as well as probably rod-shaped microemulsions have been found.

Key words: Microemulsions, linear viscoelasticity, viscosity, interfacial tension of the droplet interface, fluctuating deformed droplets.

1. Introduction

Rheological shear experiments can be divided into harmonic shear experiments and steady shear experiments. In a linear harmonic experiment, the complex viscosity $\eta^*(\omega) = \eta'(\omega) - i\eta''(\omega)$ is measured as a function of the angular frequency (ω). The real part $\eta'(\omega)$ and the imaginary part $\eta''(\omega)$ can be related to the energy dissipation and the energy storage, respectively. 'Linear' means that the measured complex viscosity is independent of the (small) amplitude of the applied shear and shear rate. This linear viscoelastic technique can be used in mechanical spectroscopy and provides information about the equilibrium structure. For many systems, one can assume that a low frequency limit $\eta'(0)$ exists and $\eta''(0) = 0$. At about an angular frequency $\omega = 1/\lambda$, where λ is the longest mechanical relaxation time of the fluid system, the dynamic viscosity $\eta'(\omega)$ starts to decrease with increasing ω and $\eta''(\omega)$ is maximal.

In a steady shear experiment, the viscosity $\eta(\dot{\gamma})$ is measured as a function of the shear rate ($\dot{\gamma}$). In general, there is a low shear rate region where the applied shear stress does not affect the equilibrium structure and the viscosity does not depend on the shear rate (Newtonian behaviour).

Departure from this Newtonian behaviour can be expected at a shear rate $\dot{\gamma}_c$, if during a typical time $\tau = 1/\dot{\gamma}_c$ the equilibrium structure is distorted such that it cannot be restored during that time. Consequently, in steady shear experiments, only the low shear rate limit of the viscosity can give information on the equilibrium structure.

In Fig. 1 a sketch of possible paths of $\eta'(\omega)$ and $\eta(\dot{\gamma})$ is given. The low frequency limit of $\eta'(\omega)$ is equal to the low shear rate limit of $\eta(\dot{\gamma})$. In polymer rheology it is often found that the characteristic time at which the behaviour departs from Newtonian behaviour coincides roughly with the inverse angular frequency at which the dynamic viscosity starts to decrease ($\lambda \approx \tau$): the Cox-Merz relationship (see, for instance, Bird et al. [1]). Within a factor of 10, this behaviour has also been found for a hard sphere colloidal dispersion [6]. In the present paper, only $\eta^*(\omega)$ and the low shear rate limit of $\eta(\dot{\gamma})$ are presented.

Microemulsions are water-in-oil or oil-in-water liquid systems, which form spontaneously in the presence of a surfactant, often in combination with a co-surfactant (alcohol). The microstructure of micro-emulsions is the subject of much research. Possible microstructures are, amongst others, spheres, rods,

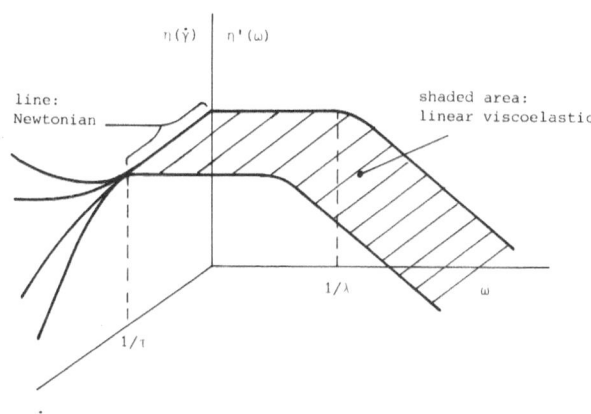

Fig. 1. Possible paths of $\eta'(\omega)$ and $\eta(\dot{\gamma})$

disks and bicontinuous shapes. In all these models, the oil and water domains are separated by a surfactant film. The interfacial energy and in particular the interfacial tension energy of these surfactant films is believed to be small. Therefore, thermal energy can cause shape fluctuations and the microstructures are also subject to shape fluctuations.

2. Materials and methods

The nonionic microemulsions were prepared from n-hexane (Baker, p.a.), deionized water and polyoxyethylene-nonylphenolether (NNP7). This surfactant has been purified by extraction of polyethyleneglycol with ethylacetate. (The NNP7 was supplied by courtesy of Servo Chemicals Co., Delden, The Netherlands). For the phase diagram of the system, see Mellema et al. [7]. The ionic microemulsion is composed of n-heptane (Merck, uvasol), brine:

deionized water containing 6.54 % wt NaCl (Baker, purity > 99 %), sodium dodecylsulphate (SDS, Merck, purity > 99 %) and 1-butanol (Merck, uvasol). The phase behaviour of this system has been studied by van Nieuwkoop and Snoei [9]. The pseudo-ternary phase diagram of this system is given in Fig. 2.

The harmonic shear measurements were performed with the help of two different apparatus: torsion pendulum resonators in the frequency range between 80 and 2500 Hz (Blom and Mellema [2] and a nickel tube resonator in the frequency range of 3.7 to 230 kHz [10].

A simplified diagram of a torsion pendulum is given in Fig. 3. A magnet is build inside the bob. The torsional motion is set up by means of a magnetic field from excitation coils positioned outside the sample holder. The measuring coils pick up inductively the motion of the magnet. For harmonic signals the equation of motion is of the forced harmonic vibration type. Analysis of the transfer function shows that for low viscous fluids, the frequency shift and the change in bandwidth of the resonance curve compared to vacuum resonance are proportional to the imaginary and real part of the damping, respectively. The damping is proportional to the plane shear impedance $R_{pl} + iX_{pl}$ applied by the fluid on the bob surface. The real and imaginary parts of the complex shear viscosity are derived from:

$$\eta' = 2R_{pl}X_{pl}/\varrho\omega, \ \eta'' = (R_{pl}^2 - X_{pl}^2)/\varrho\omega \qquad (1)$$

with ϱ the density of the fluid. Four separate resonators have been used. The resonance frequencies are about 80, 277, 720 and 2510 Hz.

The nickel tube used for the resonance experiments has a length of 0.40 m, a diameter of 4 mm and a wall thickness of 0.2 mm. Firstly, the tube is magnetized circumferentially. Torsional deformation can be set up by an axial magnetic field due to a magnetostrictive effect. When a harmonically oscillating magnetic field is produced by a coil around the tube, e. g. around the centre (see Fig. 4), torsional waves propagate up and down the tube producing in steady state a forced standing wave. By the inverse magnetostrictive effect, the motion of the tube can be detected by a measuring coil. In

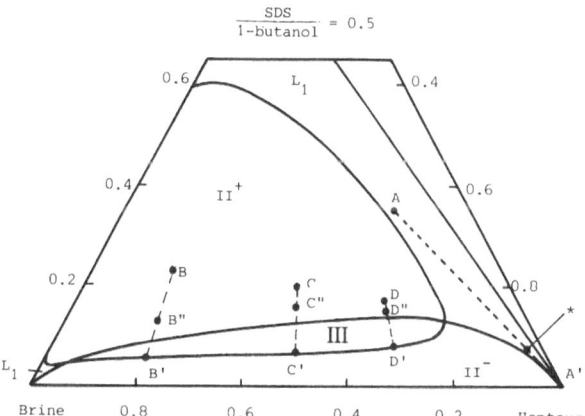

Fig. 2. Pseudo-ternary phase diagram of the SDS/1-butanol/brine/heptane system at 23 °C. L_1: single phase, II^+ and II^-: a microemulsion phase in equilibrium with an aqueous phase and an oil phase respectively, III: three phase. See text for explanation of *

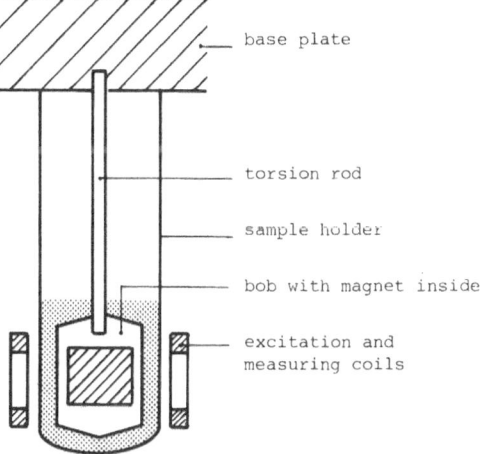

Fig. 3. Torsion pendulum resonator

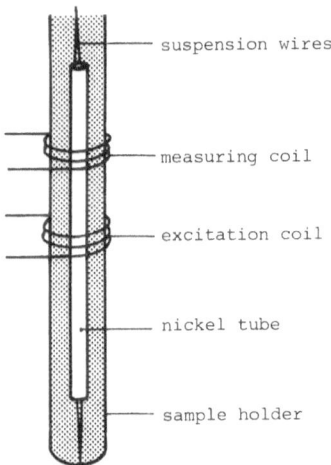

Fig. 4. Nickel tube resonator

a similar way, the torsion pendulum resonator the resonance curve in the liquid under investigation is determined. Its frequency and bandwidth change compared to vacuum is related analogously to the complex shear modulus, as for the torsion pendulum resonators. Favourable use can be made of the excitable overtones in the tube. The steady shear viscosities have partly been measured with a coaxial cylinder viscometer (Contraves low-shear) and partly with a falling ball viscometer (Haake).

3. Structure models

A simple structure model is one in which the interfacial properties of deformable spherical droplets are assigned to an infinitely thin interface. In a model given by Oosterbroek and Mellema [11], the interfacial ten-

sion Γ, the real and imaginary part of the dilatational modulus, $\varkappa$ and σ, respectively, and the real and imaginary part of the shear modulus, μ and ζ, respectively, have been incorporated. A typical relaxation time τ is approximately

$$\tau = (a\eta)/S \qquad (2)$$

where η is the effective viscosity of the emulsion and $S = \Gamma$, $\varkappa$, or μ or a linear combination of them. The steady state viscosity is smaller or equal to the steady state viscosity of hard spheres.

Another model that can be applied considers spheres with shape fluctuations due to thermal energies [8]. In Fig. 5, the results are shown for $4\pi\Gamma a^2 = 5\,kT$ and $4\pi\Gamma a^2 = 50\,kT$. As in the first model, the calculated effects are the sum of single particle effects. It appears that many relaxation times come into play, broadening the transitions in the frequency dependence of η' and η''. The magnitude of the viscoelastic effects increases compared to the first model in this section without superimposed thermal fluctuations. The steady state viscosity can easily be larger as smaller than that of hard spheres.

4. Results and discussion

SDS system

The relative steady shear viscosity as a function of the volume fraction of I-phase microemulsion systems

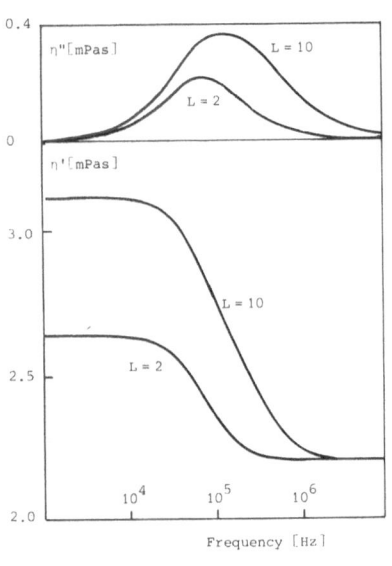

a) $4\pi\Gamma a^2 = 5\ kT$

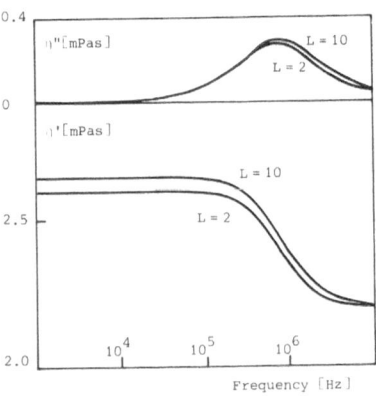

b) $4\pi\Gamma a^2 = 50\ kT$

Fig. 5. Emulsion model calculation considering thermal fluctuations: (a) $4\pi\Gamma a^2 = 5\,kT$ and (b) $4\pi\Gamma a^2 = 50\,kT$

(5 compositions on line AA' in Fig. 2) and II⁺ and III-phase microemulsions (about 35 compositions on lines BB', CC' and DD' in Fig. 2) is given in Fig. 6. The volume fractions have been calculated with help of a geometrical model [3] and composition analysis. The viscosity is given relative to the viscosity of the continuous phase. As the continuous phase of the microemulsions is either a brine-butanol or a heptane-butanol mixture, with no accurately known ratio, these viscosities have to be estimated. The used values are $\eta = 0.5 \pm 0.1$ mPas for the heptane-butanol continuous phase and $\eta = 1.15 \pm 0.15$ mPas for the brine-butanol continuous phase. Within the III-phase region there is a transition from an oil-in-water to a water-in-oil microemulsion when going from the II⁻-phase boundary towards the II⁺-phase boundary [9]. Both for the O/W and W/O system in the III-phase region, the relative steady shear viscosity as a function of the volume fraction is given in Fig. 6.

In a small area of the I-phase region (indicated with * in Fig. 2) birefringent systems which show strongly viscoelastic and non-Newtonian behaviour has been found. The investigation of these systems has not yet been finished.

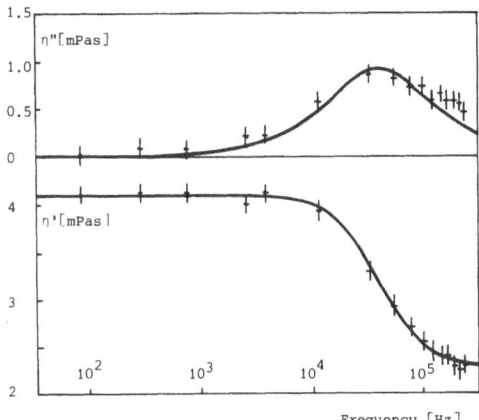

Fig. 7. Complex dynamic shear viscosity of a III-phase SDS microemulsion system

The measured viscosities are compared with the hard sphere viscosity according to de Kruif et al. [5]

$$\eta = \eta_0 \left(1 - \frac{\phi}{\phi_{max}}\right)^{-2},\qquad(3)$$

where η_0 is the viscosity of the continuous phase, ϕ is the volume fraction dispersed phase and ϕ_{max} is an adjustable parameter. From viscosity measurements of a hard sphere silica dispersion, as a function of the volume fraction, de Kruif found for the low shear limit $\phi_{max} = 0.63 \pm 0.02$.

Except for the III-phase microemulsions at high volume fractions, the measured viscosities do not differ much from the "hard sphere" viscosity given by Eq. (3). Possible causes for the small differences are the uncertainties in the calculated volume fractions and the estimated viscosities of the continuous phase.

All the investigated systems in the III and II⁺-phase region on lines B″B', C″C' and D″D' in Fig. 2 show a viscoelastic transition within the measured frequency range. A typical result of the measured complex viscosity is given in Fig. 7. All the measured steady state viscosities coincide with the low frequency limits of the real part of the complex viscosity. The deduced relaxation time changes from $4 \cdot 10^{-7}$ s to $5 \cdot 10^{-6}$ s along the lines B″B', C″C' and D″D'. At the W/O to O/W transition, the droplets have no clear preference to curve to water or to oil. One expects soft structures, i.e. easily deformating so that the viscosity is lower than for hard spheres (a completely flexible surfactant interface would lead to viscosities close to the viscosities of mixtures of shapeless flexible viscous units of oil,

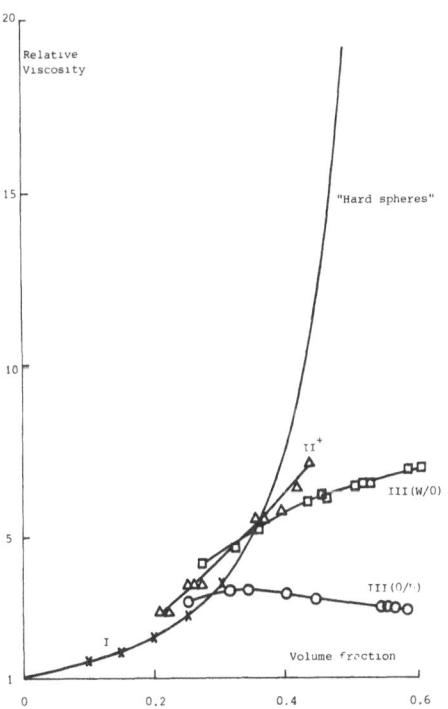

Fig. 6. Relative steady state viscosity as a function of the volume fraction for SDS microemulsions in the I, II⁺ and III-phase regions

water and surfactants). Towards the II⁻ and II⁺ boundaries, a preference for O/W at W/O emulsification is very probable. In the III-phase region, the fact that the steady state viscosities are relatively low and the longest relaxation times are small (10^{-5} s) excludes the possibility of a permanent bicontinuous structure, but instead suggests that oil or water droplets are dispersed at a relatively short length scale. This makes it worthwhile to analyse complex viscosity in terms of an individual droplet model.

If the interfacial tension is taken to be dominant in Eq. (2), its magnitude can be estimated. The resulting interfacial tension is about 3.10^{-5} N/m in points B″, C″ and D″, about 10^{-4} N/m in the middle of the III-phase region and about 10^{-5} N/m in points B′, C′ and D′. The emulsion droplet radius, calculated with help of the geometrical model and composition analysis, changes from about 10 nm in points B″, C″ and D″ to about 15 nm in points B′, C′ and D′. The interfacial energy

$4\pi\Gamma a^2$, where a is the droplet radius and Γ the interfacial tension, is about $8\,kT$ in the endpoints of lines B″B′, C″C′ and D″D′ and about $60\,kT$ in the middle of the III-phase region.

One might recall that in the III-phase region the dispersion of droplets is at high volume fraction, so that a surface shape change of one droplet cannot take place if shape changes of ambient droplets do not also take place more or less coherently. This coupling of surface motions hampers fluctuations of individual droplets with large amplitude, because the energy needed for a shape change of more droplets at the same time is much larger than for one droplet. This coupling of surface motions is not incorporated into the used model. This deficiency may explain the unexpected increase of Γ towards the O/W–W/O transition. If the observed relaxation behaviour is analysed to be descended from single particle effects, the deduced interfacial tension will be too large if assigned to a single droplet, but this (incorrect) large tension correctly reflects the suppression of thermal fluctuations of closed surfaces.

In the II⁺-phase region on lines B″, C″ and D″, no viscoelastic transition was measured. The estimated droplet radius in these areas is about 5 nm. With an interfacial tension of 10^{-5} N/m or larger, this implies that the relaxation time is too small to be measured within the frequency range of our apparatus.

NNP7 system

In Fig. 8, the relative steady shear viscosity of I-phase microemulsion systems is given as a function of the volume fraction. Two dilution series have been measured, one at the lower- and one at the higher-temperature boundary of the I-phase region. The ratio [NNP7]/hexane = 1.2 in both series. In a previous paper [8] the results of complex viscosity measurements in this I-phase region, as well as in the III-phase region, have been given [7]. When increasing the temperature from the lower to the higher temperature boundary of the I-phase region, these measurements reveal a new set of relaxation times, which are much larger than the times mentioned above for the SDS systems. It is commonly assumed that for surfactants with a polyoxyethylene moiety the driving force for microstructural changes is the decreasing solubilization of water in the hydrophilic moiety, if the temperature increases. This effect changes the curvature energy in the formation of the droplet surfaces, tending to decrease the curvature of the oil units. At the same oil amount, to be enclosed

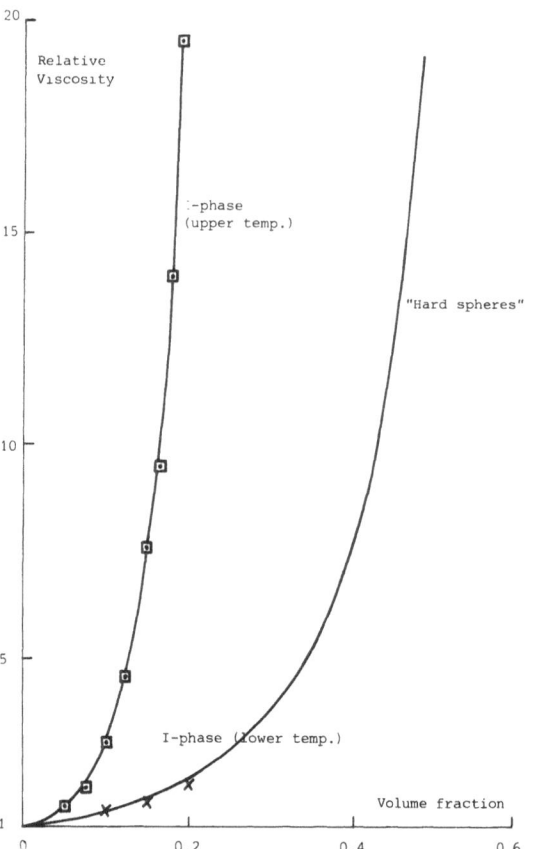

Fig. 8. Relative steady shear viscosity as a function of the volume fraction for the NNP7 system at the lower and upper temperature boundary of the I-phase region

by the surfactant layer, shape changes are necessary. Rod-like shapes can occur. Interpretation of the large low shear rate viscosities and of the longer relaxation times as due to rotational diffusion processes from rods, using the theory of Doi and Edwards [4] can be done quite satisfactorily. Disk-like shapes, however, can also occur and cannot be excluded.

References

1. Bird RB, Armstrong RC, Hassager O (1977) Dynamics of polymer liquids. vol 1, John Wiley & Sons, New York
2. Blom C, Mellema J (1984) Torsion pendula with electromagnetic drive and detection system for measuring the complex shear viscosity of liquids in the frequency range 80–2500 Hz. Rheol Acta 23:98-105
3. Blom C, Mellema J (1984) The rheological behaviour of a microemulsion in the three-phase region. J Dispers Sci Technol 5:193-218
4. Doi M, Edwards SF (1986) The theory of polymer dynamics, Clarendon Press, Oxford
5. de Kruif CG, Jansen JW, Vrij A (1987) Sterically stabilized silica colloid as a model supramolecular fluid, In: Safran SA, Clark NA (eds) Physics of complex and supermolecular fluids. John Wiley & Sons, New York, pp 315-343
6. Mellema J, de Kruif CG, Blom C, Vrij A (1987) Hard sphere colloidal dispersions: mechanical relaxation pertaining to thermodynamic forces. Rheol Acta 26:40-44
7. Mellema J, Blom C, de Kruif CG (1987) Viscoelastic behaviour of hard spheres colloidal dispersions and microemulsions. 21eme colloque annuel de groupe Français de rheologie, Strasbourg, nov 26–28, 1986
8. Mellema J, Blom C, Beekwilder J (1987) Simple linear viscoelastic models of dilute solutions of polymer molecules and emulsion droplets. Rheol Acta 26:418-427
9. van Nieuwkoop J, Snoei G (1982) Second European symposium on enhanced oil recovery. ARTEP, Paris 8–10 November
10. Oosterbroek M, Waterman HA, Wiseall SS, Altena EG, Mellema J, Kip GAM (1980) Automatic apparatus, based upon a nickel-tube resonator, for measuring the complex shear modulus of liquids in the kHz range. Rheol Acta 19:497-506
11. Oosterbroek M, Mellema J (1981) Linear viscoelasticity of emulsions. J Colloid Interface Sci 84:14-26

Received February 1, 1988;
accepted 8, 1988

Authors' address:

C. Blom
Department of Applied Physics
Twente University of Technology
P.O. Box 217
7500 AE Enschede, The Netherlands

Progress in Colloid & Polymer Science

Progr Colloid Polym Sci 76:234–241 (1988)

Comparative study of fluorinated and hydrogenated nonionic surfactants. I. Surface activity properties and critical concentrations

J. C. Ravey, A. Gherbi and M. J. Stébé

Laboratoire de Physico-Chimie des Colloïdes, Université de Nancy I, Vandoeuvre les Nancy, France

Abstract: In very dilute aqueous solutions, the behaviour of hydrogenated and fluorinated nonionics are similar but not identical. For air/water adsorption and micellisation, the main result is the equivalence in the length of the hydrophobic chains, which is about 1/1.7, but it also depends on temperature. An overview of comparative thermodynamic data is presented for both these phenomena.

Key words: Nonionic surfactants, fluorinated surfactants, surface activity, critical concentrations, thermodynamic potentials.

Introduction

A series of original nonionic fluorinated surfactants have recently been synthesized in our laboratories [1–4], in order to develop systems in which water and fluorocarbons can be cosolubilized to a large extent, and which are of low sensitivity to the eventual presence of electrolytes. More specifically, they are ethoxylated alcohols, the hydrophobic part being a perfluorocarbon chain. In one of these series, one or two SC_2H_4 groups were intercalated into the hydrophilic moiety.

Solubilization and structural properties of some of these systems have already been presented in a few papers [5–8]; as a result, the molecular structures and phase diagrams in binary and ternary oil-water-surfactant mixtures determined so far revealed a clear analogy with those of the purely hydrogenated parent compounds [8, 12, 13]. Both types of system form more or less anisometric aqueous micelles, water swollen inverse micelles, globular and bicontinuous microemulsions, isotropic phases consisting of dispersions of multilayered grains, and various liquid crystal phases. Qualitatively, the same general trends in the evolution of these systems were observed: according to the temperature, the hydrophilic-hydrophobic balance (HLB) and the chemical nature of the oil.

However, the rates of change may be quite different according to the type of surfactant. Besides, a simple equivalence rule like the one derived for ionic fluorinated soaps ($1 \times CF_2 = 1.5 \times CH_2$) is not of general validity [9, 10]: clearly, the cloud points of fluorinated compounds appear lower than those of hydrogenated ones even twice as long [11]. Moreover, a mere examination of the outlook of the binary (and ternary) phase diagrams suggests that one oxyethylene group (E) in a surfactant chain would not have identical hydrophilic properties whether the hydrophobic part were fluorinated or hydrogenated.

Therefore, in the present series of papers, we propose to report our results concerning a *quantitative comparison* of two types of nonionic emulsifiers. These investigations have concerned the evolution of the whole binary and ternary phase diagrams with the aim of phase and structure behaviour cross correlation, but for this first paper, we shall be mainly dealing with the properties of the very dilute aqueous solutions, i.e. the critical micellar concentration and the surface activity properties.

Experimental

Materials

Several series of fluorinated surfactants have been investigated. The first ones are $C_mF_{2m+1}CH_2(E)_n$, synthesized according to an original method. They will be noted $R_m^1 - E_n$. Typical compounds are such that $m = 6,7$ and $n = 3,4, 5,6$. The second ones are noted

$C_mF_{2m+1}-C_2H_4-\Sigma E_n$ or $C_mF_{2m+1}-C_2H_4-\Sigma E_p\Sigma E_q$. E stands for OC_2H_4, while Σ represents SC_2H_4. They will be abbreviated as $R_m^2 - E_n$, etc. The introduction of the Σ group facilitates the chemical synthesis or these highly pure amphiphiles. Typical values of the indices are $m=6$, $n=2,3,5,7$, $p=1,2$ and $q=2,3$. The compounds with $n=2,3$, or $p=1,2$ were purified by vacuum distillation, the others by GPC.

Methods: Surface tension

Surface tension measurements were carried out using the Wilhelmy plate method, in order to determine the critical micelle concentration (CMC) of the aqueous surfactant solutions. Temperature was monitored between 10° and 60°C. Platinum plates were used.

Results and discussion

1. Surface properties and the critical concentration of the aqueous systems

From standard surface tension investigations, a critical concentration C_1 could be measured for each solution as the location of a well marked break in the γ-log C curves. For most of the surfactants at ambient temperature, these critical concentrations are true critical micellar concentrations, since for $C > C_1$ we effectively get the classical L_1 isotropic phase, which is the aqueous micellar phase. Of course, these systems generally exhibit a liquid crystal behaviour for much higher water contents. For example, the system $R_6^2E_7$ looks like $C_{12}E_5$ [12], with the occurrence of both the hexagonal and lamellar phases. For $C > C_1$, the slope of the γ curves are practically zero, as expected for high purity amphiphiles.

However, at higher temperatures and/or for the less hydrophilic surfactants, the L_1 phase no longer exists, but the discontinuity in the slopes still exists. Moreover, there is a perfect continuity in the values of C_1 and dC_1/dt for the temperature t at which L_1 just disappears. Hence, whatever the case, C_1 corresponds to a phase transition: from a thermodynamic point of view, if C_1 is low and the micellar sizes are not too small, L_1 is a dispersion in water (W) of the micellar pseudo phase, while after the disappearance of L_1, for $C > C_1$ we get a fine dispersion of another phase into a "water phase W". According to the temperature, this dispersed phase is a lamellar (L_α), or isotropic (L_2 or L_3) one. Hence, in the case of L_α, the discontinuity in the slopes in the γ-log C curves corresponds to the formation of a very dilute dispersion of large lamellar multilayered particles. This is consistent with the current idea that the micellar sizes in the L_1 phase increase

with temperature, becoming quite large when L_1 disappears. Let us note that, whatever the case, the "W phase" is a solution of *monomeric* surfactants. Such results have been firmly derived from concomitant investigations of phase diagrams, surface tension measurements and centrifugation; they will be presented in detail elsewhere. What must be remembered is that such a behaviour is identical for both the fluorinated and hydrogenated ethoxylated nonionics.

Examples of comparative values of C_1 (or CMC) values are given in Figs. 1 and 2. The data in Fig. 1 concern amphiphiles of the type $R_m^1E_n$ as a function of temperature. The data for $C_{12}E_6$ are also indicated [14, 15]. Generally C_1 decreases with temperature, but these curves may exhibit a minimum for a temperature which depends on the hydrophilicity of the surfactant: the more hydrophobic the compound, the lower the temperature of the minimum. As will be shown in a following paper, this minimum corresponds to a change in the nature of the dilute disperse phase in W, passing from an L_α phase to the isotropic L_2 or L_3 system. Hence, this behaviour reveals an optimum tendency for the surfactants to aggregate in very dilute state for those temperatures leading to the formation of lamellar liquid crystals.

Figure 2 represents the log (CMC) variation with the number of oxyethylene groups (n) at about 20°C. As expected, the CMC increases with the length of the oxyethylene chain, at least for the present systems. A mere inspection of the curves for $R_6^1 - E_n$, $R_6^2 - E_n$ and C_mE_n immediately suggests that R_m^1 and R_m^2 are

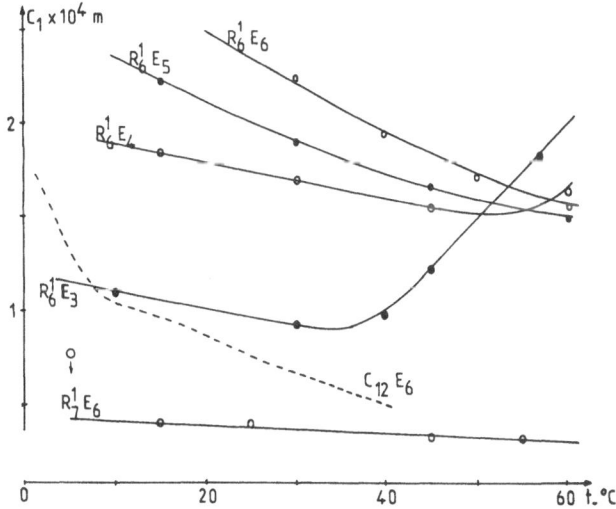

Fig. 1. Examples of critical concentrations (mole/l) for fluorinated surfactants, and for $C_{12}E_6$, as a function of temperature

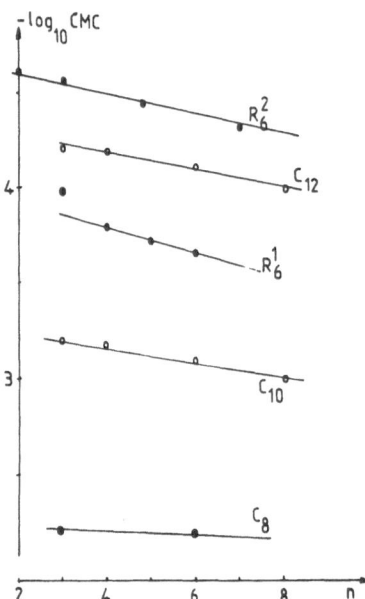

Fig. 2. Variation of $\log_{10}$CMC for different hydrogenated (C_8, C_{10}, C_{12}) and fluorinated (R_6^2, R_6^1) surfactants as a function of the number of oxyethylene groups (n) in the hydrophilic chain, at ambient temperature

roughly equivalent, respectively, to C_{11} and C_{13} chains. Then, a more quantitative comparison of the effect of fluorination of the surfactants appears very appealing and will be attempted in the following.

a) Surface film pressure (π_{CMC})

Since all these amphiphiles are more or less water soluble, they form air/water adsorbed monolayers, whose behaviour may be described by the Gibbs equation. The film pressure above the CMC (π_{CMC}) is the difference between the surface tension of pure water (γ_0) and of the solution at the CMC (γ_{CMC}). As is well known, the fluorinated (F) amphiphiles lower the surface tension of water to a larger extent than do the hydrogenated (H) ones. Typically, taking into account the change of γ_0 with temperature, π_{CMC} decreases from about 55 mN/m at 15 °C to 50 mN/m at 60 °C, whatever the present F-systems (number of oxyethylene groups $n \leq 8$). For H-compounds, π_{CMC} is in the range 40–30 mN/m for comparable hydrophilic chains: here too, Rosen et al. [16] found a decrease of π_{CMC} with rising temperature but for $C_{12}E_n$, if $n \leq 5$. Therefore, the F-substitution on the alkyl chains induces a different apparent activity of these oxyethylene groups relative to temperature effects.

On the other hand, for our F-compounds, π_{CMC} decreases very slightly with n, but only at low temperatures. Typically, at 15 °C, the rate of decrease is about 1 % per added oxyethylene group. The same trends (and the same order of magnitude of the effect) were noted for comparable H-surfactants.

Only a few results [17] have been reported on the effect of the hydrophobic chain length on π_{CMC}: for $C_{12-18}E_n$ alkanols, correlating to a decrease of the molecular area A_{CMC}, some increase of π_{CMC} may be deduced. The same variation occurs with F-amphiphiles; for example, the film pressure is 53.8 and 55.0 mN/m, respectively, for $R_6^1 - E_6$ and $R_7^1 - E_6$ at 15 °C as a consequence of increasing hydrophobic interaction between the (F)-alkyl chains. Such a result seems valid only for the lower temperatures, since at 60 °C, π_{CMC} is 50 mN/m, whatever the F-compound.

b) Molecular area (A_{CMC})

Although at the CMC the monolayer is far from being a "dilute" spread monolayer, an increase of A should correspond to a decrease of π. Effectively, the higher the temperature, the larger A_{CMC}. For example, the area is 34 and 40 Å^2 for $R_6^1 - E_3$ and 44 and 53 Å^2 for $R_6^1 - E_6$, respectively, at 15 ° and 60 °C. This behaviour parallels that of H-nonionics, as noted by Rosen [16], at least for $n < 8$. However, that trend is not general, since A_{CMC} decreases for e.g. $C_{10}E_8$ and is almost constant for $C_{15}E_8$. Hence, this increase of A must be ascribed to a larger hydrophobicity of the surfactants. Besides, the values of the product $\pi_{CMC} \cdot A_{CMC}$ increases by only about 10 % when temperature changes from 15 ° to 60 °C.

Similarly, the area A_{CMC} decreases with the length of the hydrophobic tail, as noted by Ueno [18] for H-surfactants. Thus, A is 44 and 36 Å^2, respectively for $R_6^1 - E_6$ and $R_7^1 - E_6$ at 15 °C. Let us note that the product $\pi \cdot A$ is noticeably larger for the former system (3.4 versus 2.8 kcal/mole).

Concerning the influence of the number of oxyethylene groups (n) on A_{CMC}, the classical relation

$$A_{CMC} = A_0 \sqrt{n} \tag{1}$$

derived for $C_m E_n$ amphiphiles for $n \geq 5$ ($A_0 \sim 20$ Å^2 for $m = 12$, $A_0 \sim 15$ Å^2 for $m = 16$) [19, 20], seems still roughly valid for F-compounds. As a matter of fact, a better representation would be

$$A_{CMC} = A_0 \sqrt{n+1}, \text{ for } R_6^1 - E_n, \tag{2}$$

Table 2. A few CMC values for fluorinated nonionic surfactants $(mol \cdot l^{-1})$

t	R	n	$CMC \times 10^5$	$C_mE_n: m_e^H$ (n = const)	n_e^H [m = const = m_e^H (15 °C)]
15	R_6^1	3	10.5	11.53	3
	–	4	18.4	11.19	4
	–	5	22.3	11.15	5
	–	6	25.7	11.15	6
	R_7^1	6	4.1	12.66	6
45	R_6^1	3	12.3	10.80	–
	–	4	15.5	10.69	9
	–	5	16.6	10.72	9.5
	–	6	18.4	10.72	10.5
	R_7^1	6	3.4	12.22	10

m_e^H is the equivalent chain length of hydrogenated compound, for the same value of n. n_e^H is the equivalent oxyethylene chain length of hydrogenated compound, the alkylchain of which is taken as identical to m_e^H (15 °C).

mine the H-surfactant which is equivalent to a given F-amphiphile. Results are given in Table 2 for different temperatures, 15° and 45 °C. First, if we suppose the two surfactants to be compared must have the same hydrophilic chain, we see that the equivalence rule slightly depends on the temperature: e.g. it is 1.7 × CH_2 for one CF_2 at 15 °C, and 1.6 × CH_2 for one CF_2 at 45 °C. Let us recall that, for ionic compounds, the equivalence rule would be rather 1/1.5 [9, 10]. This systematic variation of the equivalent chain length with temperature may be understood in another way: instead of assuming a dubious decrease of the intrinsic hydrophobic character (decrease of equivalent $-m$), we could state that this change is due to an increase of the hydrophilic properties (increase of equivalent $-n$). More specifically, if we assume that the equivalence in hydrophobic chain length is 1/1.7 whatever the temperature, the use of the relation given in Appendix I shows that the "effective" oxyethylene chain in F-surfactant $R_6^1 - E_n$ is approximately: $p = n + 2$ at 30 °C, $n + 4$ at 45 °C and $n + 8$ at 60 °C, as compared to $C_{11}-E_p$. In other words, using Eq. (5), we may write:

$$\log_{10} CMC = f_x(t) + g_x(t) \cdot n \qquad (6)$$

with x = F or H, according to the type of emulsifier. $g_x(t)$ is a decreasing function of t ($g_H(t) \approx -0.025\, m \cdot t$: the CMC generally decreases when temperature increases). Hence, if we fix $f_H(t) = f_F(t)$, we obtain the empirical relation:

$$g_F(t) \approx g_H(t) \cdot [1 + (1/n) \cdot 2^{(t-15)/30}]. \qquad (7)$$

As a result, the decrease of hydrophilicity is less pronounced for F-nonionics; the larger the corrective term the shorter the oxyethylene chain. This result would suggest that the effectiveness of the oxyethylene groups is enhanced by the close vicinity of a CF_2 chain (i.e. they dehydrate less easily). This is consistent with the data on surface activity, as discussed above; this is also to be compared with a former result presented elsewhere [8] and related to the view of the whole of the aqueous binary phase diagrams (indicating, for example, that each additional oxyethylene group shifts the temperature of emergence of the L_2/L_3 phase in dilute systems by 10°–15 °C and 20 °C, respectively, for H- and F-nonionics).

For $R_m^2 - E_p - \Sigma - E_q$ systems, where one -SC_2H_4 group is inserted into the hydrophilic chain, we have shown that one Σ counterbalances about 2 E-groups [29]. Moreover, by using the relation in Appendix I for $R_m^2 - E_n$, the H-chain equivalent to R_6^2 would be $C_{12.5}$. Under the assumption that one CF_2 is equivalent to 1.7 × CH_2, we find that the Σ group which acts as the bond between the true hydrophobic and hydrophilic parts would be equivalent to about 0.3 × CH_2 group, in accordance with the previous results. For example, since one CH_2 counterbalances a E_8 chain, one inserted Σ is also equivalent to 0.25 × CH_2.

As far as the thermodynamic potentials of micellisation are concerned, a few data are given in Table 1. Of course, it can be checked that the values of ΔG_m for F-amphiphiles are nearly equal to the corresponding potentials of hydrogenated compounds whose hydrophobic chains are 1.7 times longer. The values of ΔS_m

appear slightly smaller: that could be due to a larger rigidity and bulkiness of the fluorinated chains. ΔH_m also appears smaller (typically between 1 and 2 kcal/mole). As expected, ΔG_m decreases very little with n (at least for $n > 2$), while ΔS_m seems to increase somewhat and remains almost constant when adding one CF_2. Clearly, these last results are at variance with the data for the water/air adsorption, but agree with data for H-surfactant micellization. Finally, the aggregation (micellization) is favoured by an increase of the temperature, which is concomitant with a decrease of the CMC. This conclusion, valid for the present fluorinated compounds, may not be a general rule, since the corresponding trends for H-nonionics depend on the length of the hydrocarbon moiety.

Conclusion

As far as very dilute aqueous systems are concerned, the main result is the equivalence in the length of the hydrophobic chains, which is one CF_2 for $1.7 \times CH_2$. However, this equivalence in the overall hydrophilic-hydrophobic balance changes with the temperature, suggesting a different hydrophilicity of the first oxyethylene groups (i.e. in the close vicinity of the CF_2 groups). This behaviour probably results from a larger area per polar head induced by the bulkier CF_2 as compared to CH_2, at least in micellar and liquid crystalline states. For example, this area is 47 and 42 Å^2, respectively, for $R_6^1 - E_4$ and $C_{12}E_4$ in the lamellar liquid crystal. This will be discussed more thoroughly in a subsequent paper, in considering the outlook of the whole phase diagrams, i.e. for any finite surfactant concentration.

Clearly, both adsorption and micellization are entropically driven, as for H-surfactants. But it is worth noting that *both* the temperature and the number of oxyethylene groups favour air-water adsorption, while this is not the case for the micellization. Finally, we must emphasize that this 1/1.7 equivalence is valid only for the surface and CMC properties. Indeed, there is absolutely no reason to believe, a priori, that the same rule still holds for phenomena like cloud points and PIT. This will be discussed in other papers.

To conclude, let us note that this equivalence could be very crudely explained on a simplistic geometrical basis. Indeed, assuming the density of the H- and F-alcanes to be, respectively, 0.8 and 1.9, the ratio of the trans-sectional area of these linear chains is (14/0.8)/(50/1.9), i.e. 1/1.5. Hence, if the data were calculated

Table 3. Values of coefficients of Eq. (8)

a_0	=	2.87	a_1	=	-4.8
b_0	=	-1.206	b_1	=	0.135
c_0	=	0.079	c_1	=	-0.2
d_0	=	0.0046	d_1	=	-5

on a volume fraction basis, instead of the mole basis, the two types of nonionics would appear much more similar. In other words, the adsorption and micellization potentials of linear amphiphiles seem to depend essentially on the *water volume displaced by the hydrophobic chain*.

Appendix I

From data collected in literature, an empirical relation may be proposed for a crude evaluation of the CMC of (linear and saturated) nonionic polyoxyethylene alcohols C_mE_n [23]. In general terms, this function of m, n and absolute temperature allows a precision of about 25% to be achieved in the whole range of CMC $10°-10^{-6}$ mol/l and $0 < t(°C) < 50°C$. From Molyneux equation, the ΔG_m micellization potential may also be calculated to a good degree of confidence. But this may not be the case for the calculation of ΔS_m and ΔH_m.

This relation may also be applied for parent compounds. For example, one phenyl group in the hydrophobic chain is equivalent to $4 \times CH_2$. From the data of the present paper, one CF_2 is equivalent to $1.7 \times CH_2$.

The relation is written as follows:

$$\ln CMC = (T_0/T) [a + b \cdot (m-1) + c \cdot n + d \cdot n \cdot (m-1)] \quad (8)$$

with $a = a_0 [1 + a_1 (T/T_0 - 1)]$

The values of the coefficients, a_0, a_1 are given in Table 3 for $T_0 = 300$ K.

Acknowledgments

The authors wish to thank the Institut National de Recherche Chimique Appliquée (G. Thiollet) and L. Mansuy for the synthesis of the surfactants.

References

1. Selve C, Castro B, Leempoel P, Mathis G, Gartiser T, Delpuech JJ (1983) Tetrahedron 39:1313–1316
2. Gartiser T, Selve C, Mansuy L, Robert A, Delpuech JJ (1984) J Chem Res 5:292–293
3. Lampin JP, Cambon A, Szonyi F, Delpuech JJ, Serratrice G, Thiollet G, Lafosse L (1985) French Patent N° 2565:226
4. Cambon A, Delpuech JJ, Matos L, Serratrice G, Szonyi F (1980) Bull Soc Chim 965–970

5. Mathis G, Leempoel P, Ravey JC, Selve C, Delpuech JJ (1984) J Am Chem Soc 106:6162–6171
6. Ravey JC, Stébé MJ, Oberthür R (1987) In: Mittal KL, Bothorel P (eds) Surfactants in Solution. Plenum Press, New York, vol 6:1421–1430
7. Ravey JC, Stébé MJ (to be published) In: Mittal KL, Charroras DK (eds) Surfactants in Solution. Plenum Press, New York
8. Ravey JC, Stébé MJ (1987) Progr Colloid Polym Sci 73:127–133
9. Shinoda K, Hato K, Hayaski T (1972) J Phys Chem 76:909–914
10. Kunieda H, Shinoda K (1976) J Phys Chem 80:2468–2470
11. Ravey JC (1983) J Colloid Interface Sci 94:289–291
12. Mitchell DJ, Tiddy GJT, Waring L, Bostock T, Mc Donald MP (1983) J Chem Soc Faraday Trans I, 79:975–1000
13. Schick MJ (1987) Nonionic Surfactants Physical Chemistry, Marcel Dekker, Inc, New York, Basel
14. Goto A, Takemoto M, Endo F (1985) Bull Chem Soc Jpn 58:247–251
15. Herrington T, Sahi SS (1985) J Chem Soc Faraday Trans I, 81:2693–2702
16. Rosen MJ, Cohen AW, Dahanayake M, Xi-Tuan Hua J (1982) J Phys Chem 86:541
17. Galera PA, Gonzales T, Otero E (1982) Comm J Com Esp Deterg 13:285
18. Ueno M, Takasawa Y, Migashige H, Tabata Y, Meguro K (1981) Colloid Polym Sci 259:761
19. Ohba N, Takahashi A (1968) Chem Phys Appl Prat Ag Surface CR Congress Int Deterg 5th 2:481
20. Schott H (1969) J Pharm Sci 58:1521
21. Katz JL (1975) J Colloid Interface Sci 56:179
22. Lange H (1965) Kolloid Z Z Polymer 201:131
23. Buzier M (1984) Thesis, Univ of Nancy, France
24. Molyneux P, Rhodes CT, Swarbrick J (1965) Trans Faraday Soc 61:1043
25. Mcauliffe C (1963) Nature (Lond) 200:1092
26. Becher P (1984) J Dispers Sci Tech 5:81
27. Meguro K, Takasawa Y, Kawahashi N, Tabata Y, Ueno N (1981) J Colloid Interface Sci 83:50
28. Olofsson G (1985) J Phys Chem 89:1473–1477
29. Matos L, Ravey JC, Serratrice G (in press) J Colloid Interface Sci
30. Corkill JM, Goodman JF, Harrold SP (1964) Trans Faraday Soc 60:202

Received February 26, 1988;
accepted March 7, 1988

Authors' address:

J. C. Ravey
Laboratoire de Physico-Chimie des Colloides
U. A. CNRS No 406, Lesoc
Université de Nancy I
B. P. No 239
F-54506 Vandoeuvre les Nancy, France

Progress in Colloid & Polymer Science   Progr Colloid Polym Sci 266:242–246 (1988)

Study of the adsorption of organic molecules on clay colloids by means of a fluorescent probe

K. Viaene[1]), M. Crutzen[1]), B. Kuniyma[1]), R. a. Schoonheydt[2]) and F. C. De Schryver[1,2])

[1]) Department of Chemistry, Leuven, Belgium
[2]) Laboratorium voor Oppervlakte Scheikunde, Leuven, Belgium

Abstract: The adsorption and distribution of molecules on a clay surface has been investigated by means of fluorescent probes, containing pyrene as chromophore and a trimethyl ammonium head group, with varying chain length.

After adsorption, in aqueous media, these molecules form clusters on the surface, resulting in ground state interactions and an efficient excimer formation. The lack of these intermolecular interactions in organic media suggests that hydrophobic interactions are the driving force for this clustering.

Co-adsorbed detergent molecules influence the distribution of the probe molecules, at least when the chain length of the detergent is the same as the total length of the probe. For each probe, a so-called "critical chain length" exists, for the detergents to be able to destroy the probe clusters.

Excimer formation has proved to be a good tool to gain additional information on the adsorption process. It was found that excimer formation on the interlamellar surface is less efficient then excimer formation on the external surface. Also, redistribution and exchange phenomena of the adsorbed probe molecules have been observed by studying excimer emission.

Key words: Clays, fluorescence, pyrene, detergents.

Introduction

Recently, the adsorption and distribution of molecules on clay surfaces was investigated by means of fluorescent probes [1–11]. Indeed, the photophysical properties of such probes can be used to obtain information on the interaction of the adsorbed molecules and the clay surface.

The attractive features of clay minerals are their large surface area, the swelling behavior, the ion-exchange, their abundant occurrence in nature and their catalytic properties [12].

As a probe, a pyrenyl ammonium derivative was used, since these molecules are strongly bound to the negatively charged clay particles. Three such derivatives, with varying chain length, have been synthesized: (1-pyrenyl)trimethylammoniumchloride (PN), 3-(1-pyrenyl)-propyltrimethylammonium bromide (P3N) and 8-(1-pyrenyl)octylammonium chloride (P8N) [11, 14].

The intermolecular excimer formation has proven to be a particularly useful tool to investigate the adsorption phenomena.

Three clay minerals were studied: hectorite, barasym (a synthetic mica-type montmorillonite) and laponite (a synthetic hectorite). These clays are chosen because of their low iron content and their different external surfaces (Table 1). The low iron content in particular is very important because the iron, present in the crystal lattice, can quench the fluorescence of the adsorbed probes [4].

Table 1. Surface properties of the used clays

Clay	External surface, m^2	Na$^+$ CEC equiv/g	Fe$_2$O$_3$ wt %
Hectorite	63	526	0.02
Barasym	133	464	0.05
Laponite	360	733	0.06

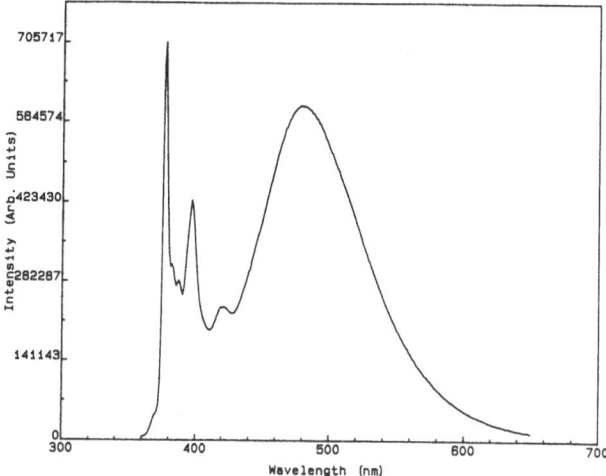

Fig. 1. The fluorescence spectrum of P3N adsorbed on laponite

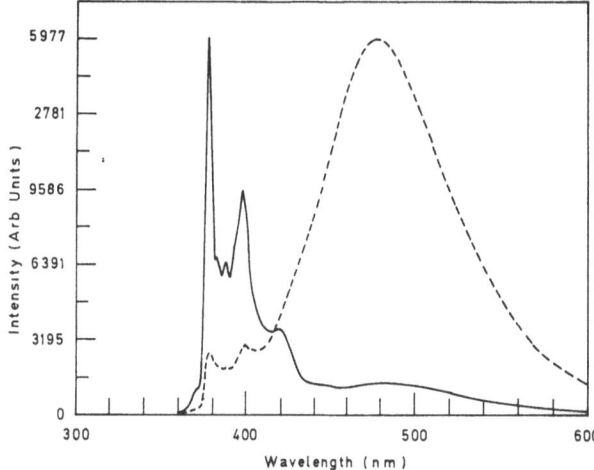

Fig. 2. The fluorescence spectra of P3N adsorbed on laponite suspended in water and methanol

Cluster formation of the adsorbed molecules

The fluorescence spectrum of P3N, adsorbed on a clay surface (Laponite) is shown in Fig. 1. Despite the low concentration (less then 0.5 % of the CEC), efficient excimer formation is observed. In addition, the excitation spectra taken in the monomer and excimer region differ in their respective maxima and vibrational fine structure. These results indicate ground state interactions between the adsorbed probes. The fact that these phenomena are observed at very low concentrations indicates a non-homogeneous distribution of the adsorbed probes on the clay surface. This was also observed by Nakamura et al. for a similar probe adsorbed on a clay surface [9]. These authors explained the non-homogeneity of the adsorption by a non-homogeneous distribution of the negative sites on the clay surface. However, an alternative explanation can be proposed, that the cluster formation is due to a minimalization of the contact surface between the hydrophobic pyrene derivatives and the surrounding water phase. This suggests that the distribution of adsorbed molecules is determined by the surrounding medium and not by the negative sites on the surface.

In order to find out which is the major determining factor, clay suspensions in different solvents were studied. The preparation of those clay samples is described elsewhere [14].

In Fig. 2, the fluorescence spectra of P3N adsorbed on Laponite and suspended in water and methanol are given. In non-aqueous media, no excimer emission is observed. This absence of excimer emission indicates the absence of clusters on the clay surface in non-aqueous media. This proves that hydrophobic interactions are the major determining factor for the distribution of organic molecules adsorbed on a clay surface.

Influence of co-adsorbed detergent molecules

The influence of detergent molecules has been reported in several studies [6, 8–11]. The major influence of these detergent molecules is decreasing the excimer emission. These coadsorbens also form clusters on the surface, and the probes are now preferentially solubilized in these detergent clusters. When enough detergents are adsorbed, no excimer emission is observed. The possibility of two pyrene molecules to form excimers is negligible.

It was further noticed that the chain length of the detergent plays an important role in affecting the distribution of the probe molecules.

For example, the distribution of P3N is not influenced by trimethylakylammonium surfactants with a chain length smaller then 8 methylene groups. For PN, however, the distribution remains unchanged for detergents with less then 6 methylene groups. So, for each probe there exists a so-called critical chain length. This critical chain length is the same as the total length of the probe, including the pyrene chromophore.

The determining factor influencing the distribution of the probe molecules is the ability of the detergent molecules to shield the pyrene chromophore from the water phase. When this shielding cannot be realized by the coadsorbens, separate clusters of surfactants and probes are formed on the clay surface. Only when the chain length of the detergent is at least as long as the length of the probe are mixed clusters formed.

For detergents with a chain length above the critical chain length, an influence of the structure, as well as the chain length, is observed. The ratio detergent over probe ([det]/[probe]), at which the excimer emission becomes negligible, decreases with increasing chain length. Also, when DDAC, a detergent with two aliphatic tails, is used instead of a one-tailed detergent, the [det]/[probe] ratio decreases.

What do these detergent clusters look like? Our observations point to the existence of many small detergent clusters on the clay surface. Indeed for CTAC, at a detergent/probe ratio of 40, the excimer emission has disappeared, whereas for DDAC this ratio equals 17. The low value of this ratio, as well as the influence on this ratio of the used detergent, indicates the existence of many small clusters, containing on average one probe molecule. The lower value of the ratio for DDAC could indicate a smaller aggregation number of DDAC, compared to the one-tail detergent molecules, on the clay surface, analogous to that in micellar solutions [15].

Influence of aggregation of clay particles

The replacement of Na^+ by Ca^{2+} on the clay surface leads to an aggregation of individual clay platelets. The effect on the photophysical properties of P3N, adsorbed on Laponite, is shown in Fig. 3 in the case of

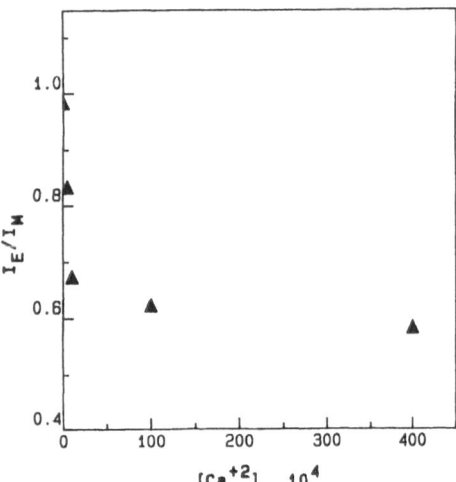

Fig. 3. Influence of Ca^{2+} concentration on excimer fluorescence

Laponite. The 480 nm emission is slightly blue-shifted with increasing Ca^{2+}, but more important, a decrease in the 480 nm fluorescence with increasing Ca^{2+} concentration is observed. When detergent is coadsorbed, no change in the fluorescence spectra is noticed when the amount of Ca^{2+} is varied.

These results indicate that, in the absence of detergent molecules, the excimer formation is less efficient when the aggregation of the clay platelets increases. This increased aggregation is due to the formation of bilayer hydrates. The P3N molecules adsorbed in the interlamellar space of these bilayer hydrates will have a restricted mobility, due to the closeness of the clay platelets, which does not allow free movement of P3N. When detergent molecules are coadsorbed, the P3N molecules adsorb preferentially in the detergent clusters and the effect of Ca^{2+} is wiped out.

The fluorescence spectra of P3N adsorbed on barasym and hectorite indicate that the former behaves as a Na^+ laponite and the latter as a Ca^{2+} laponite. Fluorescence spectroscopy is thus a tool for obtaining qualitative information about the aggregation of the clay particles in suspension.

Study of the adsorption process by means of the excimer formation

Due to the low concentration of the probe in the exchange solution, excimer formation only occurs after adsorption on the clay surface. Thus, by following the steady-state intensity of the excimer fluorescence as a function of time, an idea about the adsorption and redistribution rate can be obtained. The data are shown in Fig. 4.

An important characteristic is the immediate observation of excimer emission within the time scale of our experiment (seconds). This indicates a fast adsorption upon adding the probe to the clay suspension.

In general, excimer fluorescence decreases as a function of time in the absence of detergent. The time dependence, however, depends on the type of clay and the type of exchangeable cation.

For Ca^{2+}-exchanged laponite, the excimer fluorescence intensity at time zero after adsorption increases proportionally to the Ca^{2+} content. The excimer fluorescence decreases with time and the rate of decrease is also proportional to the Ca^{2+} content.

For barasym, the time dependence is small, while for hectorite the time dependence can be compared with the one for laponite in the presence of Ca^{2+} ions.

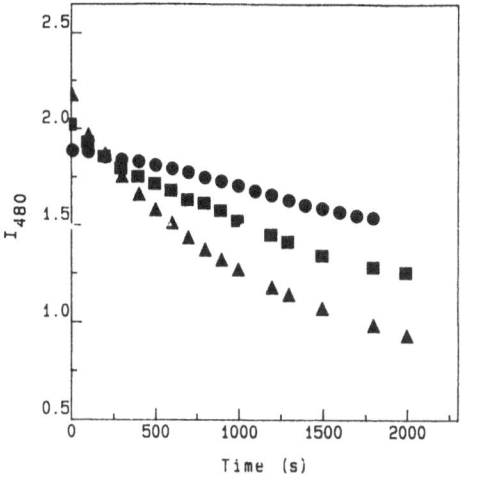

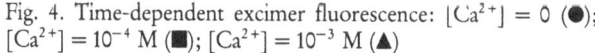

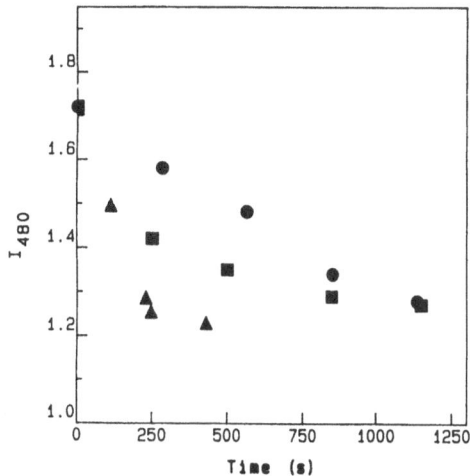

Fig. 4. Time-dependent excimer fluorescence: $[Ca^{2+}] = 0$ (●); $[Ca^{2+}] = 10^{-4}$ M (■); $[Ca^{2+}] = 10^{-3}$ M (▲)

Fig. 5. Interparticle exchange: $[NaCl] = 10^{-3}$ M (●); $[NaCl] = 5 \times 10^{-3}$ M (■); $[NaCl] = 10^{-2}$ M (▲)

To explain these results, the following adsorption model is proposed. Upon addition of the probe molecules to a clay suspension, immediate adsorption occurs on the external surface of the particles.

The second molecule which comes in, will be adsorbed preferentially next to the first one because of preferred van der Waals interactions between both. This explains the very efficient excimer formation at the clay surface.

The decrease of the excimer fluorescence with time is explained by a redistribution of adsorbed probe molecules from the external to the interlamellar surface. This decrease is more pronounced in the presence of Ca^{2+} because the interlamellar surface increases, as shown in the previous section.

A similar time dependence is noticed for pyrene adsorbed on silica [16]. The time dependence in those systems is correlated with the porosity of the particles.

When detergent molecules are adsorbed, they will be distributed over the external and the interlamellar surface by the mechanism explained above. The adsorbed probe molecules are preferentially located in the detergent clusters which are present both on the external and the interlamellar surfaces. No change of the fluorescence with time is expected as long as the detergent clusters are in their equilibrium configuration and no external perturbation is applied to them.

Exchange of adsorbed molecules between clay particles

When clay particles are added to a clay suspension with adsorbed probe molecules, after the attainment of the equilibrium excimer fluorescence, the fluorescence intensity decreases with time to reach a constant value (Fig. 5).

When these experiments are repeated in the presence of different concentrations of NaCl, the rate of decrease of the fluorescence grows with increasing NaCl concentration.

These observations can be explained by a redistribution of the probe molecules between the clay particles within the time scale of our experiment. By adding new clay particles the redistribution of probe molecules necessary leads to a decrease of the excimer fluorescence because the probability of encounter of two probe molecules decreases with an increase of surface area. When the ionic strength of the medium increases, the repulsive potential between the clay particles decreases and so does their average distance. As a consequence, the rate of redistribution of the probe molecules increases.

Acknowledgements

The authors thank the European Research Office, the University Research Fund, the FKFO, the FGWO and the Ministry of Scientific Planning for financial support to the laboratory. K. V. and M. C. thank the NFWO for a doctoral fellowship.

References

1. Abdo S, Canesson P, Cruz M, Fripiat JJ, Van Damme H (1981) J Phys Chem 85:797
2. Della Guardia RA, Thomas JK (1983) J Phys Chem 87:990
3. Della Guardia RA, Thomas JK (1984) J Phys Chem 88:964
4. Della Guardia RA, Thomas JK (1983) J Phys Chem 87:3550
5. Ghosh PK, Bard AJ (1984) J Phys Chem 88:5519
6. Habti A, Keravis D, Levitz P, Van Damme H (1984) J Chem Soc Faraday Trans 2, 80:67
7. Nakamura T, Thomas JK (1985) Langmuir 1:568
8. Nakamura T, Thomas JK (1987) Langmuir 3:234
9. Nakamura T, Thomas JK (1986) J Phys Chem 90:641
10. Schoonheydt RA, De Pauw P, Vliers D, De Schryver FC (1984) J Phys Chem 88:5113
11. Viaene K, Ji C, Schoonheydt RA, De Schryver FC (1987) Langmuir 3:107
12. Theng B (1978) The Chemistry of Clay-Organic Reactions. Adam Hilger, London
13. Mineralogical Society monograph, no 6 (1987) Chemistry of clays and clay minerals. Newman ACD, London
14. Viaene K, Schoonheydt RA, Crutzen M, Kunyima B, De Schryver FC (1988) Langmuir 4:749
15. Verbeeck A, Ph D Thesis KU Leuven (1987)
16. Wellner E, Ottolenghi M, Avnir D, unpublished results

Received December 10, 1987;
accepted December 21, 1988

Authors' address:

K. Viaene
Department of Chemistry
Celestijnenlaan 200 F
3030 Leuven, Belgium

Progress in Colloid & Polymer Science

Progr Colloid Polym Sci 76:247–250 (1988)

The behaviour of mixed hydrocarbon-fluorocarbon surface active agents at the air-water interface

S. J. Burkitt[1]), B. T. Ingram[2]) and R. H. Ottewill[1])

[1]) School of Chemistry, University of Bristol, U.K.
[2]) Procter and Gamble (NTC), Newcastle-upon-Tyne, U.K.

Abstract: The behaviour of binary mixtures of ammonium perfluorooctanoate with ammonium octanoate, ammonium decanoate and ammonium dodecanoate at the air-water interface has been investigated in detail by surface tension measurements. These were carried out both below and above the critical micelle concentration for mixtures containing mole fractions of the alkanoate ranging from zero to unity. It was found that the lowest surface tensions (~ 15 mN m^{-1}) were obtained in the mixture with ammonium octanoate. The variation of the critical micelle concentrations of the mixtures with mole fraction was found to be non-ideal and in the case of ammonium perfluorooctanoate — ammonium dodecanoate was strongly indicative of independent micellisation.

Key words: Surface tension, ammonium perfluoro octanoate, ammonium alkanoate.

Introduction

In recent studies [1, 2] we have investigated, using small angle neutron scattering, the behaviour of mixed hydrocarbon-fluorocarbon surface active agent systems. From a fundamental point of view, micelles of these materials are of considerable interest, since the fluorocarbon chain is rather stiff and contrasts with the more flexible hydrocarbon chain. Also, a question then arises as to whether hydrocarbon and fluorocarbon surface active agents can combine to form mixed micelles [2] or whether each prefers to remain in its own environment and thus form independent micelles [3].

In this contribution we have investigated the air-water interfacial tensions of mixtures of ammonium perfluoro octanoate (APFO) with ammonium octanoate (AmOct), ammonium decanoate (AmDec) and ammonium dodecanoate (AmDod). These results have enabled the limiting surface tensions of the air-water interface to be obtained for each of the mixtures. In addition, the critical micelle concentrations have been obtained as a function of the mole ratio of the two components. These data have then been compared briefly with models of mixed systems using ideal solution [4, 5] and regular solution [6] approaches. A more detailed analysis will appear in due course [7].

Experimental

Materials

Octanoic acid, decanoic acid and dodecanoic acid were Fluka puriss grade materials with an estimated g.l.c. purity > 99 %. The acids were converted to their ammonium salts by slow neutralisation with ammonium carbonate (BDH Analar grade material stabilised by ammonium carbamate). The ammonium perfluoro octanoate (APFO) was Rimar material.

Surface tension measurements

The air-water interfacial tension of the solutions of surface active agents in ammonium chloride — ammonium hydroxide buffer solutions were measured using a Krüss Interfacial tensiometer based on the ring method of du Nouy [8, 9]. The accuracy of the instrument, after calibration, was checked by measuring the surface tension of doubly distilled water. All measurements were carried out at 25 °C.

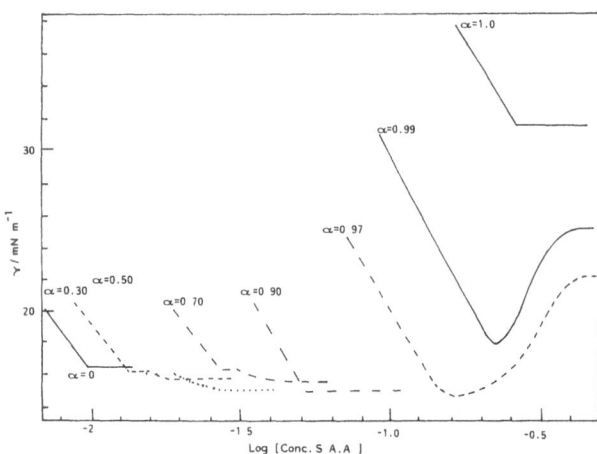

Fig. 1. Surface tension against logarithm of surface active agent concentration for mixtures of ammonium perfluoro octanoate and ammonium octanoate. pH = 8.4 and ionic strength = 0.1

Results

Mole fraction, α

The stoichiometric mol fraction, α, of the mixed surface active agents was defined by the expression,

$$\alpha = \frac{[AmAlk]}{[AmAlk] + [APFO]}$$

where [AmAlk] = the concentration of the ammonium alkanoate in mol dm^{-3} and [APFO] = the concentration of ammonium perfluoro octanoate.

Surface tension measurements

The surface tension results obtained on AmOct-APFO mixtures at various α values are shown in Fig. 1 in the form of surface tension [γ] against the logarithm of the total concentration of surface active agent. As can be seen from this figure small additions of APFO to AmOct had a dramatic effect on the surface tension and at $\alpha = 0.97$, a 3 % molar addition of APFO, the surface tension of water was reduced to ca 15.5 mN m^{-1} at the minimum of the curve.

Critical micelle concentrations

The critical micelle concentrations (CMC) were obtained from the γ against log [concentration] curves. In the case of curves having a sharp break at the CMC, for example, at α-values of 0, 0.3, 0.5, 0.7, 0.9 and 1.0 the first break was taken as the CMC. For $\alpha = 0.98$ and $\alpha = 0.99$ the value was chosen as the minimum in the curve; although open to objection it does not significantly affect the conclusions drawn from the results.

The results obtained for the CMC's of the various mixtures are listed in Table 1, together with the surface tension values obtained from the curves in the plateau region above the CMC.

Figure 2 shows a plot of the limiting surface tension values as a function of the mol ratio, α, for the three mixed systems. A number of points of interest arise from this graph. Although the longest chain hydrocarbon material, ammonium dodecanoate, produces the lowest surface tension at the air-water interface and

Table 1. CMC and surface tension values – mixed systems

Mol Ratio	AmOct-APFO		AmDec-APFO		AmDod-APFO	
	CMC/	γ/	CMC/	γ/	CMC/	γ/
α	mol dm^{-3}	mN m^{-1}	mol dm^{-3}	mN m^{-1}	mol dm^{-3}	mN m^{-1}
0.00 [APFO]	0.0096	16.6	0.0096	16.7	0.0096	16.6
0.05	0.0098	16.5	0.0098	16.7	0.0097	16.4
0.10	0.01056	16.4	0.0099	16.3	0.0109	16.7
0.30	0.131	16.0	0.0110	16.2	0.0124	16.7
0.50	0.015	15.2	0.0145	16.0	0.0102	17.2
0.70	0.027	15.7	0.0214	16.0	0.0083	–
0.90	0.052	15.1	0.0427	18.6	0.0066	22.2
0.95	0.118	15.9	0.0495	20.1	0.0064	22.4
0.97	0.171	22.4	0.0589	21.9	–	–
0.98	0.188	23.2	–	–	–	–
0.99	0.224	25.3	0.0562	24.3	–	–
1.00 [AmAlk]	0.257	31.8	0.0407	26.2	0.0059	21.8

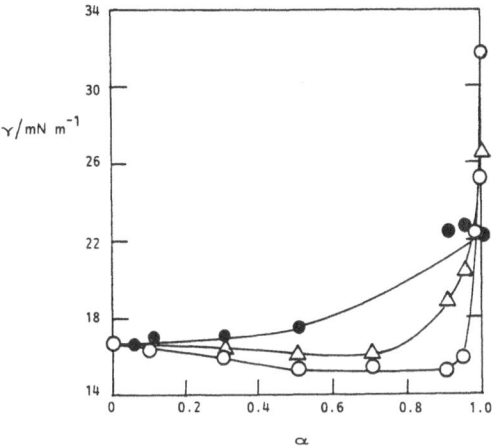

Fig. 2. Limiting surface tension of mixtures of surface active agents as a function of the mol ratio α. ($\bigcirc$) APFO-AmOct; ($\triangle$) APFO-AmDec; ($\bullet$) APFO-AmDod

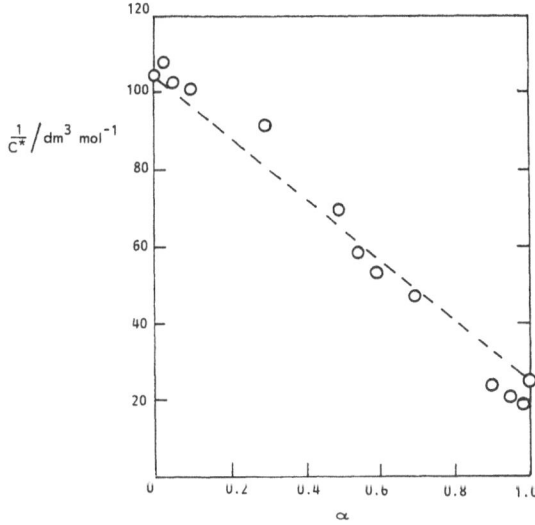

Fig. 4. $1/C^*$ against α for APFO-AmDec mixtures. (– – –) Ideal mixing; ($\bigcirc$) experimental results

ammonium octanoate the highest, the lowest overall surface tension is produced by the mixture of ammonium octanoate and ammonium perfluoro octanoate, a value of 15.1 mN m^{-1}. It should also be noted that for this system the limiting value is 22.4 mN m^{-1} at $\alpha =$ 0.97; at the minimum of the curve the values goes below 15 mN m^{-1}.

In Fig. 3, 4 and 5 the reciprocals of the CMC values, C^*, are plotted as a function of α. For comparison the value expected for ideal mixing of the surface active agents [4, 5] according to the equation,

$$\frac{1}{C^*} = \frac{\alpha}{C_{M_1}} + \frac{(1-\alpha)}{C_{M_2}} \tag{1}$$

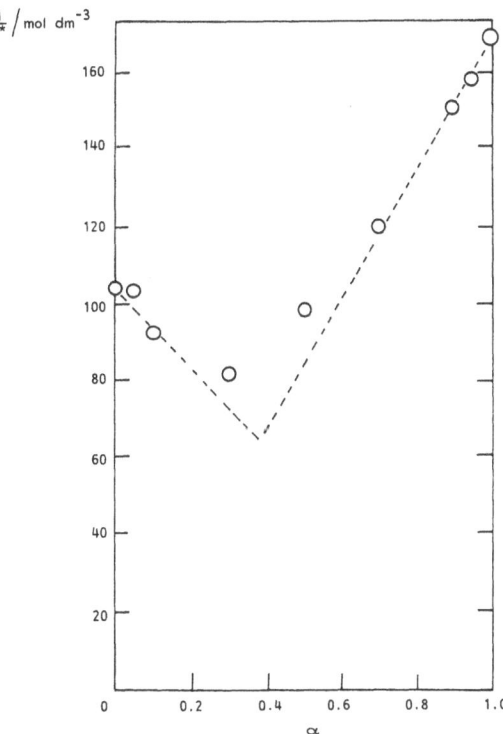

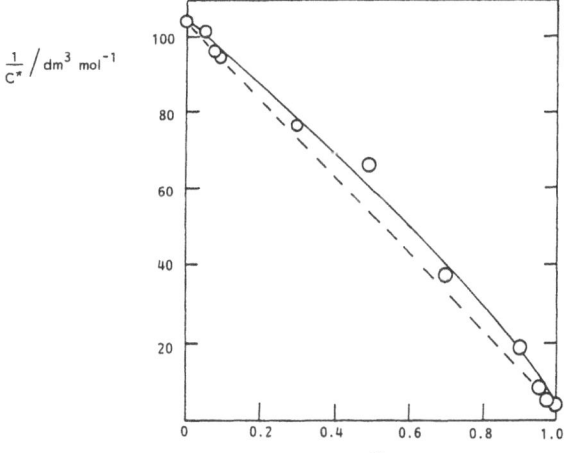

Fig. 3. $1/C^*$ against α for APFO-AmOct mixtures. (– – –) Ideal mixing; (——) regular solution theory with $\beta = -1.3$; ($\bigcirc$) experimental results

Fig. 5. $1/C^*$ against α for APFO-AmDod mixtures. (– – –) Calculated on the basis of independent micellisation; ($\bigcirc$) experimental results

is also plotted. In this equation C_{M_1} was taken as the CMC of the ammonium alkanoate and C_{M_2} the critical micelle concentration of APFO.

Discussion

The surface tensions of mixtures of APFO with respectively AmOct, AmDec and AmDod have been investigated in ammonium chloride: ammonium hydroxide solutions; the pH values of these systems were 8.4, 8.8 and 9.2. Figure 3 shows that for the AmOct-APFO systems, small negative deviations from ideality are found over the whole mole fraction range. For the AmDec-APFO system (Fig. 4) the deviations are negative at α values between zero and ca. 0.5 but become positive above this. For the AmDod-APFO system very substantial positive deviations from ideality are observed.

For the case of regular solutions it has been suggested by Rubingh [6] that Eq. (1) can be written in the form,

$$\frac{1}{C^*} = \frac{\alpha}{f_1 C_{M_1}} + \frac{(1-\alpha)}{f_2 C_{M_2}} \tag{2}$$

where

$$f_1 = \exp \beta (1-x)^2 \text{ and } f_2 = \exp \beta x^2$$

with

$$x = \frac{\alpha f_2 C_{M_2}}{\alpha f_2 C_{M_2} + (1-\alpha) f_1 C_{M_2}}$$

the mol fraction of component 1 in the mixed micelle. For regular solutions the value of β should remain constant with α.

In the case of AmOct-APFO although there was some variation of β with α it was possible to obtain a reasonable fit to the experimental data with a β value of -1.3. In the case of AmDec-APFO the values changed from being a negative quantity at low α values to being

positive at high α values, suggesting a variation from mutual mixing to weak mutual phobicity. Both of these systems have been examined by small angle neutron scattering and evidence obtained for the formation of mixed micelles. However, in the case of AmDec-APFO the results were compatible with hydrocarbon rich and fluorocarbon rich zones within the micelles [2].

In the case of AmDod and APFO the β values obtained were all positive indicating mutual phobicity between the two surface active agents. Moreover, the values of β obtained were variable although over part of the curve an average value of $\beta = 2.2$ gave a reasonable fit. However, using the assumption that the two materials micellised independently the form of the $1/C^*$ against α curve was also calculated. The results are shown in Fig. 5 and, as can be seen, quite a good fit to the experimental results was obtained supporting the contention of independence. This now requires independent experimental confirmation.

References

1. Burkitt SJ, Ottewill RH, Hayter JB, Ingram BT (1987) Colloid Polym Sci 265:619
2. Burkitt SJ, Ottewill RH, Hayter JB, Ingram BT (1987) Colloid Polym Sci 265:628
3. Mukerjee P, Mysels KJ (1975) ACS Symposium Series 9:239
4. Lange H, Beck KH (1973) Kolloid ZZ Polym 251:424
5. Clint JH (1975) J Chem Soc Faraday Trans I 71:1327
6. Rubingh DN (1979) In: Mittal KL (ed) Solution Chemistry of Surfactants. Plenum, New York
7. Burkitt SJ, Ingram BT, Ottewill RH, to be published
8. Lecomte du Nouy P (1919) J Gen Physiol 1:521
9. Harkins WD, Jordan HF (1930) J Am Chem Soc 52:1751

Received December 14, 1987;
accepted December 30, 1987

Authors' address:

Prof. Dr. R. H. Ottewill
School of Chemistry
University of Bristol
Cantock's Close
Bristol BS8 1TS, U.K.

Progress in Colloid & Polymer Science   Progr Colloid Polym Sci 76:251–259 (1988)

Contact angle hysteresis due to surface roughness

M. Bracke[1]), F. De Bisschop[2]), and P. Joos[1])

[1]) University of Antwerp, Department Biochemistry, Wilrijk, Belgium
[2]) State University of Ghent, Physics Laboratory Medical Faculty, Ghent, Belgium

Abstract: By means of the calculus of variations it is shown that even on rough surfaces the Young equation applies and that the Wenzel equation relies on a false assumption. Contact angle hysteresis, i. e. the difference in the apparent advancing and receding angles, for homogeneous rough solid substrates is due to the local slope of the solid surface at the three phase contactline.

The maximum advancing and receding contact angles were measured using the tilted plate method of Wolfram. The thermodynamic Young angle is the arithmetic mean between these advancing and receding angles. The hysteresis was constant for a given sample of a solid substrate. Furthermore, we define the rugosity of the surface as a function of this hysteresis.

By means of Young angle measurements, adsorption of Triton X-100 solutions as a function of the concentration is investigated. The adsorption at apolar surfaces (paraffin) is equal to that at the air/liquid interface. For more polar substrates, such as polyethylene (PE), this adsorption is lower and is nearly zero for polar substrates such as polyethyleneterephtalate (PET).

By coronization of PE, the rugosity as well as the polarity increases. Also, the surface is electrically charged, but these charges disappear slowly in time. The strength of this paper is that we take into account both the advancing and the receding contact angles, whereas in the literature the experimental receding values are neglected.

Key words: Contact angle hysteresis, surface roughness, Wenzel's equation, adsorption at the solid surface, coronization, surface properties.

Introduction

Measurements of the contact angle at the solid/liquid/air contact line exhibit a lack of reproducibility. Depending on the way in which the liquid is brought into contact with the solid, there are advancing and receding angles, although the Young equation requires a unique contact angle. It is known that contact angle hysteresis depends on surface roughness, since contact angle hysteresis disappears on very smooth homogeneous surfaces [1–4]. We are aware that other parameters such as surface heterogenity also affect hysteresis, but in this paper we will only pay attention to surface roughness.

For rough surfaces, at least according to the literature, the Young equation has to be replaced by the Wenzel equation [5], which differs from the Young equation by a surface roughness coefficient. The comprehensive mathematical treatment of Cox [6] makes

it clear that the Wenzel angle will be exhibited if contact angle hysteresis is absent. Also, Johnson and Dettre [11] have shown that Wenzel's equation cannot say anything about contact angle hysteresis. There are a number of theoretical studies on contact angle hysteresis but, unfortunately, experimental proof of these theories is almost non-existent. Furthermore these theoretical studies rely merely on mechanical grounds. In this paper, the calculus of variations is applied to a highly artificial system with rotational symmetry and it is found that the Wenzel equation does not apply here. In view of the work of Cox [6], which considers single and double periodic corrugations on the surface, we think that this conclusion is general. Therefore, contact angle hysteresis (by which we mean the difference between advancing and receding angles) can be understood by surface roughness, which can be obtained experimentally using a concept first put forward by Shuttleworth and Bailey [7].

The apparent contact angle will be related to the real contact angle (Young angle) and to the thermodynamically defined surface roughness.

In the experimental part, maximum advancing and minimum receding angles are measured and theoretically discussed. No dynamic contact angles due to liquid movement, are considered.

Theoretical basis

1. The Young equation

The Young equation is usually demonstrated by making use of force balance considerations, but it can also be obtained as a minimization requirement of free energy in a given equilibrium system.

In order to outline the method and to show the connection with rough surfaces, we will first consider a liquid drop in equilibrium on a horizontal, smooth, homogeneous solid surface (Fig. 1).

The liquid surface tension is σ_{lv}, the surface energy at the solid vapour interface is σ_{sv} and σ_{sl} is the interfacial energy at the solid liquid interface.

The liquid profile is expected to show rotational symmetry about the y axis, the x axis being directly along the solid surface. The free energy contribution depending on the liquid profile $y(x)$ for this system is given as:

$$F = 2\pi \int_0^{x_P} \left[\frac{1}{2} \varrho g y^2 - \sigma_{lv} y (1 + y'^{-2})^{1/2} + (\sigma_{sl} - \sigma_{sv}) \right] x \, dx \tag{1}$$

The first term in the integrand, where ϱ is the density of the liquid, accounts for the potential energy due to gravitation, the second term for the interfacial energy at the liquid vapour interface and the last term stands for the energy contribution due to partial occupa-

tion of the solid surface by liquid, the total solid surface inside being constant.

On should expect that the second term of the integrand of Eq. (1), giving the contribution of the free energy of the liquid/vapor interface, is equal to the surface tension σ_{lv} multiplied by the arc length: $(1 + y'^{+2})^{1/2}$.

This expression is correct for a contact angle $\theta \leq 90°$. For an obtuse angle ($\theta > 90°$), there is a domain (overhanging drop) for certain values of x ($x \geq x_p$) where y is not a single valued function of x. A closer inspection shows that over the whole domain the arc length can be written as $- y'(1 + y'^{-2})^{1/2}$. After integration over the correct limits, the total expression may be useful for numerical calculation of the drop profile (in the extreme value of x, i. e. x_p, y' and dx change over sign).

Furthermore, the formulation given in Eq. (1) contains only conservative forces, the reason why the present model is not adequate to describe a liquid drop in motion, where viscous dissipation should be considered.

The free energy contribution formulated in Eq. (1), must be minimal in equilibrium circumstances. This minimum however, is bound to the requirement of constant liquid volume V, written here as:

$$V = 2\pi \int_0^{x_P} xy \, dx . \tag{2}$$

This is a bound extremation problem, where the constraint – in this case the constant liquid volume – should be introduced in the extremized quantity. Therefore, the constraint is multiplied by a parameter λ, a Larange multiplier, leading to a new function Φ, to be treated formally as a freely extremized quantity:

$$\Phi = 2\pi \int_0^{x_P} \left[\frac{1}{2} \varrho g y^2 - \sigma_{lv} y (1 + y'^{-2})^{1/2} + (\sigma_{sl} - \sigma_{sv}) + \lambda y \right] x \, dx . \tag{3}$$

The integrand of this equation will further be written simply as $\Psi(x, y, y')$ and, according to the calculus of variations, the Euler equation applies:

$$\frac{\partial \Psi}{\partial y} - \frac{d}{dx} \left(\frac{\partial \Psi}{\partial y'} \right) = 0 . \tag{4}$$

It has been experimentally verified that Eq. (1) and the subsequent mathematical treatment of the extremation problem as given, lead to the exact liquid profile $y(x)$ [8], not only in the case of a sessile drop but in all cases investigated.

As was found elsewhere [8–10], the Euler equation leads to the well known Laplace formula.

However, since in Eq. (3) the limits of integration x_0 and x_P could also be varied, a complementary condition at each end point must be fulfilled

$$\left[\Psi - y' \left(\frac{\partial \Psi}{\partial y'} \right) \right] \delta x + \left(\frac{\partial \Psi}{\partial y'} \right) \delta y = 0 . \tag{5}$$

At the lower end point (x_0, y_0), the variation $\delta x_0 = 0$, but $\delta y_0 \neq 0$.

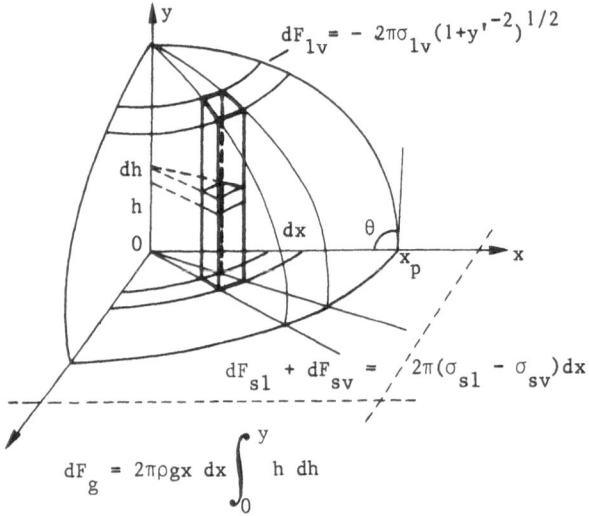

Fig. 1. Physical model for a liquid drop on a smooth solid surface

From Eq. (5), a trivial characteristic of the lower end point is obtained, i. e.: $y_0' = 0$.

Admitting with respect to the upper end point (p) that $\delta x_p \neq 0$, $\delta y_p = 0$; Eq. (5) leads to:

$$\frac{\sigma_{lv}}{y_p'(1 + y_p'^{-2})^{1/2}} = \sigma_{sl} - \sigma_{sv}. \tag{6}$$

Since $y_p' = -tg\,\theta$, for a sharp as well as for an obtuse contact angle, we can write ater substitution in Eq. (6):

$$\sigma_{sv} - \sigma_{sl} = \sigma_{lv} \cos\theta_y. \tag{7}$$

This is the well known Young equation, in which we write θ_y instead of θ, to draw attention to the fact that we are dealing with the Young contact angle. This kind of derivation, making use of variational calculus is already known [8–10]; it is repeated here to clarify where the Wenzel equation can be derived from and how the situation changes with a rough solid surface.

2. The Wenzel equation

For a rough solid surface, the profile of the solid can be written as $w(x)$, assuming rotational symmetry about the y axis (Fig. 2). In that case, one obtains for Eq. (1)

$$F = 2\pi \int_0^{x_p} \left[\frac{1}{2} \varrho g(y^2 - w^2) - \sigma_{lv}y'(1 + y'^{-2})^{1/2} \right. $$
$$\left. + (\sigma_{sl} - \sigma_{sv})(1 + w'^2)^{1/2} \right] x\,dx \tag{8}$$

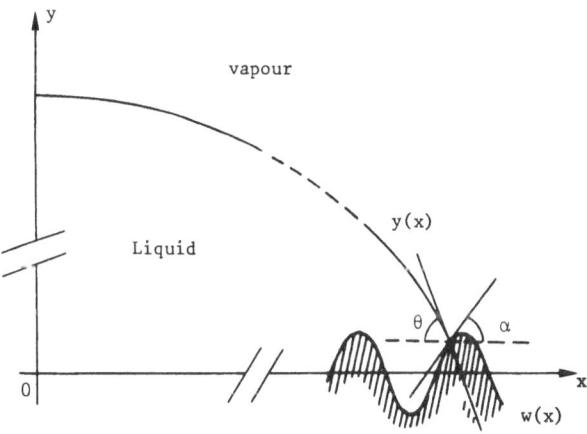

θ : apparent contact angle

$\theta_y = \theta + \alpha \quad (\theta_y$: Young angle $)$

$w_p' = tg\,\alpha$

$y_p' = tg(\alpha - \theta_y)$

Fig. 2. Liquid drop on a rough solid surface

and for Eq. (2), concerning the constant liquid volume:

$$V = 2\pi \int_0^{x_p} (y - w)\, x\,dx. \tag{9}$$

The integrand $\Psi(x, y, y')$ in Eq. (3), should now be written as:

$$\Psi = \left[\frac{1}{2}\,\varrho g(y^2 - w^2) - \sigma_{lv}y'(1 + y'^{-2})^{1/2} \right.$$
$$\left. + (\sigma_{sl} - \sigma_{sv})(1 + w'^2)^{1/2} + \lambda(y - w) \right] x \tag{10}$$

assuming the profile of the solid surface $w(x)$ to be a single valued function of x.

At the contact point (x_p, y_p) we can write $y_p - w_p = 0$. We will further examine the consequences of a deliberately false statement, i. e. $\delta y_p = 0$, in order to show where the Wenzel equation fails.

From Eq. (5) we would obtain

$$\frac{\sigma_{lv}}{y_p'(1 + y_p'^{-2})^{1/2}} = (\sigma_{sl} - \sigma_{sv})(1 + w_p'^{+2})^{1/2} \tag{11}$$

with $y_p' = -tg\,\theta$, θ representing an apparent contact angle (Fig. 3), we could write

$$\sigma_{lv} \cos\theta = (\sigma_{sl} - \sigma_{sv})(1 + w_p'^{+2})^{1/2}. \tag{12}$$

Assuming that the surface rugosity r can be defined as

$$r = (1 + w_p'^{+2})^{1/2}$$

one obtains form Eq. (12) the well known Wenzel equation [5]:

$$\sigma_{lv} \cos\theta = r(\sigma_{sl} - \sigma_{sv}). \tag{13}$$

Obviously, Eq. (13) is incorrect, since the deduction started from a manifestly false condition $(\delta y_p = 0)$. Johnson and Dettre [11] came to the same conclusion: this is one of the assumptions we have to make to derive Wenzel's equation, i. e. the variation on the solid surface has to correspond to exactly one period of the surface.

We have shown theoretically why the Wenzel equation does not apply, but only for a highly artificial system of roughness with rotational symmetry. We think the requirement of rotational symmetry is not so serious. Firstly, Cox [6] also made his analysis for periodic roughness in one direction and generalized his results to random periodic roughness. Furthermore, it can be shown that by using the calculus of variations the same results can be obtained for a smooth vertical surface at right angles to a rough horizontal surface and wetted by the same liquid drop. Experimental data will be presented to disprove once more the Wenzel equation.

3. The Young equation on rough surfaces

As already argued in the previous section, the boundary condition $\delta y_p = 0$ is incorrect, dealing with rough surfaces, and should be replaced by:

$$\left(\frac{\partial y}{\partial x} \right)_p = w_p'.$$

Inserting this into Eq. (5), one obtains:

$$\sigma_{lv} \frac{1 + y'_p w'_p}{y'_p(1 + y'^{-2}_p)^{1/2}} - (1 + w'^{+2}_p)^{1/2} (\sigma_{sv} - \sigma_{sl}) = 0.$$

(14)

Refering to Fig. 3, in which the interrelation between the thermodynamically defined contact angle and the apparent contact angle is indicated, and depending on the slope of the solid surface at the endpoint (x_p, y_p), we could write

$$w'_p = tg\,\alpha \tag{15a}$$

$$w'_p = -\,tg\,\alpha' \tag{16a}$$

and consequently

$$y'_p = -\,tg\,\theta \tag{15b}$$

$$y'_p = -\,tg\,\theta'. \tag{16b}$$

The angle θ is an apparent receding contact angle θ_R and θ' an apparent advancing contact angle θ_A. From geometrical arguments it follows that the angle θ_y the liquid profile makes with the solid surface can be written as

$$\theta_y = \theta_R + \alpha \tag{15c}$$

$$\theta_y = \theta_A - \alpha'. \tag{16c}$$

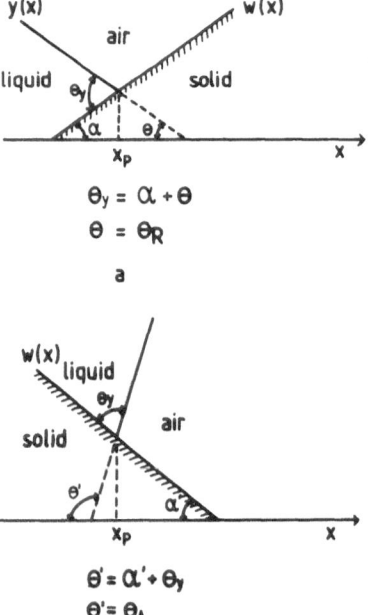

Fig. 3. Interrelation between the apparent contact angle and the thermodynamically defined contact angle (Young angle)

Substituting these Eqs. (15, 16) in Eq. (14) it follows that:

$$\sigma_{lv} \cos \theta_y = \sigma_{sv} - \sigma_{sl}. \tag{17}$$

Attention should be stressed that α and α' change continuously over the solid surface, hence an infinite set of apparent contact angles results.

Following Bailey et al. [7], one can expect that when the contact line meets the steepest slope α^M, or at least these steepest corrugations the system can sustain, one observes a maximum advancing θ_A^M, and minimum receding contact angle θ_R^m.

In this context we reformulate the equation put forward by Bailey [7] and repeated by others [11, 12]:

$$\theta_y = \frac{\theta_A^M + \theta_R^m}{2} \tag{18}$$

and

$$\alpha^M = \frac{\theta_A^M - \theta_R^m}{2} \tag{19}$$

from which the surface roughness can be defined as

$$r = (1 + tg^2\,\alpha^M)^{1/2} \geq 1. \tag{20}$$

4. Comparison with other work

There are many theoretical papers on contact angle hysteresis. However, it is beyond the scope of this paper to give a review of them. We want to discuss just a few of them, merely as an illustration, to point out the difference from our present work.

Johnson and Dettre [11] started from the same geometrical point of view, used also by Shuttleworth and Bailey [7], as we did. However, they still supported the Wenzel equation and considered the Wenzel contact angle the most probable one. Good et al. [12] extremized the free energy by a computer method. Their impression of the minimized free energy by a computer method. Their impression of the minimized free energy is fundamentally equivalent to ours. However, they did not consider the constant liquid volume as a constraint with respect to extremization, and they pointed out that the Wenzel equation is correct in a gravitational field, an assertion rejected by Johnson and Dettre [11].

Wolfram [13] correctly extremized the free energy, without considering the effect of gravitation, and considered only a flat solid surface. Neglection of gravitational potential energy is allowed, as far as it is not relevant in the complementary condition (cf. Eq. (5)), since the term concerned cancels. After this, however, he introduced the gravitational force as a gradient of potential energy, and this is incorrect. In a second step, he introduced the Wenzel equation in a somewhat obscure way.

Huh and Mason [14] obtained a modification of the Wenzel equation, and stated that the Wenzel equation is still correct for large drops. We stress the fact that these modified equations are not essentially different from the Wenzel equation. Cox [6] obtained a similar expression to ours (Eqs. (16), (17)) and confined his results to periodic surface corrugations. But he agrees that his theory fails for non-periodic surfaces. The right conclusion is obtained, i. e. that at a microscale level of surface roughness, the Young equation applies. All these results, including ours, are merely theoretical; hence, there is an urgent need for experimental confirmation.

Experimental methods

The wetting of solid substrates was examined with aqueous Triton X-100 solutions. Triton X-100 was purchased from Merck and used without further purification. Surface tensions as a function of concentration are given elsewhere [15].

We mentioned that very recently measurements have been given on advancing and receding angles on rough substrates [16] and it was found that the hysteresis H, defined as the difference between the advancing and the receding angle, is constant for all wetting solutions. [In fact our angle $\alpha^M = H/2$] For these wetting solutions we hesitate to consider also polymer surfactants, because adsorption for these solutions is more complex. The data presented by Johnson et al. [16] for non-polymer surfactants, when analysed in the way mentioned in the following, lead to the same conclusion as our experimental data.

For apolar surfaces such as paraffin and polyethylene, maximum advancing and minimum receding contact angles were measured, using the tilted plate method, as described by Wolfram [13]. The drops were photographed and projected. For polyethylene-therephtalate (PET) and cellulose triacetate (CTA), advancing and receding angles were measured, using strips of these materials as Wilhelmy plates with a Cahn eletrobalance.

Results and discussion

1. Contact angle of pure water on paraffin

Wolfram [13] gives extensive results of the maximum advancing and minimum receding contact angles for pure water drops on paraffin substrates with different surface roughness. Using Eq. (18), we calculated from his results the Young angle as $\theta_y = 104 \pm 2\,^\circ$. We did similar experiments, as discussed in the next section, and also obtained $\theta_y = 104 \pm 2\,^\circ$.

Our paraffin was simple Parafilm and its roughness (r), as defined by Eq. (20), was found to be $r = 1.005$.

Neumann [3] measured a Young angle on vacuum deposited paraffin, a smooth surface without contact angle hysteresis, and found 105°.

From these data, we conclude that the true Young contact angle for water on paraffin is: $\theta_y = 105\,^\circ$, which

results in a nice agreement with the theory of Fowkes. Fowkes [17] presented a table of surface tensions of alkanes and interfacial tensions of these alkanes with water, as a function of chain length. It gives that the surface tenison σ_{av} increases with chain length and that σ_{al} increases only weekly. We plotted the surface tension σ_{av} as a function of n^{-1} (n chain length), and extrapolated to $n = \infty$, and obtained $\sigma_{sv} = 33.6\ 10^{-3}$ Nm^{-1}. Assuming the contribution of dispersion forces to the surface tension of water to be $\sigma_{lv}^d = 22.0\ 10^{-3}$ Nm^{-1}, we calculated the interfacial tension, using the equation of Fowkes [18] for a totally apolar substrate (paraffin):

$$\sigma_{sl} = \sigma_{sv} + \sigma_{lv} - 2(\sigma_{sv}\,\sigma_{lv}^d)^{1/2} \qquad (21)$$

yielding: $\sigma_{sl} = 52.0\ 10^{-3}$ Nm^{-1} ($\sigma_{lv} = 72.8\ 10^{-3}$ Nm^{-1}).

Using these data, we calculated the contact angle and found: $\theta_y = 104.6\,^\circ$, a result in remarkable agreement with the experimental data, notwithstanding the fact that our extrapolation procedure for obtaining the surface tension of paraffin from values for liquid alkanes remains questionable.

2. Wetting of paraffin by a surface active solution (Triton X-100)

Surface roughness is a mechanical property of the solid surface, which manifests itself by contact angle hysteresis. This surface roughness must be independent of the surface tension of the wetting liquid. Using the tilted plate method, we have measured the maximum advancing and minimum receding contact angle on solid paraffin. Making use of Eq. (18), we have obtained the Young angle.

We have plotted the results, following Lucassen-Reynders [19] as $\sigma_{lv} \cos\theta$ vs. σ_{lv}, σ_{lv} being a function of the surfactant concentration. As shown by Lucassen-Reynders:

$$\frac{d}{d\sigma_{lv}}(\sigma_{lv}\cos\theta) = \frac{\Gamma_{sv} - \Gamma_{sl}}{\Gamma_{lv}} = -\frac{\Gamma_{sl}}{\Gamma_{lv}} \qquad (22)$$

where: Γ_{sv} = adsorption at the solid/vapour interface, which is zero in the present situation; Γ_{lv} = adsorption at the liquid/vapour interface; Γ_{sl} = adsorption at the solid/liquid interface.

As shown by Van Voorst Vader [20] for the paraffin/surfactant system, the slope $d(\sigma_{lv}\cos\theta_y)/d\sigma_{lv} = -1$, over the whole concentration range, indicates equal adsorption at the vapour/water as at the solid/

water interface. Our results, shown in Fig. 4, confirm these of Van Voorst-Vader, although these angles are of the advancing type. Our experiments are described by $\sigma_{lv} \cos \theta_y = - 0.99\ \sigma_{lv} + 54.4$ (linear regression).

The critical wetting tension, when $\theta_y = 0°$, is 27.4 10^{-3} Nm^{-1}. With a maximum angle $\alpha^M = 5.5°$, we can calculate $\sigma_{lv} \cos \theta_A^M$ and $\sigma_{lv} \cos \theta_R^m$. The rugosity $r = 1.005$. It is seen that the theoretical curve follows the experimental receding and advancing angles closely.

3. Wetting of polyethylene (PE) by a surface active solution (Triton X-100)

We used two PE samples of different roughness, a smooth one manufactured with a steel drum and a rougher one made with a rubber drum. The difference in roughness can be observed by electron microscopy.

According to our theory, both PE samples are expected to show the same Young angles θ_y when wetted by the same surfactant solution, although the advancing and receding angles due to different roughness, are different. As seen from Figs. 5 and 6, both Young angles are very close indeed. For pure water, we find a Young angle $\theta_y = 82°$, whereas Wolfram [13] gives $\theta_y = 85°$.

From the slope of the curve we obtain $\Gamma_{sl}/\Gamma_{lv} = 0.56$, and the experimental data are described by $\sigma_{lv} \cos \theta_y =$

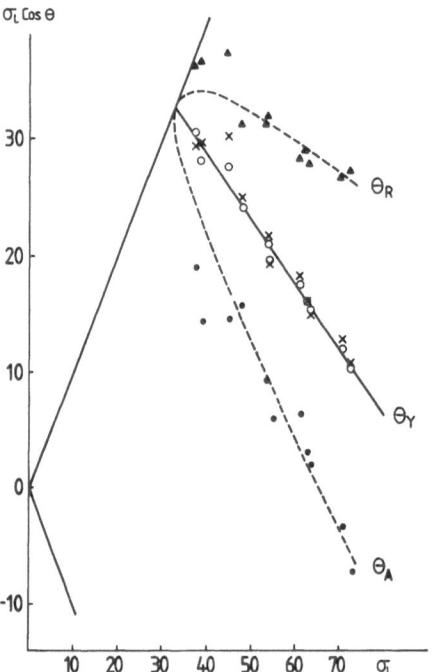

Fig. 5. Triton X-100 solutions on smooth PE. (O, ×) Young angles on smooth and rough PE; (▲) receding angles; (●) advancing angles

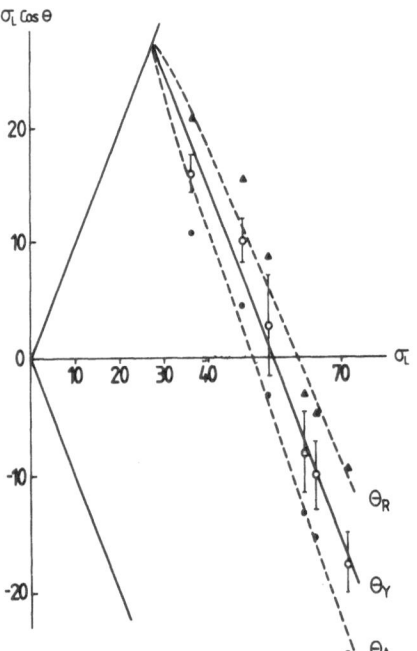

Fig. 4. Triton X-100 solutions on paraffin. (O) Young angles; (▲) receding angles; (●) advancing angles

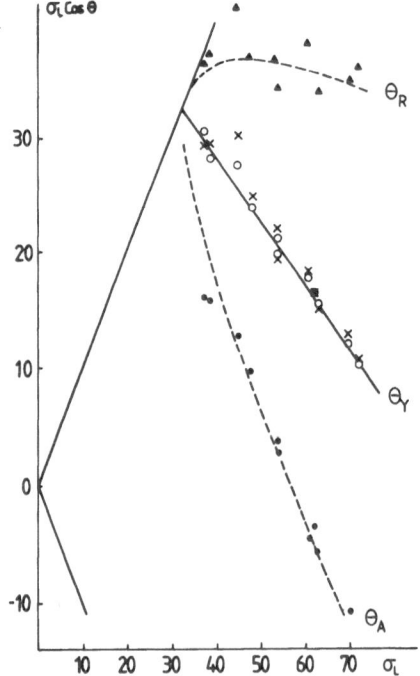

Fig. 6. Triton X-100 solutions on rough PE (same as Fig. 5)

0.56 σ_{lv} + 52.2 (linear regression); the critical wetting tension being: $\sigma_c = 32.7 \ 10^{-3} \ \text{Nm}^{-1}$.

Using this empirical equation for the Young angles, we have calculated the advancing and receding angles for the smooth ($\alpha^M = 13°$, $r = 1.026$) and the rough ($\alpha^M = 20°$, $r = 1.064$) PE sample.

The agreement between calculated and experimental values is very satisfactory (see Figs. 5 and 6).

4. The wetting of polyethyleneterephtalate (PET) by a surface active solution (Triton X-100)

Advancing and receding angles (as always, extreme values are meant) are obtained, using PET strips as Wilhelmy plates, pending from a Cahn electro balance. The results, given in Fig. 7, are described by the relation $\sigma_{lv} \cos \theta_y = -0.06 \ \sigma_{lv} + 19.3$ (linear regression), the critical wetting tension being $\sigma_c = 18.7 \ 10^{-3}$ Nm^{-1}. Advancing and receding contact angles are calculated with $\alpha^M = 8°$, corresponding with $r = 1.010$.

It is noteworthy that by increasing the polarity of the solid (paraffin, PE, PET), characterized by the Young angle of pure water ($\theta_y = 103.5°$, $82°$ and $78.5°$ respectively), the ratio Γ_{sl}/Γ_{lv} decreases ($\Gamma_{sl}/\Gamma_{lv} = 1$, 0.56 and 0.08 respectively). At the PET/water interface, the adsorption of Triton X-100 is very small indeed.

5. The wetting of cellulose triacetate (CTA) by a surface active solution

Data for CTA were obtained in a similar way as for PET; the results are given in Fig. 8. The most striking fact is that the slope of $\sigma_{lv} \cos \theta_y$ vs. σ_{lv} is positive, suggesting that Γ_{sl}/Γ_{lv} is negative. Since Γ_{lv} is positive, the adsorption of Triton X-100 at the CTA/solution interface should be negative, which is very unlikely.

However, our CTA strips were electrically charged, probably leading to this unexpected result (positive slope, not the negative adsorption). Indeed, Eq. (22) was obtained as a combination of the Gibbs and Young equations, whereas the Gibbs equation only applies for uncharged interfaces. For charged interfaces, the Lippmann equation should be considered:

$$\frac{d}{d\sigma_{lv}} (\sigma_{lv} \cos \theta_y) = -\frac{\Gamma_{sl}}{\Gamma_{lv}} - \frac{q}{\Gamma_{lv}} \cdot \frac{d\psi}{d\mu} \qquad (23)$$

where q = surface charge; ψ = potential difference.

Assuming $\Gamma_{sl} \simeq 0$, it is not well understood why the potential difference (ψ) should depend on the chemical potential (μ) of the nonionic surfactant. The fact cannot be ruled out that with the receding meniscus, some liquid film remains on the solid substrate by redrawing the drop, and hence some surfactant may be adsorbed.

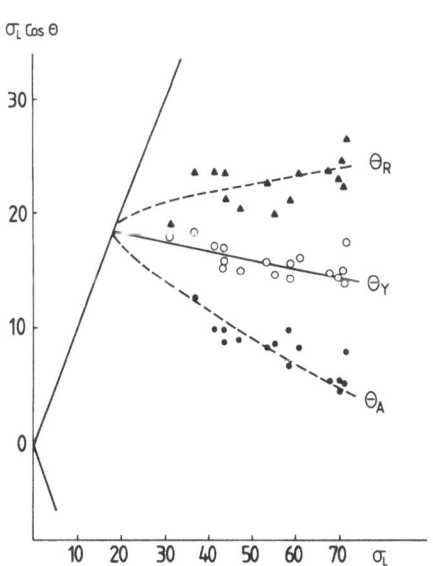

Fig. 7. Triton X-100 solutions on PET. (O) Young angles; (▲) receding angles; (●) advancing angles

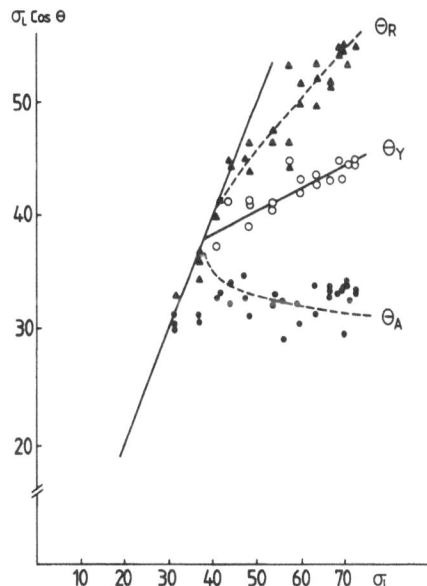

Fig. 8. Triton X-100 solutions on CTA. (O) Young angles; (▲) receding angles; (●) advancing angles

This corresponds to a rather high value of $\sigma_l \cos \theta_R$ which leads to a high value of $\sigma_l \cos \theta_y$ and also to a positive slope. The fact that the angle α^M obtained from the experiments is constant does not supplement the latter, but rather the first conclusion. The experiments with coronized PE samples, presented in the next section, may give some evidence for this.

6. Effect of coronization on the surface properties of polyethylene

Coronization is an industrial process to improve wetting. A PE sheet is drawn between two electrodes, and between these, a gas discharge takes place, the energy input being expressed as Jm^{-2}.

The PE material was the same as in Fig. 5, i.e. smooth PE with roughness $r = 1.026$, $\alpha^M = 13°$ and a critical wetting tension $\sigma_{lv} = 32.7 \ 10^{-3} \ Nm^{-1}$, the adsorption ratio being $\Gamma_{sl}/\Gamma_{lv} = 0.56$.

This PE sheet was coronized (energy input 7200 Jm^{-2}), left for 4 months before the wetting by Triton X-100 solutions was measured, the results of which are shown in Fig. 10.

It can be seen that the critical tension for wetting is increased: $\sigma_c = 37.6 \ 10^{-3} \ Nm^{-1}$ and so is the roughness: $r = 1.064$ with $\alpha^M = 20°$.

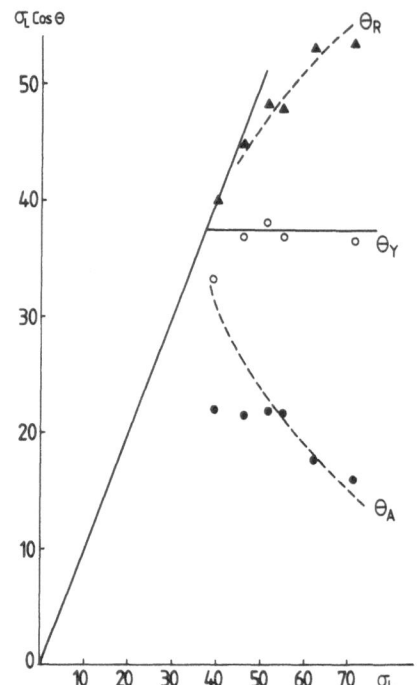

Fig. 10. Triton X-100 solutions on old coronized smooth PE. (O) Young angles; (▲) receding angles; (●) advancing angles

From Fig. 10, it follows that: $\Gamma_{sl}/\Gamma_{lv} \simeq 0$, indicating that the PE surface has undergone a chemical modification and became polar.

Another PE sample was coronized (energy 6000 Jm^{-2}) and the wetting by Triton X-100 solutions was measured, the whole procedure taking 2 days. The results are given in Fig. 9.

The critical wetting tension is high again, indicating improved wetting ($\sigma_c = 40 \ 10^{-3} \ Nm^{-1}$); the roughness is increased, $r = 1.103$ with $\alpha^M = 25°$, and the slope $d\sigma_{lv} \cos \theta_y / d\sigma_{lv}$ is positive, as for CTA, due to surface charges brought about by the coronization process.

In summary, coronization has the following consequences: (i) the surface roughness is increased, (ii) the surface is chemically modified and becomes polar, (iii) surface charges are brought onto the strip, and disappear after a sufficiently long time, whereby the critical wetting tension also decreases.

Conclusion

i. After Shuttleworth and Bailey [7], advancing and receding (extreme) angles are apparent contact angles, due to the different slopes of the corrugations on the smooth surface. Microscopically, the Young angles

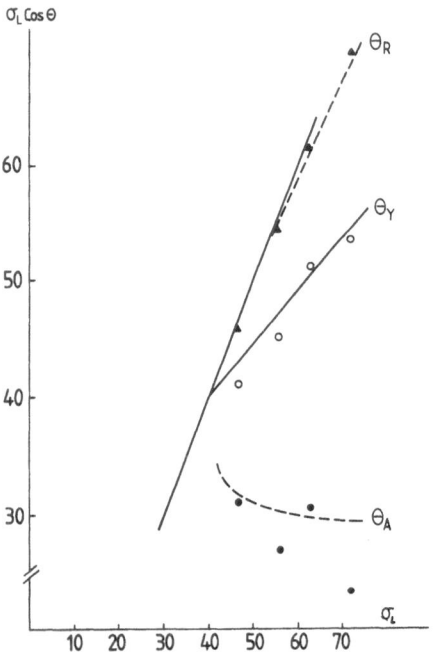

Fig. 9. Triton X-100 solutions on fresh coronized smooth PE. (O) Young angles; (▲) receding angles; (●) advancing angles

occur. Following this concept the hysteresis or the maximum angle, α^M as defined in Eq. (19), should be constant whatsoever the wetting solution. This is proven experimentally and is in agreement with experimental work of other investigators.

ii. Secondly, using this concept it is possible to obtain as a mean, the Young angle, for advancing and receding angles, which, together with the Gibbs equation, gives information about the adsorption of surfactants at the solid liquid interface. As expected, the more polar the solid interface, the less is the adsorption of surfactant at the interface.

iii. By coronization, the surface roughness is increased, the surface becomes more polar and electrical charge is set on the surface. This electrical charge disappears after a long time.

References

1. Oliver JF, Mason SG (1977) J Colloid Interface Sci 60:11
2. Ivanov IB (1977) In: Padday JF (ed) Wetting, spreading and adhesion. Academic Press
3. Neumann AW (1977) In: Padday JF (ed) Wetting, spreading and adhesion. Academic Press
4. Mason SG (1977) In: Padday JF (ed) Wetting, spreading and adhesion. Academic Press
5. Wenzel RN (1949) J Phys Colloid Chem 53:1466
6. Cox RG (1983) J Fluid Mech 131:1
7. Shuttleworth R, Bailey GLJ (1948) Discussions Faraday Soc 3:16
8. De Bisschop F (1979) Aanvullende studie van de theorieën van Young and Laplace, in verband met de statica van capillaire Fluida. Hogere aggregaatsthesis, Gent
9. Collins RE, Cooke Jr CR (1959) Trans Faraday Soc 55:1602
10. Johnson Jr RE (1959) J Phys Chem 63:1655
11. Johnson Jr RE, Dettre RH (1964) Adv Chem Ser 43:112
12. Eick JD, Good RJ, Neumann AW (1975) J Colloid Interface Sci 53:235
13. Wolfram E (1977) In: Padday JF (ed) Wetting, spreading and adhesion, Academic Press
14. Huh C, Mason SG (1977) J Colloid Interface Sci 60:11
15. Rillaerts E, Joos P (1982) J Phys Chem 86:3471
16. Johnson BA, Kreuter J, Zografi G (1986) Colloids Surf 17:325
17. Fowkes FM (1980) J Phys Chem 84:510
18. Fowkes FM (1968) J Colloid Interface Sci 28:493
19. Lucassen-Reynders EH (1963) Phys Chem 67:969
20. Van Voorst-Vader F (1977) Chem Ing-Tech 49:488

Received October 21, 1987; accepted November 23 1987

Authors' address:

M. Bracke
University of Antwerp
Department of Biochemistry
B-2610 Wilrijk, Belgium

Progress in Colloid & Polymer Science   Progr Colloid Polym Sci 266:260–264 (1988)

Interfacial activity of 1-(2′-hydroxy-5′-methylphenyl)-dodecane-1-one oxime and the interfacial mechanism of copper extraction

J. Szymanowski and K. Prochaska

Technical University of Poznań, Institute of Chemical Technology and Engineering, Poznań, Poland

Abstract: The interfacial tension isotherms for pure 1-(2′-hydroxy-5′-methylphenyl)-dodecane-1-one oxime at toluene/water and octane/water interfaces were determined and interpreted. The surface excess was calculated according to the Gibbs, Szyskowski and Temkin isotherms and the third order polynominal and used to predict reaction orders against hydroxyoxime for different mechanism versions of copper extraction from acidic sulphate solutions. The obtained results support the interfacial mechanism of copper extraction with the formation of stable 2:1 complex from intermedate 1:1 compelex and the hydroxyoxime molecule present in the aqueous layer near the interface as the slowest step.

Key words: Hydroxyoxime, copper extraction.

Introduction

In our previous papers [2–4] the interfacial tension isotherms were determined and interpreted for some model pure 2-hydroxy-5-alkyl benzophenone oximes. The surface excess isotherms were computed using different theoretical and semiempiric approaches and used to discuss the kinetics of copper extraction.

The aim of this work is to determine and interpret the interfacial tension isotherms for pure model 1-(2′-hydroxy-5′-methylphenyl)-dodecane-1-one oxime, as being a model for another group of hydroxyoxime extractants (SME 529 and Lix 84), and to discuss the kinetics and mechanism of copper extraction.

Experimental

Pure individual 1-(2′-hydroxy-5′-methylphenyl)-dodecane-1-one oxime was used. Its extraction properties were described and

interpreted previously [1]. The interfacial tension was measured by the drop volume method at 20 °C. Drops of the aqueous phase were formed in the organic phase containing oxime. Octane and toluene were used as model organic diluents. The time of surface aging was 5–25 min. Solutions were prepared and measurements were carried out in the same way as previously described [2].

Results and discussion

The interfacial tension isotherms are given in Figs. 1 and 2. In octane the oxime is much less soluble than in toluene, and the measurements were restricted only to a region of a low oxime concentration.

This interfacial data can be well matched by different adsorption isotherms, i. e. the Szyszkowski isotherm, the Temkin isotherm and the polynominal of the third order (Table 1). The errors of models (e) are relatively low and/or the correlation coefficient is high. The meanings of the regression coefficients are as follows:

The Szyszkowski isotherm,

$$\Delta \gamma = a_1 \ln(c/a_2 + 1), \tag{1}$$

the Temkin isotherm,

$$-\ln c = a_1 \sqrt{\Delta \gamma} + a_2, \tag{2}$$

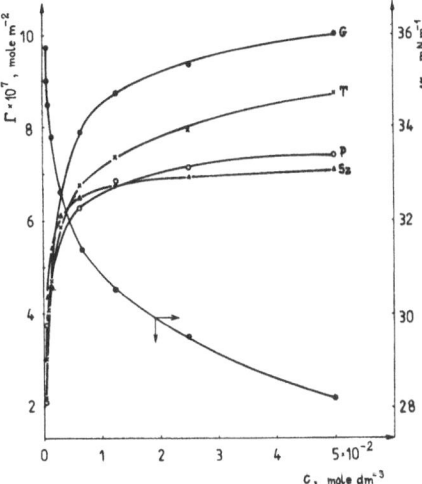

Fig. 1. Interfacial tension and surface excess isotherms for toluene/water system (G, Sz, T and P denote the Gibbs, Szyszkowski, Temkin and polynomial isotherms, respectively)

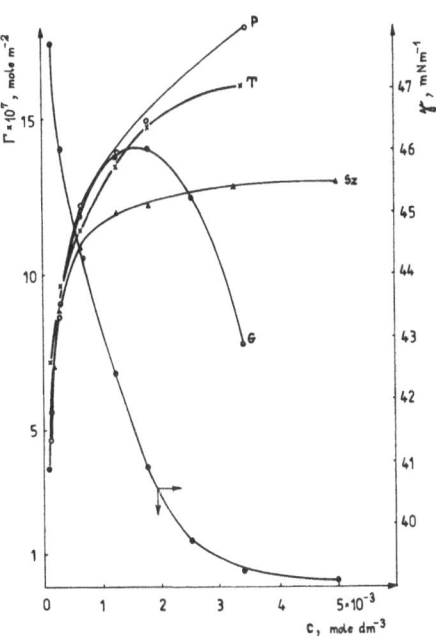

Fig. 2. Interfacial tension and surface excess isotherms for octane/water system (G, Sz, T and P denote the Gibbs, Szyszkowski, Temkin and polynomial isotherms, respectively)

the polynomial

$$y = a_1 (\ln c)^3 + a_2 (\ln c)^2 + a_3 (\ln c) + a_4, \qquad (3)$$

where y and c denote the interfacial tension and oxime concentration, respectively, and $\Delta y = y° - y$, where $y°$ is the interfacial tension for $c = 0$.

The surface excess can be easily computed by differentiating Eqs. (1)–(3) and by introducing the term dy/dc into the Gibbs isotherm.

The character of the obtained relationships surface excess versus concentration, calculated according to various adsorption isotherms, is somewhat different

(Figs. 1 and 2), although deviations observed between adsorption isotherms are relatively small for low oxime concentrations. In the direct differentiation of the experimental curve y vs c an explicit maximum is obtained, as in the case of 2-hydroxy-5-alkyl-benzophenone oximes [3, 4], and the region of low oxime concentrations (below maximum) is only interpreted.

The obtained relations Γ vs c can be further used to discuss the kinetics and mechanism of copper extrac-

Table 1. Regression coefficients (a_i) and errors of model (e) for matching of different isotherms to the interfacial tension isotherm

a_i/e	Diluent	Szyszkowski	Temkin	Polynomial
a_1	Octane	$1.377 \cdot 10^{-4}$	52.79	$-2.055 \cdot 10^{5}$
a_2		$6.512 \cdot 10^{-2}$	-11.28	$2.258 \cdot 10^{3}$
a_3		—	—	-8.1166
a_4		—	—	$4.911 \cdot 10^{-2}$
e		$9.04 \cdot 10^{-8}$	$0.9998^{a)}$	$2.79 \cdot 10^{-4}$
a_1	Toluene	$4.703 \cdot 10^{-4}$	83.99	$-7.181 \cdot 10^{-6}$
a_2		$4.771 \cdot 10^{-2}$	-10.62	$2.628 \cdot 10^{-5}$
a_3		—	—	$1.839 \cdot 10^{-3}$
a_4		—	—	$2.256 \cdot 10^{-2}$
e		$1.49 \cdot 10^{-8}$	$0.9994^{a)}$	$6.01 \cdot 10^{-6}$

[a)] Regression coefficient

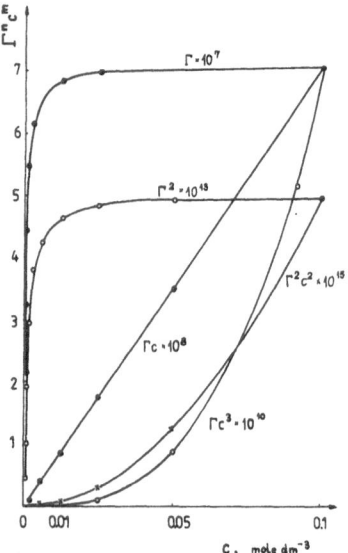

Fig. 3. Predicted effects of hydroxyoxime concentration upon extraction rates in toluene/water system

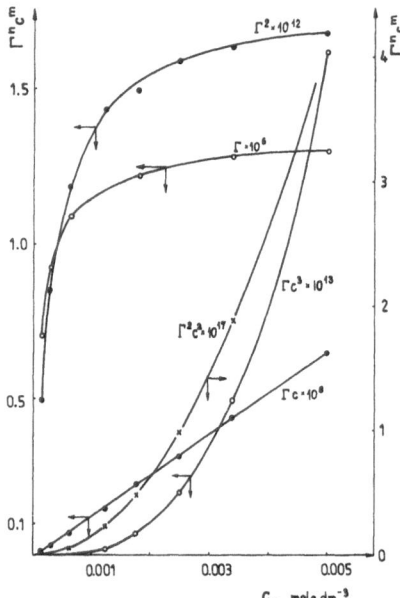

Fig. 4. Predicted effects of hydroxyoxime concentration upon extraction rates in octane/water system

tion. The following two mechanism versions can be considered:

Scheme I

$$Cu_w^{2+} + HB_{ad} = CuB_{ad}^+ + H_w^+ \qquad (4)$$

$$CuB_{ad}^+ + HB_{int} = CuB_{2\ ad} + H_w^+ \qquad (5)$$

$$CuB_{2\ ad} + 2HB_o = CuB_{2\ o} + HB_{ad} + HB_{int} \qquad (6)$$

Scheme II

$$Cu_w^{2+} + HB_{ad} = CuB_{ad}^+ + H_w^+ \qquad (7)$$

$$CuB_{ad}^+ + HB_{ad} = CuB_{2\ ad} + H_w^+ \qquad (8)$$

$$CuB_{2\ ad} + 2HB_o = CuB_{2\ o} + 2HB_{ad} \qquad (9)$$

where HB stands for hydroxyoxime, subscripts w and o denote a water phase and an organic phase, respectively; subscripts ad and int denote the molecules in the interfacial monolayer and at the border of the sublayer from which the hydroxyoxime is transferred into the monolayer without diffusion. As diffusion is neglected and the oxime interfacial concentration is assumed to be equal to the surface excess, then the following kinetic equations can be obtained: $r_{4.7} = k\Gamma$, $r_5 = k$ $[HB]_o\ \Gamma\ r_6 = k\ [HB]_o^3\ \Gamma$, $r_8 = k\Gamma^2$, and $r_9 = k\ [HB]_o^2\ \Gamma^2$, for reactions 4 and 7, 5, 6, 8 and 9 assumed as the limiting steps, respectively.

Exemplary relations r vs c, computed according to the Szyszkowski isotherm, are given in Figs. 3 and 4. Their character is quite the same, as the toluene/water and octane/water systems are compared. Relations computed according to the other adsorption isotherms have a similar character, although the $\Gamma^m c^n$ values are different.

Relations Γ vs $[HB]_o$ and Γ^2 vs $[HB]_o$ are always convex, which is in agreement with experimental data demonstrating the decrease of the extraction order against hydroxyoxime as its concentration increases [5]. The relation $\Gamma\ [HB]_o$ vs $[HB]_o$ is almost linear, and usually somewhat convex. Thus, such behaviour is either in agreement with those works in which the constant order of the extraction rate was observed, or it changed only in a limited range. Such data are usually obtained from kinetic studies [5]. Relations $\Gamma\ [HB]_o^3$ vs $[HB]_o$ and $\Gamma^2\ [HB]_o^2$ vs $[HB]_o$ are very concave, and $\Gamma\ [HB]_o^3$ and $\Gamma^2\ [HB]_o^2$ values sharply increase as the oxime concentration increases. This means that the extraction order against hydroxyoxime should sharply increase as the concentration increases, which is in full contradiction to all experimental data [5]. Thus, Eq. (6) and (9) as the limiting steps can be disregarded.

Almost all kinetic data obtained independently by different authors shows that the rate of copper extraction is proportional to the reciprocal of the hydrogen

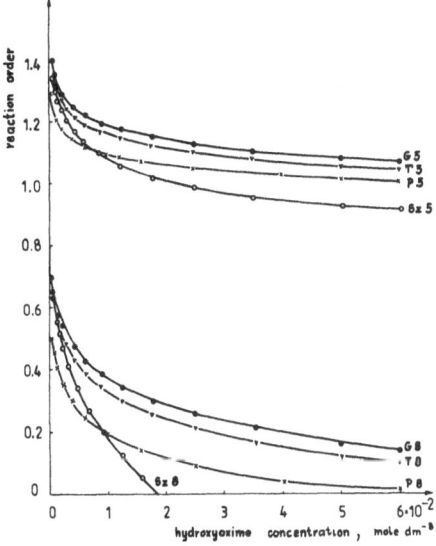

Fig. 5. Predicted reaction orders against hydroxyoxime for toluene/water system

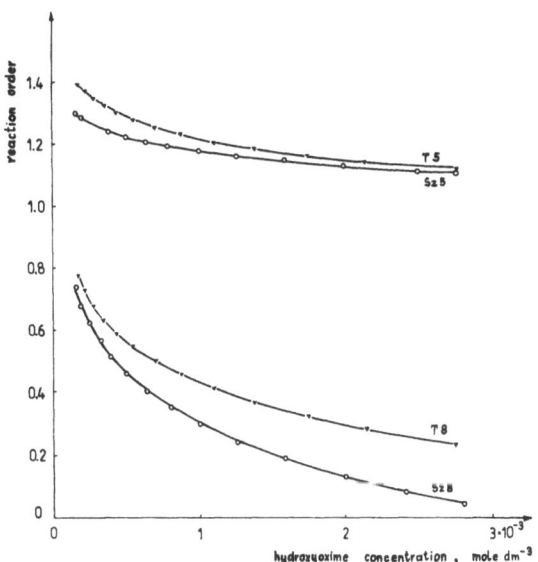

Fig. 6. Predicted reaction orders against hydroxyoxime for octane/water system

ion concentration [5]. Due to this, reactions 4 and 7 can be also disregarded as the limiting steps, because in this case the extraction rate could not depend upon the acidity of the aqueous phase, i.e. $r_{4,7} = k_{4,7}$ $[Cu^{2+}]_w[HB]_{ad}$. Thus, only reactions 5 and 8 will be further taken under consideration. In these two cases, extraction rates are proportional to the copper concentration and to the reciprocal of hydrogen ion concentration. The differences are only observed for reaction orders against hydroxyoxime. Due to this, reaction orders predicted from the interfacial data should support one of these two different versions of the copper extraction mechanism.

As relations $\Gamma^m c^n$ vs c are presented in the logarithmic form, they can be differentiated, and extraction orders against hydroxyoxime concentration, predicted from the interfacial tension data, can be calculated. However, such graphic and/or numerical differentiations are troublesome and connected with a high error rate. To avoid this, a polynomial $y = a_1 x^2 + a_2 x + a_3$ was matched with success to the discussed relations. The errors in the model for these relations are small and of the order $10^{-3} - 10^{-4}$. Thus, the polynomial can be used to calculate the reaction order against hydroxyoxime, as predicted from the interfacial tension isotherm (Figs. 5 and 6).

Results presented in Fig. 5 demonstrate that in the toluene/water system, quite similar values of the reac-

tion order were obtained for all the four adsorption isotherms considered. The differences between reaction orders computed according to the different isotherms are low, especially as scheme I with reaction 5 as the slowest one is considered. This suggests that this version of the extraction mechanism is more probable in comparison to version II, in which reaction 8 limits the process. In the octane/water system (Fig. 6), the reaction orders computed only according to the Szyszkowski and Temkin isotherms are presented, because the Gibbs isotherm and the polynomial could be interpreted in the region of a low oxime concentration, as was mentioned previously [3, 4]. This is probably connected with hydroxyoxime dimerization which, in octane, is already observed for a very low oxime concentration.

The obtained reaction orders significantly support scheme I, with reaction 5 as the slowest step. In this case only, both in toluene and octane, the predicted reaction orders against hydroxyoxime are near 1 for each adsorption isotherm considered in a quite large range of hydroxyoxime concentrations. Each isotherm gives a somewhat different relation between the reaction order and hydroxyoxime concentration, but the general character is similar. The effect of hydroxyoxime concentration upon reaction order is important only for very low bulk concentrations, i.e. below 0.01 mol dm^{-3} and 0.001 mol dm^{-3} in toluene, and octane, respectively. In this region, the reaction order significantly increases as hydroxyoxime concentration decreases, and very low concentrations of the order of 2 can be obtained. Thus, the predicted reaction orders are in quite good agreement with those values determined experimentally.

The obtained results demonstrate the usefulness of the interfacial tension measurements in discussing the kinetics and interfacial mechanism of copper extraction by hydroxyoximes. This approach gives additional data supporting the interfacial mechanism of copper extraction and demonstrates that the extraction of copper by hydrophobic hydroxyoximes, as in our case of 1-(2'-hydroxy-5'methylphenyl)-dodecane-1-one oxime, proceeds according to the proposed mechanism. The formation of the stable 2 : 1 complex from the intermediate 1 : 1 complex and the hydroxyoxime molecule present in the aqueous layers near the interface is the slowest step. Thus, the behaviour of 1-(2'-hydroxy-5'-methylphenyl)-dodecane-1-one oxime is similar to that previously reported by us [3] for hydrophobic 2-hydroxy-5-alkylbenzophenone oximes containing more than four carbon atoms in their alkyl chain.

Acknowledgement

This work was supported by Polish Research Program CPBP No. 03.08.

References

1. Szymanowski J, Atamańczuk B (1982) A Modified Couchy Distribution Model for the Extraction of Copper from Very Dilute Acidic Sulphate Solutions. Hydrometallurgy 9:29–36
2. Szymanowski J, Prochaska K (1988) Interfacial Activity of Model 2-Hydroxy-5-Alkylbenzophenone Oximes and Their Intermediates. J Colloid Interface Sci 123:456-465
3. Szymanowski J, Prochaska K (1988) Surface Excess Isotherms and Mechanism of Copper Extraction by Hydroxyoximes. J Colloid Interface Sci, in press
4. Szymanowski J, Prochaska K, Bogacki M (1987) The Correlation of Copper Extraction Rate with Surface Excess as Determined by the Gibbs Isotherm Using Spline Functions. J Colloid Interface Sci 117:293–295
5. Szymanowski J, Stepniak-Biniakiewicz D, Prochaska K, Rashid ZA (1986) The Hydrophile-Lipophile Balance of Hydroxyoximes, their Adsorption and the Rate of Copper Extraction. Preprints International Solvent Extraction Conference, München, 2:415–423

Received November 11, 1987;
accepted November 23, 1987

Authors' address:

Prof. Dr. J. Szymanowski
Technical University of Poznań
Institute of Chemical Technology and Engineering
pl. Skłodowskiej-Curie 2
60-965 Poznań, Poland

Progress in Colloid & Polymer Science   Progr Colloid Polym Sci 266:265–270 (1988)

Co(II) Extraction by sodium di(2-ethylhexyl) phosphate at the water/toluene interface: phase structure and aggregation in relation to the kinetics of complexation

J.-Y. Calves, F. Danes, E. Gentric, Y. Lijour, A. Sanfeld[1]) and P. Saumagne

Université de Bretagne Occidentale, Brest, Cédex, France
[1]) Université Libre de Bruxelles, Campus Plaine, Bruxelles, Belgium

Abstract: The kinetics of liquid-liquid extraction of aqueous Co(II) by toluenic NaD2EHP and the structural and aggregative properties of the organic phase O, a middle surfactant-rich phase B and the aqueous phase W were studied for the non-stirred system H_2O/toluene/NaD2EHP/$CoBr_2$/KBr. Interfacial turbulences also appeared in this system.

Two kinds of toluenic NaD2EHP solutions, different in their viscous behaviour, could be distinguished according to the hydration degree of the surfactant. Dry or poorly hydrated NaD2EHP gives molecular solutions of monomeric species if dissolved at more than 70 °C or polymeric species if dissolved at c.a. 30 °C. Hydrated NaD2EHP forms microemulsions. The complex formed is either octahedral with two H_2O molecules or tetrahedral. It can also form a viscous phase denser than water.

NaD2EHP enters the aqueous phase giving rise to two kinds of emulsions (A and B). The mmore surfactant-rich layer B lies below the W/O interface. The kinetic study shows that reaction takes place both at the macroscopic W/O interface and at the droplets surfaces of phases A and B. The first reaction site prevails at high cobalt concentrations and the second at low cobalt concentrations.

Key words: Co-extraction, aqueous solutions, Na-di(2-ethylhexyl)-phosphate, toluenic solutions, hydration degree, differant complexes.

Introduction

The kinetics of the Co(II) complexation by sodium di-(2-ethylhexyl) phosphate (NaD2EHP) are studied in a non-agitated system according to the following liquid-cationic exchange:

$$(Co^{2+})_{aq} + (Na^+, D2EHP^-)_{org} \rightleftharpoons Co(D2EHP)_{2, org}$$
$$2(Na^+)_{aq}. \tag{1}$$

Upon reaction, both the initial aqueous and organic phases evolved into up to four phases and interfacial turbulences developed. The kinetics results can be interpreted only if the nature of the different phases, their physical properties and the aggregation of the products and the reactants are known. Therefore, our paper will focus firstly on the phases structure before

and after reaction, studied by different techniques (UV-VIS, IR, turbidimetry, viscosimetry). The spectroscopic results take into account previous results published by Faure [1, 2], Sato [3] and Barnes [4].

1. Structure of the initial system

1.1 Organic solutions

Solid NaD2EHP has the appearance of a gum and may contain up to 40 % water (w/w) which can be eliminated upon heating at 105 °C. According to Faure [1], several mesomorphic phases exist, depending on the degree of humidity: hexagonal (0–17 % water), cubic (20–32 %), lamellar (> 37 %).

A solution of NaD2EHP was obtained by dissolving the appropriate amount of the wet or dry compound in toluene saturated with water (0.05% H_2O w/w). The organic phase O was clear. It contained NaD2EHP in a molecular form or as microemulsion (inverse micelles with 12 NaD2EHP molecules and 48 water molecules, according to Faure [1]). The viscosity of the solution was influenced by the hydration degree r_h of the surfactant (molar ratio H_2O/NaD2EHP). The relative viscosity can be expressed by:

$$\eta = \frac{\eta - \eta_o}{h_o} \qquad (2)$$

where η is the absolute of the toluenic solution and η_o is the solvent viscosity (0.59 cP at 20°C).

η depends on the volumic fraction ϕ_s of the surfactant and also on r_h.

Two classes of solutions can be distinguished, depending on whether r_h is lower or higher than 4 (Faure [1]):

a) $r_h > 4$ (sufficiently hydrated solutions)

$$\eta = 2.5 \, \phi_s \, (1 + 2r_h). \qquad (3)$$

In this case, η values are rather high and show interactions between micelles.

b) $r_h < 4$ (insufficiently hydrated solutions).

The viscosity decreases with time, presents a viscoelastic behaviour and depends on the temperature of dissolution t_d.

If $70°C < t_d < 80°C$, $\eta \simeq 0$

If $30°C < t_d < 40°C$, η is relatively high and poorly reproducible.

This behaviour demonstrates strong intermolecular associations in solution obtained by low temperature dissolution of NaD2EHP and the existence of isolated NaD2EHP molecules when the surfactant is dissolved at high temperature.

1.2 Aqueous solutions

The aqueous solutions of Co^{2+} were prepared from dehydrated $CoBr_2$ dissolved in water saturated in toluene. The electronic spectrum of the pink-coloured solution shows two peaks at 512 nm and 475 nm,

which suggests an octahedrally coordinated cobalt of the form $[Co(H_2O)_6]^{2+}$.

1.8 cm^3 of an NaD2EHP solution was gently poured over the same volume of a Co(II) solution in an non-agitated system. The reaction immediately initiated at the macroscopic W/O interface which turned to deep blue and the complex eventually diffused towards the organic phase.

In most experimental conditions, emulsification was detected and presented two different forms:

A dilute emulsion A with the appearance of milky wakes;

A concentrated emulsion B which consisted in an opaque layer located immediately below the W/O macroscopic interface.

The limit between A and B was not well defined. Both emulsions actually could differ only in the volumic fraction occupied by the surfactant droplets, as in the system investigated by Gambi et al. [5].

Both emulsions also turned blue as the reaction proceeded, which indicates that the complex also formed at the surface of the emulsion droplets.

The intensity of emulsification was particularly high for low metal ion concentrations and an inert electrolyte (KBr) sometimes had to be added in order to be able to run spectra of the aqueous phase.

Ultimately, after a few hours, viscous pink-purple droplets with various diameters of up to 3 mm were visible at the bottom of the cell.

Interfacial turbulences also quickly developed at the W/O interface and eventually spread to the whole layer B.

2. Phases structures

2.1 Infrared spectra of the toluenic complex

The infrared spectrum of NaD2EHP in a nujol mull is shown in Fig. 1A. In addition to the 1040 cm^{-1} band, representing P—OR vibrations [6] two bands at 1240 cm^{-1} and 1100 cm^{-1} are assigned to the asymmetric and symmetric P=O vibrations respectively.

The blue complex was separated from the organic phase and dried at 70°C under vacuum before recording its infrared spectrum (Fig. 1B). The P-OR bands were essentially unaffected which demonstrates that the alkoxy groups do not take part in the complexation. On the other hand, the P=O asymmetric stretch at 1240 cm^{-1} shifted to 1185 cm^{-1}, whereas its intensity relative to the P-OR bands remained constant. However, the band at 1100 cm^{-1} underwent little

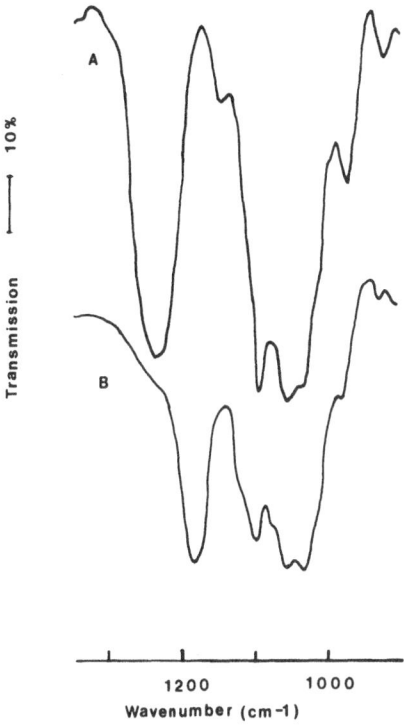

Fig. 1. (A) Infrared spectrum of NaD2EHP in jujol; (B) Infrared spectrum of the Cobalt complex after drying

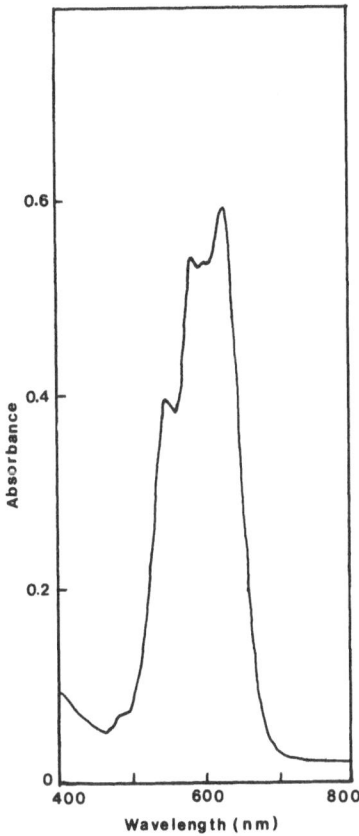

Fig. 2. Electronic spectrum of the Cobalt complex in toluene

change of frequency, while its relative intensity was significantly reduced. These data are compared in Table 1 with previous studies by Ferraro [7]. No satisfactory structure can be drawn to account for the infrared data. The IR shift of the P=O band at 1185 cm^{-1} upon complexation only suggests that the cation is bonded via the phosphoryl oxygen atoms of NaD2EHP.

The infrared spectra of the viscous purple droplets was also recorded after drying under the same conditions as for the blue complex. No differences relative to the blue complex were observed.

2.2 Electronic spectra of the toluenic complex

The electronic spectrum of the complex in the toluenic phase (Fig. 2) shows strong overlapping bands with four maxima at 542 nm (band 1), 580 nm (band 2), 597 nm (band 3) and 622 nm (band 4) plus a shoulder at 480 nm.

Sato et al. [3] have analyzed the blue complex formed between Co(II) and HD2EHP (di-(2-ethylhexyl)phosphoric acid) and have found 3 bands at c.a. 550, 590 and 630 nm which were assumed to be representative of a typical tetrahedral coordination of a monomer with two loosely bound water molecules.

Barnes et al. [4] have found a maximum at 522 nm in the visible spectrum of the pink cobalt complex and have assigned it to the hexacoordinated species Co(D2EHP)$_2$(H$_2$O)$_2$ (possibly polymeric).

Our spectra show at least two extra bands at 480 and 580 nm, which indicates a more complex situation. A distortion in the tetrahedral configuration would result in a splitting of the energy levels $T_1(P)$, $T_1(F)$ and $T_2(P)$ giving rise to six allowed transitions. With five

Table 1. Asymmetric and symmetric P=O frequencies in NaD2EHP and Co(D2EHP)$_2$

	asymmetric /cm^{-1}	symmetric /cm^{-1}	references
NaD2EHP	1243	1100	7
Co(D2EHP)$_2$	1203	1093	7
NaD2EHP	1240	1100	this work
Co(D2EHP)$_2$	1185	1100	this work

observed bands including the 480 nm one, this type of distorted T_d symmetry can be envisaged. However, it is difficult to predict whether this would occur in a monomer, a dimer or a polymer. Furthermore, the 580 nm band is much too intense to be caused by a transition in an octahedrally coordinated Co(II), but the presence of the weak 480 nm band cannot entirely rule this structure out.

The ratio $\alpha = A_4/A_1$ between the absorbances of the first and the fourth band, respectively, increased in parallel with the fraction of tetrahedral complex. Its value lay in the range 1.2–1.8, depending on the conditions of preparation of the toluenic complex. Thus: if the complex was separated by evaporation of toluene, dried (under vacuum at 50 °C) and then redissolved in toluene, varied between 1.6 for toluene saturated in water and 1.8 for toluene dried over molecular sieves, or if the complex was prepared from a highly hydrated NaD2EHP ($r_h > 4$) in toluene saturated in water, α was in the range 1.2–1.5. α increased with the proportion of remaining NaD2EHP and decreased with time.

Thererfore we assign the bands at 580, 595 and 625 nm to a tetrahedrally coordinated complex (I) and the band at 540 nm to an octahedral coordination of Co^{2+} in the dihydrated complex (II) (Scheme 1).

2.3 Visible spectrum of the NaD2EHP emulsion

A linear decrease of the absorbance A was observed when the wavelength λ increased from 400 to 800 nm. The following relation accounts for this trend:

$$A/A_{800} - 1 = K(1 - \lambda/800) \qquad (4)$$

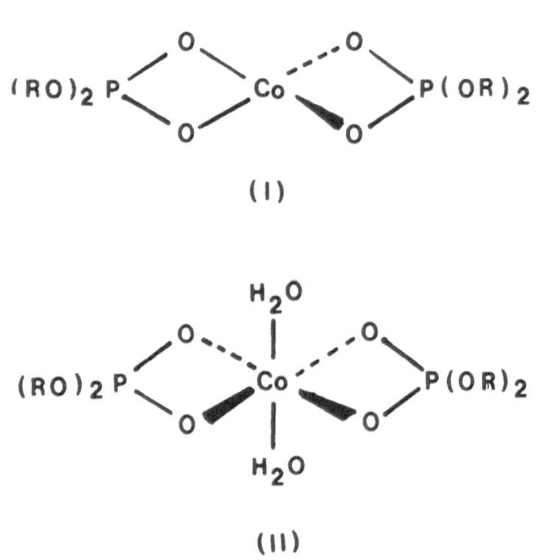

(I)

(II)

Scheme I. Complex (I) and Complex (II) (see text)

where A_{800} is the absorbance at $\lambda = 800$ nm.

K represents a measure of the mean size d of the droplets. Table 7 presents different values of K according to the NaD2EHP concentration C_d, the Co^{2+} concentration C_m and the KBr concentration C_e (Table 2).

A value of 1 for K corresponds to d values larger than $\lambda/4.7$ (in this case $d > 170$ nm) while K values of about 2 or 2.7 correspond to d values significantly smaller than $\lambda/4.7$ (in this case $d < 85$ nm).

2.4 Absorbance of the NaD2EHP emulsion

The absorbance A of the emulsion, recorded at 750 nm for $[Co(H_2O)_6]^{2+}$, does not absorb at this wavelength. A reached a steady value within 20 min and, for the range of concentrations which have been studied, did not significantly depend on C_d or C_e but strongly depended on C_m. Thus, the following dependence has been found [8], for $C_e = 1000$ mol m^{-3},

$$A \text{ (arbitrary units)} = [(C_m)^5 + 7.5 \times 10^6]^{-1}. \qquad (5)$$

3. Kinetics of reaction

The kinetics was followed on an UV-VIS Varian Cary 219 spectrophotometer "in situ" by the decrease of the 512 nm band owing to the hexahydrated Co^{2+} species in the aqueous phase. The contribution of the emulsification to the intensity of the 512 nm band was estimated from the absorbance at 750 nmm (see [8]).

3.1 Factors influencing the kinetics

The influence of the following parameters on the reaction rate w related to the unit area of the macroscopic W/O interface was studied:

a) C_d, thhe NaD2EHP concentration in the toluenic solution = 6 to 50 mol m^{-3},

Table 2. K represents a measure of the mean size of the droplets for the concentrations C_d, C_m and C_e

C_d/mol m^{-3}	C_m/mol m^{-3}	C_e/mol m^{-3}	K
3	6–100	0–1000	2.00
6–50	6–100	0–1000	2.58–2.65
6–50	0	0	2.50
6–50	0	1000	1.00

b) I_h, an arbitrary hydration index of toluenic NaD2EHP $I_h = 0$ if the hydration is less than $4\,H_2O$ per each NaD2EHP, $I_h = 1$ for complete hydration,

c) C_e, the inert electrolyte (KBr) concentration in the aqueous solution $= 0$ to $1000\,mol\,m^{-3}$,

d) C_m, the Co^{2+} concentration in the aqueous phase $= 6$ to $100\,mol\,m^{-3}$.

3.2 Results

Before the formation of phase B (i.e. during the first 20 min), the reaction was relatively faster and results were poorly reproducible. The following results refer to the period extending from 20 to 60 min:

a) C_d did not influence the reaction rate w,

b) C_e did not influence w if $C_m > 30\,mol\,m^{-3}$. If $C_m < 30\,mol\,m^{-3}$, w was reduced by an increase of C_e (for $C_e = 1000$, w was 1.7 times smaller than for $C_e = 0$ and 1.4 times smaller for $C_e = 400$),

c) I_h did not influence w if $C_m < 30\,mol\,m^{-3}$. When $C_m > 30\,mol\,m^{-3}$, the reaction rate increased by about 50% if $I_h = 1$,

d) The C_m dependence on w was more complex. Figure 3 shows v values of w for $I_h = 1$, $C_e = 1000\,mol\,m^{-3}$ and $C_d = 25\,mol\,m^{-3}$. A maximum is observed for $C_m = 20\,mol\,m^{-3}$ and a minimum for $C_m = 40\,mol\,m^{-3}$. The reaction order first decreases from 1 for $C_m = 0$ to -1 for $C_m = 30\,mol\,m^{-3}$ and then increases up to 2 at high C_m concentrations.

3.3 Mechanism

Proper reaction orders are 0 relative to NaD2EHP and 1 relative to $CoBr_2$. The reaction is superficial and located both at the macroscopic W/O interface of area s_{mi} with a partial rate v_2 and at the surface s_e of the emulsion droplets with a partial rate v_1. The total reaction rate v is:

$$v = v_1 + v_2 \tag{6}$$

with

$$v_1 = k'_1 C_m s_e / s_{mi} \tag{7}$$

and

$$v_2 = C_m/(1/k_2 + R_{de}). \tag{8}$$

K'_1 and k_2 being the chemical rate constants for the two locations and R_{de} the resistance to the diffusion of the Co^{2+} ions towards the W/O interface.

A proportionality between the intensity of emulsification E and R_{de} from one hand and the ratio s_e/s_{mi} from the other hand can be envisaged:

$$s_e/s_{mi} = k''_1 E \tag{9}$$

$$R_{de} = k'_r E \tag{10}$$

$$E = [(C_m)^5 + 7.5 \cdot 10^6]^{-1}. \tag{11}$$

Introducing the two constants k_1 and k_r as:

$$k_1 = k''_1 k'_1 \tag{12a}$$

and

$$k_r = k'_r k_2 \tag{12b}$$

one obtains v from Eqs. (6), (7) and (8).

$$v = k_1 C_m[(C_m)^5 + 7.5 \cdot 10^6]^{-1} \\ + k_2 C_m\{1 + k_r[(C_m)^5 + 7.5 \cdot 10^6]^{-1}\}^{-1}. \tag{13}$$

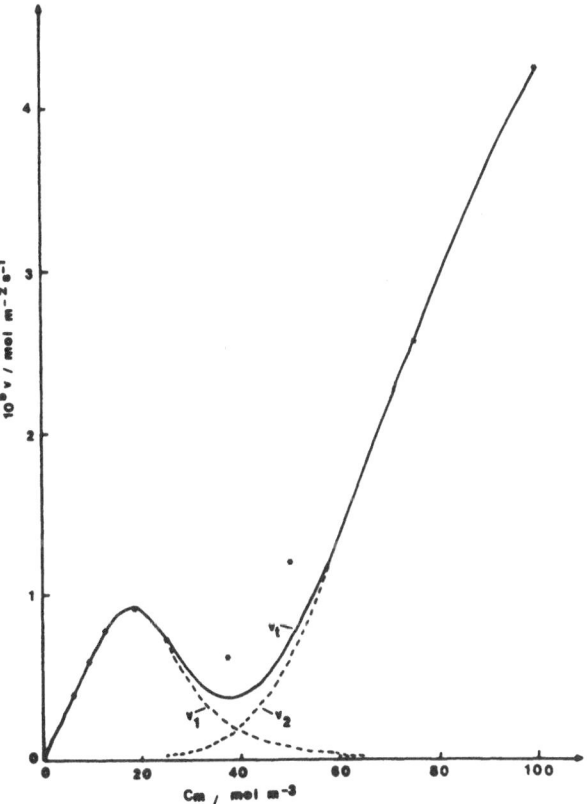

Fig. 3. Reaction rate v vs C_m. v: total rate for $I_h = 1$ and $C_e = 1000\,mol\,m^{-3}$. v_1 and v_2: partial reaction rates. * Experimental points (v_e)

The experimental points v_e in Fig. 3 are in good agreement with relation (13), except for the intermediate concentrations C_m between 30 and 60 mol m^{-3}, where v_e are significantly greater than the calculated rates. This difference is explained by spontaneous interfacial turbulences which occurred in this range of concentrations and enhanced the reaction rate [9].

It can be shown [8] that no mechanism with a single site of reaction (diffusional, kinetic or both) can account for the observed dependence w (C_m, C_e, I_h).

Conclusion

We have attempted to elucidate the structure of the complex formed upon liquid-liquid extraction of aqueous Co(II) by NaD2EHP. If a tetrahedral complex seems most probable, an octahedral complex with two coordinated water molecules may also exist. The kinetics of complexation have also been studied. Emulsification occurred in most experimental conditions. Therefore, the complexation takes place not only at the macroscopic W/O interface, but also at the overall surface of the emulsion droplets. In both mechanisms, the actual rate constants are 0 relative to NaD2EHP and 1 relative to the metal ion concentration.

Acknowledgement

We are grateful to the European Community Commission for financial support (Grant 019JC(CD)–2. 1986).

References

1. Faure A, Lovera J, Gregoire P, Chachaty C (1985) J Chimie Phys 82:779–790
2. Faure A, Tistchenko AM, Zemb T, Chachaty C (1985) J Phys Chem 89:3373–3378
3. Sato T, Nakamura T (1972) J Inorg Nucl Chem 34:3721–3730
4. Barnes JE, Setchfield JH, Williams GOR (1976) J Inorg Nucl Chem 38:1065–1067
5. Gambi CMC, Leger L, Taupin C (1987) Europhys Lett 3:213–220
6. Thomas LC (1974) Interpretation of the Infrared Spectra of Organophosphorus Compounds. Heyden, New York, pp 52–54
7. Ferraro JR (1959) J Inorg Nucl Chem 9:150
8. Danes F, Gentric E, Lijour Y, Sanfeld A, Saumagne P (19??) J Colloid Interface Sci, to be published
9. Calves J-Y, Danes F, Gentric E, Lijour Y, Sanfeld A, Saumagne P (19??) J Colloid Interface Sci, to be published

Received December 10, 1987;
accepted December 16, 1987

Authors' address:

J.-Y. Calves
Université de Bretagne Occidentale
6 Ave. Le Gorgeu
F-29287 Brest Cedex, France

Progress in Colloid & Polymer Science   Progr Colloid Polym Sci 76:271–277 (1988)

Surface enhanced Raman spectroscopy (SERS) of surfactants adsorbed to colloidal particles

J. Wiesner, A. Wokaun and H. Hoffmann

Physical Chemistry Institute, University of Bayreuth, F.R.G.

Abstract: The stabilization of colloidal silver particles has been studied by SERS. Replacement of the primary anionic stabilizer, present during colloid preparation, by cationic surfactants was monitored by the corresponding changes in the SERS spectrum. Partial information on the geometry of adsorption was obtained. Small silver particles prepared by γ-irradiation can be protected from air oxidation by a compact layer of perfluoro-surfactant stabilizer. A quantitative model for changes in the absorption spectrum caused by surface oxidation of the silver spheres is presented.

Key words: Metal colloids, surface enhanced Raman scattering, colloid stabilizers, surfactants, surface plasmon absorption.

Introduction

Colloidal dispersions of small metal particles are of interest in a variety of fields including chemical catalysis, photography, pigmentation technology, medicine, and cell biology. Prescriptions for the preparation of very stable gold colloids date back to the famous work of Faraday, but comparatively little is known about the mechanism of stabilization by the reagents added during metal salt reduction. Frequently, the stabilizers employed are polydentate organic ligands, which suggests complexation of the metal particles. Colloid stability would then be a consequence of particle repulsion by the large excess negative charge present on each complexed particle.

The discovery of surface enhanced Raman scattering (=SERS) [1, 2] has opened new ways to study colloid stabilization on a molecular scale. Briefly, the presence of amplified local fields on the surface of group IB metal particles enhanced the Raman scattering cross section of adsorbates by factors $\leq 10^6$, such that stabilizers bound to the surface can selectively be observed in the presence of a surrounding solution containing the same molecule. There are several prerequisites for achieving this enhancement [2]: The laser wavelength must match the excitation frequency of the dipolar surface plasmon; intrinsic dielectric losses of the metal should be small, which explains the exceptional role of group IB metals; and particle dimensions

must be in a defined range depending on the system. Largest enhancements are often found for particles of 10–100 nm sizes. Anisometric particles produce even stronger local fields due to the so-called lightning rod effect. In colloid work this has the consequence that the largest signals often do not arise from small monodisperse metal spheres but from strings or small aggregates which are the first step in colloid agglomeration [3]. Dynamic charge transfer processes between metal and adsorbates may lead to further resonant enhancement of the Raman scattering cross section, which has been termed the chemical effect [1].

The application of SERS to colloid studies has been pioneered by Creighton et al. [4] and by Kerker et al. [5–7]. The aim of this work is to obtain more detailed information on the mechanism of colloid stabilization. In particular, we are interested in whether the standard stabilizers used during silver salt reduction (for which apparently there is limited variability) can subsequently be replaced by other surface active molecules, e.g. surfactants. This approach would open a multitude of novel possibilities for colloid stabilization. Here we report on two successful realizations of this concept: (i) the replacement of an anionic primary stabilizer by a cationic surfactant and (ii) the protection of very small (3 nm radius) silver particles by a compact perfluoro-surfactant layer.

Experimental

Preparation of the colloids

Silver colloids were prepared by reduction of silver nitrate and perchlorate solution, either by standard chemical procedures or initiated by γ-irradiation. Fe(II)SO$_4$ [7, 8], EDTA (ethylene-diamine-tetraacetic acid, dinatrium salt) [9] or NaBH$_4$ [4] were used as reducing agents. In the Carey-Lea preparation [8], the reducing FeSO$_4$ solution contains citrate as the primary colloid stabilizer. In contrast, EDTA [9] acts both as a reductant and as a stabilizer for the growing colloid.

The colloids reduced using γ-irradiation were prepared by A. Henglein [10]. The solutions contained isopropanol as a reductant, and either polyvinylsulfate (PVS) or methylammonium perfluoro-nonanoate ($C_8F_{17}COO^-CH_3NH_3^+$) as a stabilizer.

UV absorption spectroscopy has been used to check the quality of the colloid preparation. Colloids produced by chemical reduction are characterized by absorption maxima between 390 and 410 nm. Solutions prepared by γ-irradiation in an inert gas atmosphere show an absorption maximum at 380 nm, which rapidly shifts to 400 nm upon exposure to air.

Stabilization by surfactants

When a concentrated (10^{-2} M) solution of the cationic surfactant cetyl-trimethyl-ammonium bromide (CTAB) is slowly added to a stirred silver colloid dispersion, the solution stays clear and the visible color remains unchanged; the absorption maximum experiences only a slight red shift. This process leads to a sign change of the ζ-potential (see below). If the surfactant concentration is chosen too low ($10^{-4} - 10^{-3}$ M), or if the solution is added too rapidly, aggregation of the colloidal particles followed by precipitation of the metal occurs.

ζ-potential measurements

In the instrument used (Pen Kem, Model 501) the drift velocity of the charged particles is compensated by the rotation of a prism cube. When complete cancellation oft the drift has been achieved, the ζ-potential can be read directly.

Raman measurements

Raman scattering was excited using either the 514.5 nm and 488 nm lines of an Ar$^+$ laser or the 647.1 nm line of a Kr$^+$ laser. The laser beam was focused to a 10 × 0.2 mm line onto the front face of a quartz cuvette containing the colloid. Laser power was restricted to $\leq$ 100 mW to minimize photochemical changes. A back-scattering geometry was used. Raman signals were dispersed in a SPEX 14018 double monochromator run at 3 cm^{-1} resolution, and detected using photon counting. Typical integration times of 10 s per resolution interval were employed. To ensure that the recorded signals were indeed due to SERS from molecules adsorbed to the colloidal particles, reference spectra were run on samples containing the same solution concentrations but no colloidal silver. In all cases experimental conditions were chosen such that contributions from the solution to the observed signal could be neglected.

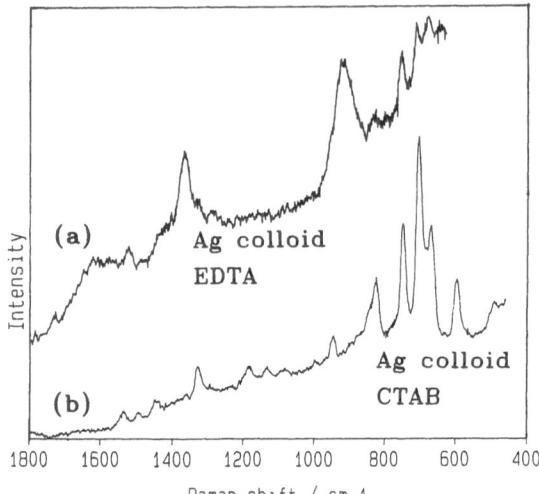

Fig. 1. (a) SERS spectrum of silver colloid prepared by reduction with EDTA, which acts as the primary stabilizer. (b) After addition of a 10^{-2} M CTAB solution, the EDTA ligands have been completely replaced by the cationic surfactant. An excitation wavelength of 647 nm was used

Results

ζ-potential

For the colloid prepared in the presence of EDTA (Ag/EDTA), a ζ-potential of -94 mV was determined. Upon addition of the CTAB solution the ζ-potential changed to $\geq +100$ mV, indicating a net positive charge of the particles stabilized by the cationic surfactant.

SERS spectra of Ag/citrate, Ag/EDTA and Ag/CTAB

For Ag colloids prepared in the presence of citrate the SERS spectra obtained agreed with the bands observed by Siiman et al. [7]. After CTAB addition, a structureless spectrum with a high inelastic background was recorded. Apparently citrate is not or incompletely replaced by CTAB such that an ill-defined first coordination sphere results. The strong background may be due to microcrystallites of CTAB present in the solution.

SERS spectra of Ag/EDTA and Ag/CTAB are shown in Fig. 1(a) and (b), respectively. In (a), the solution concentration of EDTA is 2.5×10^{-4} M. As EDTA acts both as a reductant and as a stabilizer, it is likely that some oxidation products of EDTA remain bound to the surface. This resulted in strong fluorescence with excitation in the green spectral region. An excitation wavelength of 647 nm has therefore been used.

The spectrum of Fig. 1(a) exhibits only some of the bands that have been observed for EDTA adsorbed on silver electrodes [11]. Strong enhancements are seen for the symmetric stretching mode of the carboxylate group (1360 cm^{-1}) and for a C–C stretching mode at 910 cm^{-1}. Weaker bands in the 600–800 cm^{-1} region have not been assigned yet.

After addition of the CTAB solution the SERS spectrum of Fig. 1(b) is obtained. The complete absence of EDTA bands demonstrates that the anionic stabilizer has, within our detection limits, been completely replaced by CTAB. Largest enhancements are found for vibrations involving the $-CH_2-N^+-(CH_3)_3$ head group. The intensive bands at 719 cm^{-1} and 766 cm^{-1} can be assigned [12] to the $-CH_2-N^+-$ stretching motion and are dependent on the conformation of the attached hydrocarbon chain. The peak at 967 cm^{-1} is due to the asymmetric stretch of the $N^+-(CH_3)_3$ head group [12]. Backbone motions (C–C stretches) are seen at 1130 cm^{-1}. No detailed assignments have been made for the other bands of the spectrum. Although an inelastic scattering background is commonly observed in SERS, the background intensity is unusually high in the spectrum of Fig. 1(b). This might be due to the formation of microcrystals of CTAB, which has a relatively high Krafft point (26 °C).

SERS spectra of Ag/PVS and Ag/C$_8$F$_{17}$COO$^-$

Raman spectra from silver colloids prepared by γ-irradiation are shown in Fig. 2. The colloids stabilized by PVS exhibit strong fluorescence when excited in the blue or green spectral region. By adding a trace of KI, the fluorescence is quenched, and SERS spectra are obtained with high signal-to-noise ratios. The spectrum is dominated by four intensive bands between 1200 and 1400 cm^{-1}, and a strong signal at 837 cm^{-1}. Assignments will be discussed in the next section. As mentioned above, the absorption maximum of the Ag/PVS system immediately shifts from 380 nm to 400 nm upon exposure to air [10].

In contrast, when C$_8$F$_{17}$COO$^-$ is used for stabilization, the absorption maximum (397 nm after preparation in inert gas atmosphere) remains unaltered in open cuvettes for several months. Exposure to laser radiation for several hours does not induce noticeable changes. SERS signals from this colloid are weak due to the small size of the particles. Therefore the colloid, which was contained in a rotating cuvette, had to be excited at relatively high laser power levels (~ 200 mW). Figure 2(b) exhibits, on top of a large ine-

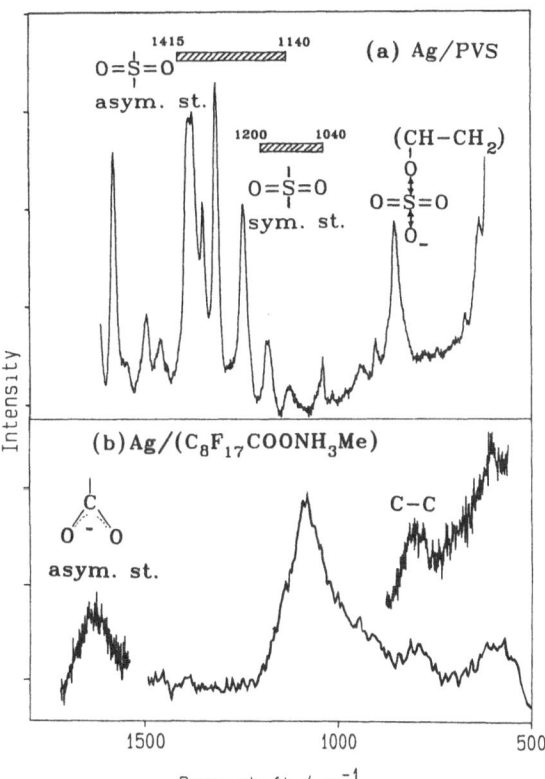

Fig. 2. SERS spectra of silver colloids prepared by γ-irradiation [10]. (a) Stabilizer polyvinylsulfate; excitation by 70 mW of 514.5 nm radiation; (b) stabilizer C$_8$F$_{17}$COO$^-$ CH$_3$NH$_3$$^+$; excitation by ~ 200 mW of 488 nm radiation in a rotating cuvette

lastic scattering background, a broad low-frequency band around 600 cm^{-1}, C–C stretching motions at 820 cm^{-1}, a very intensive peak centered at 1100 cm^{-1} with dominant contributions from C–F stretches, and a signal at 1550 cm^{-1} assigned to the asymmetric stretching motion of the carboxylate group. We have unambiguously established this peak in a separate experiment using a stationary cuvette (upper trace in Fig. 2(b)).

Discussion

SERS of Ag/EDTA and Ag/CTAB

The spectra shown in Fig. 1 demonstrate that SERS is indeed a powerful tool to elucidate the molecular environment of metal colloid particles. Upon addition of the concentrated CTAB solution to the EDTA-stabilized colloid (a), the EDTA signals disappear completely, as seen in Fig. 1(b). This observation implies that

EDTA has been replaced, rather than surrounded, by the cationic surfactant.

The implications of this observation on the nature of the colloidal surface layer should now be discussed. Comparison of the SERS spectra of adsorbates on colloids and on electrodes held at varying potentials has demonstrated that the surface charge on the colloidal particle is positive [13]. This suggests that the outermost silver layer is oxidized to Ag^+ by oxygen from the solution. Evidence for this process will be given below. The fact that a cationic surfactant can be bound to such a surface must imply the intermediacy of anions (e.g. the bromide counterion of CTAB). There is multiple evidence in SERS literature that halide ions can enhance adsorption and thereby lead to large SERS cross sections [1].

The observation of strong enhancement for the $N^+-(CH_3)_3$ and $-CH_2-N^+-$ stretching modes indicates that at least one layer of surfactants is adsorbed with the ionic head groups pointing towards the surface. The $-CH_2-N^+-$ stretching modes observed between 700 and 800 cm^{-1} are indicative of the conformation of the hydrocarbon chain close to the nitrogen. In the solid, where the chains are ordered in a perfect all-anti conformation, only one strong band at 760 cm^{-1} is observed in the solid-state Raman spectrum. In the SERS spectrum of Fig. 1(b), a second more intense band is recorded at 719 cm^{-1} which is assigned to the syn conformation of the $CH_2-CH_2-CH_2-N^+$ fragment. Furthermore, the observation of selective enhancement for the asymmetric $N^+-(CH_3)_3$ stretching mode at 967 cm^{-1} (the corresponding symmetric vibration expected at ~ 905 cm^{-1} is absent) indicates that the symmetry axis of the $N^+(CH_3)_3$ head group is tilted with respect to the surface normal. Both observations are compatible with the adsorption geometry illustrated in the photographs of molecular models shown in Fig. 3, where the head group is inclined relative to the surface. The alkyl chain is directed away from the surface, with the $-CH_2-CH_2-CH_2-N^+-$ fragment either in an anti (left) or in a syn type conformation (right).

Silver sols produced by γ-irradiation

Polyvinylsulfate, the stabilizer employed in the colloid preparation of Fig. 2(a), binds strongly to the colloid surface due to its polyelectrolyte nature. The bands between 1400 cm^{-1} and 1200 cm^{-1} can be assigned to symmetric and asymmetric stretching motions of the $O=S=O$ fragment [14]. For the solvated ion $RO-SO_3^-$ two bands are expected in the ranges 1350–1220 cm^{-1} and 1140–1050 cm^{-1}. For diesters of the type $RO-SO_2-OR'$ the frequencies are higher, i.e. 1415–1390 cm^{-1} and 1200–1185 cm^{-1}. The latter compounds might be suitable models for sulfate ions bound to the surface of the silver colloid, i.e. $RO-SO_3^-Ag^+$. For reference we have recorded IR spectra of solid potassium polyvinylsulfate in solid KBr dilution. A broad unresolved band is seen which consists of at least two intensive components at ~ 1190 and ~ 1170 cm^{-1}, as well as two weaker bands at 1130 cm^{-1} and 1115 cm^{-1}. The solid state Raman spectrum of the reference substance was not accessible due to excessive elastic scattering. From this comparison it becomes apparent that the intensive SERS band centered at 1366 cm^{-1} can be assigned to $R-O-SO_3^-$ groups bound to the cationic silver surface. The lower frequency asymmetric stretches are due to $R-O-SO_3^-$ groups of the polymer which are close to the surface but directed into solution. More detailed assignments will be possible when solution Raman spectra of oligo- or poly-vinylsulfate become available. The intensive band at 837 cm^{-1} is an S–O stretching motion [14] which experiences strong enhancement due to its close proximity to the silver surface.

If instead perfluoro-nonanoate is used as stabilizer (Fig. 2(b)), the spectrum is dominated by three vibrations corresponding to the asymmetric stretching motion of the carboxylate group, the stretching of the $-CH_2-COO^-$ bond, and a typical C–F stretching mode around 1100 cm^{-1}. Observation of this latter band can again be taken as an indication of conformational disorder in the perfluoro-alkyl chain close to the metal surface, although this point will have to be corroborated by further experimentation. Strong SERS enhancement implies [1, 2] that the vibrational dipole of the C–F bond must have a component perpendicular to the surface, which is incompatible with a chain in all-trans conformation oriented along the surface normal. The preferred anchoring of the carboxylate group might again result in a syn conformation for the $CF_2-CF_2-CF_2-COO^-$ chain fragment closest to the surface.

It has already been indicated that colloids stabilized by $C_8F_{17}COO^-$ are remarkably insensitive to oxygen, which is attributed to dense packing of the perfluoro-alkyl chains. In contrast, with PVS as a stabilizer the absorption of the silver spheres shifts from 380 nm to

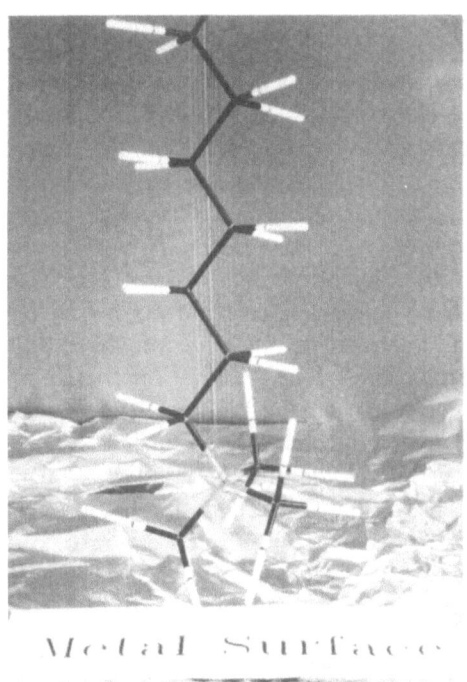
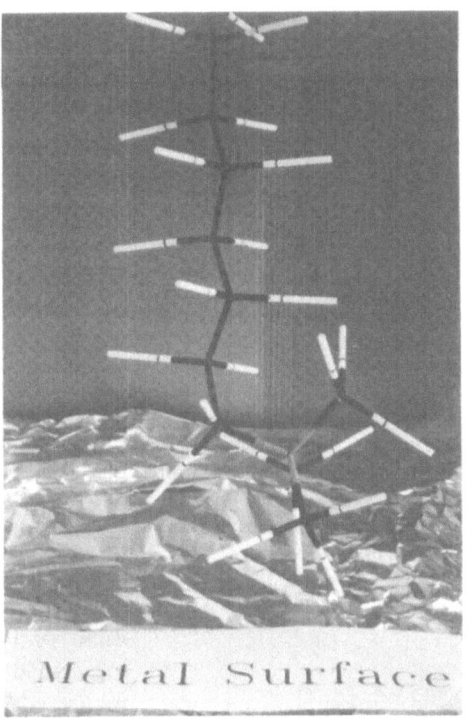

Fig. 3. Molecular models for the conformation of CTAB adsorbed to the silver surface. The head group is inclined with respect to the surface; the segments of the alkyl chain closest to the surface exhibit both anti (left) and syn (right) conformations

400 nm upon exposure to air. The accepted interpretation [10] of this phenomenon is that the colloid particles become surrounded by Ag^+ ions due to surface oxidation.

We have performed quantitative theoretical calculations to test this interpretation. To model our silver colloid we have used a radius $R = 3.6$ nm as indicated by Henglein et al. [10]. First, one could ask what fraction of the silver conduction electrons has to be taken up by oxygen to effect the above-mentioned shift. To obtain an estimate for this fraction, one can use the Drude model for the dielectric constant,

$$\varepsilon(\omega) = 1 - \frac{\omega_p^2}{\omega^2} \quad \text{with } \omega_p^2 = \frac{Ne^2}{m},$$

where N is the number density of conduction electrons and m is their mass. A shift of the absorption maximum, at constant particle size and shape, from 380 nm to 400 nm would require the number density N to decrease by 8 %. A particle with $R = 3.6$ nm consists of 11000 atoms, of which 900 atoms would lose an electron by oxidation. This number is a substantial fraction of the number of surface silver atoms, i.e. 3800 for a perfect sphere.

An independent verification of this number can be obtained as follows. The oxidized particle is modelled as consisting of a silver core of radius $R = 3.5$ nm, with the unchanged dielectric function of silver. This core is coated by an overlayer of Ag^+, of thickness 0.2 nm, which is roughly the diameter of the Ag^+ ion. To calculate the surface plasmon-induced absorption [15] for a coated sphere, we have adapted the appropriate expression for the local field of a coated sphere, as given by Craighead and Glass [16] and by Kerker et al. [6]. As an input parameter for the calculation the dielectric constant ε_D of the coating layer is required. In estimating ε_D we have been guided by geometric considerations. For the parameters given above, the area per Ag^+ ion on the surface of the metallic core is 0.17 nm^2, i.e. four times the area required by the hard sphere cross section of Ag^+. Thus there is plenty of space for solvent molecules between the adsorbed Ag^+ ions, the charge being compensated by the anionic stabilizer and a diffuse layer of counterions. The coating can then be viewed as a highly concentrated aqueous ,solution' of Ag^+ ions. If we extrapolate from the measured refractive index of concentrated $AgNO_3$ solutions, we obtain $n_D = 2$ or $\varepsilon_D = n_D^2 = 4$. The absorption spectra calculated with these parameters are shown in

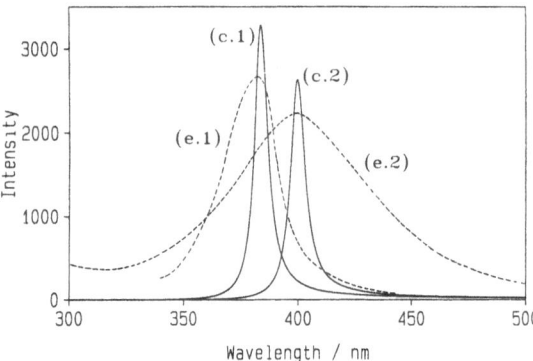

Fig. 4. Experimental (e, dashed lines) and calculated (c, solid lines) absorption curves of γ-irradiated silver colloids stabilized by PVS. The colloids prepared in an inert gas atmosphere exhibit an absorption maximum at 380 nm (e.1 [10], c.1), which shifts to 400 nm upon exposure to air (e.2, c.2). The model used for the calculation is described in the text. Calculated curves are narrower than the experimental ones because the size dependence of the dielectric constant has been neglected

Fig. 4, together with the experimental absorption curves before and after exposure to air. For the original uncoated spheres an absorption maximum at 383 nm is calculated, which shifts to 400 nm for the oxidized particle surrounded by the Ag^+ coating. Thus the experimentally observed shift is reproduced by the calculation without any further adjustable parameter. The calculated absorption curves are narrower than the experimental ones, because we have not included the increase in the dielectric losses of the metal for small particle sizes [17]. In addition, the particles in the experimental solution are not monodisperse.

Conclusions

The results presented demonstrate the potential of SERS to eludicate the mechanisms of colloid stabilization on a molecular level. The fact that an anionic stabilizer can be replaced by a cationic surfactant is indeed quite surprising and it must be kinetically controlled: with sufficiently high surfactant concentration, a transition between two metastable phases (negatively and positivley charged colloid) is effected without agglomeration of the particles. At lower CTAB concentrations, where the ligand replacement is slower, precipitation of the metal is indeed observed.

The following information on the geometry of adsorption has been obtained. (i) Strongest enhancement is observed for vibrations involving the head group, indicating that the latter is oriented towards the metal surface. (ii) Marker bands are seen for syn and anti conformations in the alkyl chain, which shows that there are at least two geometries of adsorption. (iii) With polyelectrolytes such as PVS, both free and surface-bound sulfate groups are detected. As an alternative approach, the preparation of colloids in the presence of surfactant stabilizers is presently being pursued.

For the smallest silver spheres ($R \approx 3$ nm) produced by γ-irradiation, Raman signals are comparatively weak because their size is one order of magnitude smaller than the dimensions required for maximum electromagnetic enhancement [2]. With PVS as a stabilizer, rapid oxidation of the silver surface layer is observed when the colloids are exposed to air. A quantitative model has been developed for the absorption of the silver spheres coated by Ag^+; oxidation of one quarter of the surface silver atoms gives rise to a shift in the absorption maximum as large as 20 nm. Replacement of PVS by $C_8F_{17}COO^-$ results in a marked increase of colloid stability. The silver is effectively protected against oxidation by the densely packed perfluoro-alkyl layer. This practical result illustrates the thrust of our continuing investigations of colloid stabilization using surface-enhanced Raman spectroscopy.

Acknowledgements

Colloids prepared by γ-irradiation were kindly supplied by Prof. A. Henglein. Sincere thanks are due to K. Carron, W. Fluhr, E. Guthmann and W. Häfner for their valuable contributions to various aspects of this work. Financial support by the Deutsche Forschungsgemeinschaft is gratefully acknowledged.

References

1. Otto A (1983) In: Cardona M (ed) Light-Scattering in Solids. Vol 4, Springer, Berlin, pp 289—418
2. Wokaun A (1985) Mol Phys 56:1—33
3. Creighton JA (1982) In: Chang RK, Furtak TE (ed) Surface Enhanced Raman Scattering. Plenum, New York, pp 315—337
4. Creighton JA, Blatchford CG, Albrecht MG (1979) J Chem Soc Faraday II 75:790—798
5. Kerker M, Wang DS, Chew H (1980) Appl Opt 19:4159—4174
6. Kerker M, Blatchford CG (1982) Phys Rev B 26:4052—4063
7. Siiman O, Bumm LA, Callaghan R, Blatchford CG, Kerker M (1983) J Phys Chem 87:1014—1023
8. Carey-Lea M (1986) Am J Sci 37:476
9. Fabrikanos A, Athanassion S, Lieser KH (1963) Z Naturforschung 18 b:612—617
10. Henglein A, Lilie J (1981) J Am Chem Soc 103:1059—1066

11. Bunding-Lee KA, Mc Lennan G, Gordon JG (1987) Ber Bunsenges Phys Chem 91:305–308
12. Fringeli UP, Günthard HH (1981) In: Grell E (ed) Membrane Spectroscopy. Springer, Berlin, pp 270–332
13. Koglin E, Sequaris JM, Valenta P (1983) In: Aussenegg FR, Leitner A, Lippitsch ME (ed) Surface Studies with Lasers. Springer, Berlin, pp 64–71
14. Pretsch E, Seibl J, Simon W, Clerc T (1981) Tabellen zur Strukturaufklärung organischer Verbindungen. Springer, Berlin
15. Weitz DA, Garoff S, Gramila TJ (1982) Opt Lett 7:168–170
16. Craighead H, Glass AM (1981) Opt Lett 6:248–250
17. Kreibig U, von Fragstein C (1969) Z Phys 224:307–323

Received December 10, 1987;
accepted December 23, 1987

Authors' address:

J. Wiesner, A. Wokaun
Physical Chemistry Institute
University of Bayreuth
D-8580 Bayreuth, F.R.G.

Progress in Colloid & Polymer Science

Progr Colloid Polym Sci 266:278–282 (1988)

Adsorption of cationic polyelectrolytes at clay-colloid interface in dilute aqueous suspensions – effect of the ionic strength of the medium

G. Durand, F. Lafuma and R. Audebert

Laboratoire de Physico-Chimie Macromoléculaire, Université Pierre et Marie Curie, Unité Associée au CNRS n° 278, E.S.P.C.I., France

Abstract: The behavior of dilute clay colloids/cationic copolymers systems is investigated as a function of the ionic strength I of the medium for various charge levels of the polyelectrolyte chain. Changing I not only has an influence on the interaction between the surface and the electrostatically bound segments but can also more or less affect the copolymer conformation according to its ionicity degree. The dependence of adsorption and flocculation processes on the ionic strength can be explained through these both effects.

Key words: Adsorption, flocculation, polyelectrolyte, colloid, ionic strength, clay-colloid interface, aqueous suspensions.

Introduction

The adsorption mode of polymers at a solid-liquid interface is highly dependent on their affinity for the surface. This parameter has been investigated in most theoretical works concerning polymer adsorption, for non-ionic ones [1] as well as for polyelectrolytes [2–3].

In previous publications [4–7], we examined the adsorption and flocculation properties of cationic copolymers towards aqueous suspensions of negatively charged particles at low ionic strength ($I < 10^{-3}$ M). In such systems the polymer-surface attraction can be modulated by varying the cationic monomer content (r %) in the copolymer. The results emphasized the great importance of the interfacial polymer conformation on adsorption and flocculation mechanisms [5–7]: the most charged copolymers ($r > 30$ %) display a high affinity for the surface and induce a charge neutralization mechanism for flocculation, whereas for the less charged ones ($r < 5$ %) flocculation is mainly achieved via interparticular bridging.

Another way to vary the polymer-surface interaction in charged systems is to change the ionic strength I of the medium by adding a passive electrolyte. Increasing I, for instance, has the effect of screening all the charges in presence; this leads to dramatic changes in polyelectrolyte conformation and weakens both particle-particle repulsions and particle-polymer attractions. The present paper is an experimental study of the consequences of these effects on the behaviour of Na-Montmorillonite platelet dispersions in the presence of cationic copolymers.

Experimental

Materials

Na-Montmorillonite purification and characteristics have been described elsewhere [7].

We used the same series of statistical copolymers as in our earlier study, their general formula is:

$$\text{+CH}_2-\text{CH+}_x \text{ - - - } \text{+CH}_2-\text{CH+}_y$$

$$\begin{array}{cc}
| & | \\
C=O & C=O \\
| & | \qquad\qquad\quad /CH_3 \\
NH_2 & O-(CH_2)_2-N^+-CH_3, Cl^- \\
& \qquad\qquad\quad \backslash CH_3
\end{array}$$

with various ionic content (r % = $y/(y + x) \times 100$) and molecular weight $\bar{M}_w$ [7]. In the present work, the ionic strength I of the systems was settled by adding the required amount of sodium chloride to Milli-Q water from Millipore system.

Methods

Adsorption and flocculation

Adsorption isotherm determination and flocculation tests were carried out in the same way as in our previous paper [7]. Experiments were performed at ionic strengths < 1M, since, for higher values, some spontaneous coagulation of the clay platelets is observed.

Viscometry

Intrinsic viscosities of the copolymers at various ionic strengths were measured either with an Ubbelohde capillary viscometer (shear rate $\dot{y} \simeq 1900\ s^{-1}$) or with a low-shear viscometer (Contraves LS 30) each time the viscosity appeared to be shear-dependent. This was especially the case when $r > 30\%$ and $I \leq 10^{-2}$ M: e.g. for $r = 100\%$, $I - 2.10^{-3}$ M and a molecular weight $\bar{M}_w = 2.10^6$; the viscosity of dilute solutions displays a Newtonian behaviour only for $\dot{y} < 1\ s^{-1}$. Isoionic dilutions were also performed when necessary ($r = 100\%$, $I \leq 2.10^{-3}$ M).

Determination of the persistence length

The persistence length q of a worm-like chain generally expresses its degree of stiffness well. Several methods have been described in the literature to calculate this parameter from various physico-chemical data. Among them, one of the most currently used is that of Yamakawa and Fujii [8], which is devoted to the determination of q from intrinsic viscosity results, but is inconvenient in neglecting the excluded volume effects and electrostatic contributions. We applied the Yamakawa and Fujii method to our intrinsic viscosities data, assuming that the mass per unit length of copolymer chains is the arithmetic mean of those of the two components. Another simplification was introduced in using only the linear asymptotic part of the Yamakawa function. This appears to work with polymers displaying a minimum number of statistical segments [9]. This requirement becomes rather restrictive when q increases, but even for $r = 100\%$ at the lowest ionic strength ($I = 10^{-3}$ M), four samples in our series appeared to satisfy it.

Results and discussion

Copolymers conformation

In dilute solutions, polyelectrolytes are well known to display rather extended conformations at low ionic strength because of the repulsions between charged neighbouring units along the chain. When the concentration of passive electrolyte is increased in the medium, these charges are screened, leading to an increasingly coiled conformation. When calculated with the Yamakawa and Fujii treatment (see Experimental part), the persistence lengths of our copolymers were found to vary as $I^{-1/2}$ for any value of the cationic monomer content (Fig. 1). This kind of dependence is in good agreement with many earlier experimental

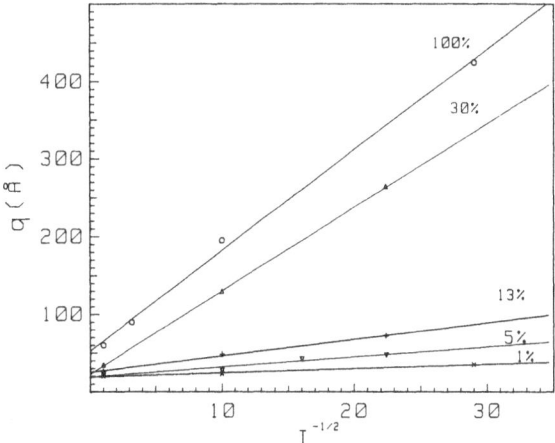

Fig. 1. Effect of the ionic strength I on the persistence length of cationic copolymers. ($\times$) $r = 1\%$, (∇) $r = 5\%$, ($+$) $r = 13\%$, ($\triangle$) $r = 30\%$, ($\bigcirc$) $r = 100\%$

works [10] together with the recent theories of Fixman [11] and Le Bret [12] while the first theoretical investigations of Odijk [13] and Skolnick [14] predicted an I^{-1} variation.

As expected, the effect of the ionic strength on q decreases with the cationic monomer content of the chain (Fig. 1). This result is better displayed in Fig. 2, where q is plotted as a function of r at different I values. For ionicities of 10% or less, the charged units stand far from each other along the chain, since the monomer distribution is random in our copolymers [15]. There-

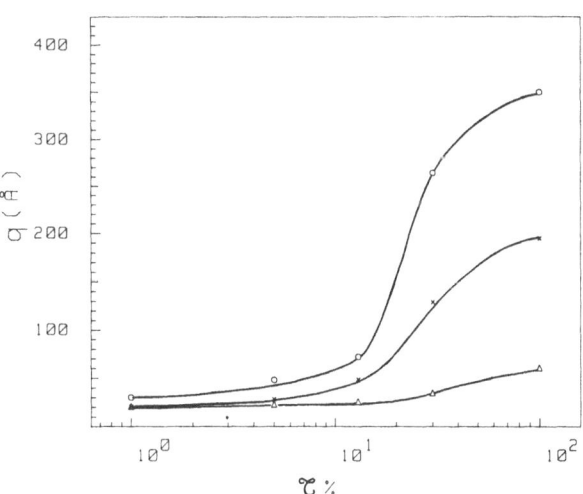

Fig. 2. Plots of the persistence length q vs $r\%$ for various ionic strengths: ($\triangle$) $I = 1$M, ($\times$) $I = 10^{-2}$ M, ($\bigcirc$) $I = 2\ 10^{-3}$ M

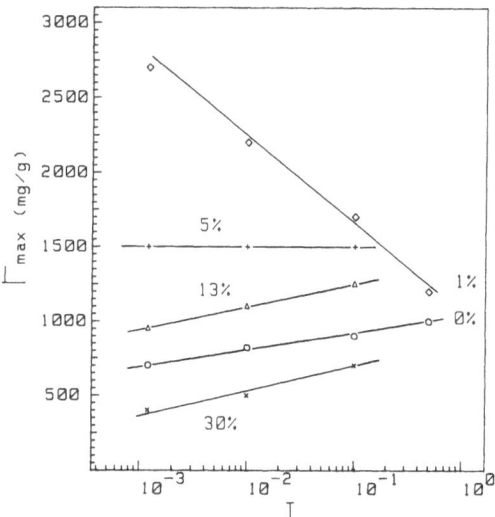

Fig. 3. Effect of I on Γ_{max} at different copolymer ionicities ($\bar{M}_w = 10^6$, montmorillonite content: $0.05\ \mathrm{g \cdot 1^{-1}}$). (O) $r = 0\%$, ($\Diamond$) $r = 1\%$, (+) $r = 5\%$ (Δ) $r = 13\%$, ($\times$) $r = 30\%$

compared to nonionic polyacrylamide [7]. So, when the ionic strength is increased, no important consequence is expected on interfacial conformation, but electrostatic interactions between the specifically adsorbed quaternary ammonium groups and negative sites of the surface are increasingly screened by the Cl^- and Na^+ ions in excess and the adsorption mode becomes similar to that of polyacrylamide. The same observations have been made by Wang and Audebert [5] when they examined the effect of I on hydrodynamic layer thickness of such a copolymer adsorbed on to silica.

For $r > 10\%$, Γ_{max} increases with I. In such cases the polyelectrolytes display a high number of sites available for electrostatic interaction with the surface. However, their adsorption at low ionic strength is limited by electrostatic repulsions occurring between charged segments near the interface [7]. Increasing I now has the effect of screening those repulsions, too, leading in the same way to a less flat and more flexible interfacial conformation with loops and tails. The latter results are the most currently observed for polyelectrolytes adsorbing onto surfaces of opposite charge [17, 18], but the former ones have also been found in some systems [19–21]. Anyway, on the theoretical point of view there is still a discrepancy concerning the effect of I on Γ [2, 3].

We think our results can help to conciliate these apparently conflicting behaviours if one considers the balance between the respective effects of the chain conformation in the plateau region and of the number of possible electrostatic chain-surface attractions. For weakly charged surfaces [20] or polymers (see our results or reference [19]), the macromolecules are flexible whatever the salt concentration, but the number of effective electrostatic attraction is low: thus, they are easily screened by the presence of small ions in the diffuse part of the double layer (Na^+) and around the polyelectrolyte charges (Cl^-) and accordingly, adding salt in such systems is unfavourable to polymer adsorption. Unfortunately, examples of this kind of behaviour are not numerous in the literature and the corresponding systems have not been well characterized. In the most encountered cases of highly charged polyelectrolyte and surfaces of high or moderate opposite charge, the number of available interacting sites is high enough to promote adsorption, although their interaction energy is diminished by screening: for low ionic strength, the surface saturation is limited by long range electrostatic repulsions which are progressively changed into steric repulsions at higher I values, leading to an increase in Γ_{max}.

fore, their mutual repulsions are not strong enough to induce a high extension of the chain and the persistence length only slightly increases with r and $I^{-1/2}$. Conversely, a spectacular jump of q is observed around $r = 20\%$ for $I < 10^{-2}$ M i.e. in the conditions for which those repulsions become effective. Above $r = 50\%$ the counter-ion condensation phenomenon limits the increase of q values. Thus, only the conformations of the most charged copolymers are highly affected when the ionic strength of the medium is lowered.

Adsorption

In Fig. 3, the maximum adsorbed amount Γ_{max} (plateau value of the adsorption) is reported as a function of I for different copolymers of identical molecular weight ($\bar{M}_w \simeq 10^6$). Different behaviours can be observed according to the polymer ionicity.

In the case of homopolyacrylamide ($r = 0\%$), the slight increase of Γ_{max} with I is generally attributed to some decrease in solvent quality [16]. A decrease of Γ_{max} is observed for $r = 1\%$. At low ionic strength, the interfacial conformation of this polymer must be flexible with loops and tails (see the results in solution). The higher level of adsorption displayed by the less charged copolymers ($r < 1\%$) has been explained by their greater affinity for the oppositely charged surface

Table 1. Effect of I on the OFC values (in ppm) for copolymers of different ionicities ($\bar{M}_w = 10^6$) – montmorillonite content: 0.5 g · l^{-1}

r %	I		
	1.2 10^{-3} M	5 10^{-3} M	10^{-2} M
1	450	250	90
5	280	180	130
30	80	65	55
100	23	20	15

Flocculation

The optimum flocculation concentrations (OFC) – taken as the dose of added polymer required to obtain the best supernatant clarification – were determined for various values of r and I (Table 1). The OFC is always lowered when I increases, but the dependence is much more pronounced for the less charged polymers, i.e., when bridging flocculation operates [7]. This can be understood, since increasing I reduces the range of double-layer repulsions without noticeable modification of the polymer conformation: therefore, the interparticular distance to be overcome becomes smaller and the minimum bridge length is shortened. Then, a smaller amount of adsorbed polymer may induce flocculation since the layer thickness increases with Γ.

For $r > 30$ %, flocculation occurs via charge neutralization at low ionic strength [7]. At higher ionic strengths, changes in polymer interfacial conformation towards a more flexible one and partial screening of the electrostatic polymer surface attraction (see preceding section) decrease the efficiency of this mechanism, counterbalancing the global reduction of particle-particle repulsions.

This explains why, for $r > 30$ %, the OFC values onyl slightly decrease with electrolyte addition. Moreover, in similar systems, the OFC have even sometimes been found to increase with I [22]. So the influence of the ionic strength in promoting a more loopy polyelectrolyte conformation at the interface probably partly modifies the contribution of the charge neutralization mechanism towards bridging.

Conclusion

Changing the ionic strength in our colloidal systems may give rise to very different results according to the copolymer composition.

At low copolymer ionicities, the ionic strength has only a weak influence on the polymer conformation and governs mainly the polymer particle and particle-particle electrostatic interactions. In such systems there are rather few electrostatic attractions between the polymer and the particle. So, when they are screened in the presence of added salt, the adsorption mode tends towards that of nonionic polyacrylamide, whereas bridging flocculation is enhanced as the interparticular repulsion range is decreased.

The most charged copolymers display high persistence lengths in aqueous solutions of low ionic strengths. Their rigid conformation must be partly kept when adsorbed from the same mediums onto particles of opposite charge. This, together with their high degree of charge, leads to flatter layers on the surface and to charge neutralization mechanism for flocculation [7]. In such cases, increasing I only partly screens the numerous polymer particle interactions and also has the effect of inducing a more flexible interfacial conformation for the adsorbed polymer with formation of some loops and tails. The increases, by the same way, the adsorption level and probably also gives rise to some briding flocculation.

Acknowledgement

We are greatly indebted to the Société Lyonnaise des Eaux (SLEE) and to the GS "Traitement Chimique des Eaux" who supported this work.

References

1. Fleer GJ, Lyklema J (1983) In: Parfitt GD, Rochester CH (eds) Adsorption of polymers, Adsorption from solution at the solid/liquid interface. Academic Press, pp 153–220
2. Hesselink FTh (1977) J Colloid Interface Sci 60:448
3. Papenhuijzen J, Van der Schee HA, Fleer GJ (1985) J Colloid Interface Sci 104:540
4. Mabire F, Audebert R, Quivoron C (1984) J Colloid Interface Sci 97:120
5. Wang TK, Audebert R (1988) J Colloid Interface Sci 21:32
6. Wang TK, Audebert R (1987) J Colloid Interface Sci, 119:459
7. Durand-Piana G, Lafuma F, Audebert R (1987) J Colloid Interface Sci 119:474
8. Yamakawa H, Fujii M (1974) Macromolecules 7:128
9. Bohdanecky M, Kovar J (1982) In: Jenkins AD (ed) Viscosit of polymer solutions. Elsevier, New York
10. Tricot M (1984) Macromolecules 17:1698 (see also Miyamoto S (1981) Makromol Chem 182:559 and Kitano T, Taguchi A, Noda I, Nagasawa M (1980) Macromolecules 13:57
11. Fixman M (1982) J Chem Phys 76:6346
12. Le Bret M (1982) J Chem Phys 76:6243

13. Odijk T (1977) J Polym Sci, Polym Phys Ed 15:477 (see also
    Odijk T, Houwart AC (1978) J Polym Sci Polym Phys Ed
    16:627)
14. Fixman M, Skolnick J (1977) Macromolecules 10:944
15. Durand G (1987) Thesis, Université Pierre et Marie Curie, Paris
16. Papenhuijzen J, Fleer GJ, Bijsterbosch BH (1985) J Colloid
    Interface Sci 104:530
17. Cosgrove T, Obey TM, Vincent B (1986) J Colloid Interface Sci
    111:409
18. Kokofuta E, Takahashi K (1986) Macromolecules 19:351
19. Pelton RH (1986) J Colloid Interface Sci 111:475
20. Froehling PE, Bantjes A (1977) J Colloid Interface Sci 62:35
21. Hendrickson ER, Neuman RD (1986) J Colloid Interface Sci
    110:243
22. Graham NJD (1981) Colloids Surf 3:61

Received December 11, 1987;
accepted December 23, 1987

Authors' address:

F. Lafuma
Laboratoire de Physico-Chimie Macromoléculaire
Université Pierre et Marie Curie
Unité Associée mau CNRS no 278
E.S.P.C.I.
10, rue Vauquelin
F-75231 Paris Cedex 05, France

Progress in Colloid & Polymer Science   Progr Colloid Polym Sci 76:283–285 (1988)

Adsorption of surfactants at the kaolinite-water interface: A calorimetric study

K. A. Wierer and B. Dobias

Surface Chemistry and Flotation, Institute of Physical and Macromolecular Chemistry, University of Regensburg, Regensburg, F.R.G.

Abstract: The adsorption of pyridinium salts on kaolinite was investigated by measuring the adsorption density, the heat of displacement and the floatability. From the results, we can conclude that the adsorption occurs in three steps and depends very much on the counter ions of the surfactants.

Key words: Adsorption, surfactants, kaolinite, enthalpy of displacement, calorimetry, pyridinium salts.

Introduction

The adsorption of a given surfactant on a mineral surface is chiefly determined by the chemical character of the lattice species in connection with the charge state and structure of the electrical double layer at the mineral-water interface. The charge formation in the electrical double layer is substantially contributed by ions which are called potential determining ions (PDI) [1, 2]. In a previous paper [3], we investigated the adsorption of H^+ and OH^- onto the kaolinite surface charge. There we found a negative surface charge on the equiadsorption point (EAP).

Now we will study the adsorption of cationic surfactants (alkyl-pyridinium salts) onto the kaolinite surface. By measuring the heat of adsorption we obtained further information for interpreting the adsorption isotherms.

Experimental

Kaolinite (> 98.4 %, specific surface: 15.8 m²/g) was purchased from Krantz Co., Bonn, the alkylpyridinium salts from Henkel, Düsseldorf.

Calorimetry

The apparatus used was a microcalorimeter, Modell 508 Tronac Co., Salt Lake City, USA. We used the method of isothermal titration of 30 ml kaolinite dispersion. The heat of reaction was deter-

mined as follows: The registered heat of reaction Q_R is composed of following single terms:

$$Q_R = \Delta H_{ads} + \Delta H_{dil} + \Delta H_{dis}$$

Isothermal reaction vessel

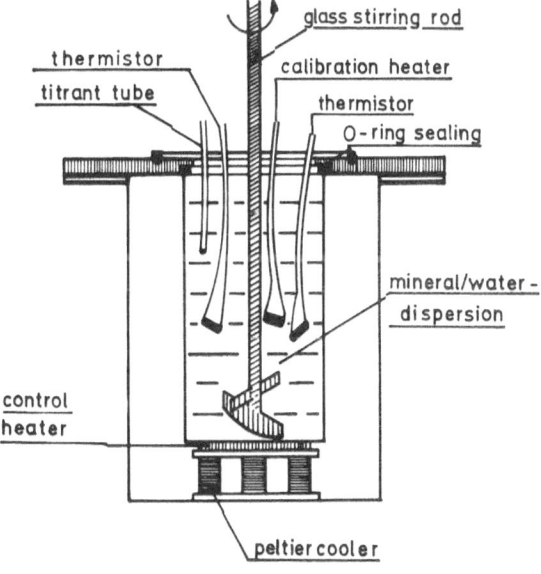

Fig. 1. Isothermal reaction vessel

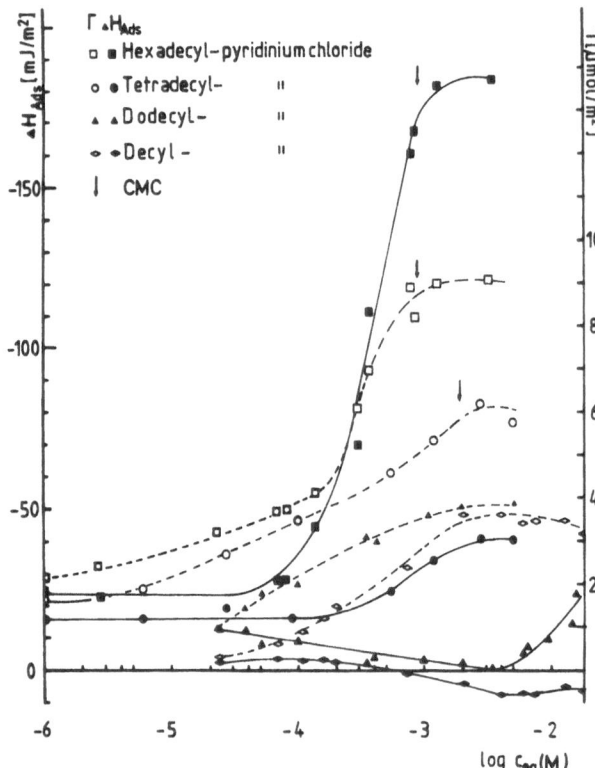

Fig. 2. Exchange enthalpy ΔH and adsorption density Γ of various alkylpyridinium chlorides on kaolinite; pH = 8.5, $I_c = 10^{-3}$M NaCl, $T = 25\,°C$

with ΔH_{ads} as enthalpy of adsorption, ΔH_{dil} as enthalpy of dilution of the titrant and ΔH_{dis} as enthalpy of dissolution of the lattice ions off the mineral surface. ΔH_{dil} and ΔH_{dis} were separately determined and substracted from ΔQ_R, to obtain ΔH_{ads}.

In Fig. 1 the reaction vessel is shown, which was put in a thermostat, and contained the mineral dispersion.

Results

In investigating the adsorption density Γ of alkylpyridiniumchloride (with different alkyl chain length: C_{10}, C_{12}, C_{14}, C_{16}) on kaolinite, we found a continuous increase of Γ until the maximum at the CMC of the surfactant was reached (Fig. 2). With shorter chain length the adsorption maxima became smaller. On measuring the heat of adsorption, we found exothermic enthalpies throughout caused by the C_{16} and C_{14}-surfactant. Remaining at first almost constant, they then increased very steeply at higher concentration (C_{16}: $c > 1 \times 10^{-4}$M; C_{14}: $c > 5 \times 10^{-4}$M). This was caused by a second, strongly exothermic reaction, as the experimental record in Fig. 3 indicates. The enthal-

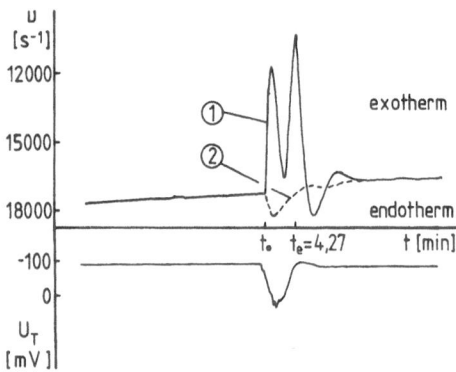

Fig. 3. Calorimetric record of the CpCl-adsorption at 2.7×10^{-3}M (initial concentration) on kaolinite; (ν) Impulses of the heater, (U_T) thermovoltage, (1) heat of reaction, (2) heat of dilution

pies of the C_{12} and C_{10} surfactants decreased with increasing concentration.

In seeking an explanation for the second peak of the enthalpy, we varied the counterions of the surfactants and used cetylpyridinium nitrate (CpNO₃). Besides the adsorption of the Cp⁺, we also investigated the coadsorption of NO_3^- and the heat of adsorption of CpNO₃ (shown in Fig. 4). In the presence of NO_3^-, Cp⁺ adsorbed at a lower equilibrium concentration than in the presence of Cl⁻. At low concentration the enthalpy was exothermic, became zero at $c = 5 \times 10^{-6}$M, and then it increased very strongly up to the CMC. The second peak occurred simultaneously with the steep increase and the adsorption of the NO_3^- started; so the second exothermic peak was caused by the coadsorption of counterions. With the coadsorption, the formation of a second inverse adsorption layer also began and decreased the hydrophobicity and the floatability (see Fig. 5).

Conclusions

From the obtained results we conclude that the adsorption of cationic surfactants on kaolinite proceeds in three steps:

1. At low concentration: Single surfactant ions adsorb at reactive, negatively charged surface sites mainly through electrostatic interactions. The enthalpy of adsorption is exothermic. There the chain length plays an inferior role.

2. At middle concentration: Further molecules adsorb on less reactive sites close to the first occupied ones. The alkyl chains start to associate due to van der

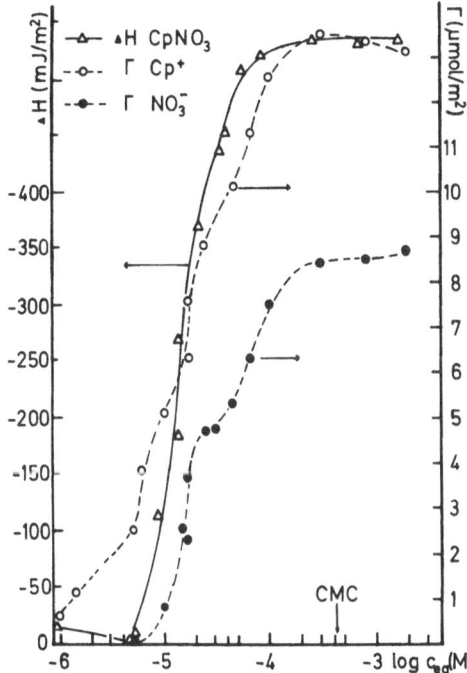

Fig. 4. Exchange enthalpy ΔH and adsorption densities Γ of cetyl-pyridinium nitrate (surfactant cation and NO_3^--anion) on kaolinite; pH = 8.5, $I_c = 10^{-3}$M NaCl, $T = 25\,°C$

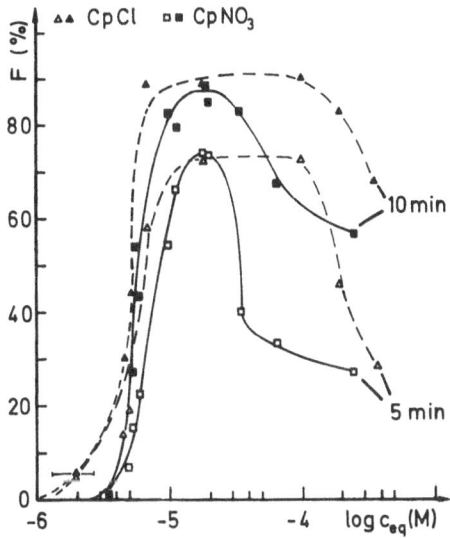

Fig. 5. Floatability F of kaolinite as dependent on the concentration of cetylpyridinium salt; 100 ml flotation cell, 10 g mineral, pH ~ 8

Waals forces, and the water molecules, which were orientated around the individual chain, are displaced and return to the free bulk phase (high entropy gain) [4]. Because of the repulsive forces between the close heads of the adsorbed surfactants, the enthalpy is very small.

3. At high concentration: This step occurs only with the C_{14} and C_{16} surfactants. Here the coadsorption of counterions starts with the increasing surfactant adsorption to eliminate the growth of the surplus charge of the positive heads of the adsorbed surfactants. Also, a second inverse layer is built, decreasing the floatability. A high exothermic enthalpy is created by the coadsorption of the counter ions (NO_3^- more than Cl^-).

References

1. Dobias B (1972) Tenside Deterg 9:322; (1976) ibid 13:131
2. Dobias B (1984) Surfactant Adsorption on Minerals Related to Flotation. Structure and Bonding 56, Springer Verlag, Berlin
3. Wierer KA, Dobias B (1988) Exchange enthalpy of H^+/OH^- on minearals with different potential determining ions. J Coll Interface Sci, 122:171
4. Schubert H (1977) Aufbereitung fester mineralischer Rohstoffe, Bd II, p 313, VEB Deutscher Verlag für Grundstoffindustrie, Leipzig

Received December 16, 1987;
accepted January 4, 1988

Authors' address:

B. Dobias
Surface Chemistry and Flotation
Institute of Physical and Macromolecular Chemistry
University of Regensburg
D-8400 Regensburg, F.R.G.

Progress in Colloid & Polymer Science   Progr Colloid Polym Sci 76:286–289 (1988)

Interactions of surfactants with polyvalent cations in solution and at the mineral-water interface

J. Oberndorfer and B. Dobias

Surface Chemistry and Flotation, Institute of Physical and Macromolekular Chemistry, University of Regensburg, Regensburg, F.R.G.

Abstract: The surfactant abstraction isotherms in the systems Na-dodecylsulfate or Na-dodecylsulfonate — fluorite were determined at constant pH and temperature. The surfactant adsorption density was calculated from the abstraction data.

Key words: Adsorption, precipitation, fluorite, Na-dodecylsulfate, Na-dodecylsulfonate.

Introduction

The abstraction of surfactants in salt-type mineral dispersions has been subject of numerous investigations [1, 7, 8]. The abstraction results from surfactant adsorption at the mineral and from precipitation of surfactant with the dissolved lattice metal ions. Even the adsorption of precipitate at the minerals has been discussed. None of these studies distinguished between these different ways of surfactant consumption in the dispersion.

Starting with the solubility and the surface structure of fluorite, this paper discribes the precipitation of the surfactants Na-dodecylsulfate and Na-dodecylsulfonate in a CaCl$_2$ solution and the abstraction in a fluorite dispersion.

The measurement of the lattice ion concentration at each point of the abstraction isotherms enabled us to derive the adsorption isotherms.

Experimental

Material

Na-dodecylsulfate was supplied by Eastman Kodak in electrophoresis grade. Na-dodecylsulfonate (Aldrich) had a purity of 99 %. CaCl$_2$ and NaOH from Merck were of analytical grade. The fluorite powder was produced by grinding fluorite crystals picked up from china (Krantz Co., Bonn) in a corundum ball mill.

The specific surface of 1.13 m^{2}g^{-1} was found using the B.E.T. method (Model 2100 D Micromeritics Instrument Coporation, Georgia, USA.)

Methods

The calcium concentrations were measured in 0.1 M KCl with a Varian 775 AAS at 422.7 μm in a reducing acetylene/nitrous oxide flame.

The fluoride concentration was analyzed with an ion sensitive fluoride electrode Ingold 15720S and a reference electrode Ingold 373-90-W using a Knick 741 pH-meter in the mV mode. All solutions were buffered with TISAB [4]. For pH measurements, a Knick 741 pH-meter with Ingold glass electrodes was used.

The surfactant concentration was determined by the two-phase titration method (DIN ISO 2271).

A Lange LPT3 light scattering photometer was used to measure the turbidity of the aqueous sols of calcium dodecyl-sulfate and calcium dodecylsulfonate after a 24-h conditioning time. The turbidity was expressed in turbidity units/Formacin [5].

The zeta potential of the sols was determined with a Pen Kem Laser Zee Meter Model 501.

Results and discussion

Figure 1 shows the turbidity of the sol and the zeta potential of the precipitate in dependence on the Na-dodecylsulfate equilibrium concentration after 24 h.

The total calcium concentration was 10^{-4}M, the pH was 10. The turbidity increased with increasing surfactant concentration owing to the formation of increasing precipitation. At about the CMC of the surfactant, the turbidity reached the maximum and started to decrease. The zeta potential was strongly negative, with its charge maximum at the CMC.

After filtration of the sols through a 0.22 μm Millipore filter, we measured the calcium and the surfactant

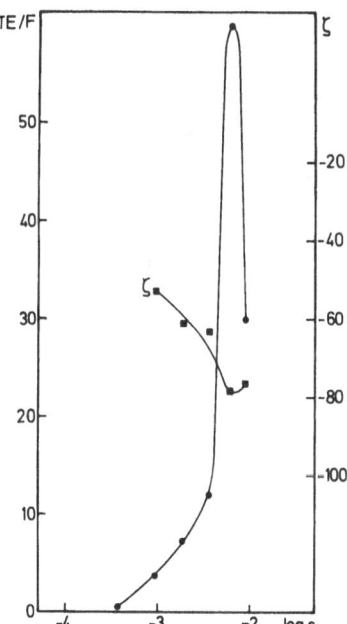

Fig. 1. Turbidity TE/F ● and Zeta-potential ζ/mv ■ of a Ca-dodecylsulfate sol as a function of the equilibrium surfactant concentration c/mol l^{-1} for $C_{Ca} = 1 \times 10^{-4}$ mol/l

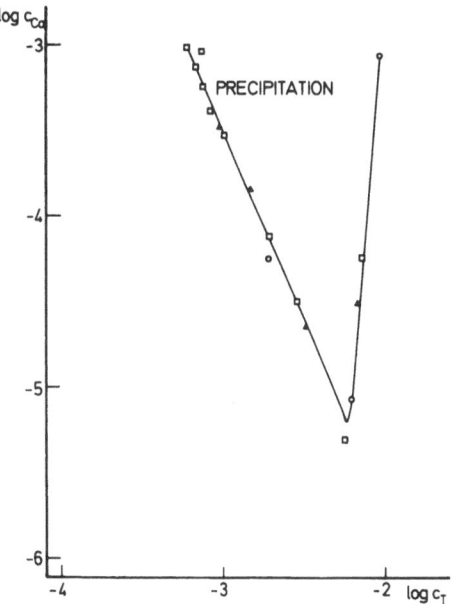

Fig. 2. The log(CaCl$_2$)- log (Na-dodecylsulfate) precipitation domain

concentrations in the clear filtrate. The results (Fig. 2) give the precipitation boundary.

It has been shown [2, 3, 6] that at surfactant concentrations below the CMC, the beginning of precipitation can be discribed by a solubility product $L = C_{Ca} \cdot C_T^2$. Above the CMC, the slope of the precipitation boundary reverses.

In Fig. 3, the calcium and fluoride concentrations and the abstracted dodecylsulfate concentrations are given in relation to the dodecylsulfate equilibrium concentration in the system fluorite/Na-dodecylsulfate at pH 10 and 20 °C after a 24-h conditioning. The calcium and fluoride concentrations are constant at low surfactant concentration. The excess of fluoride in the solution relative to the crystal stoichiometry shows an excess of calcium at the mineral surface. The excess concentration of calcium is about 1.03×10^{-5} mol m^{-2}.

When the surfactant concentration crosses the precipitation boundary, the calcium concentration decreases, and the fluorite concentration increases, indicating precipitation.

From the reactions in the system

$$Ca^{2+} + 2T^- \rightarrow CaT_2 \text{ (precipitation)}$$

$$Ca F_2 \rightarrow Ca^{2+} + 2F^- \text{ (dissolution)}$$

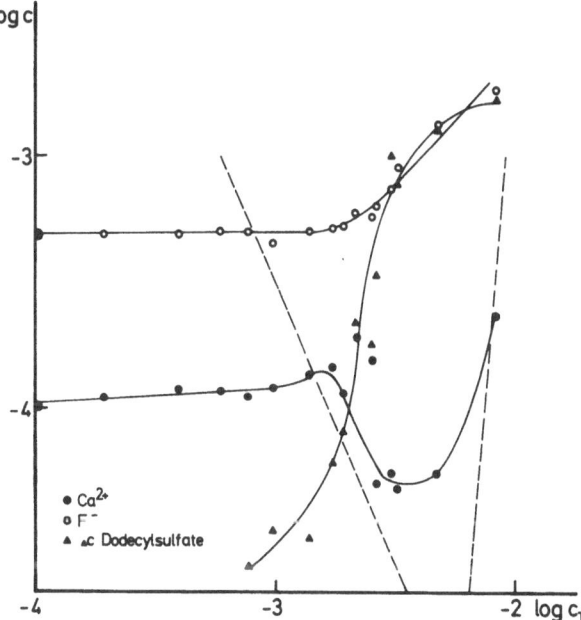

Fig. 3. The lattice ion concentrations and the decrease of surfactant concentration c/mol l^{-1} in the fluorite dispersion as a function of the Na-dodecylsulfate equilibrium concentration C_T/mol l^{-1}. (---) Ca-dodecylsulfate precipitation boundary

it follows:

$$\Delta C_{Ca} = -\frac{1}{2} \Delta C_{Tpre} + \Delta C_{Ca_{dis}}$$

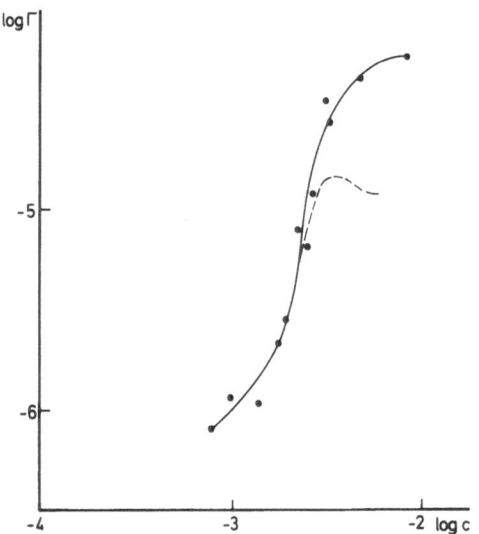

Fig. 4. Abstraction density γ/mol m^{-2} of Na-dodecylsulfate on fluorite as a function of the equilibrium concentration c/mol l^{-1} of the surfactant. (---) Adsorption density

$$\Delta C_{Ca} = -\frac{1}{2}\,\Delta C_{Tpre} + \frac{1}{2}\,\Delta C_F$$

$$\Delta C_{Tpre} = -2\,\Delta C_{Ca} + \Delta C_F$$

Using the last equation it is possible to calculate the adsorption isotherm from the abstraction data.

In Fig. 4 the abstraction and the adsorption densities are given as a function of the Na-dodecylsulfate equilibrium concentration. Most of the surfactant abstraction results from precipitation. The adsorption density decreases above the CMC.

Figures 5–8 show results of our investigations in the system fluorite/Na-dodecylsulfonate at concentrations below the CMC.

All reactions are similar in both systems, but they occur at lower surfactant concentration with Na-dodecylsulfonate. Even the adsorption density at the beginning of precipitation and at the maximum are equal in both systems.

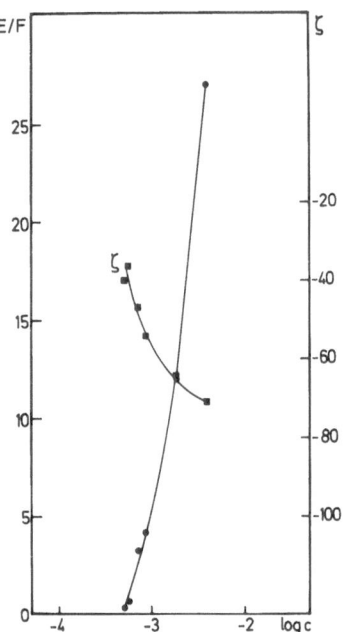

Fig. 5. Turbidity TE/F and zeta-potential ζ/mv of a Ca-dodecylsulfonate sol as a function of the surfactant concentration c/mol l^{-1} for $C_{Ca} = 1 \times 10^{-4}$ mol/l

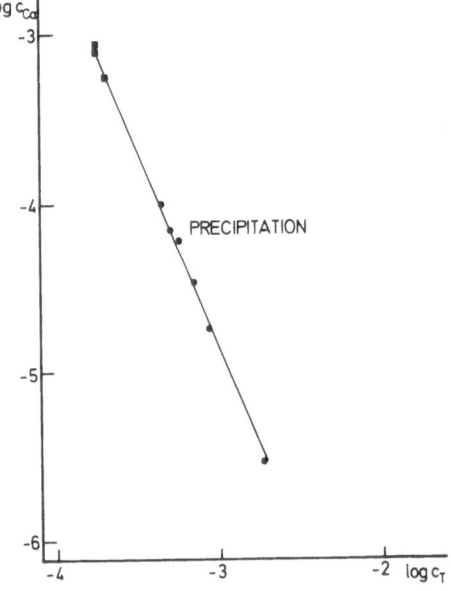

Fig. 6. The log (CaCl$_2$) – log (Na-dodecylsulfonate) precipitation domain

Conclusion

1. Na-dodecylsulfate and Na-dodecylsulfonate precipitate with Ca-ions at Ca-concentrations present in solutions of fluorite. The linear solubility boundary in the precipitation domain reverses its slope at the CMC of the surfactant.

2. The zeta potential of the precipitate is negative at all surfactant ion concentrations with a maximum at the CMC.

3. The adsorption density can be calculated from the abstraction isotherms using the proposed equation.

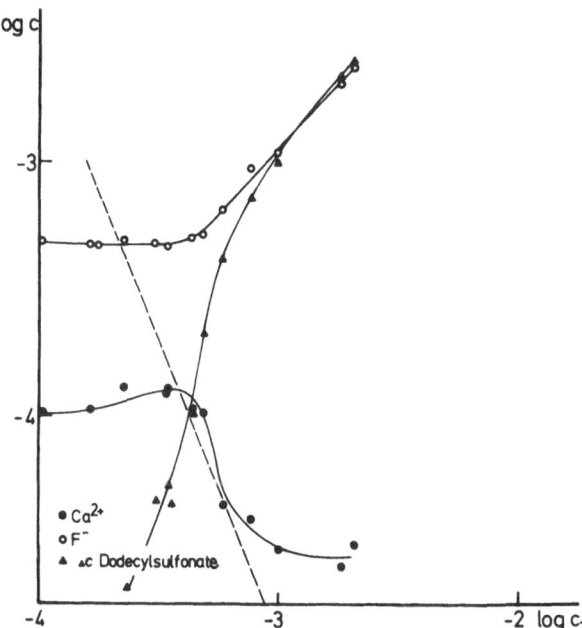

Fig. 7. The lattice ion concentrations and the decrease of surfactant concentration c/mol l^{-1} in the fluorite dispersion as a function of the Na-dodecylsulfonate equilibrium concentration C_T/mol l^{-1}. (---) Ca-dodecylsulfonate precipitation boundary

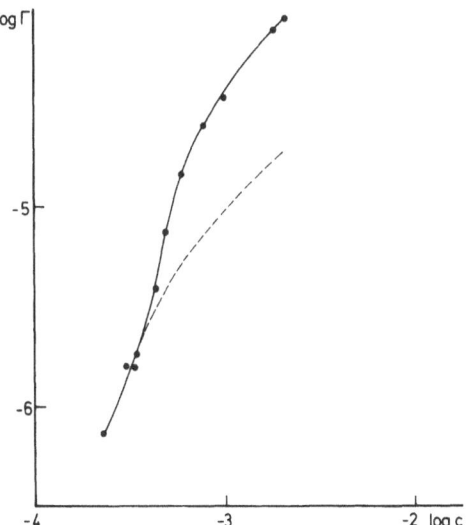

Fig. 8. Abstraction density y/mol m^{-2} of Na-dodecylsulfonate on fluorite as a function of the equilibrium concentration c/mol l^{-1} of the surfactant. (---) Adsorption density

References

1. Ananthapadmanabhan KP, Somasundaran P (1985) Surface precipitation of Inorganics and its role in Adsorption and Flotation. Colloids Surf 13:151–167
2. Baviere M, Bazin B, Aude R (1983) Calcium Effect on the Solubility of Sodium Dodecylsulfate in Sodium Chloride Solutions. J Colloid Interface Sci 92:580–583
3. Chou SI, Bae JH (1983) Surfactant Precipitation and Redissulation in Brine. J Colloid Interface Sci 96:192–203
4. Deutsche Einheitsverfahren zur Wasser-, Abwasser- und Schlammuntersuchung, Verlag Chemie, Weinheim/Bergstraße Verfahren DIN 38405 D4–1
5. Deutsche Einheitsverfahren zur Wasser-, Abwasser- und Schlammuntersuchung. Verlag Chemie, Weinheim/Bergstraße, Verfahren DIN 38404 C2–5
6. Peacock JM, Matijevic J (1980) Precipitation of Alkylbenzene Sulfonates with Metal Ions. J Colloid Interface Sci 77:548–554
7. Schulz P, Dobias B (1985) Effect of lattice ions in the selective flotation of salt-type minerals. Proc XVth Intern Minearl Processing Congr, Cannes
8. Wierer K (1987) Adsorption von H$^+$/OH$^-$-Ionen und Tensiden an Mineralen. Ph D Thesis, University of Regensburg

Received December 16, 1987;
accepted January 4, 1988

Authors' address:

B. Dobias
Surface Chemistry and Flotation
Institute of Physical and Macromolecular Chemistry
University of Regensburg
D-8400 Regensburg, F.R.G.

Progress in Colloid & Polymer Science   Progr Colloid & Polymer Sci 76:290 (1988)

Abstracts

The nature of adsorption at liquid surfaces

R. K. Thomas

University of Oxford, U. K.

Abstract: Specular reflection of either neutrons or X-ray may be used to probe the density profile of liquid surfaces. The basis of the method and its range of sensitivity have been described.

Results have been obtained from surfactants at the surface of water for soluble and insoluble layers. In the case of sodium dodecyl sulphate, both surfactant and water profiles at the air-solution interface have been measured by making use of contrast variation in neutron reflectivity. At a molecular level, the water surface appears to be "roughened" by the presence of SDS. Results on other surfactants using X-rays show that the technique can observe adsorption of ions in the Stern layer.

Adsorption of polymers at the air-solution interface has also been studied by both techniques. First results again indicate a high sensitivity. Measurements have been made of PEO-water using neutrons and of PDMS-butanone using X-rays. The possibility of obtaining the completed segment density profile at the surface has been assessed.

Author's address:

R. K. Thomas
University of Oxford
Oxford, United Kingdom

Adsorption of sodium dodecyl sulfate on alumina

A. Skauge[1]), T. Lind[2]) and H. Høiland[2])

[1]) Norsk Hydro Research Centre, Bergen, and
[2]) Department of Chemistry, University of Bergen, Bergen, Norway

Abstract: The adsorption of sodium dodecyl sulfate (NaDDS) on alumina has been studied under varying conditions. Adsorption isotherms of NaDDS in aqueous NaCl solutions have been determined both with and without added butanol. The shape of the two isotherms differ. Without any butanol the adsorption isotherm is S-shaped, with butanol it looks like a Langmuir type. The maximum NaDDS adsorbed is less with butanol added.

When cyclohexane is added to this system of NaCl, water, NaDDS, and butanol a Winsor microemulsion system is obtained. Depending upon the butanol and NaCl concentrations, an oil-in-water, a middle phase, or a water-in-oil microemulsion phase will form. The NaDDS adsorption on alumina has been measured for all three types of microemulsions. The results show that the adsorption density depends upon the type of microemulsion. It is highest for the middle phase microemulsion. However, the adsorption density from any type of microemulsion is significantly lower than from an aqueous solution of NaDDS.

Authors' address:

A. Skauge
Norsk Hydro Research Centre
N-5001 Bergen, Norway

Interfacial tension during transport of n-alkanols and n-carbonic acids between water and hexane

P. Joos, M. Vermeulen and J. van Hunsel

Department Biochemistry, University Antwerp (U.I.A.), Wilrijk, Belgium

Abstract: By interfacial tension measurements, the concentration at the interface during transport of surfactants (alkanols, carbonic acids) can be obtained. The interfacial tensions observed are not equilibrium values but steady state data due to the diffusion flux through the interface. Equilibrium data are obtained by previous equilibration of the surfactant over both phases. The steady state interfacial tensions can be predicted by diffusion penetration theory and for the alkanols by the complete diffusion equations as well. Aliphatic carbonic acids form in the organic phase dimers and this changes the evaluation drastically.

For both kind of systems the agreement between experimental data and theoretical predictions is good.

The system in which a carbonic acid is transferred from the organic phase to the aqueous phase containing NaOH is at present not well understood, but is of prime importance for technical processes.

Authors' address:

P. Joos
Department Biochemistry
University Antwerp (U.I.A.)
B-2610 Wilrijk, Belgium

Disproportionation of gas bubbles as affected by viscoelastic surface properties

A. Prins

Agricultural University of Wageningen, The Netherlands

Abstract: Disproportionation of gas bubbles as described by de Vries [1] takes place under the condition that the surface tension remains constant during the process. For most practical systems, however, this will be not the case: due to the compression of the bubble which becomes smaller, the surface tension becomes smaller too. Because the disproportionation process is driven by the Laplace pressure difference, the process is slowed down.

The extent of the surface tension lowering is determined by the value of the surface dilational viscosity η_s which is defined by

$$\eta_s = \frac{\Delta \gamma}{d\ln A/dt}$$

where $d\ln A/dt$ is the relative rate of change in surface area and $\Delta\gamma$ the connected change in surface tension. Experimental techniques are available for the measurement of η_s, from which it appears that η_s depends strongly upon the value of $d\ln A/dt$. For various practical systems it is found that a power law equation can be used to describe the rheological behaviour of the surface:

$$\log \eta_s = n + m \log^{d\ln A/dt}$$

where n and m are characteristic constants for the system.

Combining this relationship with the de Vries equation, the bubble radius as a function of the time can be calculated by means of a computer program. The results may be discussed in view of experiments in which the bubble radius is measured as a function of time.

References

1. de Vries AJ (1958) Recueil 77:209

Author's address:

Prof. Dr. A. Prins
Wageningen
Agricultural University
Laboratory of Dairying and Food Physics
Wageningen, The Netherlands

Progress in Colloid & Polymer Science   Progr Colloid & Polymer Sci 76:292 (1988)

Radiation-induced metal aggregates and deposits: Synthesis and properties

J. Belloni

Laboratoire de Physico-Chimie des Rayonnements, UA 75 du CNRS — Université Paris-Sud, Orsay, France

Abstract: Radiation-induced reduction of metal ions in solution is used to produce metal atoms originally isolated. Provided their extreme sensitivity towards reverse oxidation can be overcome, these native metal atoms aggregate whenever they meet other atoms or ions to give either subcolloidal particles stabilized by a surfactant like polyvinylalcohol or transparent deposits on specific supports (Al_2O_3, SiO_2, TiO_2, SnO_2, Si, etc.).

The extent of aggregation must be low enough to ensure a high dispersion of the metal, and at the same time high enough to ensure its stability towards corrosion by the medium. Pulse radiolysis studies provide information on the roles played by the solvent, the surfactant and the ionic strength on nucleation dynamics. A tight control of the nucleation process has enabled the synthesis of noble metals (15 Å Pt, 10 Å Ir) as well as nonnoble metals (20 Å Pb, 40 Å Ni, Co, Zn, Hg) or even bimetallic alloys (CuPd, NiPt, CuAu) exhibiting perfectly-ordered arrangements of atoms as shown by electron diffraction.

The catalytic efficiency of electron transfer reactions of subcolloidal particles or of several monolayer deposits upon semiconductors has been studied in solution and at electrodes' interface, respectively. The mechanism of the metal catalyzed electron transfer from methylviologen to protons or of the anion superoxide dismutation, as observed by pulse radiolysis, has been studied.

Author's address:

J. Belloni
Laboratoire de Physico-Chimie des Rayonnements
UA 75 du CNRS — Université Paris-Sud
F-91405 Orsay Cedex, France

Micellar aggregation of monomeric and polymeric surfactants

E. Jahns and H. Finkelmann

Institut für Makromolekulare Chemie, Universität Freiburg, Freiburg, F.R.G.

Abstract: It is well known that amphiphilic molecules form micelles in aqueous solution above a critical concentration (CMC). If the length of the hydrophobic and hydrophilic parts of the amphiphile is balanced, lyotropic liquid crystalline phases can be observed. While the lyotropic LC-phases are well determined in structure, the region of the isotropic micellar solution has been less investigated.

The existence of micellar shapes like spheres, cylinders and disks in the isotropic solution is dependent on temperature and concentration. There are several ways to influence the shape of the micelles. The hydrophilic/lipophilic balance of the amphiphile can be changed by shortening or lengthening its hydrophobic or hydrophilic parts. Another possibility is to synthesize formanisometric amphiphiles which are able to build up only one type of micellar shape. These could serve as model substances for more complicated systems. A third way is to polymerize the amphiphiles, which should result in more stable micelles. By changing the degree of polymerization the micellar shape should be influenced.

In this paper a series of non-ionic amphiphiles, starting with $C_{14}E_8$ was investigated by measuring the rotational viscosity up to high concentrations in water. A dependence of the reduced viscosity of the isotropic solution on the lyotropic LC-phases in the neighborhood of the phase diagram is observed. Changes in micellar shapes can be qualitatively shown. Preliminary measurements using the Kerr effect have also been made. A polymerized amphiphile with a $C_{14}E_8$ unit shows a stabilization of its mesophases with respect to the monomer.

Authors' address:

E. Jahns
Institut für Makromolekulare Chemie
Universität Freiburg
Stefan-Meier-Straße 31
D-7800 Freiburg, F.R.G.

Progress in Colloid & Polymer Science   Progr Colloid & Polymer Sci 76:293 (1988)

Thermodynamics of micelle formation of alkali metal perfluorononanates in water:
Role of hydrophobic hydration of the perfluoro-alkyl group

I. Johnson and G. Olofsson

Division of Thermochemistry, Chemical Center, University of Lund, Lund, Sweden

Abstract: The enthalpy and heat capacity changes for micelle formation of lithium perfluorononanate in water have been determined from calorimetric measurements of differential enthalpies of dilution of concentrated surfactant solution. The final concentrations varied between 0.001 and 0.025 $mol\,l^{-1}$ and measurements were made at four temperatures between 15°C and 43.5°C. The enthalpy of micelle formation of sodium perfluorononanate was determined at 30°C from measurements of differential enthalpies of solution of the crystalline surfactant. The volume change accompanying micelle formation of sodium perfluorononanate at 30°C was determined from density measurements. In evaluating the experimental results, the concentrations of surfactant in monomer and micellar forms in solutions above the CMC were computed using the thermodynamic model for the association of ionic amphiphiles proposed by Jönsson and Wennerström.

At ambient temperatures, the perfluorononanate salts give nearly the same enthalpy and volume changes of micelle formation as the hydrocarbon analogues. However, the heat capacity change is larger, probably stemming from a larger partial molar heat capacity contribution of the perfluoroalkyl group in the aqueous, monomeric amphiphile. The larger size of the perfluoroalkyl group means that there will be a larger number of water molecules in the first hydration layer around a perfluoroalkyl group than around the analogous alkyl group. Thus the stronger hydrophobic character of perfluorocarbon surfactants, as manifest in their low CMC and high surface activity compared to the hydrocarbon analogues, is probably due to the difference in size between the two types of hydrophobic groups.

Authors' address:

G. Olofsson
Division of Thermochemistry
Chemical Center
University of Lund
Lund, Sweden

Phase equilibria and microemulsion structure for a triple-chained soap: Sodium di-2-butyldecanoate

A. Khan, K. P. Das, L. Eberson, and B. Lindman

Physical Chemistry 1 and Organic Chemistry 3, Chemical Centre, Lund University, Lund, Sweden

Abstract: The binary and ternary (with decane) phase diagrams with the triple-chained soap sodium di-2-butyldecanoate have been determined by ^2H NMR and polarizing microscopy methods. In the binary aqueous system, two isotropic solution phases, one rich in water, L_1 (0–12 wt% surfactant) and the other rich in the surfactant, (49–61% surfactant) and a lamellar liquid crystalline phase (21–44% surfactant) are found to exist. The samples with compositions of 49–61% surfactant appear first as homogeneous lamellar liquid crystal, but within a few weeks they are converted to the second isotropic solution phase. The lamellar liquid crystalline phase solubilizes a very limited amount of decane (~1%) above which a clear solution phase is formed. The solution phase (microemulsion) in the three-component system is large and extends from very high water contents to almost pure oil. At about 1:1 weight ratio between water and decane, there is a minimum in the amount of surfactant needed to form microemulsion.

The solution structures have been studied by the ^1H FT-PGSE (pulsed-gradient spin echo) NMR diffusion method. The isotropic L_1-phase consists of normal micelles, whereas the structure of second isotropic solution appears to be of bicontinuous type. Mi-croemulsion structure depends strongly on sample composition. Microemulsion is of the W/O type in the decane-rich region and of the O/W type in the region rich in water. In the central part of the microemulsion region, where the system is rich in both water and decane, the structure is bicontinuous. The microemulsion region adjacent to the lamellar liquid crystal has a structure which is neither normal micelles nor bicontinuous. ^1H NMR linewidth measurements have been used to evaluate the size and shape of micelles and ^2H NMR quadrupolar splitting in the lamellar liquid crystal has been used to study the water binding to the interface.

Authors' address:

A. Khan
Physical Chemistry 1
Chemical Centre
Lund University
P. O. Box 124
S-22100 Lund, Sweden

Progress in Colloid & Polymer Science   Progr Colloid & Polymer Sci 76:294 (1988)

Hydrophile-lipophile balance and interfacial tensions in water-hydrocarbon-surfactant systems

D. Wielebinski, B. Föllner, K. Selcan, and G. H. Findenegg

Physikalische Chemie II, Ruhr-Universität Bochum, Bochum 1, F.R.G.

Abstract: Three-component systems consisting of water, a hydrocarbon oil, and a nonionic polyoxyethylene surfactant of the form $C_nH_{2n+1}(OCH_2CH_2)_mOH$ (denoted C_nE_m) exhibit a rich phase behaviour with liquid one-, two- and three-phase regions (and several liquid-crystal phases at higher surfactant concentrations). The capability of a surfactant to solubilize oil in water and water in oil, and to reduce the water-oil interfacial tension σ, depends on a balance of hydrophilic and hydrophobic interactions and on temperature (Shinoda and Friberg 1986). A well-balanced C_nE_m surfactant is predominantly water-soluble at low temperatures and predominantly oil-soluble at higher temperatures. This "phase inversion" usually involves the formation of a surfactant-rich third liquid phase which is described as a microemulsion, or organized surfactant phase (Shinoda and Lindman 1987), although from a phenomenological point of view these systems are closely analogous to liquid three- or four-component systems with smaller amphiphilic molecules (which are not forming micellar solutions or liquid-crystal phases). The aqueous phase α, surfactant-rich phase β, and oil-rich phase γ can coexist only over a limited temperature range between a lower critical endpoint temperature T_1 (at which α and β become identical in the presence of γ) and an upper critical end-point temperature T_u (at which β and γ become identical in the presence of α) (Kahlweit 1982; Kahlweit et al. 1983, 1984; Kahlweit and Strey 1985).

We have studied the interfaces in such two- and three-phase surfactant systems which often exhibit very low interfacial tensions. The main motivation for this work was to gain a deeper understanding of how such ultralow tensions can arise and how they are correlated with the phase behaviour of these systems. We analysed our results in terms of modern theories which link the interfacial tensions with the microscopic structure of the interfaces.

Authors' address:

G. H. Findenegg
Physikalische Chemie II
Ruhr-Universität
D-4630 Bochum 1, F.R.G.

Light scattering experiments on the microstructure of microemulsions

R. Strey and M. Teubner

Max-Planck-Institut für Biophysikalische Chemie, Göttingen, F.R.G.

Abstract: It is well established that amphiphiles give the highest mutual solubility of comparable amounts of water and oil in the vicinity of the three-phase regions [1]. The characteristic size of the water- and oil-rich domains in homogeneous microemulsions increases with decreasing the amount of amphiphile in the system [2]. We demonstrated that, by a detailed knowledge of the phase behaviour of such systems [1, 3], it is possible to choose a system with domain sizes of the order of the wavelength of visible light. These systems show beautiful color effects due to the dispersion of white light. By matching the refractive indices of aqueous and oleic moieties, the structural properties of such microemulsions have been quantitatively studied by static and dynamic light scattering. For the first time the typical broad static scattering peak — usually observed by small-angle X-ray and neutron scattering for a variety of microemulsions — was shown to be measurable by visible light. The static light scattering peak was well described by a recently published [4] intensity relation derived from a Landau free energy expansion including a negative gradient term. Dynamic light scattering experiments yield a q-dependent diffusion coefficient exhibiting a characteristic minimum in the vicinity of the static peak.

References

1. Kahlweit M, Strey R (1985) Angew Chem 24:654
2. Lichterfeld F, Schmeling T, Strey R (1986) J Phys Chem 90:5762
3. Kahlweit M, Strey R, Firman P (1986) J Phys Chem 90:671; Pospischil K-H (1986) Langmuir 2:170
4. Teubner M, Strey R (1987) J Chem Phys 87:3195

Authors' address:

R. Strey
Max-Planck-Institut für biophysikalische Chemie
Postfach 28 41
D-3400 Göttingen, F.R.G.

Structural investigation of organometallic inverse micellar systems

E. Hirsch[1]), R. Makhloufi[1]), S. J. Candau[1]), J. L. Mansot[2]) and J. M. Martin[2])

[1]) Laboratoire de Spectrométrie et d'Imagerie Ultrasonores, Unité Associée au CNRS n° 851, Université Louis Pasteur, Strasbourg, and
[2]) École Centrale de Lyon, Laboratoire de Technologie des Surfaces, Unité Associée au CNRS, Ecully, France

Abstract: Inverse micellar systems of organometallic compounds dispersed in parafffinic solvents have been studied. The disperse solutions have been investigated by light scattering and electron transmission microscopy. The measurements were performed, as a function of concentration, on Zirconium 3,5 dimethyl hexanoate dispersed either in heptane or in hexane.

Experimental evidence was found that the systems consist of two types of particles: spherical single micelles with a mean diameter of 40 Å, and aggregates of micelles with a hydrodynamic radius of the order of 850 Å. Electron transmission microscopy showed that these aggregates have the shape of a cluster.

Authors' address:

E. Hirsch
Laboratoire de Spectométrie et d'Imagerie Ultrasonores
Unité Associée au CNRS n° 851
Université Louis Pasteur
4, rue Blaise Pascal
F-67070 Strasbourg Cedex, France

Fractal structure of colloidal aggregates

R. Klein

Fakultät für Physik, Universität Konstanz, Konstanz, F.R.G.

Abstract: Recent scientific interest in the process of irreversible kinetic aggregation is to a large part due to the recognition that the highly disordered structures of aggregates can often be described by scaling concepts and that these structures appear to be fractals. The purpose of this study was to introduce some of these new concepts and to illustrate them by experimental results, mostly from scattering experiments. The structures obtained by different aggregation processes can be characterized by different fractal dimensions. The dynamics of the aggregation process (the cluster size distribution and its temporal development) is intimately related to the structure. Experimental information about these properties is obtained from static and quasi-elastic light scattering, which clearly reveal the fractal nature of the aggregates. By using the concepts of scaling, the highly disordered structures can be characterized in quantitative terms. Some specific physical properties of aggregates have been considered in more detail, such as hydrodynamic behaviour, rotational diffusion and elastic properties.

Author's address:

R. Klein
Fakultät für Physik
Universität Konstanz
Postfach 55 60
D-7750 Konstanz, F.R.G.

Interfacial analysis of pyrogenic and precipitated silica sols. Kinetics of silica coagulation

R. Zerrouk, R. Mercier and A. Foissy

U.A. CNRS 04 0436, Laboratoire d'Electrochimie des Solides — Université de Franche-Comté, Besançon, France

Abstract: The behaviour of pyrogenic and precipitated colloidal silica or sols has been studied. Their interfacial parameters (surface charge, adsorption and electrokinetic potential) were determined. Protomeric measurements of the surface charge in the presence of mono- (Na^+, K^+) and divalent (Ca^{++}) cations as a function of pH show a typical curve with a very low charge below pH 6. In alkali media the charge was determined for different cation concentrations (10^{-3} to 10^{-1} Mole/l.). In the case of precipitated silica, the charge seems to depend on the nature of the cation (Na^+, K^+) and presents an unexpected lowering with Ca^{++}.

This behaviour could be correlated to a microporosity of precipitated silica surface which prevents the approach of big ions: the bigger the hydrated radius of counterions, the lower the surface charge becomes.

The (σ_0, pH) curves were fitted using a triple layer model (pKa, pK_{M^+}, C_1) and a refinement of these three parameters simultaneously.

A study of the stability of silica sols ($0 \sim 10, 20, 70$ nm) is in progress as related to the preceeding measurements. The influence of both Mg^{++} and Ca^{++} has been determined on the evolution of the coagulation critical concentration (c.c.c.) versus pH (between 7 and 9). Stability diagrams are similar to the findings of Iler (1) and Allen and Matijević (2). The linear separation ($\log$ c.c.c. $= f$ (pH) between stable and coagulated regions enables us to write a chemical reaction including the SiO^- superficial ionized sites and leading to coagulation when a typical constant of the $[SiO^-]$ $[Ca^{++}]$ product is reached. This behaviour is similar to the laws of precipitation in homogeneous media. The coagulation of silica by divalent ions does not follow the DLVO theory because it is proved that the potential increases continuously (at constant pH) with the addition of Ca^{++} ions in the electrolyte.

The fractal dimension (SANS) of the aggregates obtained in different ways has been determined.

References

1. Iler RK (1975) J Colloid Interface Sci 53(3):476–488
2. Allen LM, Matijević E (1969) J Colloid Interface Sci 31(3):287–296

Authors' address:

R. Zerrouk
U.A. CNRS 04 0436
Laboratoire d'Electrochimie des Solides
Université de Franche-Comté
F-25030 Besançon Cedex, France

Electro-optic and rheological behaviour of aqueous dispersions of polytetrafluorethylene (PTFE) fibrils

H. Hoffmann, M. Angel, G. Huber, and H. Rehage

Lehrstuhl für Physikalische Chemie I der Universität Bayreuth, Bayreuth, F.R.G.

Abstract: For comparison of the dynamic behaviour of rodlike micelles with that of stiff rodlike particles, electro-optic and rheological measurements were carried out on PTFE fibrils. The crystalline fibrils had an axial ratio of about 100 and a thickness of 200 Å. The dispersions were stabilized by ionic perfluorosurfactants and could be kept for months without precipitation.

The rotational diffusion constant of the fibrils was determined as a function of the volume fraction. Both electric birefringence and rheological measurements gave identical results. The rotational diffusion constant did not change at the overlap concentration. The shear modulus increased in the semidilute concentration region with the same power law for both the micellar and the PTFE dispersions.

Electric birefringence measurements with AC-pulses showed a remarkable result. The stationary birefringence changed sign as a function of the frequency. This result indicates that the fibrils orient parallel to the electric field for frequencies above 1 kHz and perpendicular for lower frequencies. These conclusions could be confirmed by light scattering measurements in AC-fields.

Authors' address:

Prof. H. Hoffmann
Lehrstuhl für Physikalische Chemie I
Universität Bayreuth
Postfach 10 12 51
D-8580 Bayreuth, F.R.G.

Progress in Colloid & Polymer Science

Progr Colloid & Polymer Sci 76:297 (1988)

Nucleation, crystal growth and aggregation in model systems relevant to pathological mineralization

H. Füredi-Milhofer, D. Škrtić, M. Marković, and Lj. Komunjer

Laboratory for Precipitation Processes, "Rudjer Bošković" Institute, Yugoslavia

Abstract: Understanding of the kinetics and mechanisms of nucleation, growth and aggregation of crystals in supersaturated solutions of high ionic strength and the action of inhibitors thereon is of paramount importance for the understanding of the pathogenesis of certain forms of tissue mineralization, e.g. the deposition of mineral in kidney and urinary stones, dental calculus, etc. In our laboratory, methods for analyzing the kinetics of precipitation from supersaturated solutions have been developed and quantitative parameters have been defined which characterize the rate controlling processes, e.g. nucleation, crystal growth and aggregation. The methods were applied to the kinetics of precipitation of calcium oxalate hydrates from solutions of high ionic strength and low supersaturations, at which precipitates were formed spontaneously by heterogeneous nucleation. It has been shown that the nature of the solid phase formed and the kinetics of precipitation strongly depend on the experimental conditions employed, particularly on the reactant concentration ratio, mode of stirring and temperature. Phosphate ions inhibit crystal growth but do not affect nucleation or the rate and mechanism of aggregation of the particles. Some amino acids (glutamic acid, ornithine, tryptophan) acted as nucleation modifiers by promoting the formation of the thermodynamically stable calcium oxalate monohydrate. The type and intensity of the effect of these additives on the rates of crystal growth and aggregation has been shown to depend on the type and concentration of the amino acid.

Authors' address:

H. Füredi-Milhofer
Laboratory for Precipitation Processes
"Rudjer Bošković" Institute
P. O. Box 1016
YU-41001 Zagreb, Yugoslavia

Microcalorimetric analysis of the surface ionization of titanium dioxide and silica

J. B. Baussant, R. Mercier, A. Foissy and J. Bernard

Laboratoire d'Electrochimie des Solides — Université de Franche-Comté, Besançon, France

Abstract: The surface of mineral oxides dispersed in aqueous media is currently described in terms of ionization and complexation of superficial M–OH groups. Many measurements have been made by potentiometric acid-base surface titration, ion adsorption and electrophoretic mobility. In some favorable cases, the equilibrium constants of the chemical reactions taking place in the interface have been evaluated by extrapolation of the pH of zero surface charge. Such data are necessarily associated with a physico-chemical model of the electrical double layer (diffuse layer, Stern-Grahame or triple layer model). There is a reasonable agreement between experimental data and these models when assigning appropriate values to the fitting parameters. More experimental and independant measurements are needed, however, to obtain a precise and unequivocal description of interfacial reaction.

We here report our attempts to obtain calorimetric data on surface ionization reactions in aqueous dispersions of titanium dioxide (anatase) and silica. The apparatus and procedure have already been described (A. Foissy, Doctoral Thesis, 1985).

The accuracy of measurements depends on parameters such as stirring rate, particle aggregation, initial pH of the suspension and reaction rate. For titanium dioxide, a good reproducibility was obtained by adding the reagents (acid, base or dissolved salts) into a dispersion previously equilibrated at zero superficial charge pH. Assuming the following reactions on TiO$_2$:

$$TiOH + H^+ \longleftrightarrow TiOH_2 \tag{1}$$

$$TiOH + OH^- \longleftrightarrow TiO^- + H_2O. \tag{2}$$

The measured mean ionization enthalpies were -24 kJ/mole and -29 kJ/mole for positive and negative charge formations, respectively.

A decrease in the absolute value of the ionization enthalpies was observed when the surface charge density increased, but overall measurement accuracy did not allow a quantitative analysis of this phenomenon. The same comment can be made for the change in enthalpy with ionic strength and the nature of the counter ions.

Our results are comparable to the rare measurements of the same type in the literature. They can be compared to the free energies calculated from our potentiometric measurements of the charge density (chemical equilibria 1 and 2 above) but the experimental accuracy, again, does not lead to reliable values for the resulting thermodynamic constants.

In the case of silica, we investigated the acidic reaction of the surface as described by the following reaction:

$$SiOH + OH^- \longleftrightarrow SiO^- + H_2O$$

between pH 7 and pH 9 the mean molar ionization enthalpy varied more or less linearly between -33 ± 2 kJ/mole and -20 ± 2 kJ/mole. This decrease in absolute value is consistant with the additional electrical energy needed to remove a proton from the surface when the latter is increasingly negative.

These measurements have to be improved to obtain more reliable and accurate data. Nevertheless, this work has permitted a better understanding of surface ionization and adsorption reactions.

Authors' address:

J. B. Baussant
Laboratoire d'Electrochimie des Solides
Université de Franche-Comté
U.A. CNRS 04 0436
F-25030 Besançon Cedex, France

Progress in Colloid & Polymer Science   Progr Colloid & Polymer Sci 76:298 (1988)

Properties of hydrous iridium oxide colloids

A. Harriman[1]), G. R. Millward[2]), and J. M. Thomas[1])

[1]) Davy Faraday Research Laboratory, The Royal Institution, London, and
[2]) Department of Physical Chemistry, University Chemical Laboratory, University of Cambridge, Cambridge, U.K.

Abstract: Colloidal particles consisting of hydrous $IrO_2 \cdot xH_2O$ have been prepared by a number of different methods and characterised by a variety of instrumental techniques. The colloids, which are formed initially as spheres of about 30 Å diameter, aggregate into large but loosely-bound agglomerates with average diameter about 1000 Å. The colloids are deep blue, having a structureless absorption band centred at 580 nm, and readily undergo redox reactions. Reduction occurs to form lower valent hydrous oxides and, with strong reducing agents, to metallic iridium. Such processes can be initiated by electrochemical, photochemical and radiation chemical reactions and, in some cases, it is possible to resolve the kinetic parameters. Since protons accompany the electron transfer steps, the rates of redox reactions observed with these colloids show a strong pH dependence.

In neutral solution, the colloids possess an overall negative surface charge. Positively-charged dyes are bound to the surface by electrostatic forces and, upon irradiation, charge injection into the colloid can be observed by flash photolysis methods.

Authors' address:

A. Harriman
Davy Faraday Research Laboratory
The Royal Institution
21, Albemale Street
London W1X 4BS, England

Surface properties of carboxylated latex particles

P. Baglioni, R. Cocciaro and L. Dei

Department of Chemistry, University of Florence, Florence, Italy

Abstract: The interfacial properties of a styrene-butadiene-acrylic acid polymeric latex and their variation by the addition of a surfactant, sodium dodecylsulfate (SDS), able to be adsorbed at the latex interface, were studied by means of the ESR spectroscopy of nitroxide probes.

The parameters analyzed in this study were the nitrogen coupling constant, $\langle A_N \rangle$, the correlation time, τ, and the ESR lineshape.

We demonstrated that the latex interface presents, to a foreign molecule that can interact with the latex, two different "sites" of interaction: apolar "sites" directly exposed to the solvent interaction and polar "sites" mainly located in the water phase.

The addition of SDS decreased the exposition of the latex hydrophobic polymeric chains to the solvent only for high latex surface coverage (more than 70%), indicating that the latex "stabilization" can occur only when its interface is almost fully covered by SDS molecules.

The ESR spectra obtained for high SDS concentrations added to the latex indicated that SDS micelles and latex particles, fully covered by SDS monomers, coexisted in solution.

Authors' address:

P. Baglioni
Department of Chemistry
University of Florence
Florence, Italy

Surface pressure and surface potential studies of poly(isobutyl cyanoacrylate)-ampicillin association

A. Baszkin, M. Deyme, P. Couvreur, and G. Albrecht

Physico-Chimie des Surfaces et Innovation en Pharmacotechnie, U.A. CNRS 1218. Université Paris-Sud, Chatenay-Malabry, France

Abstract: Nanoparticles of poly(isobutyl cyanoacrylate) have recently found an application as carriers of biologically active substances. However, until now, little has been known about the nature of the association between these carriers and a particular drug.

Studies on poly(isobutyl cyanoacrylate) monolayers, using surface pressure and surface potential techniques, were performed in order to get an insight into the effect of polymer interfacial structure on its interactions with ampicillin.

Four polymerization processes were used to obtain the nanoparticles: monomer alone, monomer with polymerization adjuvants (glucose and dextran), monomer with ampicillin and monomer with adjuvants and drug.

The monolayers of poly(isobutyl cyanoacrylate) were studied on the aqueous support at pH varying from 2.7 to 8.8.

The shape of the obtained isotherms was that of expanded-type polymer monolayer curves. The limiting areas were lower than those reported in the literature for poly(acrylate) derivatives. The hydrolysis of polymer groups, followed by the submergence of the polar parts so produced, explained this behaviour.

The influence of ampicillin was significant only when associated with polymerization adjuvants. An increase in surface areas occupied by a monomer unit revealed the interactions of ampicillin with the polymer. This interaction was weak enough to disapear at the collapse point.

Surface potential and surface pressure measurements, with ampicillin injected under the polymer monolayers, reveal the existence of a limiting density of monomer/Å^2 above which no penetration of the drug into the monolayer may occur.

Authors' address:

A. Baszkin
Physico-Chimie des Surfaces et Innovation en
Pharmacotechnie U.A. CNRS 1218
Université Paris-Sud
5, rue J. B. Clément
F-92290 Chatenay-Malabry, France

Critical behaviour in microemulsions

A. M. Bellocq and D. Gazeau

Centre de Recherche Paul Pascal (CNRS), Domaine Universitaire, Talence, France and Greco "Microemulsions"

Abstract: Light scattering results obtained for the four-component system water-dodecane-pentanol-SDS in several distinct positions of a critical line have provided evidence for a singular behavior of the critical exponents ν and γ. Their values vary continuously from the Ising values to largely smaller ones as the critical end point is approached. In a previous study, the critical points were approached in the single phase domain either by increasing the water to surfactant ratio (path I) or by increasing temperature (path II). In order to gain information on the phase transition mechanism, we performed new measurements along paths different from paths I and II. Data were obtained in both the single and two-phase regions. Furthermore, measurements of the exponents β and μ, re-spectively, related to the shape of the coexistence curve and the interfacial tension were also carried out. Relations between the critical exponents ν, γ, β and μ were considered.

Authors' address:

A. M. Bellocq
Centre de Recherche Paul Pascal (CNRS)
Domaine Universitaire
and Greco "Microemulsions"
F-33405 Talence Cedex, France

A light scattering and thermodynamic study of the micellar solution of a nonionic surfactant

H.W. Richtering, W. Burchard, E. Jahns, and H. Finkelmann

Institute of Macromolecular Chemistry, University of Freiburg, F.R.G.

Abstract: This contribution deals with combined static and dynamic light scattering from the $C_{14}E_8$ [$CH_3(CH_2)_{13}(OCH_2CH_2)_8OH$] surfactant in the isotropic aqueous phase. Concentrations up to 36% were studied in a temperature range from 15 °C to 50 °C. The isotropic phase showed very complex behaviour.

A strong temperature and concentration dependence was observed and structure determination was difficult, since all measurable quantities were strongly influenced by thermodynamic interaction. However, the intricate variation of the mutual diffusion coefficient D was reduced to a very smooth increase of the friction coefficient when D was corrected for the osmotic compressibility. Determination of particle scattering and structure factors was not possible in most cases because of a low angular dependence of the scattered light. Structure determination became possible, however, from the analysis of the concentration dependence of the osmotic compressibility. The main results were: 1. At low temperatures up to 30 °C, micelles were found containing 119 molecules. This structure was stable up to concentrations of about 30%. 2. Above 30 °C, these spherical micelles started to associate and form random percolation clusters. 3. Near the phase boundary to the hexagonal liquid crystalline phase, the random clusters started to form more dense and geometrically anisotropic clusters, which were clearly seen in the time correlation function and the angular dependence of the static light scattering.

Authors' address:

H.W. Richtering
Institute of Macromolecular Chemistry
University of Freiburg
D-7800 Freiburg, F.R.G.

Singular behavior of the refractive index in a critical microemulsion

N. Rebbouh[1]), J. Buchert[2]), and J.R. Lalanne[1])

[1]) CNRS Paul Pascal, Domaine Universitaire, Talence, France
[2]) Non-Linear Topics Division, Institute of Physics, A. Mickiewicz University, Poznan, Poland

Abstract: It has been known for more than 20 years, both theoretically [1] and experimentally [2, 3], that the refractive index n of binary mixtures exhibits an anomalous temperature behavior near the critical mixing point. Two singular contributions from the density ϱ [$\Delta n_\varrho \propto t^{1-\alpha}$ where t is the reduced temperature and $\alpha \# 0.11$ the critical exponent of the constant pressure heat capacity] and the so-called local field correction F [far from the critical temperature T_c, $\Delta n_F \propto t^{-\nu}$ where $\nu \# 0.63$ is the critical exponent of the correlation length] have been considered generally.

The refractive index of the critical microemulsion 77.65% (by weight) n-dodecane, 5.15% water, 12.22% n-pentanol and 4.98% SDS has been investigated in the vicinity of the critical temperature T_c, both by refractometric and interferometric techniques. It exhibits a singular decrease near T_c. The fit of such a singularity by the density contribution leads to systematic residuals and an unrealistic value of $\alpha[1-\alpha \# (1.046 \pm 0.008) 10^{-3}]$ in agreement with the two-scale-factor universality prediction. The one by the local field gives $\nu = (0.29 \pm 0.05)$, a value which is inconsistent with hyperscaling. Moreover, our observation cannot be explained by gravity or Soret-induced gradients in the sample.

The exact origin of this singularity is, as far as we know, not understood at the present time. It can be connected to a singular increase of the volume of the dispersed phase near T_c and needs both experimental and theoretical further investigations which are now in progress in our laboratory.

References

1. Larsen SY, Mountain RD, Zwanzig R (1965) J Chem Phys 42:2187
2. Hartley CL, Jacobs DT, Mockler RC, O'Sullivan WJ (1974) Phys Rev Lett 33:1129
3. Houessou C, Guenoun P, Gastaud R, Perrot F, Beysens D (1985) Phys Rev A 32:1818

Authors' address:

N. Rebbouh
CNRS Paul Pascal
Domaine Universitaire
F-33405 Talence, France

Progress in Colloid & Polymer Science   Progr Colloid & Polymer Sci 76:301 (1988)

Lyotropic liquid crystals — shape of micelles and molecular geometry

M. A. Schafheutle and H. Finkelmann

Institut für Makromolekulare Chemie, Universität Freiburg, Freiburg, F.R.G.

Abstract: Amphiphiles investigated up to now have basically differed in the chemical constitution of their hydrophilic parts. A broad variety of non-ionic or ionic hydrophilic groups have been attached to hydrophobic groups which are mainly linear or branched alkyl chains. Insufficient attention has been focused on surfactants, which systematically induce a defined micellar aggregation of rod-like or disk-like micelles by their chemical constitution.

In principle, two possibilities exist for determining the anisometric dimension of micelles by the chemical constitution of the monomeric surfactants: A. The hydrophobic (hydrophilic) part of the surfactant forms a disk which is laterally surrounded by hydrophilic (hydrophobic) groups. This type of surfactant preferentially enables the formation of rod-like (reverse rod-like) micelles. B. The hydrophobic (hydrophilic) part of the surfactant molecule is formed by a rigid rod-like moiety, to which the hydrophilic (hydrophobic) group(s) is (are) attached at the end(s) of the rod. This type of surfactant promotes the formation of disk-like (reverse disk-like) micelles.

Some selected examples of new monomeric and polymeric non-ionic surfactants have been studied, in which the shape of the micelles and the liquid crystalline phases in aqueous solutions were determined by the molecular constitution of the surfactants.

Authors' address:

M. A. Schafheutle
Institut für Makromolekulare Chemie
Universität Freiburg
Stefan-Meier-Straße 31
D-7800 Freiburg, F.R.G.

Rheological properties of viscoelastic surfactant systems

H. Rehage and H. Hoffmann

Lehrstuhl für Physikalische Chemie I der Universität Bayreuth, Bayreuth, F.R.G.

Abstract: Dynamic rheological results on viscoelastic surfactants of cetylpyridinium salicylates were obtained. The solutions were prepared from cetylpyridinium chlorides and sodium salicylate.

For surfactant concentrations between 10 and 100 mM, we observed five different concentration regions with increasing Na Sal concentration. For a 30 mM surfactant concentration, the zero-shear viscosity (η_0) began to rise abruptly at a salicylate/surfactant ratio $R_I \simeq 0.6$ and reached a maximum at $R_{II} = 0.73$. At the maximum η_0 was between five and six orders of magnitude higher than the viscosity of water. Upon further increase of R, the viscosity passed through a minimum at $R_{III} = 2$ and a second maximum at $R_{IV} = 6.6$ and for even higher salicylate/surfactant ratios the viscosity decreased finally monotonously. The viscoelastic properties of the solution in the four different concentration regions II-V ($R_{II} < R < R_{IV}$) are determined by a shear modulus $G°$ and a single relaxation time τ. The complicated concentration dependence of the zero-shear viscosity is due to the different dynamic behaviour of the rod-like micelles. The shear modulus in the four different concentration regions is constant and independent of R. It follows a simple power law for the elastic properties of the surfactant concentration. This is clear evidence that the structures present in these solutions do not vary with the ratio R.

In region II the structural relaxation time τ is determined by entanglement processes; in region III it is determined by kinetic processes in which the breaking of the rods is rate limiting. In this region, the relaxation time becomes faster with increasing R until the rod-like micelles are completely uncharged. For even higher salicylate concentrations the charge on the rods is reversed. The charge slows down the dynamic properties of the surfactant solutions.

Authors' address:

H. Rehage
Lehrstuhl für Physikalische Chemie I
Universität Bayreuth
Postfach 10 12 51
D-8580 Bayreuth, F.R.G.

Oil penetration into surfactant monolayers

K. Jakobsen

Rogaland Research Institute, Ullandhaug, Stavanger, Norway

Abstract: Some phenomena of oil solubilization appear to require the assumption that hydrocarbons, some more than others, do penetrate into the surfactant monolayer at the oil-water interface. From experiments it appears that oil penetration is longer for small alkanes and as a rule of thumb, it appears that chains of length less than the surfactant length penetrate completely, whereas for larger oils, alkane uptake drops dramatically.

It is believed that oil penetration, or the role of oil, is crucial in determining phase behaviour. The curvature of the oil-water interface reflects in the main a balance between oil penetration into the surfactant tails, and head group repulsion. Therefore it is necessary to have a better model of oil penetration in order to improve the understanding of the concept of optimum salinity. Adding a cosurfactant or a cosolvent, the mechanism of oil penetration is somewhat modified. A more complex mechanism then takes place.

The phenomenon of oil penetration has been studied and quantified, both with and without a nonsurfactant. The effect of oil molecule length, temperature, roughness at the surfactant tail, cosurfactant partition were considered. The results could be used to predict phase behaviour, effect on interfacial tension with increasing oil molecule length, and natural curvature of microemulsion aggregates, and could also be used in a thermodynamic model of phase behaviour.

Author's address:

K. Jakobsen
Rogaland Research Institute
P. O. Box 25 03, Ullandhaug
N-4004 Stavanger, Norway

A field-dependent NMR-relaxation study of a sodium dodecyl sulfate/butanol/toluene/brine microemulsion system

M. Jonströmer[1]), K. Nagai[2]), U. Olsson[1]), and O. Söderman[1])

[1]) Physical Chemistry 1, Chemical Center, Lund University, Sweden
[2]) Dept. of Industrial Chemistry, Faculty of Engineering, Nagasaki University, Nagasaki, Japan

Abstract: Field dependent NMR relaxation data were obtained for a microemulsion system containing sodium dodecyl sulfate (SDS)/butanol/toluene/water/sodium chloride. By increasing the salinity, the system changed from a Winsor I to a Winsor III and then finally to a Winsor II system.

Multicomponent self-diffusion studies of these systems have been carried out on the microemulsion phase at different salinities [1]. In this study it was concluded that a change from an o/w structure via a bicontinuous structure to a w/o structure takes place as the salinity is increased. To obtain further information about the state of the aggregated surfactant in this system, a specifically deuterated SDS-molecule has been synthesized and NMR spin-lattice (T_1) and spin-spin relaxation (T_2) times have been measured at three different magnetic field strengths, *viz.* 2.35 T, 5.99 T and 8.48 T, using the same conditions as in the diffusion study. The data show the following features: i. T_1 is much longer than T_2 for all samples studied. ii. There is a frequency dependence on T_1 for all samples studied. iii. T_2 varies as the salinity changes. The data may be interpreted in terms of the conformational and dynamic state of the aggregated SDS.

References

1. Guéring P, Lindman B (1985) Langmuir, 1:464

Authors' address:

M. Jonströmer
Physical Chemistry I
Chemical Center
Lund University
P. O. Box 124
S-22100 Lund, Sweden

The mechanism of demicellisation studied by the dynamic surface tension of aqueous Brij 58 solutions

P. Joos and J. van Hunsel

Department Biochemistry, University of Antwerp (U.I.A.), Wilrijk, Belgium

Abstract: The CMC of Brij 58 solutions is very low, allowing us to observe dynamic surface tension effects in the time range of about 1 s when the equilibrium of micellisation is finished. Phenomenologically, the dynamic surface tension is diffusion controlled, yielding an apparent diffusion coefficient. This apparent diffusion coefficient does not follow the theory of Lucassen, but, as shown theoretically, gives information about the reaction of demicellisation.

Diffusion equations with chemical reaction terms have been considered using local equilibrium between monomers and micelles. A shorter way has been determined for obtaining the rela- tion of Lucassen and also for other types of demicellisation reac- tions. Our experimental data are qualitatively consistent with first order reaction kinetics.

Authors' address:

P. Joos
Department Biochemistry
University of Antwerp (U.I.A.)
B-2610 Wilrijk, Belgium

A form of Rayleigh Taylor instability during curtain coating

P. Joos and Y. Balbaert

Department Biochemistry, University Antwerp (U.I.A.), Wilrijk, Belgium

Abstract: During curtain coating, liquid is extended with a certain flow rate through a narrow slit and fills as a sheet between two verti- cal guide wires. If the flow rate is reduced, at the extension, the film breaks up as several liquid jets. This is an example of Rayleigh Tay- lor instability (liquid of a higher density, here water, above liquid which of a lower density (air)). The distance between the liquid jets has been measured and related to the unstable mode of the Rayleigh Taylor instability. The effect of surface tension may be clearly demonstrated, but curiously the observed distances between the jets correspond to the onset of the instability mode and not to the mode of the fast growing disturbance.

Authors' address:

P. Joos
Department Biochemistry
University Antwerp (U.I.A.)
B-2610 Wilrijk, Belgium

Mechanisms of action of threshold-active substances with respect to calcium carbonate precipitation in the washing process

C. P. Kurzendörfer and W. von Rybinski

Laboratories of Henkel KGaA, Düsseldorf, F.R.G.

Abstract: This investigation focused on laundry detergent additives that inhibit the formation of salt deposits on textiles. The mecha- nisms of action of these additives have been determined regarding the formation of calcium carbonate in model experiments under molar ratios of Ca^{2+}/CO_3^{2-} and $CaCO_3$/additive relevant to practi- cal conditions. The investigation of the additives (sodium salts of ethylenediaminetetrakis (methylenephosphonic acid) (EDMP), hydroxyethanebis (phosphonic acid) (EHDP), Na-triphosphate (STP), Na-ethylenediaminetetraacetate (EDTA), Na-nitrilotriaceta- ate (NTA) and Na-citrate) was carried out by scattering light meas- urements, X-ray analysis, as well as by microcalorimetry and potentiometric determination of the calcium activity.

The additives EDMP, EHDP and STP that were used in under- stoichiometric quantities with respect to calcium carbonate, effect a stabilization of the colloidal distribution and amorphous modifica- tion of calcium carbonate over a longer period to different extents. These mechanisms of action are characteristic of substances that are threshold-active vs. calcium carbonate. Comparable determina- tions of the effect of the additives with equal molar concentration showed the following sequence of threshold-activity: EDMP >> EHDP > STP > EDTA, NTA, citrate. Key threshold-active addi- tives induce a secondary effect regarding inhibition of incrustation in the presence of Zeolite NaA.

Investigations of the mechanisms of action are important in the experimental examination of substances with respect to their thre- shold-activity and to the formulation of laundry detergents with properties of inhibiting formation of salt deposits on textiles.

Authors' address:

C. P. Kurzendörfer
Laboratories of Henkel KGaA
P. O. Box 11 00
D-4000 Düsseldorf, F.R.G.

Progress in Colloid & Polymer Science   Progr Colloid & Polymer Sci 76:304 (1988)

Local structure analysis in organometallic reverse micellar systems

J. L. Mansot[1])[2]) and P. Lagarde[2])

[1]) Laboratoire de Technologie des Surfaces de l'Ecole Centrale de Lyon, Ecully, and
[2]) Laboratoire pour l'Utilisation du Rayonnement Electromagnétique, Université Paris-Sud, Orsay, France

Abstract: Organometallic micelles dispersed in hydrocarbons are currently being studied.

Metallic compounds (lead oxide and calcium carbonate) are dispersed in light hydrocarbons (hexane) by means of aliphatic β branched acids (3–5 dimethyl hexanoic acid). The small mineral cores (1 to 5 nm diameter) were investigated at the LIII edge of lead and the K edge of calcium using extended X-ray absorption fine structure spectroscopy (EXAFS) to determine the local order in the finely dispersed material. We demonstrated that for stoichiometric and overbased lead salts, local order exists due to the presence of three shells in the radial distribution function corresponding to Pb-O and Pb-Pb distances. A structure could be proposed for the stoichiometric lead salt.

In the case of the calcium carbonate dispersion, it was shown that the core structure is fully amorphous, the radial distribution function presenting only one peak, corresponding to the Ca-O distance with a coordination number lower than in the calcium carbonate crystalline forms.

Authors' address:

J. L. Mansot
Laboratoire de Technologie des Surfaces
de l'Ecole Centrale de Lyon
U.A. CNRS 855, B.P. 163
F-69131 Ecully Cedex, France

A neutron scattering study of the "anomalous isotropic phase" in a ternary surfactant/alcohol/brine system

G. Porte[1]), P. Bassereau[1]), J. Marignan[1]), and R. May[2])

[1]) Groupe de Dynamique des Phases Condensées and Greco "Microemulsions", U.S.T.L. Place Eugène Bataillon, Montpellier, and
[2]) Institut Laue Langevin, Grenoble, France

Abstract: The so-called "anomalous isotropic phase" is often observed in water-rich solutions of surfactants. It is usually called L_3 in non-ionic surfactant/water binary systems. It also occurs in water-rich ionic surfactant/alcohol/brine ternary systems, as first shown by Benton and Miller. This phase shows special physical properties, such as streaming birefringence and unusual turbidity, which suggest a structure different from that of the classical L_1 micellar phase.

This structure has been investigated using a variety of techniques: neutron and X-ray scattering, magnetic birefringence and conductivity measurements. The quantitative treatment of the high q-range of the scattering profiles shows that *on a local scale* the phase consists of randomly oriented surfactant-alcohol bilayers with the same thickness as in the neighbouring L_α phase. *On a larger scale* the low q-range of the scattering profiles shows that the overall structure swells linearly with the degree of dilution ϕ_W (ϕ_W is the volume fraction of the brine).

We have considered four different alternative structural models for this phase: large disconnected disks, foam-like structure, disordered lamellar stacking and bicontinuous structure where the infinite bilayer separates two interwoven equivalent self-connected infinite brine domains. Only the bicontinuous topology is quantitatively consistent with the magnetic birefringence, the conductivity and the low q neutron scattering data.

Authors's address:

G. Porte
Groupe de Dynamique des Phases
Condensées and Creco "Microemulsions"
U.S.T.L. Place Eugene Bataillon
F-34060 Montpellier Cedex, France

Progress in Colloid & Polymer Science   Progr Colloid & Polymer Sci 76:305 (1988)

A mode-mode coupling theory of electrolyte friction on charged polyions

M. Medina-Noyola[1]), R. Klein[1]), H. Ruiz-Estrada[2]), and A. Vizcarra-Rendon[2])

[1]) Fakultät für Physik, Universität Konstanz, Konstanz, F.R.G.
[2]) Cinvestav, Mexico

Abstract: A theory of the tracer-diffusion properties of an isolated polyion in an ionic solution has been formulated. Such a theory is based on an extension of the mode-mode coupling approach to self-diffusion, and its main formal result is an expression of the time-dependent electrolyte friction $\Delta \zeta^{el}(t)$ in terms of static properties, such as the interionic correlation functions of the supporting solution and the radial distribution function of the small ions around the polyion.

Explicit results, neglecting hydrodynamic interactions, and based on the use of the primitive model of the system (charged hard spheres in a uniform dielectric background) and on the mean sphe-

rical approximation were obtained. The current status on the extension of the theory to incorporate hydrodynamic interactions, and of its comparison with experimental results, was considered.

Authors' address:

Prof. R. Klein
Fakultät für Physik
Universität Konstanz
D-7750 Konstanz, F.R.G.

Ionic-strength dependence of the self-diffusion coefficient in polyball systems: theory vs. experiment

G. Nägele[1]), M. Medina-Noyola[1]), R. Klein[1]), J. L. Arauz-Lara[2]), and W. Dozier[3])

[1]) Universität Konstanz, Konstanz, F.R.G.
[2]) Syracuse University, U.S.A.
[3]) Exxon, Annandale

Abstract: The experimental measurements of the self-diffusion coefficient D_s in a suspension of polystyrene spheres by forced Rayleigh scattering have been considered in terms of the one-component macrofluid model (hard-sphere plus screened coulombic potential) in the absence of hydrodynamic interactions. The self-diffusion properties were calculated using two theories of self-diffusion. The first involves a mode-mode coupling approximation for the time-dependent self-friction function, and the second an exponential model (with exact short-term behaviour) for the memory of the self-diffusion propagator. Both theories write D_s in terms only of the static structure factor $S(k)$ of the macrofluid, calculated within the rescaled mean spherical approximation. The inaccuracies intro-

duced by the use of this approximation were estimated, and found to be significant in the strong-coupling regime. A general agreement was observed between the experimental results for the ionic strength dependence of D_s and those of the theories considered, which involve no adjustable parameter.

Authors' address:

Prof. R. Klein
Fakultät für Physik
Universität Konstanz
D-7750 Konstanz, F.R.G.

Progress in Colloid & Polymer Science   Progr Colloid & Polymer Sci 76:306 (1988)

Adsorption of hydrolyzed polyacrylamides on TiO_2 and $CaCO_3$ influences of electrical interaction and measurement procedure

G. Girod, J. M. Lamarche, and A. Foissy

Laboratoire d'Electrochimie des Solides-Université de Franche-Comté, Besançon, France

Abstract: Hydrolyzed polyacrylamides (HPAM) are currently used as thickeners in enhanced oil recovery. Adsorption of the polyelectrolyte on the reservoir rocks, however, determines the feasibility and economic interest of the technique. This study dealt with adsorption of HPAM on $CaCO_3$ (calcite) and TiO_2 (anatase), the latter being used as a model substrate to analyze the effects of electrical interactions and dissolved calcium ions on the binding of the polymer to the surface. We examined the influence of ionic strength, pH, dissolved Ca^{++} and experimental methodology such as liquid-to-solid ratio, stirring time and order of addition of the components to the system.

Experimental results showed similar surface coverages on TiO_2 and $CaCO_3$, taking into account the fact that the latter generates Ca^{++} ions in the solution. Adsorbed amounts as a function of pH, ionic strength and Ca^{++} concentration on both substrates showed that surface coverage is mainly dependent on the electrical interactions between adsorbed molecules. At higher ionic strength (1M/l

NaCl), adsorption of HPAM resembles adsorption of neutral polyacrylamide. We showed that erratic data can originate from the fact that adsorption is very fast compared to other reactions of the dissolved polymer, when performing the experiment. As an example, an increase of ionic strength can be correlated either to a decrease or to an increase of adsorption, depending on the experimental methodology. In such cases, measurements as a function of time give examples of a slow reorganisation of the adsorbed layer to return to equilibrium and give, then, coherent data.

Authors' address:

G. Girod
Laboratoire d'Electrochimie des
Solides-Université de Franche-Comté
U.A. CNRS 04 0436
F-25030 Besançon, France

Adsorption of polynaphtalene sulfonic acid onto titanium dioxide

A. Pierre, J. M. Lamarche, R. Mercier, and A. Foissy

Laboratoire d'Electrochimie des Solides – Université de Franche-Comté, Besançon, France

Abstract: Polynaphtalene sulfonic acid is a polymeric compound widely used as a fluidifying agent in cement pastes, since it is readily adsorbed onto the mineral surface. Using titanium dioxide as a non-reactive and preliminary substrate model, this study was aimed at an analysis of the influence of some practical adsorption parameters, namely, pH, ionic strength and ion nature. The solid surface was characterized by measurements of superficial charge density, electrophoretic mobility and ion adsorption as functions of the aforesaid parameters. The polymer was analyzed potentiometrically in the dissolved state to determine the ionization of the sulfonic groups. The polymer adsorption was measured and expressed by means of adsorption isotherms, and then correlated to the macroscopic properties of the dispersions, electrophoretic mobility and rheological behaviour.

The reactivity of titanium dioxide surface groups can be summarized by the following equations for a simple univalent electrolyte:

$$TiOH + H^+ \rightleftharpoons TiOH_2^+ \tag{1}$$

$$TiOH + OH^- \rightleftharpoons TiO^- + H_2O \tag{2}$$

$$TiOH + H^+ + Cl^- \rightleftharpoons TiOH_2^+ ... Cl^- \tag{3}$$

$$TiOH + Na^+ + OH^- \rightleftharpoons TiO^- ... Na^+ + H_2O. \tag{4}$$

Protometric titrations have been carried out to determine the influence of pH and ionic strength. It has been shown that a strong enhancement of the surface charge is induced by ions such as Ca^{++} and SO_4^{--}.

In acidic media (pH 3.5), adsorption is of the high affinity type. Addition of calcium ions to the solution (3.10^{-3} M/l) increases the amount of polymer adsorbed.

In alkali media, these is no adsorption without addition of ions. Addition of calcium ions at a concentration of 10^{-2} M/l leads to an adsorption isotherm similar to that obtained in acidic media. Ionic strength also increases adsorption of polymer.

Electrophoretic mobility of the particles is sensitive to the presence of polymer and calcium ion concentration. In acidic media (pH 3), for example, originally positive TiO_2 particles become negative (by adsorption of about 20% of the amount at the plateau level).

The surface charge (with or without adsorption of polymer) can be correlated to the stability of the particles' dispersion and to the viscosity of the suspension.

Authors' address:

A. Pierre
Laboratoire d'Electrochimie des Solides
Université de Franche-Comté
U.A. CNRS 04 0436
F-25030 Besançon, France

Modification of crystal size and shape by surface active agents

P. Sjödin and P. Persson

ACO Läkemedel AB, Solna, Sweden

Abstract: Addition of a surface active agent to a solution containing growing crystals of a pharmaceutical substance modifies the crystal shape significantly. The effect of several low molecular weight surface active agents on the crystal habit has been studied. The relative efficacy of a surface active agent can be estimated from considerations of the energy released in the adsorption process.

Authors' address:

P. Sjödin
ACE Läkemedel AB
Box 30 26
S-17103 Solna, Sweden

Size separation of polydisperse microspheres or microcapsules

J.-E. Proust and S. D. Tchaliovska

Physico-Chimie des Surfaces et Innovation en Pharmacotechnie, U.A. CNRS 1218, Université Paris-Sud, Chatenay-Malabry, France

Abstract: The inverse flotation or immersion of floating particles by application of external force is proposed as a size separation method for homologous polydisperse collection of microbeads, microspheres or microcapsules. The microparticles are treated in order to obtain hydrophobic surfaces, and to ensure their homogeneous wetting angles when attached to the liquid-gas interface. The monodisperse size separation is thus achieved. In the case of empty spheres (microcapsules), the method gives a relationship between the wall thickness and the radii of microcapsules.

Authors' address:

J.-E. Proust
Physico-Chimie des Surfaces et Innovation en Pharmacotechnie
U.A. CNRS 1218
Université Paris-Sud
5, rue J. B. Clémment
F-92290 Chatenay-Malabry, France

Small angle light scattering patterns from micrometer-sized spheroids

J.-C. Ravey

Laboratoire de Physico-Chimie des Colloides, U.A. CNRS N° 406, LESOC, Université de Nancy I, Vandoeuvre lés Nancy, France

Abstract: Two-dimensional small angle light scattering may be used for the characterization of optical and morphological properties of larger colloidal particles. For that purpose, we have derived the simple but useful following calculations. By using a Physical Optics Approximation, we assume that, at any point on the surface of the scatterer, the unknown field is the same as it would be if Fresnel's laws were applied to the plane tangent at that point, and several reflections/transmissions of the elementary incident beams are taken into account.

Numerical calculations compare quite favourably to exact results available for spheres (MIE) and smaller spheroids (ASANO) when there is no radius of curvature R less than $2 \pi R/\lambda = 6$.

According to the optical, orientational and morphological parameters, the small angle light scattering patterns (two-dimensional patterns) can be calculated. Their outlooks may be very dependent on the exact value of each parameter. Conversely, their experimental study can be used as a means of particle size, orientation and/or shape determination.

More specifically, these patterns present sets of successive maxima and minima, whose angular/azimuthal location can be measured and related to the properties of the spheroidal scatterer. Quantitative comparisons with RGD and diffraction approximations have been performed.

From a practical point of view, these results have been applied to suspensions of red blood cells in a Couette flow apparatus.

Author's address:

J.-C. Ravey
Laboratoire de Physico-Chimie des Colloids
U.A. CNRS N° 406, LESOC
Université de Nancy I
BP 239
F-54506 Vandoeuvre lès Nancy Cedex, France

PCPS A 036 + 037 + 038

Progress in Colloid & Polymer Science   Progr Colloid & Polymer Sci 76:308 (1988)

Non-aqueous microemulsions. An NMR relaxation study

A. Ceglie[1]), M. Monduzzi[2]), O. Söderman[3]), and B. Lindman[3])

[1]) Dipartimento di Chimica, Universita' di Bari, Bari, Italy
[2]) Dipartimento di Scienze Chimiche, Universita' di Cagliari, Cagliari, Italy
[3]) Physical Chemistry I, Chemical Center, Lund University, Lund, Sweden

Abstract: Recently, non-aqueous solutions of surfactants have been given increasing attention because of the theoretical and practical interest in them. Many studies have been carried out in order to understand both the mechanisms involved in the aggregation processes and the structural and dynamic features of these aggregates in aqueous microemulsions. In particular, important information has been inferred from the analysis of self-diffusion coefficients measured in several two-, three- and four-component aqueous systems [1–3]. On the other hand, in non-aqueous systems, significant differences have been found in comparison with the corresponding aqueous ones, both in the phase diagram and in the structural features [4, 5]. However, many uncertainties still remain regarding knowledge of the microstructure and dynamics of these systems at the molecular level.

NMR relaxation is considered to be one of the most suitable techniques for giving realistic pictures of motional and structural features, even in complicated systems. Therefore, we have undertaken a ^{13}C- and ^{2}H-NMR relaxation study of some non-aqueous microemulsions of sodium dodecyl sulphate (SDS) — pentanol or octanol — p-xylene in formamide (FM) and N-methyl-formamide (NMF) [6].

SDS is known to form spherical, or very nearly so, micelles in water. This system has been extensively characterized with field dependent NMR relaxation methods [7, 8]. A natural question then is, what kind of aggregates are formed in different polar solvents such as FM or NMF [9, 10].

Two-component systems: The ^{2}H relaxation data display a slight but significant frequency dependence which cannot be accounted for with a single motion. It follows that motions in the time scale of 10^{-9} s are present. Moreover, since at $6\,T\,R_2 = R_1$, no slower motions are involved.

Applying the "two-step" model [11] to frequency-dependent ^{2}H-NMR relaxation data from spherical lithium dodecyl sulphate micelles gave $\tau_c = 4 \times 10^{-9}$ s: if the same kind of micelles are supposed to form in FM, $\tau_c = 10^{-8}$ s is expected because of the higher viscosity of FM. Moreover, the observation that ^{2}H R_1's depend rather strongly on the concentration of SDS, implies that a substantial fraction of SDS molecules are present as monomers.

In the two-component systems, SDS-FM and SDS-NMF, the observed frequency dependence in the ^{13}C relaxation rates can be well explained if a motion with a correlation time of 10^{-9} s is present. A possible candidate for this motion is the tumbling of monomers in a solvent with large viscosity.

It can be concluded that SDS micelles in FM or NMF, if present, are much smaller than in water.

Three- and four-component systems: For the three- and four-component systems, using octanol as cosurfactant, the frequency dependence in ^{2}H R_1 is considerably stronger as compared to the two-component systems. Furthermore $R_2 > R_1$ at $6\,T$. This is due to the presence of slow motions, thus implying that at least a fraction of SDS molecules are aggregated into some sort of interface.

Due to problems with signal overlap, octanol was replaced with pentanol in the ^{13}C study of three- and four-component systems. The frequency dependence of the ^{13}C R_1's in these systems is almost of the same order of magnitude as that observed for the two-component systems. It follows that the conclusions drawn above about the two-component systems are also valid for the three- and four-component systems with pentanol as cosurfactant.

The degree of association into an interface, whatever its shape may be, while almost negligible in the two-component systems, seems to be strongly dependent on the chain length of the alcohol used as cosurfactant in the three- and four-component systems.

References

1. Lindman B, Stilbs P (1987) In: Friberg S, Bothorel P (eds) Microemulsions. CRC Press, Boca Raton, Ch 5
2. Stilbs P, Rapacki K, Lindman B (1983) J Colloid Interface Sci 95:583
3. Ceglie A, Das KP, Lindman B (1987) J Colloid Interface Sci 115:115 and references therein
4. Das KP, Ceglie A, Lindman B (1987) J Phys Chem, 91:2938
5. Das KP, Ceglie A, Lindman B, Friberg S (1987) J Colloid Interface Sci 116:390
6. Das KP, Ceglie A, Monduzzi M, Söderman O, Lindman B (1987) Progr Colloid Polym Sci 73:167
7. Ahlnas JT (1985) Thesis, Lund
8. Ellena JF, Dominey RN, Cafiso DS (1987) J Phys Chem 91:131
9. Almgren M, Swarup S, Lofroth JE (1985) J Phys Chem 89:4621
10. Rico I, Lattes A (1986) J Phys Chem 90:5870
11. Wennerstrom H, Lindman B, Söderman O, Drakenberg T, Rosenholm JB (1979) J Am Chem Soc 101:6860

Authors' address:

A. Ceglie
Dipartimento di Chimica
Universita' di Bari
Via Amendola 173
I-70126 Bari, Italy

Progress in Colloid & Polymer Science   Progr Colloid & Polymer Sci 76:309 (1988)

Detection of microemulsion structure variety by electroconductimetry, viscosimetry, densimetry and microcalorimetry

N. Azemar[1]), C. Solans[1]), P. Erra[1]), J. L. Parra[1]), M. Clausse[2]), D. Touraud[2]), and D. Clausse[3])

[1]) Instituto de Tecnología Química y Textil. (C.S.I.C.), Barcelona, Spain
[2]) U.A. CNRS N° 858, Département de Génie Biologique
[3]) DGTE. Département de Génie Chimique, Université de Technologie de Compiègne, Compiègne, France

Abstract: The microemulsion domain of the pseudo-ternary system water/sodium dodecylsulfate (SDS)/n-pentanol/n-dodecane (SDS/n-pentanol mass ratio, equal to 0.5) spans a wide range of water mass fractions (from 1 to 0.1 or so). Owing to the continuity existing between the water-rich and the water-poor composition regions it may be predicted that microemulsion structure is highly diverse and varies progressively with composition.

The aim of this study was to achieve a better understanding of the microstructures present in the microemulsion domain of the water/SDS/n-pentanol/n-dodecane system. In order to probe into microemulsion structure, the evolution of macroscopic transport properties (conductivity, viscosity) and thermodynamic properties (volumes and heat capacities) as a function of composition have been investigated.

Authors' address:

N. Azemar
Instituto de Tecnología Química
y Textil (C.S.I.C.)
Jorge Girona Salgado, 18–26
E-08034 Barcelona, Spain

Adsorption of some oligooxyethylene amine derivatives at the toluene/water interface

J. Szymanowski and K. Prochaska

Poznań Technical University, Institute of Chemical Technology and Engineering, Poznań, Poland

Abstract: The equilibrium interfacial tension in toluene/water systems containing individual model oligooxyethylene amine derivatives of the following structures:

$$RO\frown O\frown O\frown N\frown O\frown OR$$
$$\diagdown O\diagdown OR$$

$$HO\frown N\frown O\frown OR$$
$$\diagdown O\diagdown OR$$

where: $R = C_4H_9$, C_6H_{13} and C_8H_{17}, were determined and interpreted. The influence of the compounds structures and of the length of the alkyl group upon the interfacial activity and the probable configurations of the investigated compounds at the interface were considered. Different adsorption isotherms were matched to the experimental data and used to calculate the surface excess.

Compounds having three terminal alkyl groups and a hydrophilic group present in the molecule centre exhibit higher interfacial activity than compounds having two terminal alkyl groups. This high interfacial activity of the compounds having three terminal alkyls can only be explained by a change in these compound configurations at the interface connected with the bending of the oligooxyethylene chain.

The isotherms of Gibbs, Szyszkowski and Temkin and the polynomial of the third order can be used to match the experimental data. However, the relations between the surface excess and the bulk concentration are significantly different for various adsorption isotherms. As a result, the interfacial tension measurements cannot give enough strong quantitative evidence supporting the interfacial mechanism of metal extraction, and if used to interpret the kinetic data of metal extraction they can lead to ambiguous conclusions.

Authors' address:

J. Szymanowski
Poznań Technical University
Institute of Chemical Technology and Engineering
Pl. Skłodowskiej-Curie 2
P-60-965 Poznań, Poland

Colloid species as mass fractal aggregates in aluminium and iron hydroxide diluted solutions

P. Quienne[1]), J. Y. Bottero[2]), and D. Tchoubar[1])

[1]) Laboratoire de Cristallographie, U.F.R.-Faculté des Sciences, Orleans, France
[2]) Centre de recherche sur la Valorisation des Minerais de 1 E.N.S.G., Vandoeuvre, France

Abstract: The Al and Fe hydroxide particles which are formed under conditions far from equilibrium are highly reactive with the surrounding medium. This property can be explained by their very high specific surface and charge resulting from a specially divided state. The colloid species considered in this work were produced from partial neutralization of Al chloride and Fe nitrate aqueous diluted solutions (0.1 M of Al or Fe) by NaOH.

We have essentially considered two parameters:
  i. The neutralization ratio:

$$R = {}^{(NaOH)}/[Al \text{ or } Fe] \text{ with } \begin{cases} 2 < R < 3 \text{ for Al} \\ 1 < R < 2.5 \text{ for Fe} \end{cases}$$

ii. The ageing time T of resulting solutions with:

$$T < 1\,h.$$

The kinetics of the growth process first steps were studied using small angle X-ray scattering (SAXS) data. The experimental spectra were fitted with theoretical curves, obtained from simulation mass fractal models. These models describe the process of cluster growth by aggregation of diffusing clusters in different conditions of subunit cluster sticking.

Previous works [1–4] have enabled us to explain the structure of colloid species in Al solutions for R varying from 2 to 3.

The present work concerned principally the Fe solutions for R varying from 1 to 2.5. The fractal structures of resulting aggregates may be discussed by comparison with Al solution in terms of "polarizability effects" inducing different kinds of subunit clusters and different degrees of aggregate ramification. The restructuring process of the initial aggregates may be shown as a function of ageing time T.

References

1. Rakotonarivo E, Bottero JY, Cases JM, Fiessinger F (1984) Colloids Surf 9:273–292
2. Bottero JY, Axelos MAV, Tchoubar D, Cases JM, Fripiat JJ, Fiessinger F (1987) J Colloid Interface Sci, in press
3. Jullien R, Bottet R (1987) Aggregation and Fractal aggregates. World Scientific Publishing Co, PTE ltd
4. Axelos MAV, Tchoubar D, Jullien R (1986) J Phys 47:1843–1847

Authors' address:

P. Quienne
Laboratoire de Cristallographie
U.A. 810, B.P. 6759
Rue de Chartres
U.F.R. Faculté des Sciences
F-45067 Orleans Cedex, France

Micellar solubilization of alcohols in binary surfactant mixtures

C. Treiner

Laboratoire d'Electrochimie, Université Pierre et Marie Curie, Paris, France

Abstract: The partition coefficient P of 1-pentanol between micelles and water has been determined in a variety of mixed surfactant solutions in the whole surfactant composition range using gas chromatography. The following systems have been studied: (I) lithium dodecylsulfate (LiDS) + lithium perfluorooctylsulfonate (LiFOS); (II) sodium dodecylsulfate (SDS) + dodecyloxyethylene(23) (POE23); (III) SDS + dodecyloxyethylene(4) (POE4); (IV) sodium decylsulfate (SDeS) + trimethyldecylammonium bromide ($C_{10}Br$). The solubilization of the solute in the mixed surfactant solutions is calculated from the regular solution theory (applied to a three-component solution) using the interaction parameter β deduced from the same theory as applied to a binary surfactant solution. A positive departure from partition ideality is observed for system (I); an almost ideal behaviour is observed for systems (II) and a larger departure from ideality for the system (III); a strongly negative deviation from ideality is displayed with system (IV). Thus the sign and magnitude of β governs the variation of P with micelle composition only in cases (I) and (IV). In these binary surfactant solutions, 1-PeOH must be distributed throughout the micellar structure; however, in systems (II) and (III), the experimental results suggest a preferential solubilization in the mixed micelles hydrocarbon core. An application of these findings to microemulsion formulation has been suggested, and difficulties with the notion of micellar volume considered.

Author's address:

C. Treiner
Laboratoire d'Electrochimie
Université Pierre et Marie Curie
4, place Jussieu, Bat. F.
F-75005 Paris, France

Phase diagrams of the systems water, sodium dodecylsulfate, heptane and 1–, 2– and 3–pentanol

K. Veggeland[1]), E. Gilje[2]), and H. Høiland[3])

[1]) Rogaland Research Institute, Box 2503, Stavanger, Norway
[2]) Statoil, Box 300, Stavanger, Norway
[3]) Department of Chemistry, Bergen University, Norway

Abstract: The phase diagrams of the systems water, sodium dodecyl sulfate, pentanol and heptane have been determined. Pentanol includes 1–, 2– and 3–pentanol. In general terms, the extension of the homogeneous microemulsion phase depends on the pentanol, a larger microemulsion domain for 1-pentanol compared to 2– and 3–pentanol.

The electrical conductivities and the partial molar volumes and compressibilities of heptane and water have been determined at a surfactant-to-pentanol ratio of 1:2. The partial molar volumes of heptane and water are practically equal to those of pure water and pure heptane and even the partial molar compressibilities remain very close to those of pure components throughout the entire phase.

Authors' address:

K. Veggeland
Rogaland Research Institute
Box 2503
N-4004 Stavanger, Norway

Non-equilibrium states of a binary mixture with miscibility gap in the vicinity of its critical point in the homogeneous one-fluid phase region

W. Mayer and D. Woermann

Institute of Physical Chemistry, University of Köln, F.R.G.

Abstract: The temperature of a homogeneous 2,6-dimethyl pyridine/water mixture of critical composition is raised by a fast temperature jump (Joule heating pulse; characteristic heating time $\tau_h \simeq 0.4$ m s, amplitude of temperature jump ($0.04\,K < \delta\,T_h < 0.2\,K$) bringing the system closer to its lower critical temperature. The delayed increase in intensity of scattered light is measured as a function of time at different scattered angles ($\theta = 20°, 30°, 40°$) and different temperature differences ($T_c - T_f$) (T_c, T_f: critical temperature and final temperature of the sample after the temperature jump, respectively). This corresponds to a change of the scaling variable ($k\xi_f$) (k: absolute value of scattering wave vector; ξ_f: correlation length of local concentration fluctuations at T_f) in the range $0.1 < (k\xi_f) < 1.5$. The measured relaxation curves are always single exponentials (τ_{exp}: experimental relaxation time). For ($k\xi_f$), < 0.3 the structure factor relaxes as expected on the basis of linear response ($\tau_{hyd}/\tau_{exp} = 1$; τ_{hyd}: hydrodynamic relation time calculated from light scattering experiments). At larger values of ($k\xi_f$) in ($k\xi_f > 0.4$), the structure factor relaxes more slowly: τ_{hyd}/τ_{exp} decreases with increasing ($k\xi_f$). Plots of τ_{hyd}/τ_{exp} versus ($k\xi_f$) shows no dependence on the parameter (ξ_f/ξ_i) and ($T_c - T_f$) within the error limits over the entire ($k\xi_f$) range.

Authors' address:

W. Mayer
Institute of Physical Chemistry
University Köln
D-5000 Köln, F.R.G.

Electron transfer reactions of In$_2$Se$_3$ and In$_2$S$_3$ semiconductor colloids

N. M. Dimitrijević[1][2]) and P. V. Kamat[1])

[1]) Radiation Laboratory, University of Notre Dame, Notre Dame, Indiana, USA and
[2]) Institute of Physical Chemistry, Faculty of Science, Belgrade University, Belgrade, Yugoslavia

Abstract: Size quantization effect which increases the effective band-gap of the semiconductors also enhances the instablility of such semiconductor particles. In order to investigate the photochemical properties of small In$_2$Se$_3$ and In$_2$S$_3$ colloidal particles and to probe the mechanism of photocorrosion processes, we have undertaken laser flash photolysis and pulse radiolysis study of these semiconductor colloids.

Colloidal particles of In$_2$Se$_3$ and In$_2$S$_3$, $D_p < 50$ Å were prepared in aqueous and acetonitrile solutions. Their optical properties have shown hypsochromic shift in absorption edge and emission band as compared to the bulk material. Pronounced absorption peaks (or shoulders) due to excitonic transitions appeared during the growth of crystalline colloids which can be monitored spectrophotometrically. Ultraband excitation of these semiconductor colloids with a 355 nm laser pulse led to the charge separation followed by the trapping of charge carriers and/or electron transfer reactions at the semiconductor/electrolyte interface. The anodic corrosion of these semiconductor colloids was elucidated by recording time resolved transient absorption spectra. Addition of hole scavengers such as I$^-$ or triethanolamine to the colloidal suspension has been found to retard the anodic (oxidative) corrosion process.

Detailed kinetic analysis of one electron oxidation of colloidal In$_2$Se$_3$ and In$_2$S$_3$ colloids has been elucidated from the reactions with OH, Br$_2^-$, I$_2^-$ and (SCN)$_2^-$ radicals using pulse radiolysis technique. The rate constants determined for the reactions of X$_2^-$ radicals with colloids were less than predicted for the diffusion-controlled reactions and depended on the standard redox potentials of the oxidants.

Authors' address:

N. M. Dimitrijević
Institute of Physical Chemistry
Faculty of Science
Belgrade University
YU-11001 Belgrade, Yugoslavia

Surface properties of the bovine casein components: Influencing factors, chemical improving treatments

B. Closs, J. L. Courthaudon, and D. Lorient

Université de Bourgogne, École Nationale Supérieure de Biologie, Appliquée á la Nutrition et á l'Alimentation, Dijon, France

Abstract: Milk proteins and especially caseinates are very often used in the food industry as surface agents: their adsorption properties at water/oil and water/air interfaces, their ability to form emulsions and foams are directly related to their unfolded structure and their amphipatic character (hydrophobic residues located at the C-terminal side of the α_{sl} and β caseins molecules, polar residues at the N-terminal one).

In order to optimize the use of caseins as surfactants, the surface tension, the foaming capacity and stability were studied on the casein major components (α_{sl} and β) as a function of different factors: electric charge (modified by changing pH and ionic strength), protein concentration, polarity (modified by covalent binding of carbohydrates or by heat denaturation).

According to the results, the ability of a protein to diffuse to the interface and to adsorb appears to be directly related to the solubility and proteins charge (pHi optimal value). The caseinates constituted of several components with different surfactant properties should have properties directly depending on composition changes.

Authors' address:

B. Closs
Université de Bourgogne
École Nationale Supérieure de Biologie
Appliquée á la Nutrition et á l'Alimentation
F-21000 Dijon, France

Progress in Colloid & Polymer Science   Progr Colloid & Polymer Sci 76:313 (1988)

Minimal surfaces and the cubic liquid crystalline phase in microemulsions

P. Chieux[1]), J. O. Rädler[2]), and C. Toprakcioglu[2])

[1]) Institute Laue Langevin (ILL), Grenoble, France
[2]) Cavendish Laboratory, Cambridge, UK

Abstract: Microemulsions are stable mixtures of water, oil and amphiphilic additives (surfactants). At high concentrations of surfactant, liquid crystalline phases are known and believed, in some cases, to be bicontinuous. Infinite periodic minimal surfaces (IPMS) have been discussed as a possible microstructure (Scriven 1976). The ternary system containing water, alkanes and the surfactant mixture benzyltetradecyldimethylamoonium chloride and tetradecyltrimethylammonium bromide (3:1) has been found in this study to form a cubic liquid crystalline phase, which has been investigated by neutron scattering. A contrast experiment has been carried out on a system with equal volume fractions, which showed three Bragg reflections varying in agreement with theoretically calculated scattering amplitudes (D. M. Andersson, 1986) for the simple cubic Schwarz minimal surface. The position of the peaks shifts as the volume fraction is changed towards water-rich phases. This effect makes it possible to determine the interfacial area per unit cell as a function of the volume fraction and gives therefore a characterisation of the topology of the macrocrystal.

Authors' address:

C. Toprakcioglu
Cavendish Laboratory
Cambridge, U.K.

Author Index

Subject Index